Lecture Notes in Artificial Intelligence 11526

Subseries of Lecture Notes in Computer Science

More information about this series at http://www.springer.com/series/1244

David Riaño · Szymon Wilk ·
Annette ten Teije (Eds.)

Artificial Intelligence in Medicine

17th Conference on Artificial Intelligence in Medicine, AIME 2019
Poznan, Poland, June 26–29, 2019
Proceedings

Editors
David Riaño
Universitat Rovira i Virgili
Tarragona, Spain

Szymon Wilk
Poznan University of Technology
Poznan, Poland

Annette ten Teije
VU Amsterdam
Amsterdam, The Netherlands

ISSN 0302-9743 ISSN 1611-3349 (electronic)
Lecture Notes in Artificial Intelligence
ISBN 978-3-030-21641-2 ISBN 978-3-030-21642-9 (eBook)
https://doi.org/10.1007/978-3-030-21642-9

LNCS Sublibrary: SL7 – Artificial Intelligence

This Springer imprint is published by the registered company Springer Nature Switzerland AG
The registered company address is: Gewerbestrasse 11, 6330 Cham, Switzerland

Preface

The European Society for Artificial Intelligence in Medicine (AIME) was established in 1986 following a very successful workshop held in Pavia, Italy, the year before. The principal aims of AIME are to foster fundamental and applied research in the application of artificial intelligence (AI) techniques to medical care and medical research, and to provide a forum at biennial conferences for discussing any progress made. Thus, the main activity of the society is the organization of a series of biennial conferences, held in Marseilles, France (1987), London, UK (1989), Maastricht, The Netherlands (1991), Munich, Germany (1993), Pavia, Italy (1995), Grenoble, France (1997), Aalborg, Denmark (1999), Cascais, Portugal (2001), Protaras, Cyprus (2003), Aberdeen, UK (2005), Amsterdam, The Netherlands (2007), Verona, Italy (2009), Bled, Slovenia (2011), Murcia, Spain (2013), Pavia, Italy (2015), and Vienna, Austria (2017). This volume contains the proceedings of AIME 2019, the 17th Conference on Artificial Intelligence in Medicine, held in Poznan, Poland, June 26–29, 2019.

The AIME 2019 goals were to present and consolidate the international state of the art of AI in biomedical research from the perspectives of theory, methodology, systems, and applications. The conference included two invited lectures, full and short papers, tutorials, workshops, and a doctoral consortium.

In the conference announcement, authors were invited to submit original contributions regarding the development of theory, methods, systems, and applications for solving problems in the biomedical field, including AI approaches in biomedical informatics, molecular medicine, and health-care organizational aspects. Authors of papers addressing theory were requested to describe the properties of novel AI models potentially useful for solving biomedical problems. Authors of papers addressing theory and methods were asked to describe the development or the extension of AI methods, to address the assumptions and limitations of the proposed techniques, and to discuss their novelty with respect to the state of the art. Authors of papers addressing systems and applications were asked to describe the development, implementation, or evaluation of new AI-inspired tools and systems in the biomedical field. They were asked to link their work to underlying theory, and either analyze the potential benefits to solve biomedical problems or present empirical evidence of benefits in clinical practice.

AIME 2019 received 161 abstract submissions, 134 thereof were eventually submitted as complete papers. Submissions came from 36 countries, including 15 outside Europe. All papers were carefully peer-reviewed by experts from the Program Committee with the support of additional reviewers. Each submission was reviewed in most cases by three reviewers, and at least by two reviewers. The reviewers judged the overall quality of the submitted papers, together with their relevance to the AIME conference, originality, impact, technical correctness, methodology, scholarship, and quality of presentation. In addition, the reviewers provided detailed written comments on each paper, and stated their confidence in the subject area.

A small committee consisting of the AIME 2019 scientific chair, David Riaño, the local organization chair, Szymon Wilk, and the doctoral consortium chair, Annette ten Teije, made the final decisions regarding the AIME 2019 scientific program. This process began with virtual meetings starting in October 2018. The process ended with a two day face-to-face meeting of the committee in Poznan to assemble the final program.

As a result, 22 long papers (an acceptance rate of 21%) and 31 short papers (including demo papers) were accepted, two papers were withdrawn. Each long paper was presented in a 25-minute oral presentation during the conference. Each regular short paper was presented in a 5-minute presentation and by a poster. Each demo short papers was presented in a 5-minute presentation and by a demo during the poster session. The papers were organized according to their topics in the following main themes: (1) Deep Learning; (2) Simulation; (3) Knowledge Representation; (4) Probabilistic Models; (5) Behavior Monitoring; (6) Clustering, Natural Language Processing, and Decision Support; (7) Feature Selection; (8) Image Processing; (9) General Machine Learning; and (10) Unsupervised Learning.

AIME 2019 had the privilege of hosting two invited speakers: Anthony Chang, from Pediatric Cardiology, CHOC Children's, Orange, USA, and Ivana Bartoletti, Head of Privacy and Data Protection at Gemserv, London, UK. In his keynote entitled "Common Misconceptions and Future Directions for AI in Medicine: A Physician-Data Scientist Perspective," Dr. Anthony Chang discussed ten common misconceptions that both clinicians and data scientists have about the application of AI in medicine. Ivana Bartoletti's keynote focused on the ethical and privacy challenges involved in the deployment of AI in health care.

The doctoral consortium received six PhD proposals that were peer reviewed. AIME 2019 provided an opportunity for the best ones to present their research goals, proposed methods, and preliminary results. A scientific panel consisting of experienced researchers in the field (John Holmes, Beatriz Lopez, Mar Marcos, Paola Mello, Lucia Sacchi, Allan Tucker, Alfredo Vellido, Blaz Zupan) provided constructive feedback to the students in an informal atmosphere. The doctoral consortium was chaired by Annette ten Teije.

Two workshops were organized after the AIME 2019 main conference. These included the 11th International Workshop on Knowledge Representation for Health Care (KRH4C) and the 12th International Workshop on Process-Oriented Information Systems in Health Care (ProHealth), joined together for the third time at AIME. This workshop was chaired by Mor Peleg, Mar Marcos, and Manfred Reichert. The second workshop was the First Workshop on Transparent, Explainable and Affective AI in Medical Systems, chaired by Grzegorz J. Nalepa, Gregor Stiglic, Sławomir Nowaczyk, Jose M. Juarez, and Jerzy Stefanowski.

In addition to the workshops, four interactive half-day tutorials were presented prior to the AIME 2019 main conference:

(1) Evaluation of Prediction Models in Medicine (Ameen Abu-Hanna); (2) The Clinician–Data Scientist Dyad: Understanding Both for an Exceptional Convolution (Anthony Chang), (3) Medical Information Retrieval (Lynda Tamine, Lorraine Goeuriot); (4) Argumentation Technology in Medicine (Philipp Cimiano, Laura Moss, Olivia Sanchez-Graillet, Basil Ell).

We would like to thank everyone who contributed to AIME 2019. First of all, we would like to thank the authors of the papers submitted and the members of the Program Committee together with the additional reviewers. Thanks are also due to the invited speakers as well as to the organizers of the workshops, the tutorials and doctoral consortium panel. Many thanks go to the local Organizing Committee, who managed all the work making this conference possible. The free EasyChair conference system (http://www.easychair.org/) was an important tool supporting us in the management of submissions, reviews, selection of accepted papers, and preparation of the overall material for the final proceedings. We would like to thank Springer for sponsoring the conference and the Committee on Informatics of the Polish Academy of Sciences for its patronage. Finally, we thank the Springer team for helping us in the final preparation of this LNAI book.

June 2019

David Riaño
Szymon Wilk
Annette ten Teije

Organization

Program Committee

Pedro Henriques Abreu	University of Coimbra, Portugal
Laura Barnes	University of Virginia, USA
Riccardo Bellazzi	Università di Pavia, Italy
Alessio Bottrighi	Università del Piemonte Orientale, Italy
Jacques Bouaud	AP-HP/DRCI and LIMICS, UMR_S 1142
Carlo Combi	Università degli Studi di Verona, Italy
Arianna Dagliati	University of Manchester, Manchester, UK
Stefan Darmoni	Rouen University Hospital, University of Rouen, France
Kerstin Denecke	Bern University of Applied Sciences, Switzerland
Barbara Di Camillo	University of Padova, Italy
Michel Dojat	Grenoble Institut des Neurosciences, France
Georg Dorffner	Medical University Vienna, Austria
Paulo Felix	Universidade de Santiago de Compostela, Spain
Jesualdo Tomás Fernández-Breis	Universidad de Murcia, Spain
Chiara Ghidini	Fondazione Bruno Kessler, Italy
Graciela Gonzalez-Hernandez	University of Pennsylvania, USA
Natalia Grabar	STL CNRS Université Lille 3, France
Adela Grando	Arizona State University, USA
Milos Hauskrecht	University of Pittsburgh, USA
Zhe He	Florida State University, USA
John Holmes	University of Pennsylvania, USA
Arjen Hommersom	Open University, The Netherlands
Zhengxing Huang	Zhejiang University, China
Jose M. Juarez	University of Murcia, Spain
Rafał Jóźwiak	Warsaw University of Technology, Poland
Eleni Kaldoudi	Democritus University of Thrace, Greece
Elpida Keravnou-Papailiou	University of Cyprus, Cyprus
Haridimos Kondylakis	Institute of Computer Science, FORTH, Greece
Pedro Larranaga	University of Madrid, Spain
Nada Lavrač	Jozef Stefan Institute, Slovenia
Michael Liebman	IPQ Analytics, LLC, USA
Helena Lindgren	Umeå University, Sweden
Pawel Liskowski	Poznan University of Technology, Poland
Peter Lucas	Radboud University, The Netherlands

Beatriz López	University of Girona, Spain
Mar Marcos	Universitat Jaume I, Castellón, Spain
Roque Marín	University of Murcia, Spain
Paola Mello	University of Bologna, Italy
Martin Michalowski	University of Minnesota, USA
Silvia Miksch	Vienna University of Technology, Austria
Diego Molla	Macquarie University, Australia
Stefania Montani	University of Piemonte Orientale, Italy
Robert Moskovitch	Ben-Gurion University, Israel
Laura Moss	University of Aberdeen, UK
Fleur Mougin	Université de Bordeaux, France
Goran Nenadic	The University of Manchester, UK
Anthony Nguyen	The Australian e-Health Research Centre, Australia
Øystein Nytrø	Norwegian University of Science and Technology, Norway
Dympna O'Sullivan	National College of Ireland, Ireland
Barbara Oliboni	University of Verona, Italy
Enea Parimbelli	University of Pavia, Italy
Niels Peek	The University of Manchester, UK
Mor Peleg	University of Haifa, Israel
Pedro Pereira Rodrigues	University of Porto, Portugal
Christian Popow	Medical University of Vienna, Austria
Jedrzej Potoniec	Poznan University of Technology, Poland
Cédric Pruski	Luxembourg Institute of Science and Technology, Luxembourg
Silvana Quaglini	University of Pavia, Italy
David Riaño	Universitat Rovira i Virgili, Spain
Lucia Sacchi	University of Pavia, Italy
Aleksander Sadikov	University of Ljubljana, Slovenia
Brigitte Seroussi	Assistance Publique, Hôpitaux de Paris, France
Yuval Shahar	Ben-Gurion University, Israel
Erez Shalom	Ben-Gurion University, Israel
Constantine Spyropoulos	NCSR Demokritos, Greece
Gregor Stiglic	University of Maribor, Slovenia
Annette Ten Teije	Vrije Universiteit Amsterdam, The Netherlands
Paolo Terenziani	Università del Piemonte Orientale Amedeo Avogadro, Italy
Allan Third	The Open University, UK
Samson Tu	Stanford University, USA
Allan Tucker	Brunel University, UK
Ryan Urbanowicz	University of Pennsylvania, USA
Frank Van Harmelen	Vrije Universiteit Amsterdam, The Netherlands
Alfredo Vellido	Universitat Politècnica de Catalunya, Spain
Bartosz Wieloch	Poznan University of Technology, Poland

Szymon Wilk	Poznan University of Technology, Poland
Blaz Zupan	University of Ljubljana, Slovenia
Pierre Zweigenbaum	LIMSI, CNRS, Université Paris-Saclay, France

Additional Reviewers

Amorim, José P.
Barren, Matthew
Barrowman, Michael
Belissen, Valentin
Bordea, Georgeta
Bueno, Marcos
Cattelani, Luca
Chesani, Federico
Da Silveira, Marcos
de Paula Bueno, Marcos Luiz
Di Francescomarino, Chiara
Dragoni, Mauro
Giannakopoulos, George
Groznik, Vida
Hulme, William
Jenkins, David
Katzouris, Nikos
Kirinde Gamaarachchige, Prasadith
Kocbek, Primoz
Koukourikos, Antonis
Krawiec, Krzysztof
Lamberti, Claudio
Le Sueur, Helen
Leonardi, Giorgio
Lin, Lijing
Liu, Manxia
Liu, Siqi
Magge, Arjun
Malakouti, Salim
Marini, Simone
Martin, Glen
Martinez-Millana, Antonio
Massana, Joaquim
Oliveira, Bruno
Pala, Daniele
Palmerini, Luca
Payrovnaziri, Neelufar
Payrovnaziri, Seyedeh Neelufar
Piovesan, Luca
Salvi, Elisa
Santos, Miriam Seoane
Sarker, Abeed
Schut, Martijn
Shi, Boming
Striani, Manuel
Troullinou, Georgia
Viani, Natalia
Visani, Giorgio
Vogiatzis, Dimitrios
Wang, Wenjuan
Weissenbacher, Davy
Williams, Richard
Zamborlini, Veruska

Contents

Probabilistic Models

Behavior Monitoring

Invited Speakers

Common Misconceptions and Future Directions for AI in Medicine: A Physician-Data Scientist Perspective

Anthony Chang(✉)

Medical Intelligence and Innovation Institute (MI3), Children's Hospital of Orange County, Orange, CA 92868, USA
AChang007@aol.com

"*Healthcare is an information industry that continues to think that it is a biological industry.*" (Laurence McMahon at the AAHC Thought Leadership Institute meeting, August, 2016)

The following are common misconceptions (by both clinicians as well as data scientists) about AI in medicine that are understandable and human:

Clinicians will be replaced by AI. There is a fundamental lack of understanding of what clinicians do even amongst august data scientists and seasoned venture capitalists, and this deficiency renders it easy to think that computer vision and interpretation of medical images alone is sufficient to replace image-intensive subspecialties like radiologists, pathologists, ophthalmologists, dermatologists, and cardiologists. The doctor's tasks can be divided into perception (visual image interpretation and integrative data analytics), cognition (creative problem solving and complex decision making), and operation (procedures) (see Fig. 1). The computer is much stronger in perception tasks but not yet facile with cognition nor operation parts of the clinician's tasks. It is also notable that all those who proclaim an end to radiologists are not radiologists (or even physicians) themselves.

AI can be applied to every aspect of healthcare to bring value. While AI can improve workflow and increase accuracy of diagnoses, there are certain technologies that will not necessarily benefit from AI application. For example, applying AI to an older technology (like auscultation for heart murmurs) may not increase the value proposition of such an application. There is a myriad of workflow deficiencies in healthcare, however, that can be improved with AI but these deficiencies are often neglected. It is therefore important to remember design thinking principles in applying AI ("design AI") in delineating problems first.

AI, in conquering the game Go, will be successful for medicine and healthcare. AI was indeed successful in defeating the human champion in the ancient game of Go. Medicine and healthcare, especially in arenas such as a busy intensive care or emergency room setting or in domains such as chronic disease management and population health, is more akin to the real time strategy game of *Starcraft* where real time decisions on multiple fronts need to be made in a complex milieu that is different for each individual (so essentially it is even more akin to playing hundreds of *Starcraft* games simultaneously). It should be noted, however, that AI was recently successful in defeating a human real time strategy player.

D. Riaño et al. (Eds.): AIME 2019, LNAI 11526, pp. 3–6, 2019.
https://doi.org/10.1007/978-3-030-21642-9_1

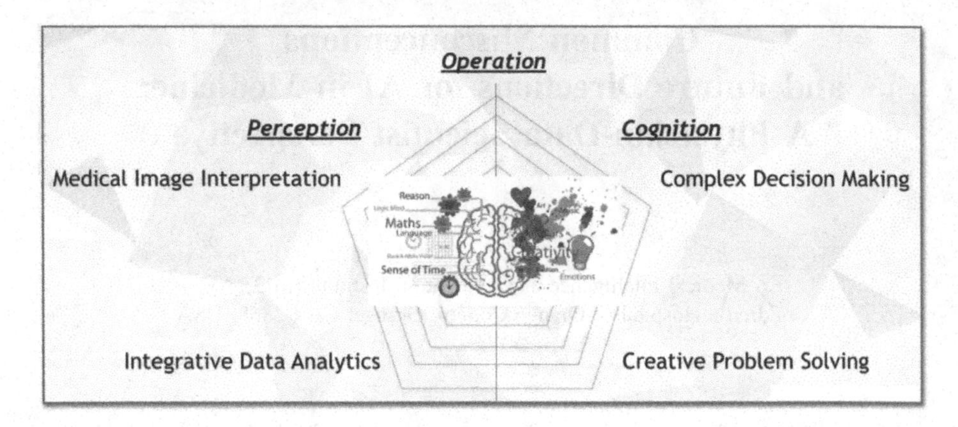

Fig. 1. The Clinician's Brain. The clinician has perception tasks such as medical image interpretation and integrative data analytics. In addition, the clinician often have cognition tasks such as complex decision making and creative problem solving, especially in chronic disease management and rare medical conditions. Lastly, some clinicians like surgeons, cardiologists, anesthesiologists, and intensive care physicians often perform procedures and this involves special spatiotemporal and manual skill part of the brain.

Deep learning, especially convolutional neural network (CNN), will be the preferred AI tool for a long time. While deep learning is in its hype and is indeed very effective for computer vision and medical image interpretation as well as decision support, the future of these areas will need even more sophisticated tools involving cognitive architecture, which is the third wave of AI. Even deep learning gurus such as Geoff Hinton feels that cognitive elements such as capsular networks will be needed to improve deep learning performance [1]. Deep learning with CNN will also need to be even more sophisticated in the future with tools such as recursive cortical network and transfer learning as there are limitations to the amount of available medical data. In addition, while CNN has been a major contribution in medical image interpretation, recurrent neural network (RNN) with robust natural language processing can also be equally as valuable by extracting information and knowledge from hospital and clinic records. Variations of CNN with RNN can also be used to examine videos which are commonplace in medical imaging [2].

We need more biomedical data for deep learning in healthcare. There are several instances that big data in medicine and healthcare are not feasible. One situation involves patients with rare diseases as there is a limit in the sheer number of patients with the rare disorder with equal limit in the number of medical images. Another situation would involve a very sophisticated or invasive test or a test with excessive risks and/or costs; these tests would result in very few samples in a population. In all of these cases, creative uses of generative adversarial networks (GANs) or one shot learning to neutralize the lack of big data can obviate the absolute necessity of big data in biomedicine.

The area under the curve (AUC) of the receiver operating curve (ROC) is a good indicator of the performance of the algorithm. First, similar to parents having higher expectations of their children than they have of themselves, clinicians and data scientists can have relatively high (and perhaps unfair) expectations of AI. It is not uncommon for human clinicians to have not much better than 50% accuracy of certain medical diagnoses and yet we have such higher (albeit understandable) expectations for machine intelligence [3]. To use the AUC of the ROC as the sole determinant of the accuracy of machine intelligence for a test, however, is problematic [4]. Some of this problem lies in the fact that the labels on images are often not entirely accurate to begin with (humans are labeling and are not infallible). Additional issues in the large datasets include: lack of precise terminology (consolidation vs pneumonia), time element in diagnoses (early vs late manifestation of pneumonia), presence of multiple labels (diagnoses are often not exclusionary or dichotomous), and variability of the dataset (quality of image). Three key elements are necessary for there to be an accurate conclusion of performance derived from this ROC AUC assessment: accuracy, threshold, and prevalence of disease, the latter is a critical element for the analysis but often not included in the overall study description.

You have to be able to program to make a contribution to AI in medicine. There are many ways other than actually programming and coding that anyone in medicine and healthcare can contribute to the overall paradigm shift of AI in medicine. The most glaring deficiency in the AI in medicine domain is not the lack of AI tools, but the quality and management of biomedical data. First, any clinician can provide domain expertise for an AI project or idea that can be misdirected. In addition, any healthcare worker can also delineate the workflow inadequacies so pervasive in healthcare for an AI project. Lastly, anyone in healthcare can also contribute to the foundation of the data-information-knowledge-intelligence pyramid by focusing more on the quality of data and integrity of data infrastructure.

AI is mainly for selected subspecialists like radiologists and pathologists. While AI and deep learning have made significant contributions in these fields, the use of other AI methodologies such as cognitive computing as well as robotic process automation (RPA) and natural language processing are helpful for almost all other subspecialists. In addition, these tools are essential in reducing the administrative burden of healthcare systems irrespective of the clinical domain. AI, therefore, has much more to offer other than CNN and deep learning with medical image interpretation; the portfolio of AI tools will all provide a new resource to alleviate the burden of healthcare delivery in all areas.

AI will make clinicians less human. With the appropriate application of AI, especially natural language processing (NLP) and understanding (NLU) tools, clinicians will be in a position to be more human as the burden and distraction of electronic medical record would be lessened. It would be a laudable vision that the future sanctuary of the physician-patient setting will have no visible machines in the environment but will be the venue for only human-to-human bonding [5].

AI devices will be difficult to be understood or regulated. There is the possibility of a self fulfilling prophecy if advanced AI tools lack explainability. Even if we treat AI and its panoply of tools as "software-as-a-device", how we can effectively and expediently approve all these upcoming AI tools as these emerge will be a challenge. Perhaps we need to match this exponential paradigm shift in technology with a parallel trajectory in how we regulate. One potential solution is to not regulate AI devices per se but rather teams and/or individuals working on the AI tools in a specific program or institution. Another possible answer lies in the sagacious Turing philosophy of "machines to machines" and devise regulatory algorithms that will overlook algorithms on a continuous basis in addition to periodic checks by regulatory agencies.

A final one: AI in medicine will be here in the future. As the computer scientist William Gibson so eloquently stated: "the future is already here, it is just not evenly distributed". The trajectory of medicine needs to change given the exponentially increasing amount of data and information to attain precision medicine and chronic disease management of our population. All stakeholders in both clinical medicine and data science have a special opportunity to create a special synergy for a once-in-a-generation transformative paradigm shift in medicine.

It is more than 100 years since the Flexner report that shaped our present medical school education strategy, and it is now more important than ever to reassess our medical educational strategy. The advent of AI is a precious gift from our technological colleagues, and while AI is not necessarily going to replace clinicians, it should be part of every medical student's educational curriculum as well as every physician's clinical portfolio from this point forward.

References

1. Hinton, G.: Deep learning – a technology with the potential to transform health care. JAMA **320**(11), 1101–1102 (2018)
2. Yu, F., et al.: Assessment of automated identification of phases in videos of cataract surgery using machine learning and deep learning techniques. JAMA New Open **2**(4), e191860 (2019)
3. Hill, A.C., Miyake, C.Y., Grady, S., Dubin, A.M.: Accuracy of interpretation of pre-participation screening electrocardiograms. J. Pediatr. **159**(5), 783–788 (2011)
4. Mallett, S., Halligan, S., Collins, G.S., Altman, D.G.: Exploration of analysis methods for diagnostic imaging tests: problems with ROC AUC and confidence scores in CT colonography. PLoS ONE **9**(10), e107633 (2014)
5. Verghese, A., Shah, N.H., Harrington, R.A.: What this computer needs is a physician: humanism and artificial intelligence. JAMA **319**(1), 19–20 (2018)

AI in Healthcare: Ethical and Privacy Challenges

Ivana Bartoletti(✉)

Head of Privacy and Data Protection, Gemserv and Founder, Women Leading in AI Network, London, UK
ivanabartoletti@hotmail.co.uk

Abstract. The deployment of Artificial Intelligence in healthcare is extremely promising and although AI is no panacea, harnessing patient data will lead to precision medicine, help detect disease before they manifest and support independent living for the elderly, amongst many other things. However, this progress will not be without challenges from both an ethical and privacy standpoint. These issues need understanding from policy makers and developers alike for AI to be embraced responsibly.

Keywords: AI · Healthcare · Privacy · Ethics

1 AI in Healthcare

Healthcare represents one of the most interesting and promising areas for the deployment of AI systems [1]. We can now detect cancer faster and earlier than before, identify diseases before they manifest and spot genetic disorders which may affect us later down the line.

Likewise, algorithms can streamline back office processing thus improving patient hospital experience as well as saving considerable resources by reducing waste and inefficiency – resources that can then be invested into better patient care.

However, healthcare is a sector where technology must meet the law, regulations, and privacy principles to ensure innovation is for the common good. This is because the deployment of AI in the medical field brings a plethora of challenges from a privacy and ethics standpoint, namely: the safeguarding of patient data, the ethical boundaries of innovation and the actual impact of technology on medics and patients alike.

2 Privacy Challenges

Privacy is a complex concept: it is culturally bound, and it evolves with time. It is culturally bound as it is influenced by the cultural norms of the specific country it inhabits, thus varying across territories, generations and backgrounds. Therefore, it is conceivable that it will also evolve alongside technological development, too. To an extent, we are already seeing the transition of our 'personhood' into data citizenship as we become increasingly monitored with data collection points placed at every corner of our daily life; through our smartphones as well as the tracking of our movements online.

D. Riaño et al. (Eds.): AIME 2019, LNAI 11526, pp. 7–10, 2019.
https://doi.org/10.1007/978-3-030-21642-9_2

Within this transformation, we are becoming acquainted with the fact that our data is of great value – as well as becoming a real currency that we can use to access entertainment and sometime services at no cost. Alongside this, we are also witnessing a public awakening regarding the malicious use of data for microtargeting, online manipulation and behavioral advertising that is leading to a decrease in public trust towards the way private and public organizations handle our data. This distrust ranges from fear of CCTV cameras and state monitoring to the devastating effects of online manipulation which last year culminated in the Cambridge Analytica scandal.

It is therefore of no surprise that citizens are becoming wearier of collection points, unethical use of data and transparency fallacies, namely the privacy notices that – due to their nature as long and often unintelligible documents – are often brushed off.

Understanding the wider attitude and changes towards privacy is key when it comes to healthcare, as trust becomes paramount. Health data is the most private information about one's self whilst being the most valuable resource to improving wellbeing, defeating diseases and supporting the elderly, those with disabilities and in social care.

3 Building Trust in Algorithms Deployment in Healthcare

Challenges around privacy emerging from the use of AI in healthcare must not be underestimated, and it is essential to recognise that the distinction between personal data and sensitive information is increasingly blurred [2] as we can now infer health information from behavior patterns and other data which is not sensitive in the first place.

The deployment of algorithms requires a clear roadmap, and that includes:

- *Data Privacy Impact Assessments* to assess the risks of privacy harms, identify the privacy enhancing technologies that need to be deployed to safeguard patient data.
- *Algorithmic Impact Assessments* to ensure algorithms do not have bias embedded into them. As algorithms use historic data, a level of bias may be inherent, or it can emerge via proxy, class labels or associations. It is crucial to ensure all tools are in place to identify risks of bias harms as well as to debias algorithms at the onset as well as later down the line.
- *Audit trails* must be accurately kept to ensure logs are kept of who is doing what, which data are used and what changes are being made to the systems.
- Procurement law in healthcare setting must ensure purchase of AI systems from third parties needs to adhere to strict procedures including assessments of how the algorithms have been trained, whether they have been audited and whether they have been assigned a trustmark recognizing due process in their algorithms processing.
- The establishment of a clear governance framework overseeing the use of AI, including the set up of the three lines of defense to mitigate risks.

4 Wider Ethical Issues Surrounding the Deployment of AI in Healthcare

There is no doubt that the interplay between data, data analytics, robotics and AI is extremely complex, nor that it will challenge the current regulatory establishment.

The key underlying issue is that 'big data' challenges the core principle of privacy – which is to collect as little data as possible. Yet big data is about collecting all available information. This dichotomy is hard to reconcile and stands as the core challenge to privacy as we have known it so far. The use of mobile medical apps, wearables, chatbots, connected devices etc., shows the amount of data that can be collected. For this data to be harnessed, clear codes of privacy and ethics need to be established and adhered to. Equally, it is perhaps necessary for these codes to be enforced, and for regulators to conduct audits into the algorithms to ensure they follow due process.

But AI is much more than technology and the deployment of AI in healthcare is likely to be challenging in other areas.

One challenge, for example, will be the impact on the workforce and how medics will be trained and learn how to cooperate and work alongside the machine. How will the machine augment a doctor's capabilities? How will they develop the confidence in it and, similarly, the detachment should they feel the need to intervene? And, equally, what will happen to medical knowledge should decisions and diagnostics be increasingly made by machines? These are all incredibly important questions we cannot ignore if we want to exploit the opportunities offered by artificial intelligence.

Lastly, one key challenge will be accountability, and namely who is responsible in case a mistake is made [3]. Interestingly, patients are used to relating to doctors, and understand an error can be made by virtue of their humanity. How will patients react to errors made by machines? This consideration is in addition to the legal issues around accountability, namely who will hold the responsibility for a mistake. Courts will have to intervene, and the law will need to catch up to identify what is the best normative answer to this problem.

5 Conclusion

The deployment of AI in health is promising and very welcome. Cooperation between doctors and machines could represent a turning point with regards to our ability to tackle diseases and improve our wellbeing.

From precision and targeted medicine to back office operations and leaner processes, from support with independent living for the elderly to greater diagnostic ability, the benefits will certainly be invaluable.

However, it is likely that AI in health will also challenge the boundaries of both current regulatory systems and privacy principles [4]. It is therefore essential to adopt a cautious approach in order to maximise the positive whilst reducing the risks of privacy, bias and ethics harms.

References

1. Bartoletti, I.: Algorithms may outperform doctors, but they're no healthcare panacea. The Guardian. https://www.theguardian.com/commentisfree/2018/jul/26/tech-healthcare-ethics-artifical-intelligence-doctors-patients. Accessed 26 Mar 2019
2. Marks, M.: Artificial intelligence based suicide prevention. Yale J. Health Policy Ethics. https://ssrn.com/abstract=3324874. Accessed 26 Mar 2019
3. Calo, R.: Robots and privacy. In: Lin, P., Beckey, G., Abney, K. (eds.) Robot Ethics: The Ethical and Social Implications of Robotics, pp. 187–202. MIT Press, Cambridge (2012)
4. Terry, N.: Of regulating healthcare AI and robots. https://ssrn.com/abstract=3321379. Accessed 26 Mar 2019

Deep Learning

Recent Context-Aware LSTM for Clinical Event Time-Series Prediction

Jeong Min Lee(✉) and Milos Hauskrecht

University of Pittsburgh, Pittsburgh, PA 15260, USA
jlee@cs.pitt.edu, milos@pitt.edu

Abstract. In this work, we propose a novel clinical event time-series model based on the long short-term memory architecture (LSTM) that can predict future event occurrences for a large number of different clinical events. Our model relies on two sources of information to predict future events. One source is derived from the set of recently observed clinical events. The other one is based on the hidden state space defined by the LSTM that aims to abstract past, more distant, patient information that is predictive of future events. We evaluate our proposed model on electronic health record (EHRs) data derived from MIMIC-III dataset. We show that the combination of the two sources of information implemented in our method leads to improved prediction performance compared to the models based on individual sources.

Keywords: Recurrent neural network · Event time series prediction

1 Introduction

Successful modeling of complex multivariate event time series and their ability to predict future events is important for applications in various areas of science, engineering, and business. In clinical settings our ability to predict future events for a patient based on clinical events observed in past, such as past medication orders, past labs and their results, or past physiological signals can help us to anticipate the occurrence of a wide range of future events that would let health care practitioners intervene ahead of time or prepare resources to get ready for their occurrence. All of this can in turn improve the quality of patient care.

One of the challenges of modeling clinical event time series is their complexity, that is, clinical event time series for hospitalized patients may consist of thousands of different types of events corresponding to administration of many different medications, lab orders, arrivals of lab results, or various physiological observations, etc. This complexity may not fit very well standard Markov time series models [19] with either observed or hidden state and transition models.

To alleviate the event complexity problem we propose to develop a new more scalable event time series model based on the long-short-term-memory (LSTM) [14] that relies on two sources of information to predict future events. One source is derived from the set of recently observed clinical events. The other one is based

D. Riaño et al. (Eds.): AIME 2019, LNAI 11526, pp. 13–23, 2019.
https://doi.org/10.1007/978-3-030-21642-9_3

on the hidden state space defined by the LSTM that aims to abstract past, more distant, patient information that is predictive of the future events. In the context of Markov state models, the next state in our models and the transition to the next state is defined by a combination of the recent state (most recent events) and the hidden state summarizing more distant past events.

In order to evaluate the proposed model, we use data derived from electronic health records (EHRs) of critical care patients in MIMIC-III dataset [16]. The clinical events considered in this work correspond to multiple types of events, such as medication administration events, lab test result events, physiological result events, and procedure events. These are combined together in a dynamically changing environment typical of intensive care units (ICUs) with patients suffering from severe life-threatening conditions.

Through extensive experiments on MIMIC-III data we show that our model outperforms multiple time series baselines in terms of the quality of event predictions. To provide further insights to its prediction performance we also divide the results with respect to different types of clinical events considered (medication, lab, procedure and physiological events), as well as, based on their repetition patterns, again showing the superior performance of our proposed model.

2 Related Work

2.1 Event-Time Series Models

The majority of discrete time-series models are based on Markov processes [24, 25]. Markov process models rely on Markov property that assumes that the state captures all necessary information relating future and past. In other words, the next state depends only on the most recent state, and is independent of the past states. In this case the joint distribution of an observed sequence is modeled as chain of conditional probabilities: $p(y_1, y_2, ..y_T) = p(y_1) \prod_{t=2}^{T} p(y_t|y_{t-1})$

For Markov process models, the conditional probability defining a transition is parameterized by an $e \times e$ transition matrix where e denotes all possible states: $A_{i,j} = p(y_t = j|y_{t-1} = i)$. Standard Markov processes assume all states of the time series are directly observed. However, the states of many real-world processes are not directly observable. One way to resolve the problem is to define the state in terms of a limited number of past observations or features defined on past observations [11,12,31].

Hidden State Models. Another is to use Hidden Markov models (HMM) [29] that introduce hidden states z_t of some dimension d. Now the observations y_t is defined in terms of the hidden states and an $e \times d$ emission table B with components: $B_{i,j} = p(y_t = j|z_t = i)$. Briefly, the transition table A is used to update the hidden states and the emission table is used to generate observations.

HMM has been shown to reach good performance in many applications such as stock price prediction [10], DNA sequence analysis [15], and time-series clustering [28]. However, classic HMM model comes with drawbacks when applied to real-world time series: the hidden state space is discrete, and the transition model is restricted to transitions in between the discrete states. Linear dynamical models (LDS) [17] remedy some of the limitations by defining real-valued hidden state-space with linear transitions among the current and next hidden state. One problem with HMM and LDS models is that the dimensionality of their hidden state space is not known a priori. Various methods for hidden state space regularization, such as work by Liu and Hauskrecht [21,22] for LDS have been able to address this problem.

Continuous Time Models. We would like to note that in addition to discrete time series models, the researchers have explored also methods permitting continuous time models. Examples are various version of Gaussian process models for predicting multivariate time series in continuous time, including those used for representing irregularly sampled clinical time series [20,23].

Neural-Based Models. Recent advances in neural architectures and their application to time-series offer end-to-end learning framework that is often more flexible than standard time-series models. In neural-based approaches, the discrete time series are typically modeled using recurrent neural network (RNN) which provides a more flexible framework for modeling time-series. Similarly to HMM and LDS, RNN uses hidden states to abstract and carry information from past history but with more flexible hidden state defined by real-valued vectors and transition rules. At each time step, hidden state is updated given the previous time step's hidden states and a new information from the current time step's input. Although its limitations on vanishing and exploding gradient problems [13], its variants such as long short-term memory (LSTM) [14] unit and gated recurrent units (GRU) [2] allow wide adoptions in event time-series modeling. They have been applied to prediction and modeling time series [1,9], vision [8], speech [7], and language [30] problems.

2.2 Clinical Event Time-Series Modeling

Modeling and prediction of discrete event time series in the healthcare area have been influenced greatly by advances in various neural architectures and deep learning. [3] used Skipgram [26] to represent and predict next visit in outpatient data. But they evaluated their model on the prediction task at the level of hospital visit, which can be of a very coarse granularity for real-world clinical applications that encompass event-specific time information. [4] modeled clinical time series with RNN and attention mechanism. However, the model is only able to perform binary classification on a whole-sequence level. Our model is able to predict fine-grained future event at the level of each time step of a sequence. [6] also used neural network models to predict the sequence of clinical events. In their approach, the patient pool was limited to patients with kidney failure and organ transplant. On the other hand, our model is tested and shows superior performances over baselines across general clinical time series that were not limited to a specific patient cohort.

3 Methodology

In this section, we first introduce state-space Markov and LSTM-based event time series models and then present our model combining the two models.

State-Space Markov Event Prediction. Given an observed events sequence $\mathbf{y} = y_1, y_2, ..., y_T$, we can model $\mathbf{y}$ by defining a Markov transition model relating the current event state y_t with the next event states y_{t+1}. In this case, we assume the event space is formed by a multivariate binary vector reflecting the occurrence of many different events (encoded as 1) over some time-window. One way to parameterize the transition between two consecutive event states is to use a transition matrix W with a bias vector b. As we want to predict multivariate binary vector, we can use sigmoid function $\sigma(x) = \frac{1}{1+e^{(-x)}}$ as the output activation function:

$$\hat{y}_{t+1} = \sigma(W \cdot y_t + b) \tag{1}$$

LSTM-Based Event Prediction. LSTM models are being successfully used to model time series with the help of hidden state vector, allowing one to summarize in the hidden state information from more distant past. At a glance, at each time step of a sequence, LSTM gets current (event) input and updates its hidden states. The hidden state then generates signals for the next hidden state, as well as, predictions for the occurrence of events in the next time-step.

In detail, at each time step t, events in the input sequence represented as multi-hot vector m_t is processed to a real-valued vector x_t through linear embedding matrix W^{emb}: $x_t = W^{(emb)} \cdot m_t$. Then, given processed input x_t and previous hidden states h_{t-1}, LSTM updates hidden states h_t:

$$\begin{aligned} f_t &= \sigma(W^{(f)} \cdot [h_{t-1}, x_t] + b^{(f)}) & i_t &= \sigma(W^{(i)} \cdot [h_{t-1}, x_t] + b^{(i)}) \\ o_t &= \sigma(W^{(o)} \cdot [h_{t-1}, x_t] + b^{(o)}) & \tilde{C}_t &= \tanh(W^c \cdot [h_{t-1}, x_t] + b^{(c)}) \\ C_t &= f_t \cdot C_{t-1} + i_t \cdot \tilde{C}_t & h_t &= o_t \otimes \tanh(C_t) \end{aligned}$$

f_t, i_t, and o_t are forget, input and output gates and $\otimes$ denotes element-wise multiplication. With these parameters ready, we can update hidden states:

$$h_t = \text{LSTM}(x_t, h_{t-1})$$

Future event occurrence prediction is generated through a fully-connected layer W^q with output activation function sigmoid:

$$\hat{y}_{t+1} = \sigma(W^{(fc)} \cdot h_t + b^{(fc)}) \tag{2}$$

This parameterization links to the state space based event predictor. When y_t of Eq. 1 is replaced to hidden states h_t, it becomes Eq. 2.

Recent Context-Aware LSTM-Based Event Predictor. When properly trained, hidden states in LSTM can be sufficient to represent and model future behaviors of event time-series by abstracting dependencies of past and future events. However, to be trained properly, LSTM (or any deep-learning based models) requires large amounts of training instances. In the clinical domain, obtaining large amounts of clinical cases (e.g., rarely ordered medication or lab tests) is hard in general. This constraint may deter us to train LSTM for predicting rare clinical cases. Meanwhile, for certain clinical event category such as medications, the future occurrence of an event may highly depend on recent previous or current occurrence of the event type and incorporating this information may help to resolve the data deficiency constraint.

Therefore, to address the problem, we propose and develop an adaptive mechanism that refers to both abstracted information of past sequence through hidden states of LSTM and concrete information about event occurrences in very recent context window. Different from the preliminary LSTM-based output generation in Eq. 2 that only depends on abstracted hidden states of LSTM, we directly refer to recent event occurrence information. The recent event at the current time step t is in multi-hot vector m_t and it is incorporated into the model through a linear transformation to model:

$$b^{(u)} = W^{(s)} \cdot m_t + b^{(s)}$$

$b^{(s)}$ can be seen as additional bias term that reflects recent event occurrence information and final prediction for event occurrence is made as follows:

$$\hat{y}_{t+1} = \sigma(W^{(fc)} \cdot h_t + b^{(fc)} + b^{(u)})$$

The proposed predictor also can be seen as combining the LSTM based predictor with state-space based Markov predictor. Especially, in context of Markov state models, the next state in our models and the transition to the next state is defined by a combination of the recent state (most recent events) and the hidden state summarizing more distant past events.

Loss Function. To measure the performance of the event prediction, $\mathcal{L}$ is defined as binary cross entropy between label vector y_t and prediction vector $\hat{y}_t$ over all sequences in the training set and $\mathbf{1}$ denotes a vector filled with 1s:

$$\mathcal{L} = \sum_t -[y_t \cdot \log \hat{y}_t + (\mathbf{1} - y_t) \cdot \log(\mathbf{1} - \hat{y}_t)]$$

Parameter Learning. The parameters of the model is learned by back propagation through time (BPTT) [32] with adaptive stochastic gradient descent based optimizer [18]. Hyper-parameters are tuned by F1-score performances on validation set with following ranges: embedding ($W^{(emb)}$) size in $\{128, 256, 512\}$; hidden states size in $\{512, 1024, 2048\}$ and learning rate $= 0.005$ batch size $= 512$. To prevent over-fitting, early stopping and dropout ($p = 0.5$) are applied.

4 Experimental Evaluation

4.1 Clinical Data

We test the proposed model on MIMIC-III, a clinical database generated from real-world EHRs of intensive care unit patients [16]. We extract 21,897 patients whose records are generated from Meta Vision system that is one of the systems used to create records in the MIMIC-III database. We extract patient in age between 18 and 99 and whose length of stay in ICU is between 3 and 20 days. We randomly split patients into the train, test, and validation sets with the ratio of 7:2:1 and generate multivariate event time-series by segmenting sequences with both input-window and future window with size $W = 24$. At the end of each input-window, its future-window is generated.

We consider the following types of events in our models: medication administration events, lab results events, procedure events, and physiological result events. **Medication administration events** indicate records of specific kind of medication administered to the patient. **Lab results events** indicate lab test and its results represented as normal, abnormal-high, or abnormal-low. **Procedure events** indicate records of procedures patient received during hospitalization. For medication, lab, and procedure event categories, we select those events observed in more than 100 different patients. **Physiological result events** consist of 23 cardiovascular, routine vital signs, respiratory, and hemodynamics signals selected by a critical care expert. Similarly to the lab result events, numeric physiological results are discretized to normal, abnormal-high, and abnormal-low. Table 1 shows the basic data statistics.

Table 1. Clinical data statistics by event categories

Category	Medication	Procedure	Lab test	Physio signal
Cardinality	136	79	1197	102
Num. of occurrences	803K	257K	4266K	8378K
Proportion of positive label	5.9%	3.2%	3.6%	83.1%

4.2 Evaluation Metrics

We evaluate the quality of time series predictions using area under precision-recall curve (AUPRC) and area under the receiver operating characteristic curve (AUROC). Although AUROC is commonly used to present result for binary classification problems, it can provide misleading information when applied to highly imbalanced dataset. On the other hand, AUPRC provides more accurate profile on performances of models under such circumstances [5,27]. As shown in Table 1, our dataset is severely skewed to negative examples. Therefore, we use AUPRC as our primary evaluation measurement.

4.3 Baseline Models

We compare our proposed model to the dense logistic regression models defined upon the following inputs (predictors):

Current Markov state (Markov) as defined in Eq. 1.
Binary History (LR-binary): Unlike the current Markov state information, this model considers the occurrence of all past events (not just the most recent one) and encodes them into one multi-hot vector.
Count History (LR-count): This model, similarly to Binary history, summarizes all past events (not just the most recent ones), but instead of multi-hot vector representation it uses a vector of event counts.
Current LSTM state (LSTM): The model uses the hidden state of the LSTM to summarize information from distant past important for prediction.

4.4 Results

All our evaluations were performed on the test set, that was not touched during the training and validation steps. Prediction results in Fig. 1 summarize the performance of our model and baselines on 24-h prediction window. The results show that our model outperforms all baselines in terms of both AUROC and AUPRC statistics. Moreover, the Markov state model is better than pure LSTM in terms of AUPRC. This shows the information from the most recent time window is most of the time the most important source for predicting the next step events. This is not surprising given the fact that many events (such as drug administrations or lab orders) are repeated every 24-h, hence once they are observed they are most likely to occur also in the next time window.

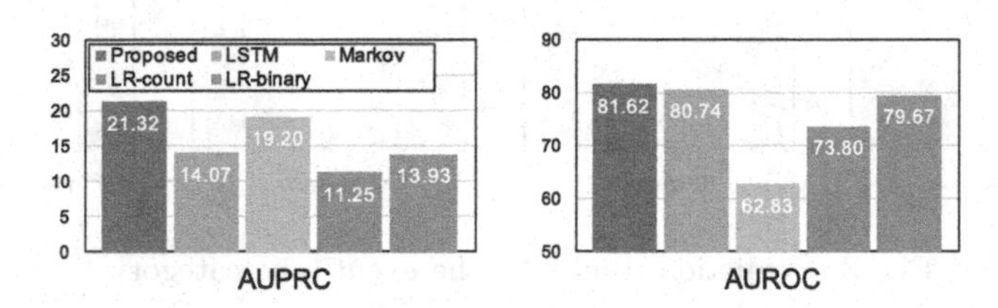

Fig. 1. Overall time-series prediction results on the 24-h window segmentation

To verify the above reasoning, and to provide further insights into the predictive performance of our models, we break the above results by considering separately predictions when the same events occurred in the previous time step and when they did not. We refer to these as to repetitive and non-repetitive patterns. The results are given in Fig. 2. From the results, we can clearly see that predicting non-repetitive events is significantly more difficult than predicting repetitive ones. However, despite this, we can also see that our model consistently outperforms other baselines across both repetitive and non-repetitive scenarios. Remarkably for non-repetitive event prediction, our model's AUROC is 32% higher than average of all baseline models in AUPRC and 11% in AUROC.

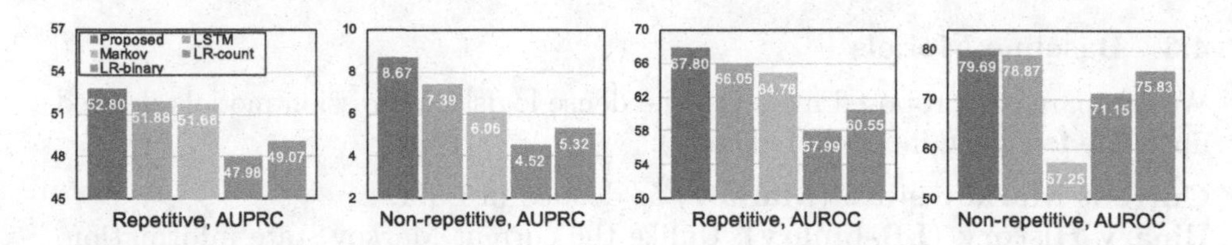

Fig. 2. Prediction results on repetitive and non-repetitive events

To analyze our results further, we next break the evaluation down by inspecting predictive performances of the models for the different event categories. The results are shown in Fig. 3. Clearly, our model consistently outperforms baseline models across all event categories in both AUROC and AUPRC statistics.

So far, all our results were obtained by considering the window size of 24 h. Next, we investigate the predictive performance of the models by varying the prediction window size. More specifically we will consider the window size W of length 6, 12 and 24 h. Due to space limits, we will consider and compare the methods only using AUPRC statistics. As shown in Fig. 4, our model shows superior performance across all time-resolutions.

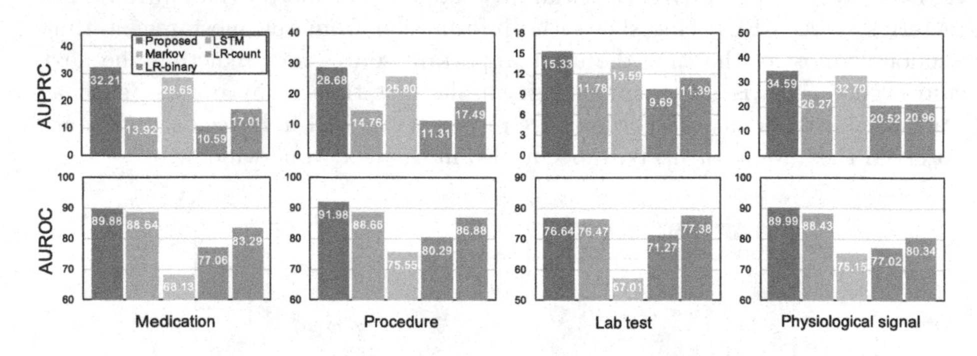

Fig. 3. Prediction results by the event type category

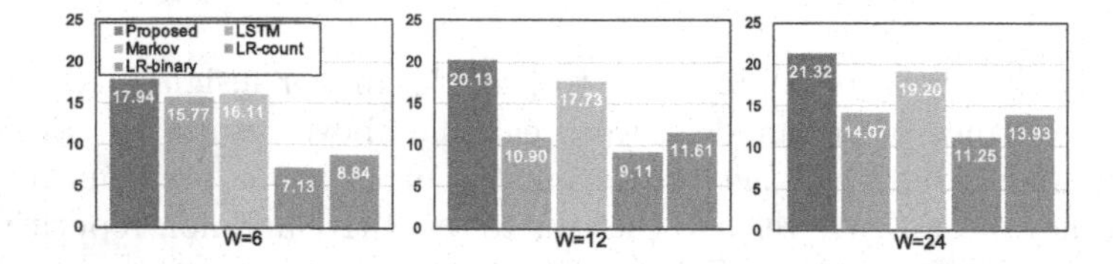

Fig. 4. AUPRC prediction statistics for the different window sizes

To dig deeper into the time segmentation results, in Fig. 5 we show the predictive performance of lab test and physiological result events. We can see that on lab test event prediction, our model dominates at larger window sizes ($W = 12, 24$): it outperforms baseline models by 27%. In smaller window size ($W = 6$), the LSTM performs slightly better than ours by 2%. On physiological event prediction, our model surpasses all baselines across all time resolutions.

Interestingly, on lab event prediction, overall predictability is high at $W = 24$ and deteriorates for smaller window sizes. This reflects the recurrent characteristic of lab events at a cycle of 24 h, that is, lab tests and their results are ordered and observed most of the time once daily. Inversely, the overall predictability of physiological events decreases with increasing window length. It indicates a recurrent characteristic of clinical events but in different recurring interval that is much shorter. Most physiological result events are automatically generated from bedside monitoring devices at short intervals, typically at a scale of seconds to minutes. Therefore, the variability of observation on a time series generated from smaller windows should be less than those of larger windows. Hence, overall predictability on smaller time resolution is consistently higher than larger ones as seen in Fig. 5.

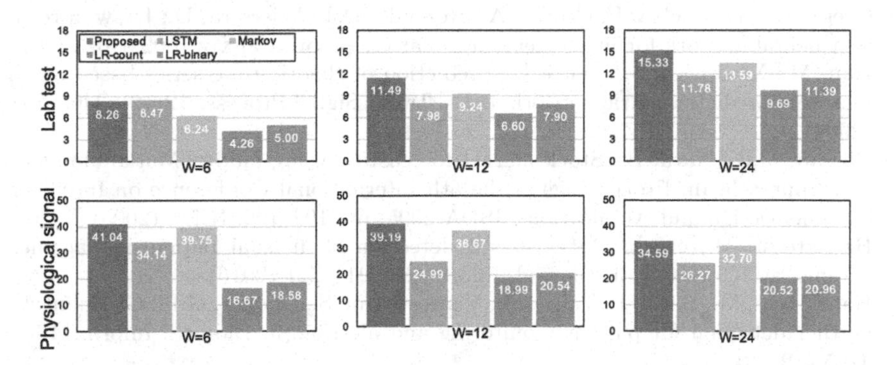

Fig. 5. Prediction result for lab and physiological events for the different window sizes

5 Conclusion

In this work, we show the importance of two sources of information for event-time series modeling. One source is derived from the set of recently observed clinical events and the other is based on the hidden states of LSTM that aims to abstract past, more distant, patient information that is predictive of future events. We show that the combination of the two sources of information implemented in our method leads to improved prediction performance on MIMIC-III clinical event data when compared to models that rely only on individual sources.

Acknowledgement. The work in this paper was supported by NIH grant R01GM-088224. The content of the paper is solely the responsibility of the authors and does not necessarily represent the official views of NIH.

References

1. Chen, P.A., Chang, L.C., Chang, F.J.: Reinforced recurrent neural networks for multi-step-ahead flood forecasts. J. Hydrol. **497**, 71–79 (2013)
2. Cho, K., et al.: Learning phrase representations using RNN encoder-decoder for statistical machine translation. arXiv preprint arXiv:1406.1078 (2014)
3. Choi, E., et al.: Multi-layer representation learning for medical concepts. In: The 22nd ACM SIGKDD International Conference on Knowledge Discovery and Data Mining (2016)
4. Choi, E., Bahadori, M.T., Sun, J., Kulas, J., Schuetz, A., Stewart, W.: Retain: an interpretable predictive model for healthcare using reverse time attention mechanism. In: Advances in Neural Information Processing Systems (2016)
5. Davis, J., Goadrich, M.: The relationship between precision-recall and the ROC curves. In: Proceedings of the 23rd International Conference on Machine Learning, pp. 233–240. ACM (2006)
6. Esteban, C., Schmidt, D., Krompa, D., Tresp, V.: Predicting sequences of clinical events by using a personalized temporal latent embedding model (2015)
7. Graves, A., Jaitly, N.: Towards end-to-end speech recognition with recurrent neural networks. In: International Conference on Machine Learning, pp. 1764–1772 (2014)
8. Gregor, K., Danihelka, I., Graves, A., Rezende, D.J., Wierstra, D.: Draw: a recurrent neural network for image generation. arXiv preprint arXiv:1502.04623 (2015)
9. Han, M., Xi, J., Xu, S., Yin, F.L.: Prediction of chaotic time series based on the recurrent predictor neural network. IEEE Trans. Signal Process. **52**(12), 3409–3416 (2004)
10. Hassan, M.R., Nath, B.: Stock market forecasting using hidden Markov model: a new approach. In: Proceedings of the 5th International Conference on Intelligent Systems Design and Applications, ISDA 2005, pp. 192–196. IEEE (2005)
11. Hauskrecht, M., et al.: Outlier-based detection of unusual patient-management actions: an ICU study. J. Biomed. Inform. **64**, 211–221 (2016)
12. Hauskrecht, M., Batal, I., Valko, M., Visweswaran, S., Cooper, G.F., Clermont, G.: Outlier detection for patient monitoring and alerting. J. Biomed. Inform. **46**(1), 47–55 (2013)
13. Hochreiter, S., Bengio, Y., Frasconi, P., Schmidhuber, J., et al.: Gradient flow in recurrent nets: the difficulty of learning long-term dependencies (2001)
14. Hochreiter, S., Schmidhuber, J.: Long short-term memory. Neural Comput. **9**(8), 1735–1780 (1997)
15. Hughey, R., Krogh, A.: Hidden Markov models for sequence analysis: extension and analysis of the basic method. Bioinformatics **12**(2), 95–107 (1996)
16. Johnson, A.E., et al.: MIMIC-III, a freely accessible critical care database. Sci. Data **3**, 160035 (2016)
17. Kalman, R.E.: Mathematical description of linear dynamical systems. J. Soc. Ind. Appl. Math. **1**(2), 152–192 (1963)
18. Kingma, D.P., Ba, J.: Adam: a method for stochastic optimization. arXiv preprint arXiv:1412.6980 (2014)
19. Lipton, Z.C., Berkowitz, J., Elkan, C.: A critical review of recurrent neural networks for sequence learning. arXiv preprint arXiv:1506.00019 (2015)
20. Liu, Z., Hauskrecht, M.: Clinical time series prediction: toward a hierarchical dynamical system framework. Artif. Intell. Med. **65**(1), 5–18 (2015)
21. Liu, Z., Hauskrecht, M.: A regularized linear dynamical system framework for multivariate time series analysis. In: The 29th AAAI Conference on Artificial Intelligence, pp. 1798–1804 (2015)

22. Liu, Z., Hauskrecht, M.: Learning linear dynamical systems from multivariate time series: a matrix factorization based framework. In: SIAM International Conference on Data Mining (2016)
23. Liu, Z., Wu, L., Hauskrecht, M.: Modeling clinical time series using Gaussian process sequences. In: SIAM International Conference on Data Mining (2013)
24. MacDonald, I.L., Zucchini, W.: Hidden Markov and Other Models for Discrete-Valued Time Series, vol. 110. CRC Press, Boca Raton (1997)
25. McKenzie, E.: Ch. 16. discrete variate time series. Handb. Stat. **21**, 573–606 (2003)
26. Mikolov, T., Sutskever, I., Chen, K., Corrado, G.S., Dean, J.: Distributed representations of words and phrases and their compositionality. In: Advances in Neural Information Processing Systems, pp. 3111–3119 (2013)
27. Saito, T., Rehmsmeier, M.: The precision-recall plot is more informative than ROC plot when evaluating binary classifiers on imbalanced datasets. PLoS ONE **10**(3), e0118432 (2015)
28. Smyth, P.: Clustering sequences with hidden Markov models. In: Advances in Neural Information Processing Systems, pp. 648–654 (1997)
29. Stratonovich, R.L.: Conditional Markov processes. Theory Probab. Appl. **5**(2), 156–178 (1960)
30. Sutskever, I., Vinyals, O., Le, Q.V.: Sequence to sequence learning with neural networks. In: Advances in Neural Information Processing Systems, pp. 3104–3112 (2014)
31. Valko, M., Hauskrecht, M.: Feature importance analysis for patient management decisions. In: International Congress on Medical Informatics, pp. 861–865 (2010)
32. Werbos, P.J.: Backpropagation through time: what it does and how to do it. Proc. IEEE **78**(10), 1550–1560 (1990)

Deep Learning Approach for Pathogen Detection Through Shotgun Metagenomics Sequence Classification

Ying-Feng Hsu[1(✉)], Makiko Ito[2], Takumi Maruyama[3], Morito Matsuoka[1], Nicolas Jung[4], Yuki Matsumoto[4], Daisuke Motooka[4], and Shota Nakamura[3]

[1] Cybermedia Center, Osaka University, Osaka, Japan
{yf.hsu,matsuoka}@cmc.osaka-u.ac.jp
[2] Fujitsu Laboratories Ltd., Kanagawa, Japan
maki-ito@jp.fujitsu.com
[3] Fujitsu Limited, Tokyo, Japan
takumi_maruyama@jp.fujitsu.com,
nshota@gen-info.osaka-u.ac.jp
[4] Genome Information Research Center,
Research Institute for Microbial Diseases, Osaka University, Osaka, Japan
{nicolasj,matsumoto,daisukem}@gen-info.osaka-u.ac.jp

Abstract. Studies have shown that shotgun metagenomics sequencing facilitates the evaluation of diverse viruses, bacteria, and eukaryotic microbes and assists in exploring their abundances in complex samples. Due to the challenges of processing a substantial amount of sequences and overall computational complexity, it is time-consuming to analyze these data through traditional database sequence comparison approaches. Deep learning has been widely used to solve many classification problems, including those in the bioinformatics field, and has demonstrated its accuracy and efficiency for analyzing large-scale datasets. The purpose of this work is to explore how a long short-term memory (LSTM) network can be used to learn sequential genome patterns through pathogen detection from metagenome data. Our experimental result showed that we can obtain similar accuracy to the conventional BLAST method, but at a speed that is about 36 times faster.

Keywords: Shotgun metagenomics sequencing · Sequence classification · Deep learning · LSTM · GPU acceleration · Parallel computing

1 Introduction

Shotgun metagenomics sequencing allows the extraction of a large number of both genes and species information from a given ecosystem. As compared to other types of DNA sequencing, it produces many short reads of 20 to 500 base pairs that are long and sensitive enough for clinical pathogen detection. Studies have shown that this method enables the evaluation of the diversity of viruses, bacteria, and eukaryotic microbes, and can help to estimate their abundances in given complex samples. Due to

D. Riaño et al. (Eds.): AIME 2019, LNAI 11526, pp. 24–30, 2019.
https://doi.org/10.1007/978-3-030-21642-9_4

challenges that come from computational complexity from millions of reads within a relatively short time, traditional sequence comparison approaches may require a much longer time to obtain the results of a given analysis. Those methods, such as the Smith-Waterman algorithm [1], BLAST [2], and BWA [3] use sequence alignment to measure the similar sequences from the reference database which is highly time-consuming process and may not be suitable for the massive amount of shotgun metagenomics sequencing studies. In this paper, we use a deep learning method (LSTM model) to discover the sequential patterns of given reference databases and have adopted it as a disease discriminator to evaluate the given shotgun metagenomics samples.

2 Methodology

To develop a reliable deep learning model, it is essential to have high-quality training data. Our data collection and preprocessing method include both an NCBI public database as ground true reference data and shotgun sequence data from 28 clinical samples as for our evaluation dataset. Those clinical samples include 5 HCV-positive (hepatitis C viruses) patients and nasal swab samples from 13 influenza-infected patients. For negative control samples, we used blood samples from 10 healthy donors. The RNA/DNA extraction of those 28 clinical samples was performed by MagNA Pure 96 System. Each library was prepared using a Nextera Library Prep Kit. For each sequence in the NCBI database, we generate more subsequences by the shifting sequence generation function below, with a multiple randomly generated *overlap_ratio*.

Shifting sequence generation = (sequence_length/subsequence_length) * (overlap_ratio)

In this manner, those shifting sequences facilitate the sequence alignment operation, while deep learning extract characteristics of those shifting sequences and support the polymorphism and indel type of sequence comparison. Please note that the *subsequence_length* is determined by target testing data set. In our case, we find the suitable length to be 50 for our pathogen detection use case of HCV or Influenza. For traditional sequence alignment approaches such as BLAST, the size of the reference database has a significant impact on overhead processing time. In contrast, for deep learning, it only increases the model training time, and there is not much difference when considering the task of sequence classification. Deep learning is considered to be an alignment-free approach [4], and it uses mathematical functions to learn the distance between the input sequence and its distance to the class boundary.

2.1 The Model: LSTM Model

An LSTM network is a type of recurrent neural network (RNN) that is designed to learn the long-term dependencies between time steps of sequential data, which meets the concept of analyzing DNA genomic sequences. To cope with the complicated scenario of sequence classification from both sequence alignment and polymorphism, we enhanced the above model with the bagging type of ensemble learning method which is a meta-algorithm that combines several machine learning prediction results and uses the meta-classifier to make the final decision. Figure 1 describes a scenario where we divide a sequence of 151 bp to 5 subsequences with a shifting sequence window of 25 and two discrimination functions. Different subsequence predictions capture information from different angles with different advantages and disadvantages. By adequately leveraging the uniqueness of them, in most cases, it is possible to obtain a higher prediction accuracy than a single classification [5]. The discrimination function is customized by different pathogen detection use cases such as for detecting a different type of virus or bacteria. Since the preceding nucleotide has a higher base calling quality to determine the sequence classification result, we usually impose its importance in our discrimination function. From our observation, in many cases, there is a no obvious effect for those later nucleotides, such as subsequence 03 or 04 in this example. In other words, our approach can maintain the same level of accuracy when increasing the processing speed by dynamically eliminating the number of evaluation subsequences.

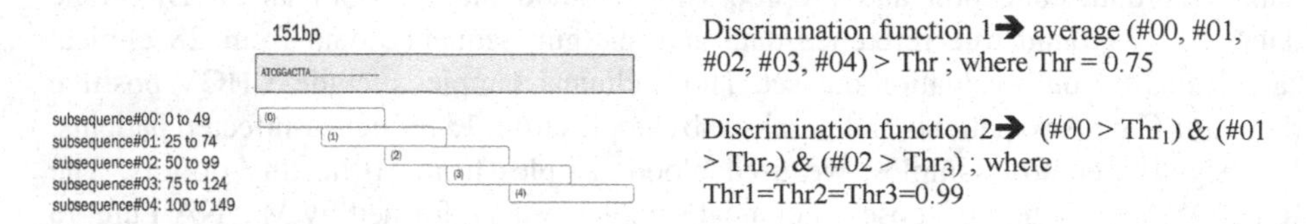

Fig. 1. Ensemble prediction results with discrimination function.

Depending on the reading capability of NGS technology, a clinical sample may produce from a million to hundreds of millions of shotgun metagenomics sequences. In a conventional computation use case, sequence comparison normally runs under a single CPU. Analyzing this type of big-data task is highly challenging and time-consuming without efficient sequence comparison algorithms. Depending on the data size and parameter setting of the LSTM model, our model training time varies from 12 to 47 h. However, for the task of sequence classification, by using multi-GPU acceleration, our proposed method is able to split hundred million of shotgun DNA sequences into many smaller batches of DNA sequences and concurrently process them within minutes.

2.2 Sequencing Data Preprocessing

DNA sequences and reads commonly have varying length. In our implementation, we use a fixed sequence length, because of two underlying reasons determined from the shotgun sequences sample. First, shotgun sequencing is relatively shorter when compared to other types of genome sequencing datasets, and it is relatively easy to adjust the sequence length instead of directly inputting various amounts of base pairs (bp) into the model. Second, sequence base calling is the process that converts the raw image data to nucleotide sequences through the sequencing instrument. The quality of nucleotide reads decreases along with the length of sequence reads, which are caused by both the sample quality and the capability of the DNA sequencer.

In this study, we obtained the sequencing data with paired-end within 200 bp read length by using an Illumina HiSeq 2500. The average quality of reads distribution from our 28 clinical samples which have approximately of 103 million shotgun sequences. About 86.34% of these samples have an average Phread quality score greater than 30. From the aspect of base calling quality, a significantly decreasing happens around a base position of 110. To ensure that the input subsequences always included a portion of high-quality reads as the input of the proposed ensemble LSTM model, we ignored those low-quality base reads and used a fixed 50 bp subsequences.

3 Experimental Results and Discussion

3.1 Pathogen Detection (True Positive Test)

Influenza Test Result

Figure 2 shows the result of all sequences selected from the 13 influenza-infected clinical samples which have about 4.8 million shotgun sequences per sample. By comparing the number of overlapping sequences with our proposed method, it seems that our result appears almost like a subset of BLAST. To further discover the sensibility behavior, we zoom in on the quality of the BLAST output by using the expected value (e-value). Figure 3 shows our result with the BLAST threshold of an E-value less than 1e-30, which is considered to be a high-quality hit for homology matches. Overall, our result is almost identical with BLAST in all samples. Precisely, it reports only slightly more influenza-related sequences than BLAST. Those sequences do not necessarily mean a false positive result, since BLAST does also generate erroneous outputs.

HCV Test Result

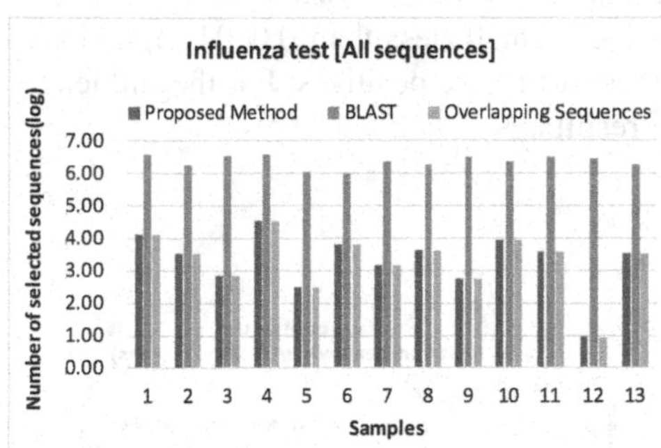

Fig. 2. Influenza test (all sequences)

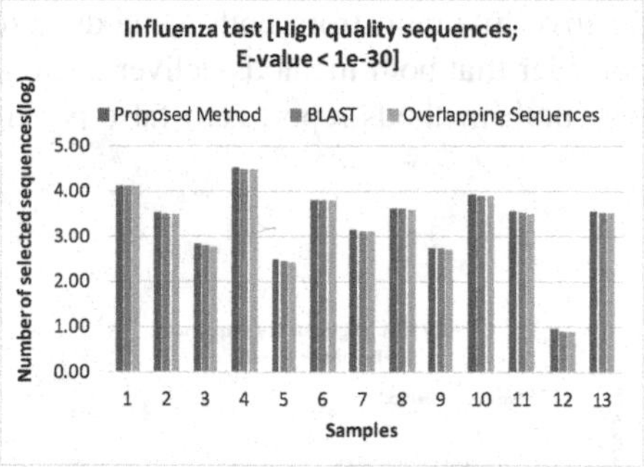

Fig. 3. Influenza test (high-quality sequences)

Figure 4 shows the sequences selection result from the 5 HCV-positive samples, which have about 2.1 million shotgun sequences per sample. In this test, our method does not have any overlapping sequences with BLAST in both sample 1 and sample 2. However, even the BLAST method only reports 2 and 1 HCV-related sequences from about 2 million sequences, respectively. From a clinical viewpoint, this could be due to the patient being in the early stage of hepatitis C. For sample 3, 4, and 5, our approach demonstrates a high degree of overlapping with BLAST, and both methods report a significantly larger amount of HCV-related sequences than those reported with samples 1 and 2. The BLAST method seems much more sensitive than our approach in this case, since it reports more possible HCV-related sequences. As compared to the test result of influenza, our method seemed to struggle to identify the HCV viral sequences. We consider that this effect is due to the higher diversity of human-genome–related sequences from the blood sample, which increase the noise level and may cause difficulty in discovering the HCV viral sequences.

3.2 False Positive Tests from Healthy Samples

Other than the accuracy of successfully detecting Influenza or HCV-related sequences, a pathogen detection system should also avoid the chance of false positive detection. In this study, we collected blood samples from 10 healthy individuals and we consider a false positive occurs when the system reports any influenza or HCV sequences in those ten samples. Figure 5 shows the false positives of those ten healthy clinical samples from both our proposed method and the BLAST method. The BLAST method produced false prediction sequences in all ten samples with about ten false positive sequences per sample, while our method reported false positive HCV sequences from 9 healthy human samples with an average of only 2 errors per sample. Apparently, our method behaves better because it produces the fewer false positive sequence. Since those healthy samples have an average of about 2.9 million sequences, the false-positive hits ratio from both methods is relatively small (less than 0.0004%), and we consider that both methods deliver a similar result of false positives. For the Influenza test, both methods report zero false positive results.

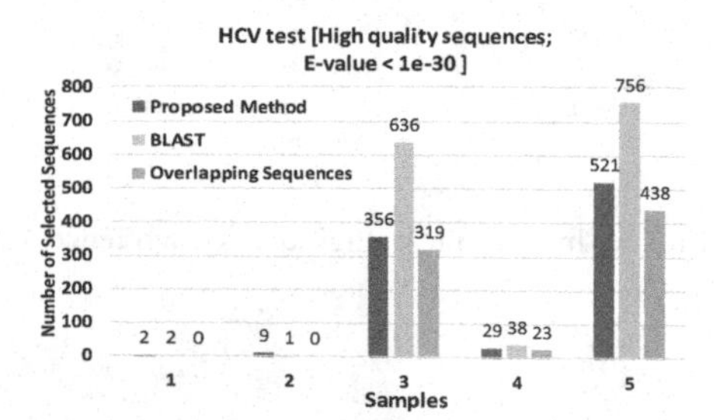

Fig. 4. HCV test (high-quality sequences)

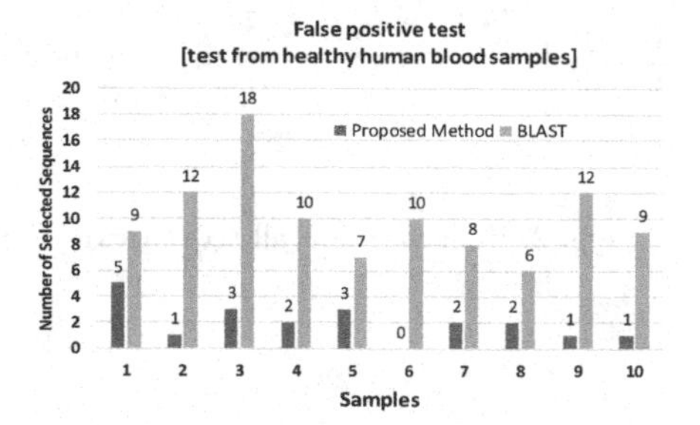

Fig. 5. False positive test from healthy samples

3.3 Time Complexity Test

Other than the accurate identification of the viral-related sequences, the computational speed is as important as the accuracy. Figure 6 shows the processing speed comparison for both our proposed deep learning approach and the conventional BLAST method. The y-axis describes how many millions of sequences are processed and analyzed per minute, and the x-axis shows the testing platform. We compared the processing speed in various scenarios from a single CPU (Intel(R) Xeon(R) E5-2690 v4 @ 2.60 GHz) and a single GPU to a multi-GPU (Tesla P100-PCIE-16 GB). From a single CPU test, our method can process 0.13 million sequences, which is slower than the 0.37 million sequences from BLAST. The proposed method is ordinarily designed for maximizing GPU acceleration with multiple GPUs for detecting target DNA sequences. We observe the linear increase pattern when the number of GPUs increase. We observed that as the number of GPUs increases, the number of processing sequences per minute linearly increases. Our best case (8 GPUs) shows that our method can evaluate about 13.38 million sequences per minute, which is about 36 times faster than the BLAST method.

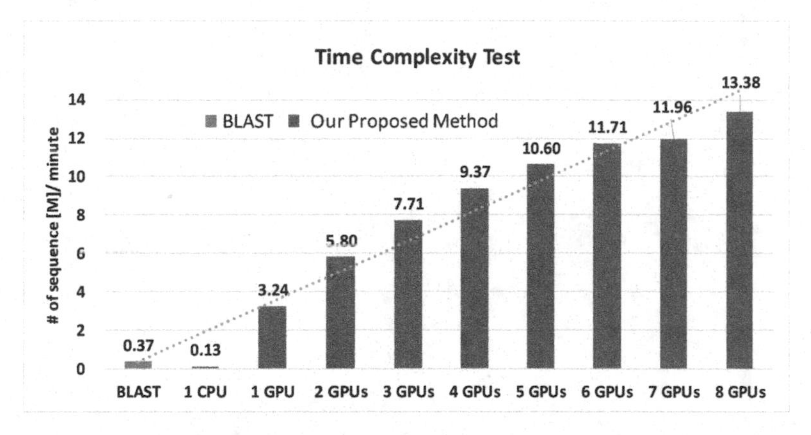

Fig. 6. Sequence processing speed (million sequences/minute).

4 Conclusions

In this work, we explored how an LSTM network can be used to learn sequential genome patterns through pathogen detection from metagenome data. We collected and conducted case studies that analyzed influenza, HCV viral sequences, and healthy samples. We evaluated the accuracy of pathogen detection with GPU acceleration. Our experimental result shows that we obtained similar accuracy to the conventional BLAST method, but at a speed that was about 36 times faster.

Acknowledgments. The authors are members of Fujitsu next generation Cloud Research Alliance Laboratory (FCRAL). This research and development work was partially supported by the MIC/SCOPE #172107106 and by Fujitsu Ltd.

References

1. Gotoh, O.: An improved algorithm for matching biological sequences. J. Mol. Biol. **162**, 705–708 (1982)
2. NCBI: BLAST: Basic Local Alignment Search Tool. https://blast.ncbi.nlm.nih.gov/Blast.cgi
3. BWA: Aligner Burrows-Wheeler (BWA). http://bio-bwa.sourceforge.net/
4. Zielezinski, A., Vinga, S., Almeida, J., Karlowski, W.M.: Alignment-free sequence comparison: benefits, applications, and tools. Genome Biol. **18**, 186 (2017)
5. Sill, J., Takacs, G., Mackey, L., Lin, D.: Feature-Weighted Linear Stacking, arXiv:0911.0460

Fully Interactive Lungs Auscultation with AI Enabled Digital Stethoscope

Tomasz Grzywalski[1], Riccardo Belluzzo[1(✉)], Mateusz Piecuch[1], Marcin Szajek[1], Anna Bręborowicz[2], Anna Pastusiak[1,3], Honorata Hafke-Dys[1,3], and Jędrzej Kociński[1,3]

[1] StethoMe®, Poznań, Poland
{grzywalski,belluzzo,piecuch,szajek}@stethome.com

[2] Department of Pediatric Pneumonology, Allergology and Clinical Immunology, Poznań University of Medical Sciences, Poznań, Poland
abreborowicz@wp.pl

[3] Institute of Acoustics, Faculty of Physics, Adam Mickiewicz University, Poznań, Poland
{anna.pastusiak,h.hafke,jen}@amu.edu.pl

Abstract. Performing an auscultation of respiratory system normally requires the presence of an experienced doctor, but the most recent advances in artificial intelligence (AI) open up a possibility for the laymen to perform this procedure by himself in home environment. However, to make it feasible, the system needs to include two main components: an algorithm for fast and accurate detection of breath phenomena in stethoscope recordings and an AI agent that interactively guides the end user through the auscultation process. In this work we present a system that solves both of these problems using state-of-the-art machine learning algorithms. Our breath phenomena detection model was trained on 5000 stethoscope recordings of both sick (hospitalized) and healthy children. All recordings were labeled by a pulmonologist and acousticians. The agent is able to accurately assess patient's lung health status by auscultating only 3 out of 12 locations on average. The decision about each next auscultation location or end of examination is made dynamically, after each recording, based on breath phenomena detected so far. This allows the agent to make best prediction even if the auscultation is time-constrained.

Keywords: AI in healthcare · Deep learning · E-health · Telemedicine · Digital stethoscope · Lung sounds auscultation

1 Introduction

Lung sounds auscultation was made popular by Leannec, who invented the stethoscope in 1816. While the most notable features of auscultation are non-invasiveness, simplicity and low cost associated with the device, its major drawback is its subjectivity, since the examination results depend on physician's

D. Riaño et al. (Eds.): AIME 2019, LNAI 11526, pp. 31–35, 2019.
https://doi.org/10.1007/978-3-030-21642-9_5

abilities to interpret the respiratory sounds [2]. A way to overcome this inherent limitation is by digital recording and subsequent computerized analysis [1].

Many efforts have been reported in literature to automatically detect lung sound pathologies using digital signal processing and simple time-frequency analysis [1]. In recent years, however, machine learning techniques have gained popularity because of their potential to find significant diagnostic information on statistical distribution of data itself. Currently, benchmark results for breath phenomena detection are obtained by implementation of deep neural networks (DNNs) and their variants [3]. DNNs consist of stacked layers of neurons, that process raw data by multiple non-linear transformations, thus incorporating the feature extraction itself in the training phase. If sufficient training examples are shown to the network, these learned features are much more distinctive and descriptive in comparison to hand-crafted features by experts [3].

When it comes to design a fully interactive lungs auscultation system, breath phenomena detection information gathered in the single auscultation point (AP) must be fused with the rest of the context, in such a way that decisions about how to proceed with the auscultation can be taken. A mathematical framework that could allow to do that is Reinforcement Learning (RL). In the common RL problem, the algorithm (*agent*) learns to solve complex problems by interacting with an environment, which in turn provides rewards or penalties depending on the results of the actions taken. The objective of the agent is thus to find the best action to take when being in a certain state (*policy*), in order to maximize the received reward [4]. We believe that adoption of the RL framework to model the problem of conducting lung sounds auscultation, can be the key to develop a fully interactive lung sounds analysis system. Also, since no literature about it was found, we contribute by presenting the first adoption of RL to this problem.

This paper presents a novel method of automatic and interactive lung sounds analysis based on DNNs and RL, which has been implemented in a system that uses a digital stethoscope for capturing respiratory sounds.

2 Fully Interactive Lungs Auscultation

Our solution consists of two main algorithms, combined together: the first one is a RL agent, trained to determine the status of the patient as fastest as possible, i.e. using the lowest number of APs as possible. The second one is a convolutional recurrent neural network (CRNN) that serves the agent as a feature extractor, providing as input to the agent – a fixed number of values representative of detection and intensity level of critical phenomena. In Fig. 2 a systematic overview of the application is given, showing how these two components interact. The entire application is enabled by a digital stethoscope (connected to a smartphone) that captures the respiratory sounds.

Breath Phenomena Detection. Our neural network architecture is a modified implementation of the one proposed in [5], i.e. a CRNN designed for polyphonic sound event detection. Convolutional layers act as pattern extractors, recurrent layers integrate the extracted patterns over time providing context information, and finally feedforward layers produce the activity probabilities for each class [5]. The network, trained with 5000 recordings is able to detect 7 types of sound events: inspirations (i), expirations (e), wheezes (w), rhonchi (r), fine and coarse crackles (fc, cc) and noise (n) [9] (Fig. 1).

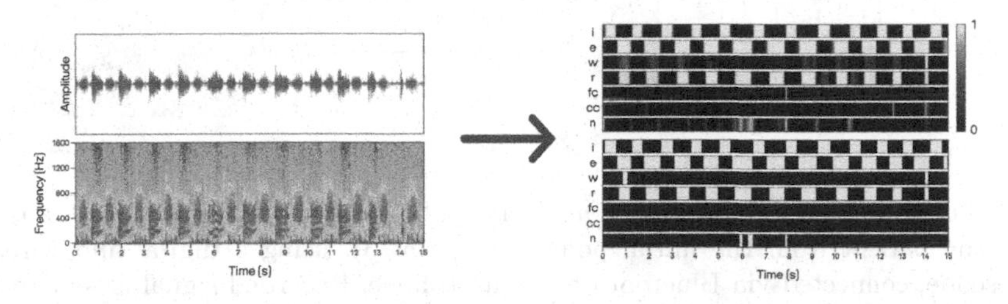

Fig. 1. The signal is transformed into spectrogram and analyzed by the CRNN. The output of the model is presented in form of raster where rows represent time, framed in windows of 10 ms each; and columns show the probability of positive detection of each phenomenon.

Reinforcement Learning Agent. The states for the agent are represented by the list of APs already auscultated, each one described with a set of features, obtained by post-processing the binarized raster coming from the breath phenomena detection module. When receiving the input state, the agent can decide either to auscultate another point or predict status of the patient if confident enough. Rewards are given if the predicted diagnosis is correct, penalties in the opposite case. A small penalty is given for each auscultated point, in order to discourage the agent from using too many points. Best policy for the agent is thus embodied in the best auscultation path, described as sequence of most informative APs. The agent itself is a direct implementation of the Q-learning algorithm [6], where Q-values are approximated by a DNN (deep Q-network) whose weights are updated through stochastic gradient descendent optimization algorithm, with the objective of maximizing the expected future reward.

3 Experiments

Dataset. The database used for the research was based on a large amount of actual auscultation recordings batched in more than 400 so-called *visits* (set of 12 recordings taken in different locations) captured in realistic conditions by experienced pediatricians using StethoMe® and Littmann 3200® digital stethoscopes [7,8]. The two devices used for data gathering were working in the same frequency range and the data was eventually normalized before being sent to the algorithms. Each sample was described by one to three experienced pediatricians and acousticians in terms of presence of adventitious sounds in certain phases of breathing cycle. For this, taggers were provided by an interactive, proprietary web-based interface that allows fast labelling of the time series data.

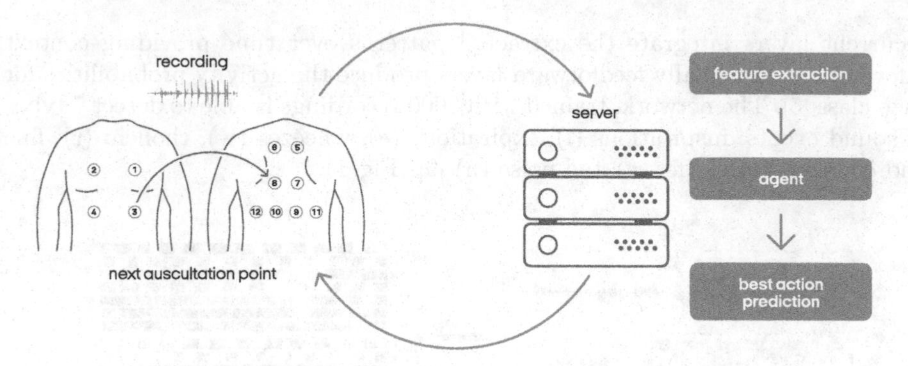

Fig. 2. Fully interactive lung sounds analysis application: the examiner starts auscultating the patient from the initial point (e.g. point 3), using a digital and wireless stethoscope, connected via Bluetooth to a smartphone. Recorded signal is sent to the server where a fixed set of features are firstly extracted based on breath phenomena detection algorithm. Then, these features are given as input to the agent that predicts next best to be taken. This is sent back to device and shown to the user (e.g. to auscultate point 8). The auscultation goes on until the agent declares an alarm value.

Moreover, for each visit an alarm label was assigned considering severeness of pathological sounds found: 0, when zero or minor (innocent) pathologies are detected; 1, when severe pathologies are found. In the first case, there is no need to consult a doctor (*no alarm*), while in the second case the patient should consult a specialist (*alarm*). In case there was more than one label for the same visit, the highest label value was taken. The database was build with signed consent from the parents of children and was approved by bioethical commission.

Evaluation. We evaluated the accuracy of the two components of the proposed solution separately. After properly tuning its main hyper-parameters (learning rate, number of layers, number of neurons per layer, etc.), the CRNN was trained with the whole 5000 recordings and tested on 2142 unseen recordings (*hidden* test set), gathered in the same way as described in the previous section, but left apart for testing purposes [9]. Then, we compared the performance of the RL (*interactive*) agent against its *static* counterpart, i.e. an agent that always performs auscultation using all 12 APs. Note that both of the agents were using the trained CRNN as feature extractor. In order to access this comparison, we performed 5-fold cross validation (CV) for 30 different random splits of the dataset into training and validation sets, simulating the auscultation procedure as depicted in Fig. 2. All best selected models for each iteration were finally tested on the hidden test set. Averaged scores are reported in Table 2. For completeness, in Table 1 results of breath phenomena detection are presented, where N_{GT} and N_{BPD} indicate the number of 10 ms frames that had positive occurrence of the event (phenomenon), for ground truth and model predictions respectively.

Table 1. Breath phenomena detection

Event	N_{GT}	N_{BPD}	Precision	Recall	Bacc
i	86907	85438	93.2%	91.7%	93.2%
e	112796	115230	91.5%	93.5%	91.2%
fc	15892	16817	58.9%	62.3%	79.2%
cc	9717	7719	40.9%	32.5%	65.0%
w	7448	6530	62.3%	54.6%	76.7%
r	11683	13050	64.8%	72.4%	85.0%
n	73705	82409	73.1%	69.3%	78.3%

Table 2. Interactive agent versus static agent

Agent	Bacc	$F1_{alarm}$	$F1_{not\ alarm}$	APs
Static	84.8%	82.6%	85.1%	12
Interactive	82.3%	81.8%	82.6%	3.2

4 Conclusions

We presented a system being able to perform lungs auscultation in a fully interactive way. This system, enabled by a digital stethoscope capturing respiratory sounds, relies on a RL agent that uses features extracted by a CRNN in order to estimate which is the best action to take at every stage of the auscultation. The agent is able to assess patient's lungs health auscultating only 3 out of 12 locations on average, keeping an acceptable diagnosis accuracy if compared to an exhaustive auscultation using all possible locations. We believe that this system, once deployed on large scale, will reduce number of unneeded visits to the doctors, convert a lot of real visits into telemedicine, but also support practitioners by providing more objective tool for breath phenomena detection.

References

1. Palaniappan, R., Sundaraj, K., Ahamed, N., Arjunan, A., Sundaraj, S.: Computer-based respiratory sound analysis: a systematic review. IETE Tech. Rev. 248–256 (2013). https://doi.org/10.4103/0256-4602.113524
2. Mangione, S., Nieman, L.Z.: Pulmonary auscultatory skills during training in internal medicine and family practice. Am. J. Respir. Crit. Care Med. **159**, 1119–1124 (1999). https://doi.org/10.1164/ajrccm.159.4.9806083
3. Kilic, O., Kiliç, Z., Kurt, B., Saryal, S.: Classification of lung sounds using convolutional neural networks. EURASIP J. Image Video Process. (2017)
4. Dayan, P., Niv, Y.: Reinforcement learning: the good, the bad and the ugly. Curr. Opin. Neurobiol. **18**, 185–196 (2008). https://doi.org/10.1016/j.conb.2008.08.003
5. Çakir, E., Parascandolo, G., Heittola, T., Huttunen, H., Virtanen, T.: Convolutional recurrent neural networks for polyphonic sound event detection. CoRR, abs/1702.06286
6. Watkins, C.J.C.H., Dayan, P.: Q-learning. Mach. Learn. 279–292 (1992). https://doi.org/10.1007/BF00992698
7. StethoMe®. https://stethome.com/. Accessed 30 Jan 2019
8. Littmann 3200®. https://www.littmann.com/. Accessed 30 Jan 2019
9. Grzywalski, T., et al.: Practical implementation of artificial intelligence algorithms in pulmonary auscultation examination. Eur. J. Pediatr. 1–8. https://doi.org/10.1007/s00431-019-03363-2

An Automated Fall Detection System Using Recurrent Neural Networks

Francisco Luna-Perejon(✉), Javier Civit-Masot, Isabel Amaya-Rodriguez, Lourdes Duran-Lopez, Juan Pedro Dominguez-Morales, Anton Civit-Balcells, and Alejandro Linares-Barranco

Robotics and Computer Technology Laboratory, University of Seville, 41012 Seville, Spain
fralunper@atc.us.es

Abstract. Falls are the most common cause of fatal injuries in elderly people, causing even death if there is no immediate assistance. Fall detection systems can be used to alert and request help when this type of accident happens. Certain types of these systems include wearable devices that analyze bio-medical signals from the person carrying it in real time. In this way, Deep Learning algorithms could automate and improve the detection of unintentional falls by analyzing these signals. These algorithms have proven to achieve high effectiveness with competitive performances in many classification problems. This work aims to study 16 Recurrent Neural Networks architectures (using Long Short-Term Memory and Gated Recurrent Units) for falls detection based on accelerometer data, reducing computational requirements of previous research. The architectures have been tested on a labeled version of the publicly available SisFall dataset, achieving a mean F1-score above 0.73 and improving state-of-the-art solutions in terms of network complexity.

Keywords: Fall detection · Deep Learning · Recurrent Neural Networks · Long Short-Term Memory · Gated Recurrent Units · Accelerometer

1 Introduction

According to the World Health Organization [12], unintentional falls are one of the most frequent causes of injuries in people over 65 years. Approximately 28%–35% of this cohort suffer at least one fall per year. This topic is gaining importance due to the progressive elderly population increase. Major injuries pose significant risk for postfall morbidity and mortality. This risk has been shown to be closely correlated to the delay in assist with first aid after the fall [7].

Fall detection systems (FDS) are devices that monitor the user's activity, and ideally alert when a fall has occurred. They allow sending an accident notification immediately to medical entities, caregivers and family members for a quick assistance. Among all the different FDS types, wearable devices allow a

D. Riaño et al. (Eds.): AIME 2019, LNAI 11526, pp. 36–41, 2019.
https://doi.org/10.1007/978-3-030-21642-9_6

continuous monitoring without dependence on the environment. This kind of devices usually use accelerometers and different algorithms to distinguish between daily activities and fall events [4]. Although threshold based algorithms show very high performance in terms of detection effectiveness and low computational complexity, they present many difficulties when trying to adapt them to new types of falls and user complexion. Machine learning methods are considered more sophisticated approaches to try to solve this problem. However, traditional supervised classification algorithms are not suitable due to the sequential nature of fall events, that implies a large computational cost, and the scarcity of datasets to study these events [5].

Recurrent Neural Networks (RNN) such as Long Short-Term Memory units (LSTM) and Gated Recurrent Units (GRU) are Deep Learning networks specifically designed to process sequences. Recent studies shed some light on the potential of RNNs for accelerometers [6]. This work focuses on finding a cost-effective RNN architecture in terms of computational complexity and effectiveness for fall detection in real-time.

2 Materials and Methods

2.1 Dataset

SisFall dataset [10] is used in this study. This dataset is composed of several simulated activities mainly classified in falls and activities of daily living (ADL). Each sample contains accelerometer measurements obtained from a device fixed to the user's waist. The measurement's sampling frequency is 200 Hz. The dataset was complemented with labeling criteria proposed in [6] dividing each activity into segments with a width of 256 samples and a stride of 128. Each segment can be classified as Fall, Alert or Background (FALL, ALERT and BKG) considering if that segment recorded part of a fall event, a fall hazard status or an ADL state without danger. A subset of 20% approximately (all activities related to 5 adult subjects and 3 elderly, randomly chosen from each category) was extracted from the total set for evaluating the effectiveness of models trained in this study. The dataset includes 94K samples for training (90K BCK, 1K ALERT, 3K FALL) and 23K for testing (22K BCK, 0.3K ALERT, 0.7K FALL).

2.2 Gated RNN

Gated RNNs introduce some memory-like cells in the architecture that hold information separated from the rest of the neural network. The information is managed through a set of gates. During the training of the network, the cells learn to close or open their gates according to the relevance of the information that comes from the sequence and the information currently stored.

LSTM units [3] contain three gates. Input and forget gates evaluate the addition of new information into memory and the deletion of part of the stored information. The output gate controls what information is provided to the next

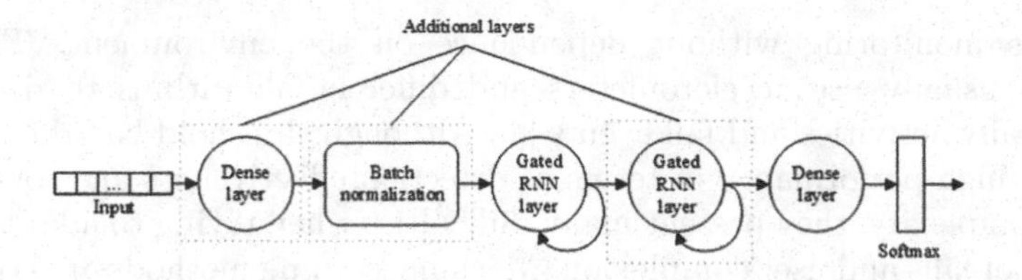

Fig. 1. Scheme representing all architectures trained in the first study stage, differentiated between them by having distinct combinations of the highlighted layers.

step. GRUs [1] are more recent cells similar to LSTM that lack of the output gate, dumping what is stored in the cell's memory during the entire training process. Both alternatives have shown to be similarly effective [2], although GRUs are more economical in terms of computation cost.

2.3 Training and Testing

This work aims to identify the most ideal RNN in order to be implemented in wearable devices. To this end, due to the high computational cost, the architecture should be simplified. Therefore, our study consists of two stages. First one is focused on comparing architectures with different layers combinations (See Fig. 1). The study includes 16 architectures (8 uses LSTM and 8 uses GRU). These are obtained by including/removing the dotted layers in Fig. 1, and are simplified versions of the proposed solution in [6] with the exception of number 8, that consist of the same architecture without dropout. Each architecture was tested by using both GRU and LSTM as RNN layers.

Due to the dataset being highly unbalanced, the overall classification accuracy is not an appropriate way to measure the effectiveness of the system. We compared the effectiveness employing the F1-score [8] for each class and average, which measures the relations between data's positive labels and those given by a classifier through a combination of precision and accuracy. Regarding the architecture complexity, the observed metric was the number of trainable parameters.

The second stage aimed to optimize the architectures that obtained the best results. Firstly, in order to deal with overfitting, dropout technique [9] was applied in the dense and recurrent layers, with the exception of the last dense used to classify the input. It was tested with 0% (without dropout), 20% and 35% values for each layer. Secondly, we used the best results combinations obtained previously to adjust batch size and learning rate hyperparameters by grid search with {32,64} and {0.0015, 0.001, 0.0005} values respectively.

3 Results and Discussion

Main results of first stage are presented in Table 1. F-1 score did not reach 0.33 for the ALERT class. Some reason for this can be the scarcity of this class examples in the dataset [5] and the falling conditions. The application of batch

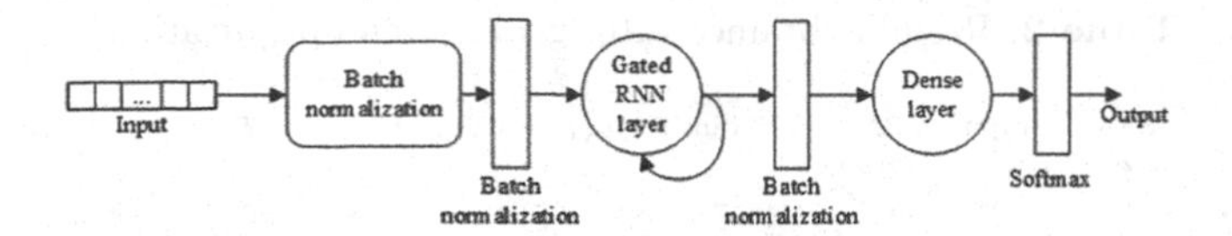

Fig. 2. Scheme representing the architecture with best results obtained in the first stage in terms of number of trainable parameters and mean F1-score.

Table 1. Results obtained with proposed architectures. First 8 rows corresponds to training results using LSTM as RNN layers and second ones using GRU. The *avg* subscript means the average between metrics values obtained. The *prec.* term consists of precision metric. Codes d1, bn and rnn2 indicates the presence in the architecture of first dense layer, batch normalization and second RNN layer respectively.

	Additional layers	Loss	$F1_{BKG}$	$F1_{ALE}$	$F1_{FALL}$	$Prec._{avg}$	$Recall_{avg}$	$F1_{avg}$	Param.
1	-	1.83	0.88	0.09	0.58	0.49	0.76	0.52	4835
2	rnn2	1.56	0.94	0.21	0.69	0.55	0.87	0.61	13283
3	**bn**	**1.04**	**0.95**	**0.22**	**0.83**	**0.61**	**0.91**	**0.66**	**4847**
4	bn, rnn2	1.02	0.96	0.29	0.82	0.63	0.91	0.69	13295
5	d1	2.07	0.85	0.08	0.45	0.45	0.73	0.46	8675
6	d1, rnn2	1.37	0.94	0.20	0.71	0.56	0.86	0.62	17123
7	d1, bn	0.97	0.95	0.25	0.82	0.62	0.91	0.67	8803
8	d1, bn, rnn2 [6]	1.31	0.96	0.26	0.82	0.63	0.89	0.68	17251
9	-	1.75	0.87	0.09	0.64	0.51	0.78	0.53	3651
10	rnn2	1.63	0.95	0.23	0.74	0.58	0.85	0.64	9987
11	**bn**	**1.10**	**0.97**	**0.32**	**0.82**	**0.64**	**0.90**	**0.70**	**3663**
12	bn, rnn2	1.20	0.96	0.29	0.80	0.62	0.90	0.69	9999
13	d1	1.89	0.84	0.09	0.43	0.44	0.75	0.45	6563
14	d1, rnn2	1.11	0.94	0.21	0.71	0.56	0.89	0.62	12899
15	d1, bn	0.97	0.96	0.28	0.87	0.65	0.92	0.70	6691
16	d1, bn, rnn2	1.28	0.97	0.34	0.82	0.64	0.90	0.71	13027

normalization presents the best contribution to performance results. Although the second RNN layer improves the results, the amount of parameters that it adds to the network implies a higher computational cost. Due to the low number of parameters and good results in comparison with the rest of the models, it was considered to optimize results of both architectures 3 and 11 (Fig. 2).

Table 2 shows the best results obtained after the parameter optimization. Effectiveness results are very similar to [6], of which is estimated an F1-score of 0.71 considering that they had used approximately the same number of samples in each class as in this study. However, the model obtained in this study has a much lower complexity since it lacks a dense layer and an additional LSTM layer. This implies a direct impact on the energy saving of the real-time capable device in which the model would be integrated, being able to have an autonomy greater than 20 h of [11].

Table 2. Results obtained after grid search optimization.

RNN type	Learn. rate	Batch size	Input drop	RNN drop	Param	$F1_{BKG}$	$F1_{ALE}$	$F1_{FALL}$	$Prec_{avg}$	$Recall_{avg}$	$F1_{avg}$
lstm	0.001	32	0.35	0	4847	0.98	0.42	0.85	0.69	0.88	0.75
gru	0.001	64	0.35	0.2	3663	0.98	0.37	0.85	0.69	0.88	0.73

4 Conclusions

In this paper, we present the use of Gated RNN based in LSTM and GRU layers as a method implementable in wearable devices with accelerometer to detect falls in real-time. After a study of 16 architectures, the best results in terms of computational complexity and classification are formed by a batch normalization layer receiving the input, a RNN layer and a dense end layer for the classification of the event. FALL and BKG classes are well classified, with F1-scores above 0.98 and 0.85 respectively. Mean F1-score obtained was 0.75 and 0.73 for LSTM and GRU versions, respectively. In future studies the authors will implement this model in a hardware device suitable for use as a wearable FDS.

Acknowledgment. This work was supported by the excellence project from the Spanish government grant (with support from the European Regional Development Fund) COFNET (TEC2016-77785-P).

References

1. Cho, K., et al.: On the properties of neural machine translation: encoder-decoder approaches. arXiv preprint arXiv:1409.1259 (2014)
2. Chung, J., et al.: Empirical evaluation of gated recurrent neural networks on sequence modeling. arXiv preprint arXiv:1412.3555 (2014)
3. Hochreiter, S., et al.: Long short-term memory. Neural Comput. **9**(8), 1735–1780 (1997)
4. Igual, R., et al.: Challenges, issues and trends in fall detection systems. Biomed. Eng. Online **12**(1), 66 (2013)
5. Khan, S.S., et al.: Review of fall detection techniques: a data availability perspective. Med. Eng. Phys. **39**, 12–22 (2017)
6. Musci, M., et al.: Online fall detection using recurrent neural networks. arXiv preprint arXiv:1804.04976 (2018)
7. Noury, N., et al.: Fall detection-principles and methods. In: 2007 29th Annual International Conference of the IEEE Engineering in Medicine and Biology Society, EMBS 2007, pp. 1663–1666. IEEE (2007)
8. Sokolova, M., et al.: A systematic analysis of performance measures for classification tasks. Inf. Process. Manag. **45**(4), 427–437 (2009)
9. Srivastava, N., et al.: Dropout: a simple way to prevent neural networks from overfitting. J. Mach. Learn. Res. **15**(1), 1929–1958 (2014)
10. Sucerquia, A., et al.: SisFall: a fall and movement dataset. Sensors **17**(1), 198 (2017)

11. Torti, E., et al.: Embedded real-time fall detection with deep learning on wearable devices. In: 2018 21st Euromicro Conference on Digital System Design (DSD), pp. 405–412. IEEE (2018)
12. World Health Organization: Ageing; life course unit. Who global report on falls prevention in older age. World Health Organization (2008)

A Semi-supervised Learning Approach for Pan-Cancer Somatic Genomic Variant Classification

Giovanna Nicora[1(✉)], Simone Marini[2], Ivan Limongelli[3], Ettore Rizzo[3], Stefano Montoli[1], Francesca Floriana Tricomi[1], and Riccardo Bellazzi[1]

[1] Department of Electrical, Computer and Biomedical Engineering, University of Pavia, Via Ferrata 1, 27100 Pavia, Italy
giovanna.nicora01@universitadipavia.it
[2] Department of Computational Medicine and Bioinformatics, University of Michigan, Ann Arbor, MI, USA
[3] enGenome S.r.l, Via Ferrata 1, 27100 Pavia, Italy

Abstract. Cancer arises from the accumulation of particular somatic genomic variants known as drivers. New sequencing technologies allow the identification of hundreds of variants in a tumor sample. These variations should be classified as driver or passenger (i.e. benign), but functional studies could be time and cost demanding. Therefore, in the bioinformatics field, machine learning methods are widely applied to distinguish drivers from passengers. Recent projects, such as the AACR GENIE, provide an unprecedented amount of cancer data that could be exploited for the training process of machine learning algorithms. However, the majority of these variants are not yet classified. The development and application of approaches able to assimilate unlabeled data are needed in order to fully benefit from the available omics-resources.

We collected and annotated a dataset of known 976 driver and over 84,000 passengers from different databases and we investigated whether unclassified variants from GENIE could be employed in the classification process. We characterized each variant by 94 features from multiple omics resources. We therefore trained different autoencoder architectures with more than 80000 GENIE variants. Autoencoder is a type of neural network able to learn a new features representation of the input data in an unsupervised manner. The trained autoencoders are then used to obtain new representations of the labeled dataset, with a reduced number of meta-features with the aim to reduce redundancy and extract the relevant information. The new representations are in turn exploited to train and test different machine learning techniques, such as Random Forest, Support Vector Machine, Ridge Logistic Regression, One Class SVM. Final results, however, does not show a significant increase in classification ability when meta-features are used.

Keywords: Somatic variant classification · Semi-supervised learning · Autoencoder

D. Riaño et al. (Eds.): AIME 2019, LNAI 11526, pp. 42–46, 2019.
https://doi.org/10.1007/978-3-030-21642-9_7

1 Introduction

It is now assessed that cancer originates from the accumulation, over life-time, of genomic mutations known as drivers. These variants are able to confer an advantage to the cell in which they occur, which starts to proliferate without control and gives rise to the neoplasm [1]. However, the majority of variants appearing in somatic cells during time is not malignant and they are referred to as passengers.

To distinguish driver from passenger mutations, functional studies should be performed, but their high cost makes them often unfeasible. Thanks to the great availability of cancer data resulting from international and national research projects, machine learning (ML) methods are often exploited to solve the task of "driver/passenger classification" [2]. To assemble training and test set, known drivers and passengers are characterized by a set of features, that include structural, evolutionary and genetic information [3]. Several features could be collected to characterize each variants, but many of them are often correlated: for instance, in silico tools to predict the damaging impact of the variant on the gene product can rely on different variant characteristics already collected as features. In similar circumstances, reducing the dimensionality of the feature space have showed to improve ML performance, by transforming the input space into a lower set of features that preserves the relevant information, thus removing redundant data [4].

Another issue is the that the majority of available data, such as those collected by the AACR Project Genomics Evidence Neoplasia Information Exchange (GENIE) [5], are not labeled and therefore cannot be exploited to train standard supervised learning approaches. Since in this field data-labeling process is high, it is desirable the development of approaches able to take into account also unlabeled data within the training procedure. Specifically, semi-supervised learning techniques exploit unlabeled data in addition to labeled data to improve model performance. The assumptions is that there is some structure to the underlying distribution of data that the unlabeled data will help elucidate [6].

2 Methods

We investigated whether unlabeled genomic data coupled with dimensionality reduction could be exploited to improve performance when classifying somatic variants from pan-cancer data into drivers or passengers. We chose to perform dimensionality reduction through autoencoders for two main reasons. It has been shown that classification performance increases when meta-features extracted from autoencoders are exploited [7]. Moreover, autoencoder could be pretrained with unlabeled data, making them feasible for the development of semi-supervised learning approaches.

2.1 Preprocessing: Labeled and Unlabeled Dataset

First we collect a labeled dataset from three public sources [8–10]. The total number of driver mutations is 976, while the number of passengers is 84440. The great unbalanced between the two classes reflects the real proportions of drivers and passengers that has been sometimes observed in tumor samples [11]. The unlabeled dataset is represented by 80032 GENIE variants (downloaded in June 2017). Each variant is identified by a genomic coordinate (chromosome, start position, reference allele and alternate allele). We characterized each variant with 94 features, including population allele frequency, Boolean features indicating the presence of the variant in cancer databases, *in silico* prediction of damaging impact, evolutionary profiling information.

2.2 Supervised and Semi-supervised Learning

We selected 4 different ML methods implemented in the scikit-learn Python package: Random Forest (RF), Support Vector Machine (SVM), One Class Support Vector Machine (OneClassSVM) and Ridge Regression (RR). RF is an ensemble classifier where a set of decision trees are trained, and final classification is the result of a voting procedure among them. SVM represents examples as points in space and try to discover a hyperplane able to divide two different classes. One Class SVM is an adaptation of standard SVM, applied to one class classification problems, i.e. when data from only one class (the target class) are available. The target class represents normal examples, while the outliers are unusual observations. In our case, we can assume that passenger variants are the usual variations occurring in the genome, while drivers are the abnormal that we want to detect. Finally, RR is a classification method where classes probabilities are calculated according to a logistic function.

We selected the 70% of labeled dataset for training, leaving 30% for test. The proportions of passengers and drivers in the two datasets are kept. We select the best parameters after a 10-fold cross validation on the training set for each of the four methods. Best models are trained on the whole training dataset and then tested.

We implemented an autoencoder through the Keras library. The architecture of the network is the following: the input layer (94 neurons) is followed by 5 encoded layers (with number of neurons equal to 90, 85, 80, 75 and 70), followed by five symmetric decoded layers. Therefore, the reduced dimension of features is 70. The size of the meta-features is selected empirically, and further analysis should be performed with different meta-features size. We decide to first investigate a relatively high hidden dimension (70) since it has been shown that for classification purpose a very small number of meta-features does not improve performance [12]. We trained the autoencoder with GENIE unlabeled dataset (10 number of epochs, size of batch 32). The trained network is exploited to reduce the dimensionality of the labeled dataset. The four ML models are trained and tested also with the meta-features dataset.

3 Results

The model parameters selected after a 10-fold cross validation are the following: RF is implemented through sklearn.ensemble.RandomForestClassifier (n_estimators = 10, min_sample_split = 2, criterio = “gini”), SVM through sklearn.svm.SVC (kernel rbf, gamma “scale”, C = 1), OneClass SVM through sklearn.svm.OneClassSVM (kernel rbf, nu = 0.1, gamma = 0.1) and Ridge Regression thorugh sklearn.linear_model. RidgeClassifier (alpha = 1.0, tol = 0.001). In Table 1 we reported some performance metrics on raw features test set, while Table 2 shows the same metrics for the 70 meta-features dataset. We select metrics, for instance Matthews Correlation Coefficient (MCC) and Balanced Accuracy, able to assess the effectiveness even with highly imbalanced dataset [13].

Table 1. Results on 94 raw features labeled dataset.

Model	MCC	TPR	TNR	Balanced accuracy
Random forest	0.903	0.999	0.878	0.93
OneClass SVM	0.5	0.9	1	0.95
SVM	0.846	0.998	0.813	0.9
RR	0.813	0.998	0.79	0.89

Table 2. Results on 70 meta-features labeled dataset.

Model	MCC	TPR	TNR	Balanced accuracy
Random forest	0.916	0.999	0.91	0.95
OneClass SVM	0.494	0.898	0.996	0.94
SVM	0.901	0.998	0.901	0.95
RR	0.847	0.999	0.78	0.89

RF and RR perform better on the meta-features dataset, while OneClass SVM and SVM show better results on the raw features dataset. However, performance differences are not clearly marked.

4 Conclusions

The distinction between passenger and driver mutations in tumor sample is a critical step to understand patient individual cancerogenesis. Thanks to the great availability of public cancer data, bioinformaticians have often trained ML models to classify somatic variants into drivers or passengers. However, many examples could not be used since the ground truth labels are missing. Semi-supervised learning approaches could allow to fully benefit from genomic sequencing data reported by public resources.

Our work represents a preliminary attempt to use unlabeled data and meta-features representation to improve performance of standard supervised methods. After the collection of two datasets (labeled and unlabeled), four ML techniques are trained and tested with the labeled dataset. Then, we implemented an autoencoder architecture pre-trained with unlabeled data, that performed dimensionality reduction of labeled dataset. The meta-features dataset is again used to train and test the same ML models. Classification performance are high, but there is not a strong improvement when using the meta-features dataset. Source code is available at https://github.com/GiovannaNicora/semi_supervised_learning-somatic_variant_classification.

Future works will include further investigation, through the testing of different autoencoder architecture, with different number of meta-features involved. Moreover, other dimensionality reduction techniques, for instance PCA, must be compared.

References

1. Stratton, M.R., Campbell, P.J., Futreal, P.A.: The cancer genome. Nature **458**, 719 (2009). https://doi.org/10.1038/nature07943
2. Gonzalez-Perez, A., Mustonen, V., Reva, B., et al.: Computational approaches to identify functional genetic variants in cancer genomes. Nat. Methods **10**, 723–729 (2013). https://doi.org/10.1038/nmeth.2562
3. Agajanian, S., Odeyemi, O., Bischoff, N., Ratra, S., Verkhivker, G.M.: Machine learning classification and structure-functional analysis of cancer mutations reveal unique dynamic and network signatures of driver sites in oncogenes and tumor suppressor genes. J. Chem. Inf. Model. **58**, 2131–2150 (2018). https://doi.org/10.1021/acs.jcim.8b00414
4. Khalid, S., Khalil, T., Nasreen, S.: A survey of feature selection and feature extraction techniques in machine learning. In: 2014 Science and Information Conference, pp. 372–378 (2014). https://doi.org/10.1109/SAI.2014.6918213
5. AACR Project GENIE. https://www.aacr.org:443/Research/Research/pages/aacr-project-genie.aspx
6. Chapelle, O., Schlkopf, B., Zien, A.: Semi-supervised learning. The MIT Press, Cambridge (2010)
7. Van Der Maaten, L., et al.: Dimensionality reduction: a comparative review. J. Mach. Learn. Res. **10**, 66–71 (2009)
8. Martelotto, L.G., Ng, C.K., De Filippo, M.R., et al.: Benchmarking mutation effect prediction algorithms using functionally validated cancer-related missense mutations. Genome Biol. **15**, 484 (2014). https://doi.org/10.1186/s13059-014-0484-1
9. Tamborero, D., Rubio-Perez, C., Deu-Pons, J., et al.: Cancer genome interpreter annotates the biological and clinical relevance of tumor alterations. Genome Med. **10**, 25 (2018). https://doi.org/10.1186/s13073-018-0531-8
10. Sherry, S.T., Ward, M.-H., Kholodov, M., et al.: dbSNP: the NCBI database of genetic variation. Nucleic Acids Res. **29**, 308–311 (2001). https://doi.org/10.1093/nar/29.1.308
11. Wood, L.D., Parsons, D.W., Jones, S., et al.: The genomic landscapes of human breast and colorectal cancers. Science **318**, 1108–1113 (2007). https://doi.org/10.1126/science.1145720
12. Wang, Y., Yao, H., Zhao, S.: Auto-encoder based dimensionality reduction. Neurocomputing **184**, 232–242 (2016). https://doi.org/10.1016/j.neucom.2015.08.104
13. Akosa, J.: Predictive accuracy: a misleading performance measure for highly imbalanced data. In: Proceedings of the SAS Global Forum (2017)

ChronoNet: A Deep Recurrent Neural Network for Abnormal EEG Identification

Subhrajit Roy(✉), Isabell Kiral-Kornek, and Stefan Harrer

IBM Research - Australia, Level 22, 60 City Rd, Southbank 3006, Australia
{subhrajit.roy,isabeki,sharrer}@au1.ibm.com

Abstract. Brain-related disorders such as epilepsy can be diagnosed by analyzing electroencephalograms (EEG). However, manual analysis of EEG data requires highly trained clinicians, and is a procedure that is known to have relatively low inter-rater agreement (IRA). Moreover, the volume of the data and the rate at which new data becomes available make manual interpretation a time-consuming, resource-hungry, and expensive process. In contrast, automated analysis of EEG data offers the potential to improve the quality of patient care by shortening the time to diagnosis and reducing manual error. In this paper, we focus on one of the first steps in interpreting an EEG session - identifying whether the brain activity is abnormal or normal. To address this specific task, we propose a novel recurrent neural network (RNN) architecture termed ChronoNet which is inspired by recent developments from the field of image classification and designed to work efficiently with EEG data. ChronoNet is formed by stacking multiple 1D convolution layers followed by deep gated recurrent unit (GRU) layers where each 1D convolution layer uses multiple filters of exponentially varying lengths and the stacked GRU layers are densely connected in a feed-forward manner. We used the recently released TUH Abnormal EEG Corpus dataset for evaluating the performance of ChronoNet. Unlike previous studies using this dataset, ChronoNet directly takes time-series EEG as input and learns meaningful representations of brain activity patterns. ChronoNet outperforms previously reported results on this dataset thereby setting a new benchmark.

Keywords: Machine learning · Recurrent neural networks · Electroencephalography

1 Introduction

Electroencephalography (EEG) is a noninvasive method to measure brain activity through the recording of electrical activity across a patient's skull and scalp and is frequently used for the diagnosis and management of various neurological conditions such as epilepsy, somnipathy, coma, encephalopathies, and others. As symptoms are not guaranteed to be present in the EEG signal at all times, the diagnosis of a neurological condition via EEG interpretation typically involves

D. Riaño et al. (Eds.): AIME 2019, LNAI 11526, pp. 47–56, 2019.
https://doi.org/10.1007/978-3-030-21642-9_8

long-term monitoring or the recording of multiple short sessions. In this process, large amounts of data are generated that subsequently need to be manually interpreted by expert investigators. The relatively low availability of certified expert investigators and high volume of data make EEG interpretation a time-consuming process that can introduce a delay of hours to weeks in the patient's course of treatment. Introducing a certain level of automation to the EEG interpretation task could serve as an aid to neurologists by accelerating the reading process and thereby reducing workload. It is these reasons why automatic interpretation of EEG by machine learning techniques has gained popularity in recent times [4,16].

When interpreting an EEG recording, first, an assessment is made as to whether the recorded signal appears to show abnormal or normal brain activity patterns [13]. This decision can influence which medication is being prescribed or whether further investigation is necessary. Typically, both, patterns in the recording and the patient's state of consciousness are being considered when deciding whether a recording shows a abnormal or normal EEG. Highly trained clinicians typically follow a complicated decision chart to make this distinction [13]. The motivation behind our work is to automate this first step of interpretation. We do so using a recently released dataset known as the TUH Abnormal EEG Corpus, which is the largest of its type to date [14] and freely available at [1]. Inspired by successes in time-domain signal classification, we explore recurrent neural network (RNN) architectures using the raw EEG time-series signal as input. This sets us apart from previous publications [12,13,16], in which the authors used both traditional machine learning algorithms such as k-nearest neighbour, random forests, and hidden markov models and modern deep learning techniques such as convolutional neural networks (CNN), however, did not use RNNs for this task.

Compared to the original studies using hand-engineered features [12,13], we show that the combination of raw time series and RNNs eliminates the need to extract hand-crafted features and allows the classifier to automatically learn relevant patterns, surpassing their results by 3.51%. Taking inspiration from 1D convolution layers [5], gated recurrent units [3], inception modules [17], and densely connected networks [8], we build a novel deep gated RNN named ChronoNet which further increases accuracy by an additional 4.26%, resulting in an overall 7.77% improvement over results reported in [12,13]. Moreover, compared to a recently published study showing state-of-the-art performance on this dataset, ChronoNet achieves 1.17% better results thereby setting a new benchmark for the TUH Abnormal EEG Corpus.

2 Background and Theory

Raw EEG signals are temporal recordings that may exhibit patterns and periodicities at various time scales. A method that has successfully been used to classify time signals, for example speech, is the use of recurrent neural networks (RNNs). In recent times, two types of recurrent units have become popular namely Long short-term memory [6] and Gated Recurrent Units [3] which we will be using in this article. For more details on the theory and applications of these architectures, we invite the reader to refer to [9].

We shall now discuss the concept of inception modules and densely connected neural networks (concepts used in convolutional neural networks) which we will use for EEG data analysis to account for patterns emerging at different scales and for mitigation of vanishing gradients, respectively. This collection of principles and modules provides the essential basis for understanding the ChronoNet architecture proposed in Sect. 3.

The inception module was proposed by Szegedy et al. [17] as a building block for the GoogLeNet architecture. Unlike traditional convolutional neural networks, the inception module uses filters of varied size in a convolution layer to capture features of different levels of abstraction. Processing visual information at different scales and aggregating them allows the network to efficiently extract relevant features. Typically, the module uses three filters of sizes 1×1, 3×3, and 5×5. Moreover, an alternative parallel path is also included which implements a 3×3 max-pooling operation. However, naively introducing more filters in convolutional layers increases the number of parameters.

DenseNet is a deep convolutional neural network architecture recently proposed in [8]. The main idea of DenseNet is that it connects each layer with every other layer in a feed-forward fashion. Each layer uses the feature maps of all its preceding layers as input and passes its own feature maps as input to all subsequent layers. Hence, while a traditional CNN with L layers has L connections, in DenseNet there are $L(L+1)/2$ direct connections. DenseNet mitigates the problem of vanishing/exploding gradients that is observed in very deep networks [10]. It achieves that by providing short-cut paths for the gradients to pass during backpropagation.

3 Methods

Inception modules (see Sect. 2) were originally proposed to enable a convolutional neural network to account for different abstraction layers in the context of image processing. Similarly, densely connected networks (see Sect. 2) were developed to address vanishing gradients due to backpropagation in deep convolutional neural networks. As described previously, EEG data contains information across different scales in the time-domain. Furthermore, using deep RNN architectures might lead to the problem of vanishing or exploding gradients. If designed properly, advantages of inception modules and densely connected layers may, thus, equally apply to problems in the time-domain. In this section, we will use the concepts of inception modules and densely connected networks to build novel recurrent neural networks architectures for time-series classification.

3.1 Convolutional Gated Recurrent Neural Network (C-RNN)

Considering the input is a time series, an obvious first approach is to stack multiple GRU layers as shown in Fig. 1a. This popular architecture for handling sequential input data has led to state-of-the-art accuracy in various pattern recognition tasks, especially in natural language processing [18,19].

However, when applied to relatively long input time-series data (as opposed to embedding vectors [15] in the case of natural language processing), this approach turns out to be computationally very intensive and time consuming to train. To solve this problem, data can be downsampled to an acceptable length before it is given as input to RNNs. However, using fixed values means that networks will not be able to adapt to the data at hand. In order to mitigate these problems, we used multiple 1D convolution (Conv1D) layers with strides larger than 1, enabling the network to learn to appropriately reduce the input signal automatically.

The resulting architecture (C-RNN) is a combination of Conv1D layers followed by stacked GRU layers. Conv1D layers have two advantages. First, they learn to sub-sample the signal and, thus, reduce the input vector's length as we move towards higher layers. This becomes particularly relevant when reaching GRU layers, which during training constitute the most computationally expensive part of the network. Second, Conv1D layers extract local information from neighbouring time points, a first step towards learning temporal dependencies. Following Conv1D layers, the GRU layers are responsible for capturing both short- and long-term dependencies.

The specific network used in this paper is presented in Fig. 1b. The formats used throughout the paper to describe Conv1D and GRU layers are (layer name, filter length, number of filters, stride size) and (layer name, number of filters) respectively.

3.2 Inception Convolutional Gated Recurrent Neural Network (IC-RNN)

In the previous C-RNN architecture, each Conv1D layer had the capability to extract local information at only one time scale determined by a single fixed filter size, limiting the flexibility of the model. Since the rate of change of information in a time series depends on the task at hand, the filter size for each Conv1D layer would have to be hand-picked to fit the particular data.

To address this problem, taking inspiration from [17], we designed an architecture which expands upon C-RNN by including multiple filters of varying sizes in each Conv1D layer. This allows for the network to extract information over multiple time-scales. However, unlike [17], in IC-RNN, filter lengths used in the Conv1D layers were drawn from a logarithmic instead of a linear scale, leading to exponentially varying filter lenghts. Our experiments demonstrated that for the dataset considered in this paper, exponentially varying filter lengths lead to better performance. We speculate that this is because compared to images where relevant features vary in the same order of magnitude, in time series the range of timescales in which features exist is much wider. Note that, to the best of our knowledge, inception modules with exponentially varying filter sizes are reported for the first time in this paper. The specific configuration used in our experiments is shown in Fig. 1c. A Filter Concat layer concatenates the incoming features along the depth axis.

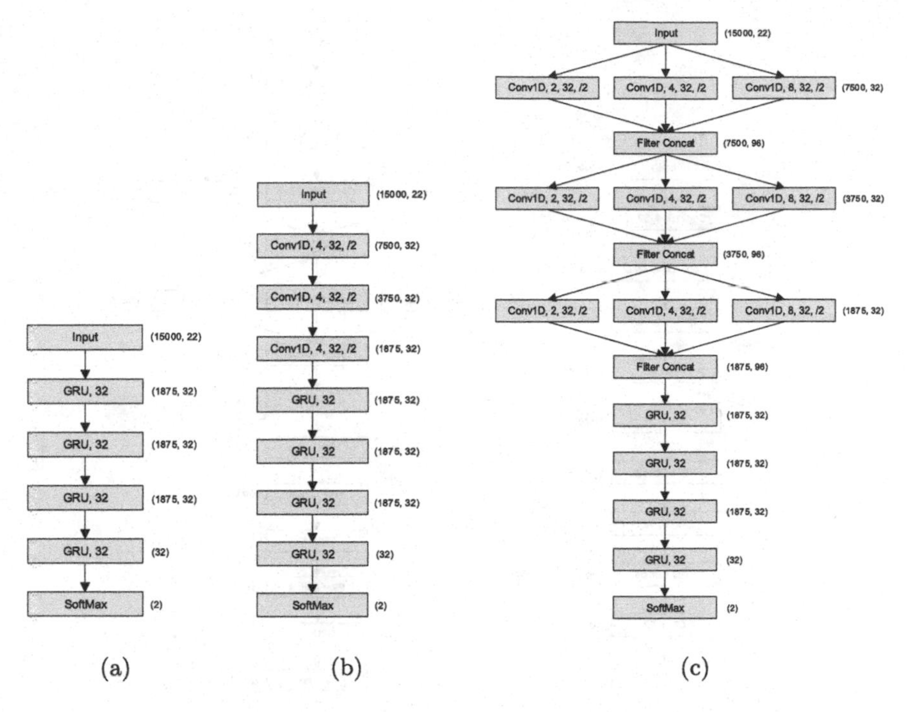

Fig. 1. (a) A deep gated recurrent neural network which feeds the input directly into stacked GRU layers. (b) The convolutional gated recurrent neural network (C-RNN) which stacks multiple 1D convolution layers followed by a small number of GRU layers. (c) The Inception Convolutional Gated Recurrent Neural Network (IC-RNN) modifies the C-RNN (see Fig. 1b) architecture by including multiple filters of exponentially varying lengths in the 1D convolution layers.

3.3 Convolutional Densely Connected Gated Recurrent Neural Network (C-DRNN)

The C-RNN architecture is not immune to the problem of *degradation* which sometimes impedes the training of very deep neural networks [10]. For simpler problems that do not need the full potential of the model complexity offered by a C-RNN, the optimization procedure may lead to higher training errors.

To tackle this issue, inspired by the DenseNet architecture proposed by [8] for CNNs, we incorporate skip connections in the stacked GRU layers of C-RNN to form the C-DRNN architecture. Each GRU layer is connected to every other GRU layer in a feed-forward fashion. Intuitively, skip connections will lead to GRU layers being ignored when the data demands a lower model complexity than offered by the entire network. The details of the network are shown in Fig. 2a.

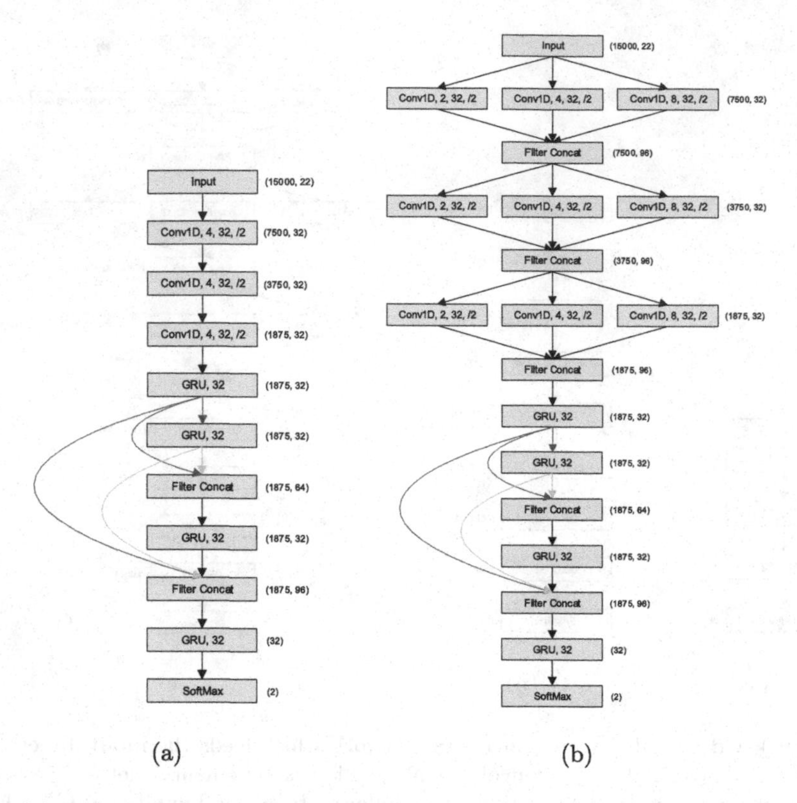

Fig. 2. (a) The Convolutional Densely Connected Gated Recurrent Neural Network (C-DRNN) architecture includes dense connections in the recurrent stage i.e. the output of each GRU layer is given as input to every other GRU layers in a feed-forward manner. (b) Proposed ChronoNet architecture which includes both multiple filters of exponentially varying lengths in the 1D convolution layers and dense connections within the GRU layers.

3.4 ChronoNet: Inception Convolutional Densely Connected Gated Recurrent Neural Network

Finally, we combine both modifications introduced for the previous two networks (IC-RNN and C-DRNN) with C-RNN to form the ChronoNet architecture. To the best of our knowledge, this is the first time this architecture has been reported. To summarize, ChronoNet is created by stacking multiple Conv1D layers followed by multiple GRU layers where each Conv1D layer has multiple filters of varying sizes and the stacked GRU layers are densely connected in a feed-forward manner.

The presence of multiple filters in the Conv1D layers allows ChronoNet to extract and combine features from different time scales. The optimum filter size for a Conv1D layer usually depends on both the task at hand and its relative position in the network. ChronoNet has the flexibility to explore multiple filter lengths for each Conv1D layer. On the other hand, densely connected GRU layers allow ChronoNet to mitigate the problem of *degradation* of training accuracy caused by vanishing or exploding gradients. This potentially enables the creation of very deep variants of ChronoNet for more complex tasks. Moreover, dense connections also strengthen feature propagation and encourage feature reuse in the GRU layers. The network we designed for the abnormal EEG classification task considered in this paper is depicted in Fig. 2b.

4 Experiments

4.1 Data Selection and Preparation

In this paper, we primarily focused on the TUH Abnormal EEG Corpus [12], which contains EEG records that are annotated as either clinically abnormal or normal. The TUH Abnormal EEG Corpus is a subset of the TUH EEG Corpus [14] which is the world's largest publicly available database of clinical EEG data. The TUH EEG Corpus comprises 23257 EEG sessions recorded over 13551 patients. In the entire dataset, almost 75% of the data represent abnormal EEG sessions. The TUH EEG Abnormal Corpus was formed by selecting a demographically balanced subset of the TUH EEG Corpus through manual review that consisted of 1488 abnormal and 1529 normal EEG sessions, respectively. These sets were further partitioned into a training set (1361 abnormal/1379 normal), and a test set (127 abnormal/150 normal). TUH Abnormal EEG Corpus consists of EEG sessions recorded according to the 10/20 electrode configuration [7]. We converted the recorded EEG signal into a set of montages or differentials based on guidelines proposed by the American Clinical Neurophysiology Society [2]. In this paper, we used the transverse central parietal (TCP) montage system for accentuating spike activity [12]. Note that we did not extract any hand-engineered features from the dataset because we envisioned that the deep RNNs used in this paper will be able to automatically extract relevant features and learn meaningful representations.

In the original study [12], the authors noted that neurologists can accurately classify an EEG session into either abnormal or normal by only examining the initial portion of the signal. This motivated the authors to build machine learning algorithms that can classify an EEG signal by taking only the first minute of data as input. Hence, training and test set were generated by extracting the first minute from the available EEG sessions. Note that during testing only the first minute was used to enable a fair comparison of the classifier to human-level performance. Using only the first minute to create the training set, on the other hand, was a design choice motivated by the fact that the first minute might be most representative of the test set. Once electrodes are placed on the scalp and data recording starts, impedances and therewith the signal will gradually change due to external factors such as slowly drying conductive paste. To have a fair comparison with [12], we trained our model only on the first minute and report the obtained results in Sect. 4.2.

However, using the above method significantly limits the amount of data that can be used for training. This results in two problems. First, deep learning is a data hungry technique and performance substantially increases as more data is included in the training set. Second, when applied to small datasets, RNNs have a tendency to quickly overfit, an effect that intensifies as networks become deeper, as is the case for those considered in this paper. To not limit ourselves unnecessarily, we analyzed the effect of including more than just the first minute from the training sessions. This was done by choosing a random subset of sessions from the original training by exclusion of any samples later used for testing. The resulting sets were then further divided into smaller training and test sets. Separate models were trained for each minute of these training sets. We analysed the performance of these models on the first minute of the intermediate small test sets.

The outcome of this experiment demonstrated that we can use up to 11 min of data from the training EEG sessions without performance degradation. This led to a 11-fold increase in our training data as compared to the method used in [12]. The sizes of the final training and test set used were 14971 abnormal/15169 normal and 127 abnormal/150 normal respectively.

In this dataset, most recordings were done with a sampling frequency of 250 Hz. Where this was not the case, sessions were resampled to 250 Hz. An input vector to the network was 1 min long, thus, consisting of 15000 time points.

4.2 Results

We used the dataset described above to train the four deep recurrent neural network architectures presented in Sect. 3. Networks were trained using the adaptive moment estimation optimization [11] algorithm with a learning rate of 0.001. Moreover, we used a batch size of 64 and trained the networks for 500 epochs. Table 1 lists mean accuracies of 5 repetitions of these experiments. Results reported to date on this dataset are included for comparison. In [12], the author explored various machine- and deep learning algorithms and observed that best performance is obtained when frequency features extracted from the input time-series signal are fed into a convolutional neural network [12] (CNN-MLP in Table 1). Furthermore, in [16] the authors used a deep convolutional neural network built by automatic hyperparameter search (DeepCNN in Table 1) and reported the best accuracy to date.

Table 1 clearly depicts that the deep recurrent neural architectures explored in this paper outperform the results shown in the original study [12] using CNN-MLP. It is important to note that in contrast to CNN-MLP, the proposed architectures do not rely on hand-crafted features. Moreover, we see that C-RNN, IC-RNN, C-DRNN, and ChronoNet are surpassing best accuracy reported in [12] by 3.51%, 5.31%, 5.09%, and 7.77%, respectively. Furthermore, compared to the recently published state-of-the-art performance [16], ChronoNet shows 1.17% better accuracy. Out of the four recurrent architectures, ChronoNet achieves both the best training and testing accuracy. This shows that the combined positive effect of including multiple filters in Conv1D layers and incorporating dense connections in the GRU

Table 1. Performance comparison of the four deep recurrent neural networks described in Sect. 3 and results reported in [12] (see CNN-MLP) and [16] (see DeepCNN).

–	Training accuracy	Testing accuracy
C-RNN	83.58%	82.31%
IC-RNN	86.93%	84.11%
C-DRNN	87.20%	83.89%
ChronoNet	**90.60%**	**86.57%**
CNN-MLP	N/A	78.80%
DeepCNN	N/A	85.40%

layers is more pronounced than using either one or none of them. Moreover, our experiments showed that ChronoNet yields similar performance (86.64%, averaged over 5 runs) when GRUs are replaced by LSTMs, however, networks with LSTM units took longer time to train than their GRU counterpart.

The training and test dataset was pre-split in the TUH Abnormal EEG corpus in a way such that each set is demographically balanced (gender and age) and no patient appears in both the training and testing set. To demonstrate that the network is not overfitting the hyper parameters on the test set, we combine the training and test set provided in the TUH Abnormal EEG Corpus and perform a 5-fold cross-validation to provide test accuracy for the proposed architecture. We achieve a 86.14% accuracy with the 5-fold cross-validation approach.

Note that the number of EEG records used in the training set is the same as the number used in other works on this dataset. While the original study [12] used only the first minute, we discovered that more than the first minute can be included in the training set. If trained on just the first minute i.e. when the training set is exactly same as used in other work, ChronoNet achieves an accuracy of 85.27% (averaged over 5 runs) which is 6.47% better than [12].

To demonstrate that exponentially varying filter sizes in the Conv1D layers of ChronoNet are a necessary component, two experiments were performed. First, shorter (compared to the longest 1D convolution filter used in ChronoNet) linearly varying filters of lengths 3, 5, and 7 were implemented. As a result, training and testing accuracies fall to 89.15% and 85.12%, respectively. Second, longer but linearly varying filters of lengths of 14, 16, and 18 were implemented. While training accuracy increased to 91.25%, the testing accuracy was reduced to 85.92. We speculate that in both cases features extracted by the network are not sufficiently diverse, and furthermore, in the later case, the increased model complexity leads to overfitting.

5 Conclusion

Determining whether an EEG recording shows abnormal or normal brain activity is often the first step in the diagnosis of a neurological condition. Since manual interpretation of EEG is an expensive and time-consuming process, any classifier that automates this first distinction will have the potential to reduce delays in treatment and to relieve clinical care givers. We introduce ChronoNet, a novel network architecture that is designed to be flexible and adaptable and, thus, uniquely suited for the analysis of EEG time-series data. This novel RNN architecture outperforms the best previously reported accuracy on the dataset used by 1.17%, setting a new benchmark.

References

1. Temple university EEG corpus. https://www.isip.piconepress.com/projects/tuh_eeg/
2. Acharya, J.N., et al.: American clinical neurophysiology society guideline 3: a proposal for standard montages to be used in clinical EEG. J. Clin. Neurophysiol. **33**(4), 312–316 (2016)
3. Cho, K., et al.: On the properties of neural machine translation: encoder-decoder approaches. arXiv:1409.1259v2, September 2014
4. Golmohammadi, M., et al.: Deep architectures for automated seizure detection in scalp EEGs. arXiv:1712.09776 [cs, eess, q-bio, stat], December 2017
5. Goodfellow, I., et al.: Deep Learning. The MIT Press, Cambridge (2016)
6. Graves, A.: Generating sequences with recurrent neural networks. arXiv:1308.0850 [cs], August 2013
7. Homan, R.W.: The 10-20 electrode system and cerebral location. Am. J. EEG Technol. **28**(4), 269–279 (1988)
8. Huang, G., et al.: Densely connected convolutional networks. arXiv:1608.06993, August 2016
9. Chung, J., Gulcehre, C.: Empirical evaluation of gated recurrent neural networks on sequence modeling. arXiv:1412.3555 [cs], December 2014
10. He, K., et al.: Deep residual learning for image recognition. arXiv:1512.03385 [cs], December 2015
11. Kingma, D.P., Ba, J.: Adam: a method for stochastic optimization. arXiv:1412.6980 [cs], December 2014
12. López, S.: Automated interpretation of abnormal adult electroencephalograms. MS thesis, Temple University (2017). https://www.isip.piconepress.com/publications/ms_theses/2017/abnormal/
13. López, S., et al.: Automated Identification of Abnormal Adult EEGs. IEEE Signal Process. Med. Biol. Symp. **2015** (2015)
14. Obeid, I., Picone, J.: The temple university hospital EEG data corpus. Front. Neurosci. **10** (2016). https://doi.org/10.3389/fnins.2016.00196
15. Pennington, J., et al.: GloVe: Global vectors for word representation. In: Empirical Methods in Natural Language Processing (EMNLP), pp. 1532–1543 (2014)
16. Schirrmeister, R.T., et al.: Deep learning with convolutional neural networks for decoding and visualization of EEG pathology. CoRR abs/1708.08012 (2017)
17. Szegedy, C., et al.: Going deeper with convolutions. In: 2015 IEEE Conference on Computer Vision and Pattern Recognition (CVPR), pp. 1–9, June 2015
18. Tang, Y., et al.: Question detection from acoustic features using recurrent neural network with gated recurrent unit. In: 2016 IEEE International Conference on Acoustics, Speech and Signal Processing (ICASSP), pp. 6125–6129, March 2016
19. Yin, W., et al.: Comparative study of CNN and RNN for natural language processing. arXiv:1702.01923 [cs], February 2017

Convolutional Recurrent Neural Network for Bubble Detection in a Portable Continuous Bladder Irrigation Monitor

Xiaoying Tan(✉), Gerd Reis, and Didier Stricker

DFKI, German Research Center for Artificial Intelligence,
Kaiserslautern, Germany
{Xiaoying.Tan,Gerd.Reis,Didier.Stricker}@dfki.de

Abstract. Continuous bladder irrigation (CBI) is commonly used to prevent urinary problems after prostate or bladder surgery. Nowadays, the irrigation flow rate is regulated manually based on the color (qualitative estimation of the blood concentration) of the drainage fluid. To monitor the blood concentration quantitatively and continuously, we have developed a portable CBI monitor based on the Lambert-Beer law. It measures transmitted light intensity via a camera sensor and deduces the blood concentration. To achieve high reliability, we need to guarantee that the measurement is conducted when there is no air bubble passing through the view of the camera. To detect bubble occurrences, we propose a convolutional recurrent neural network with a sequence of images as input: the convolutional layers extract spatial features from 2D images; the recurrent layers capture temporal features in the image sequence. Our experimental results show that our network has smaller scale and higher accuracy compared with conventional convolutional and recurrent neural networks.

Keywords: Continuous bladder irrigation ·
Blood concentration measurement · Bubble detection ·
Convolutional neural network · Recurrent neural network

1 Introduction

Continuous bladder irrigation (CBI) is commonly used after prostate or bladder surgery, like transurethral resection of the prostate, open prostatectomy and transurethral resection of bladder tumor [1,2]. It can prevent postoperative complications, like blood clot formation and retention, urinary tract obstruction, cystospasm and hemorrhage [3,4]. In some cases, CBI is also used to dissolve bladder stones or treat infected bladder lining. CBI uses a three-way Foley catheter to allow the irrigation fluid (sterile saline) to flow into and the drainage fluid to flow out of the bladder simultaneously [5]. Nowadays, to improve the efficacy of CBI, nurses regulate the flow rate of the irrigation fluid based on the qualitative estimation of the blood concentration of the drainage fluid in the

D. Riaño et al. (Eds.): AIME 2019, LNAI 11526, pp. 57–66, 2019.
https://doi.org/10.1007/978-3-030-21642-9_9

evacuating catheter. That is, nurses observe the color of the drainage fluid: if it is heavily bloodstained, they turn the irrigation faster, as the likelihood of blood clot formation is increased due to the high blood concentration; if it is lightly bloodstained, they let the irrigation run slower to inflict less pressure upon the bladder of patient. Typically, the patient prescribed with CBI needs frequent irrigation monitoring and regulation for around two days [5]. This is a heavy burden for nurses since they have to inspect patients back and forth in wards. Additionally, since the inspection by nurses is discontinuous, the flow rate regulation might be incorrect. Moreover, the color estimation by eyes under uncontrolled illumination conditions in wards is prone to be erroneous.

We have developed a portable CBI monitor which uses a camera sensor to measure the blood concentration of the drainage fluid quantitatively. It helps nurses monitor irrigation situations of patients continuously and accurately. However, occasionally an air bubble passes through the view of the camera, which deteriorates the blood concentration measurement. Therefore, we need to detect the bubble occurrences and discard the corresponding measurements. The main contribution of this paper is a small-scale convolutional recurrent neural network (CRNN) to detect the bubble occurrences. The designed CRNN takes a sequence of images as input, which is obtained by the same camera sensor used for blood concentration measurement. It firstly uses convolutional layers to extract spatial features from the 2D images, and then recurrent layers to capture temporal features in the image sequence. Our approach provides reliable bubble detection without introducing an extra sensor to our CBI monitor.

The remainder of this paper is organized as follows: Sect. 2 presents related works that study automatic CBI monitoring as well as convolutional and recurrent neural networks; Sect. 3 describes the problem we are facing and our methodology to solve it; Sect. 4 reports the experiments and results while Sect. 5 draws the conclusions and suggests some directions for future work.

2 Related Work

Nowadays, CBI is manually monitored and regulated in the hospital and is important practical training for nurses. In 2016, Ding et al. [6] designed an automatic flow rate controller for CBI. Their clinical experiments suggested that their automatic controller could promote the recovery of patients and hence improve the clinical outcome. In 2017, Arun et al. [7] published a patent about an autonomous CBI device. In both works, a color sensor was utilized to monitor blood concentration, which used a white light source to illuminate the drainage fluid in the catheter and an RGB color sensor to determine the color of the drainage fluid. The measured color might be quite different from the one observed by nurses, because of the spectral difference between the adopted white light source and natural/artificial light sources in wards, as well as the different spectral responsivities between the used RGB photo-diodes and human eyes. Moreover, the conversion from measured color to blood concentration is unclear, because the color of the drainage fluid is not only related to blood, but also urinary pigments, such as urobilin, carotin, and betanin. A bubble in the drainage fluid is prone to cause erroneous color measurements, which has not been considered in the previous works.

Convolutional neural networks (CNNs) have recently achieved great success in many computer vision tasks, such as image classification, object detection and pixel-wise prediction [8–10]. CNNs use feed-forward architectures to extract features from an image at different levels, which is similar to a human brain. Recurrent neural networks (RNNs) use the idea of processing sequential information and have shown great success in natural language processing tasks, such as language modeling, machine translation and speech recognition [11–13]. RNNs have memory over previous computations and use this information in current processing. In recent years, some works have combined convolutional and recurrent neural networks to model the spatial dependencies between image pixels and integrate context information on a single image [14–17]. They have proven extremely capable of performing image segmentation, object detection and scene labeling in single-frame images. However, there has been a comparatively small amount of work on combining convolutional and recurrent neural networks to integrate information from a sequence of images.

3 Problem Statement and Methodology

3.1 CBI Monitor

Instead of the color sensor used in [6,7], we used a camera sensor in our CBI monitor to measure transmitted light intensity. Then, blood concentration was deduced using the Lambert-Beer law. The schematic diagram of our monitor is shown in Fig. 1a: a LED provides incident light propagating through the evacuating catheter filled with drainage fluid perpendicularly; a monochrome camera captures the transmitted LED spot; a black mounting box blocks the environmental illumination.

According to the Lambert-Beer law,

$$I_t = I_0 \cdot 10^{\epsilon \Delta z c}, \tag{1}$$

where I_t and I_0 denote the transmitted and incident light intensities respectively, ϵ and c denote the molar absorptivity and concentration of the light absorptive compound, and Δz is the optical length through the drainage fluid.

By selecting proper power supply (constant current) and wavelength (800 nm) for the LED, we obtain constant I_0 and ϵ. When the catheter tube is full of drainage fluid, Δz is also constant and determined by the diameter of the tube. Therefore, we can infer the blood concentration c solely based on I_t. The value of I_t is measured by the camera sensor. Namely, the median value of the pixels in the center of the LED spot is used as I_t, as shown in Fig. 1b.

In this way, our CBI monitor measures the blood concentration with good accuracy when $c \in [0\%,\ 20\%]$, and triggers an alarm when $c > 20\%$, which means massive bleeding and immediate medical intervention. It can be easily attached

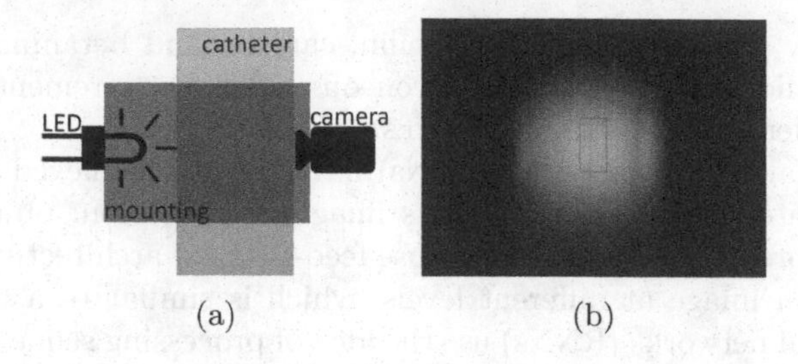

Fig. 1. (a) the schematic diagram of the CBI monitor; (b) the transmitted LED spot, the median value of the pixels in the red rectangle is the measured transmitted light intensity. (Color figure online)

to the existing CBI system and send nurses accurate blood concentration data and alarms via wireless communication. Therefore, nurses can monitor irrigation situations of patients continuously and accurately. When irrigation regulation is necessary for a specific patient, they can conduct quick and precise intervention.

3.2 Bubble Problem

Since the evacuating catheter is connected to a non-vacuum urine bag, bubbles occasionally pass through the view of the camera of our CBI monitor. The bubbles have very different appearances in the image, depending on their sizes, relative positions with camera, velocities, and the drainage fluid composition. For example, when the blood concentration is low, a bubble is distinguishable by its dark edge (Fig. 2 top row); however, when the blood concentration is high, a bubble is distinguishable by its bright center (Fig. 2 bottom row).

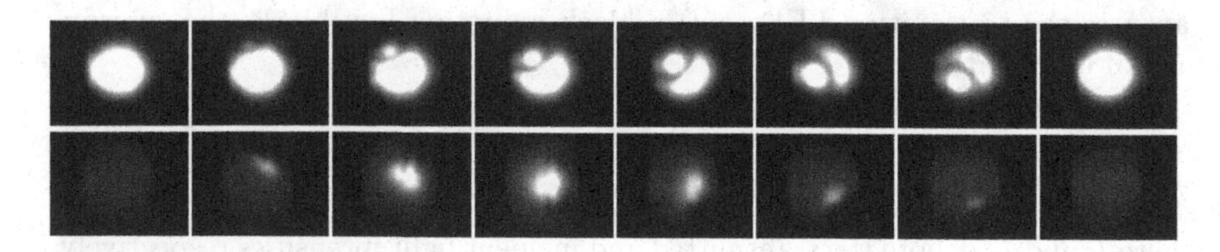

Fig. 2. Images of two bubbles passing through the view of the camera, in the solution with 0.3% blood (top) and with 12% blood (bottom) respectively.

In presence of a bubble, Δz in Eq. 1 is not constant anymore. Consequently, the exponential relationship between I_t and c becomes variable. Hence, the measured blood concentration becomes unreliable. In Fig. 3, we show how a bubble deteriorates the blood concentration measurements. For instance, in Fig. 3a without the interference of bubble, the measured blood concentration $\tilde{c} = 3.06\%$, error $\Delta c = +0.06\%$; while in Fig. 3c with bubble, $\tilde{c} = 2.35\%$, error Δc increases to -0.65%.

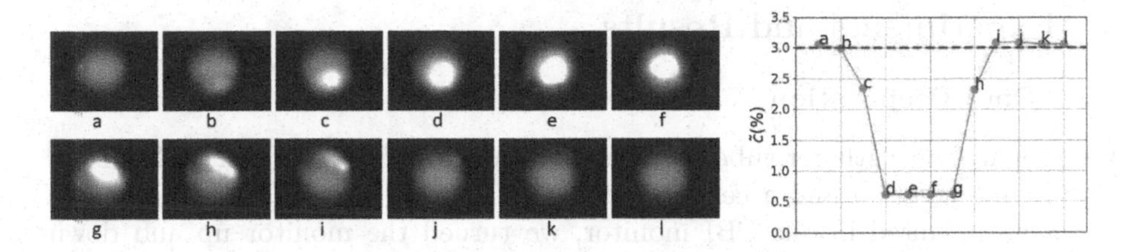

Fig. 3. Images taken with a bubble flowing in the solution with 3% blood (left) and the corresponding measured blood concentrations (right).

3.3 Neural Network Architecture

To avoid unreliable blood concentration measurements from images with bubbles, e.g., Fig. 3c–h, we need to detect the presence of bubbles and discard the corresponding images. There are commercial sensors on the market to detect bubble occurrence in fluid-filled tubes, based on ultrasonic, capacitive or optical property difference between the fluid and air. However, it is difficult to find one which can be easily integrated with our CBI monitor and provide reliable bubble detection for variable fluid composition and bubble velocity.

To detect the bubble occurrence in our CBI monitor, we designed a small-scale convolutional recurrent neural network (CRNN). As shown in Fig. 4, this CRNN takes a sequence of frames as input. It uses a spatial feature extractor (two convolutional layers) to extract a feature map from each image. The feature map is flattened and fed to a temporal feature extractor (two layers of unidirectional LSTM). The temporal feature extractor integrates information from the sequence of extracted feature maps and outputs a new feature vector which is then used to compute class probabilities by a fully-connected layer. This CRNN fuses spatial and temporal features from a sequence of frames to improve the bubble detection in the current frame.

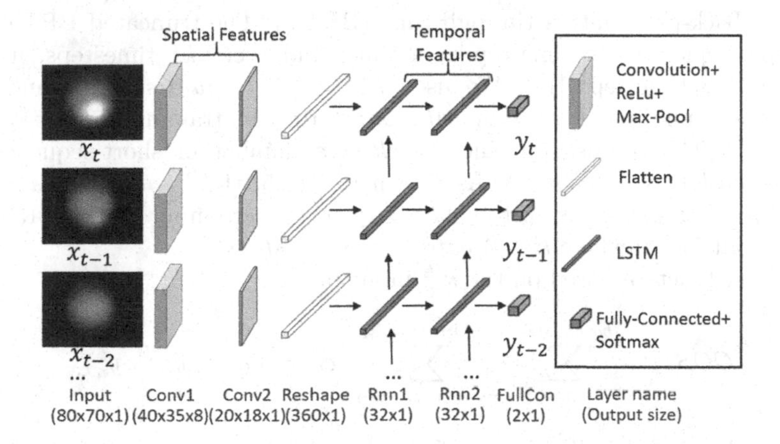

Fig. 4. The convolutional recurrent neural network (CRNN).

4 Experiments and Results

4.1 Data Preparation

We prepared 46 catheter tube segments, in each of which solution with known blood concentration and a certain volume of air were sealed. With these tube segments mounted in our CBI monitor, we turned the monitor up and down and gathered 46 image sequences. In every image sequence, there were bubbles passing with different velocities. These image sequences are referred to as sub-sequences $S(c)$ hereafter, where c denotes corresponding blood concentrations.

The training sequence consisted of 28 sub-sequences $S(c)$. Each sub-sequence was used twice and they were stitched together to a whole image sequence with c gradually rising and then decreasing: $c = 0\%, 0.1\%, ..., 0.4\%, 0.6\%, ..., 1.2\%, 1.6\%,$ $2\%, 2.2\%, ..., 2.8\%, 4\%, 6\%, ..., 22\%, 50\%, 70\%, 90\%, 90\%, 70\%, ..., 0\%$. The ratio of images with bubble in the training sequence was $22754/99536 = 23\%$. In the same way, the validation sequence was generated with other 10 sub-sequences $S(c)$: $c = 0.5\%, 1.5\%, 1.8\%, 3\%, 7\%, ..., 19\%, 40\%, 80\%, 80\%, 40\%, ..., 0.5\%$. The bubble image ratio in the validation sequence was $6502/23340 = 28\%$. In the test sequence, each of the remaining 8 sub-sequences $S(c)$ was used twice, and they were stitched together in the reverse order: $c = 100\%, 60\%, 21\%, 17\%, ..., 5\%, 1.4\%, 1.4\%,$ $5\%, ..., 100\%$. The bubble image ratio in the test sequence was $4120/15454 = 27\%$. The data sets were generated according to the fact that the blood concentration changes gradually in reality. To exclude overfitting of the CRNN to the long-term changing trend of blood concentration in training and validation data, we gave the test sequence the reverse blood concentration order.

4.2 Network Training

In our case, the outputs of the LSTMs might depend on arbitrarily distant inputs, because size and speed of bubbles are unpredictable. This makes back-propagation computation difficult. To tackle this problem, we implemented the truncated back-propagation through time (BPTT). The truncated BPTT processes the sequence one timestep at a time, and every $k1$ timesteps, it runs BPTT for $k2$ timesteps [18]. We used $k1 = k2 = num\_steps(8)$ and processed data in mini-batches of size $batch\_size(80)$. The training image sequence of length $N(99536)$ was split into $batch\_size$ number of short-sequences of length $batch\_len(N/batch\, size)$. As shown in Fig. 5, the $k \cdot num\_steps + 1$ to $(k + 1) \cdot num\_steps$ frames of the short-sequences were input to the CRNN as the kth input block, where $k \in [0, batch\_len/num\_steps)$.

The cost function based on the kth input block:

$$COST_k = \sum_{n=1}^{batch\_size} \sum_{t=k \cdot num\_steps+1}^{(k+1) \cdot num\_steps} cross\_entropy(y_n^{'t}, y_n^t), \tag{2}$$

where $y_n^{'t}$ denotes the tth output of the nth short-sequence from the CRNN, y_n^t is the corresponding true label.

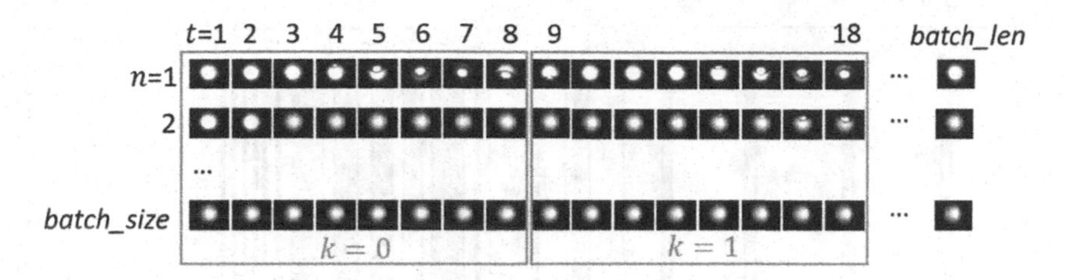

Fig. 5. The input block for CRNN training.

The BPTT based on the kth input block was conducted using the cost calculated in Eq. 2. After the last frames of the short-sequences were processed and the BPTT based on the $(batch\_len/num\_steps - 1)^{th}$ input block was conducted, one whole epoch training was finished. The validation sequence was used for hyperparameters selection and early stopping.

4.3 Test Results

The test sequence of $M(15454)$ images was input to the trained CRNN. The CRNN output two probabilities for each input image. One was the probability of the presence of a bubble in the image, denoted as P, and the other one equaled $1 - P$. An image was classified as a bubble image when $P \geq threshold$. We used the typical threshold 0.5. By comparing the classification results of the CRNN and the true labels, we obtained the confusion matrix shown in Table 1. Our CRNN achieved 97.4% accuracy, 97.5% precision, and 92.6% recall. For our CBI monitor, high recall is more important than high precision. In practice, we can increase the recall at the cost of lower precision by decreasing the threshold for bubble detection.

Table 1. Bubble detection results of the CRNN ($threshold = 0.5$).

M = 15454	Predicted NOT bubble	Predicted IS bubble	
Actual NOT bubble	TN = 11236	FP = 98	11334
Actual IS bubble	FN = 304	TP = 3816	4120

Figure 6 depicts two parts of the bubble detection results of the test sequence. It shows that the predicted bubble occurrences (blue bars) overlap well with the labeled ones (red bars). The miss-classification happens mostly at the transitional frames. As shown in Fig. 7a, when the frame index is around 2953 to 2962, a bubble passes through the camera view. Our CRNN successfully detects six frames of this bubble but misses one frame where the bubble enters the view.

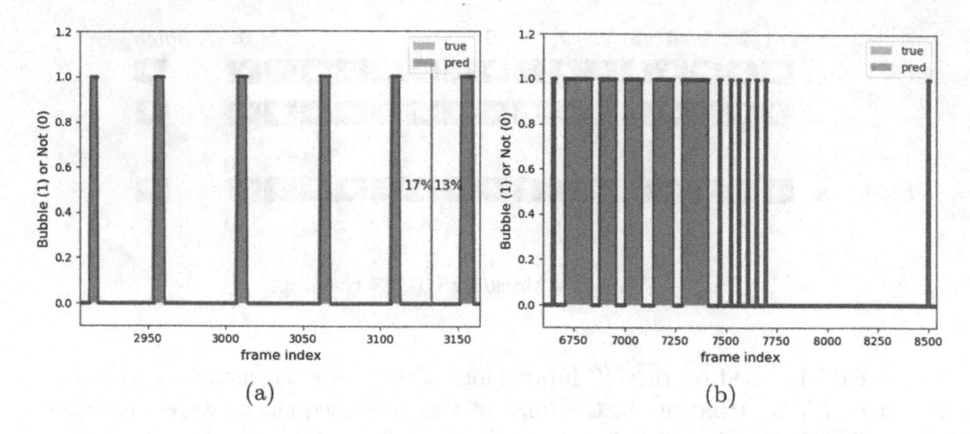

(a) (b)

Fig. 6. Comparison between predicted and true labels of the test sequence. The green line separates two image sub-sequences with different blood concentrations. (Color figure online)

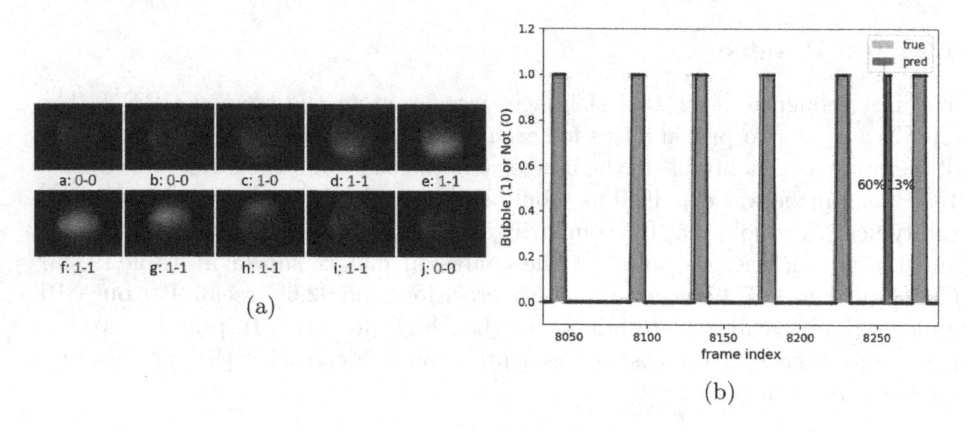

(a) (b)

Fig. 7. (a) the images corresponding to the second bubble occurrencein Fig. 6a: a–j are frames 2953–2962, red/blue digits are true/predicted labels; (b) false positive around frame 8255 after blood concentration decreases sharply. (Color figure online)

The trained CRNN also showed good performance in some extreme cases, like very long bubble/non-bubble sequence, a bubble jittering in the view of the camera and bubbles next to each other. The only severe miss-classification happened in the extreme case when blood concentration decreased sharply. For instance, in Fig. 7b, after blood concentration decreases from 60% to 13% in a single frame, the first four non-bubble frames are miss-classified as bubble frames. Since in reality the blood concentration unlikely decreases so sharply and the number of the miss-classified frames are small, the impact of this type of false positive classification is bearable.

For comparison, small-scale conventional CNN and RNN were designed, as shown in Fig. 8. With the same data sets, they were trained and tested. The comparison between them and our CRNN is shown in Table 2. The CRNN shows significant advantage over conventional CNN and RNN.

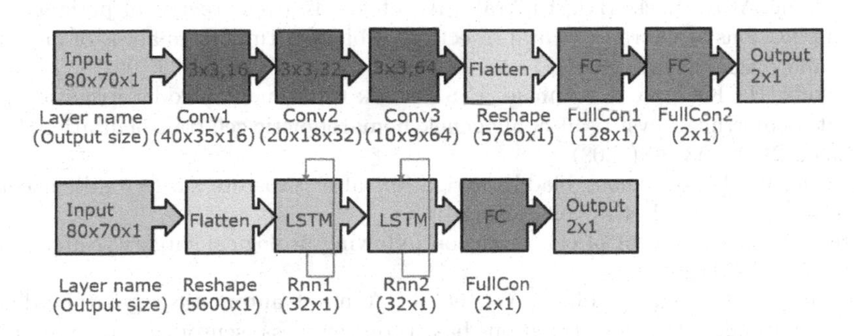

Fig. 8. The conventional CNN (top) and RNN (bottom).

Table 2. Comparison between CRNN, conventional CNN and conventional RNN.

	No. of trainable parameters	Accuracy	Precision	Recall
CRNN	59099	97.4%	97.5%	92.6%
CNN	760962	93.3%	88.1%	89.4%
RNN	729410	95.7%	97.1%	86.7%

5 Conclusion

In this paper, we have solved the bubble problem in our portable CBI monitor with a bubble detector which uses a camera sensor and a small-scale CRNN. The camera sensor has low cost and is also used for blood concentration measurement, which is the crucial function of our portable CBI monitor. The CRNN outperforms conventional CNN and RNN at bubble occurrence detection in a sequence of images, because it learns not only the spatial features of bubbles on 2D images, but also temporal features of bubbles in the image sequence. The small scale of our CRNN is beneficial for low latency, especially when implemented on mobile and embedded processors as well as SoCs. In the future, we plan to implement the trained CRNN on a suitable processor/SoC and conduct latency tests. With its capability to detect the presence of bubbles in fluid-filled clear tubes non-invasively, this bubble sensor based on camera sensor and CRNN might find its applicability in other fields, like medical treatments that use extracorporeal blood circuits and micro-fluid control.

Acknowledgments. This work is partially supported by the BMBF project VisIMON under grant No. 16SV7861R.

References

1. Nojiri, Y., Okamura, K., Kinukawa, T., et al.: Continuous bladder irrigation following transurethral resection of the prostate (TURP). Nihon Hinyokika Gakkai Zasshi **98**(6), 770–775 (2007)
2. Avallone, M.A., Sack, B.S., El-Arabi, A., et al.: Ten-year review of perioperative complications after transurethral resection of bladder tumors: analysis of monopolar and plasmakinetic bipolar cases. J. Endourol. **31**(8), 767–773 (2017)
3. Molouk, H., Kamran, A., Ahmad, R.S., et al.: Continuous bladder irrigation prevents hemorrhagic cystitis after allogeneic hematopoietic cell transplantation. Urol. Oncol. **26**(1), 43–46 (2008)
4. Okorie, C.: Is continuous bladder irrigation after prostate surgery still needed? World J. Clin. Urol. **4**(3), 108–114 (2015)
5. Steve, S.: Management of clot retention following urological surgery. Nurs. Times **98**(28), 48–50 (2002)
6. Aimin, D., Huling, C., Lihua, W., et al.: A novel automatic regulatory device for continuous bladder irrigation based on wireless sensor in patients after transurethral resection of the prostate: a prospective investigation. Medicine **95**(52), e5721 (2016)
7. Rai, A., Yadav, N.: Autonomous continuous bladder irrigation devices, colorimeters, and methods of monitoring bladder pressure (2017)
8. Alex, K., Ilya, S., Geoffrey, E.H.: ImageNet classification with deep convolutional neural networks. Adv. Neural Inf. Process. Syst. **25**(2), 1915–1929 (2012)
9. Ali, S.R., Hossein, A., Josephine, S., et al.: CNN features off-the-shelf: an astounding baseline for recognition. In: Conference on Computer Vision and Pattern Recognition Workshops. IEEE, USA (2014)
10. Vijay, B., Alex, K., Roberto, C.: SegNet: a deep convolutional encoder-decoder architecture for image segmentation. IEEE Trans. Pattern Anal. Mach. Intell. **39**(12), 2481–2495 (2017)
11. Tomas, M., Martin, K., Lukas, B.: Recurrent neural network based language model. In: 11th Annual Conference of the International Speech Communication Association, Japan (2010)
12. Michael, A., Michel, G., Chris, Q., et al.: Joint language and translation modeling with recurrent neural networks. In: Conference on Empirical Methods in Natural Language Processing, pp. 1044–1054. Association for Computational Linguistics, USA (2013)
13. Alex, G., Navdeep, J.: Towards end-to-end speech recognition with recurrent neural networks. In: 31st International Conference on Machine Learning, China (2014)
14. Haiqiang, Z., Heng, F., Erik, B., et al.: Combining convolutional and recurrent neural networks for human skin detection. IEEE Signal Process. Lett. **24**(3), 289–293 (2017)
15. Zhiwei, D., Arash, V., Hexiang, H., et al.: Structure inference machines: recurrent neural networks for analyzing relations in group activity recognition. In: Conference on Computer Vision and Pattern Recognition. IEEE, USA (2016)
16. Ming, L., Xiaolin, H.: Recurrent convolutional neural network for object recognition. In: Conference on Computer Vision and Pattern Recognition. IEEE, USA (2015)
17. Russell S., Mykhaylo A., Andrew Y.N.: End-to-end people detection in crowded scenes. In: Conference on Computer Vision and Pattern Recognition. IEEE, USA (2016)
18. Sutskever I.: Training recurrent neural networks. Ph.D. thesis, University of Toronto (2012)

Mammogram Classification with Ordered Loss

Rami Ben-Ari(✉), Yoel Shoshan, and Tal Tlusty

IBM Research, Haifa, Israel
{ramib,yoels}@il.ibm.com, ttlusty@ibm.com

Abstract. Breast radiologists inspect mammograms with the utmost consideration to capture true cancer cases. Yet, machine learning models are typically designed to perform a binary classification, by joining several severities into one positive class. In such scenarios with mixed gradings, a reliable classifier would make less mistakes between distant severities such as missing a true cancer case and calling it as normal or vise versa. To this end, we suggest a simple yet elegant formulation for training a deep learning model with ordered loss, by increasingly weighting the loss of more severe cases, to enforce importance of certain errors over others. Training with the ordered loss yields fewer severe errors and can decrease the chances of missing true cancers. We evaluated our method on mammogram classification, using a weakly supervised deep learning method. Our data set included over 16 K mammograms, with a large set of nearly 2,500 biopsy proven cancer cases. Evaluation of our proposed loss function showed a reduction in severe errors of missing true cancers, while preserving overall classification performance in the original task.

Keywords: Mammography · Deep learning · Weakly supervised · Ordered loss

1 Introduction

Nearly 40 million mammography exams are performed on a yearly basis in the US alone. These arise predominantly from screening programs implemented to detect breast cancer at an early stage. All this data has to be inspected for signs of cancer by one or more experienced readers. This is a time-consuming, costly, and most importantly error-prone endeavor. In the drive for improved health care, AI systems are being developed to assist radiologists in this task.

When it comes to screening, the goal is to separate suspiciously malignant cases for recall and further diagnosis. A strong use case for an AI model is *triage*, where an AI system ranks the tests for radiologist inspection according to their severity. In breast radiology, the confidence for malignancy is graded by BIRADS (Breast Imaging-Reporting and Data System), ranging from 0–6. BIRADS assessment categories are (0) Incomplete, (1) Negative, namely normal

D. Riaño et al. (Eds.): AIME 2019, LNAI 11526, pp. 67–76, 2019.
https://doi.org/10.1007/978-3-030-21642-9_10

(2) Benign finding, (3) Probably benign finding, (4) Suspicious abnormality, (5) Abnormality, highly suggestive for malignancy and (6) Known biopsy-proven malignancy. Suspicious abnormalities go under biopsy test (usually BIRADS 4 and 5) to distinguish between cancerous and yet benign pathologies. The biopsy result indicates positive for cancerous lesion and negative for a benign finding. Figure 1 presents examples of mammograms of four different types, BIRADS 1 and 2 (abbreviated by B1 and B2) and cases of biopsied breasts with positive and negative results (abbreviated by PB and NB respectively). This figure demonstrates the challenge in detecting lesions and recognizing the cancerous cases as it is often the case even for expert radiologists.

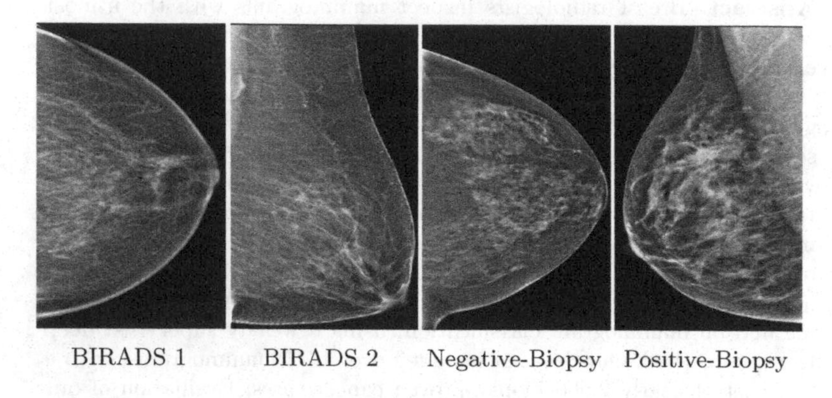

Fig. 1. Examples of Mammograms of four different types. Findings are subtle and differentiation is a challenge often for expert radiologists.

Existing machine learning and deep learning based methods in mammography mostly target a binary classification task [1–4,6], often using cross-entropy (CE) as loss function. In this loss function there is no order or importance associated with the samples in a class. While in natural images this is often a satisfactory criterion, in the medical domain, images in the same class may correspond to diverse stages of a disease, where errors may have different consequences.

Mammograms are often classified in literature into two categories: normal and benign grouped together, versus suspiciously malignant cases. A common practice in the classification and triage of mammograms is to set BIRADS 1,2 as a negative set and BIRADS 3–6 as positive *e.g.*, [2,3]. Yet, there are many classifiers that share the same triage operation point, indicated by the tuple, true-positive rate (TPR) and false-positive rate (FPR), but yield a different error distribution. For example, a positive set composed of 80 NB and 20 PB cases can yield 80% true-positive rate (TPR) with classifier C_1 by categorizing all 80 NB and no PB as positive (meaning that all 20 PB are missed). Yet, a second classifier C_2 can reach the same TPR by categorizing 60 NB and 20 PB cases as positive. While both classifiers yield the same triage performance (the FPR is ignored for sake of simplicity), C_2 classifier is a highly preferred model, due to it's significantly lower crucial errors such as missing of true cancers. This emphasizes the importance for the types of mistakes obtained by a certain binary classifier. Naturally, we would prefer a classifier that is less likely to miss a true cancer. This can be achieved by training a model with an ordered loss.

In this paper, we suggest a generalization of CE to an ordered loss, by suggesting a new formulation that diverges from the standards CE, and increasingly weights the losses of more severe errors. This modeling enforces a preference on certain errors over others. For instance, false-negatives of malignant cases (PB) are more heavily penalized compared to false-negatives of a biopsied, yet benign cases (NB). A model trained with such ordered loss is less likely to miss true cancers.

We evaluate our model with a weakly labeled classification task similar to [1,2,10], where no local annotations around findings are needed for training. Images in this setup are labeled by their global BIRADS assessment and biopsy outcome. In our testbed, we grouped B1, B2 as negative and NB, PB as positive set, according to a typical triage task. Our data set consists of a large scale mammography corpus with over 16 K mammograms, 9,774 biopsy verified labels, including 2,466 true cancers. To the best of our knowledge, tests on such a large scale of full field digital mammography with biopsy validated tags are rarely found in literature. We demonstrate the advantage of our ordered loss by measuring two performance values, one for the original task of triage and the other for a true cancer classification (or alternatively number of missed cancers). We show that a model trained for triage with our ordered loss is less likely to miss a true cancer, without jeopardizing the triage performance.

Our work entails four major contributions: (1) We suggest a simple yet powerful and easy to implement, ordered loss that generalizes the commonly used cross-entropy. This loss mimics a human-like behavior, where more "attention" is paid to find the true cancer (2) Our model was verified on a test bed with a large set of biopsy verified cases (3) Our evaluation is based on extensive tests, including k-fold cross validation, as well as a held out set. (5) The suggested model is general and can be implemented in other medical domains.

Previous Work. Two obvious approaches for handling discrete ordinal labels are (1) treating the different severities as unrelated classes and learning to predict them as in a multiclass classification settings, and (2) treating them as real-valued responses and using a standard regression setting with a loss function such as sum-squared error. However, option (1) suggests an undesirable loss in terms of a task originally defined as binary classification. In a multi-class classification model, miss-classification between sub-classes in the positive set, such as NB and PB will be penalized in the same manner as miss-classification between PB and B1. Such penalizing pattern is prone to yield poor results in the original binary classification task of triage. The sum of squared error loss function in (2) is often inferior to the commonly used cross-entropy (CE) loss.

Ordered labels has been used in different applications [7,8]. These types of labels arise when preferences are specified by several raters for each item, such as movie ratings. To this end, Rennie and Srebro [8] employ a discrete ordinal regression loss, in movie rating, to fit rating levels by ordered labels. They study various generalizations of logistic and hinge loss, to learn a real valued predictor. Niu *et al.* [7] address the age estimation in human faces, by solving an ordinal regression problem in a convolutional neural network (CNN) framework. In their work, they reduce the ordinal regression problem into a set of multiple binary classification tasks. These approaches address non-medical use cases and define the problem as a regression task. In this paper, we address a different problem of binary classification with various sub-classes, carrying a severity label. We suggest a new loss function to enforce a prior for the types of errors made between different sub-classes. We validate the impact of this loss in mammogram classification with a weakly supervised deep CNN.

2 Mammogram Classification

Considering a binary classification task, the most common deep learning loss function is cross entropy:

$$L_{CE} = y \log(p) + (1 - y) \log(1 - p) \tag{1}$$

where $y \in \{0, 1\}$ denotes the image label and $p \in [0, 1]$ stands for the positive class probability. In medical applications, several types of abnormalities are often grouped together establishing the positive set while the rest are grouped to a single negative set.

Ordered Loss. Let us map the BIRADS monotonically, according to their severity (or importance) to a new set $\mathcal{B}$, which we call *pseudo*-BIARDS (P-BIRADS). Often, the positive and negative sets are split according to a threshold in the ordered set $\mathcal{B}$, dividing it into two mutually exclusive and ordered sets: $\mathcal{B}_N$ and $\mathcal{B}_P$. We can now define the following ordered loss:

$$L_O(B) = y(B - \max \mathcal{B}_N)^{\gamma_{FN}} \log(p) + (1 - y)(\min \mathcal{B}_P - B)^{\gamma_{FP}} \log(1 - p) \tag{2}$$

where $B \in \mathcal{B}$ is a **pseudo-BIARDS** and $\gamma_{FN} \geq 0, \gamma_{FP} \geq 0$ define the amplification of false negative and false positive losses, according to the order in $\mathcal{B}$. Note that $B > \max \mathcal{B}_N$ for $y = 1$ since $B \in \mathcal{B}_P$ in this case. Similarly, $B < \min \mathcal{B}_P$ for $y = 0$ since the sample belongs to the negative set. In our study, we use the following mapping (without loss of generality): BIRADS 1 $\rightarrow$ 1, BIRADS 2 $\rightarrow$ 2, negative biopsy $\rightarrow$ 3 and positive biopsy $\rightarrow$ 5. The biopsied cases correspond to BIRADS 0 or BIRADS $\geq$ 3 at the screening stage. We therefore obtain $\mathcal{B} = \{1, 2, 3, 5\}$ where, $\mathcal{B}_N = \{1, 2\}$ and $\mathcal{B}_P = \{3, 5\}$. This assignment is similar to BIRADS prediction except that for pseudo-BIRADS 3 & 5, the diagnosis is based on biopsy results. An example for the ordered loss function is shown in Fig. 2. At a given score (probability) for a positive sample, $L_O(B = 5) \geq L_O(B = 4) \geq L_O(B = 3)$, creating a preference for less severe errors. This formulation further suggests a loss with the following properties:

$$\lim_{\gamma \to 0} L_O = L_{CE}, \forall B \in \mathcal{B}, \gamma := |\gamma_{FN}| + |\gamma_{FP}| \tag{3}$$

$$L_O(B) \equiv L_{CE}, for B = \max \mathcal{B}_N + 1 or B = \min \mathcal{B}_P - 1 \tag{4}$$

The property in (3) ensures convergence of the ordered loss to the standard CE loss. Varying γ_{FN} and γ_{FP} continuously changes the CE loss in an ordered manner. Property (4) indicates that the loss function is bounded from below with the bound equal to the CE loss (see also Fig. 2). The loss for errors between neighboring ordered labels *e.g.*, $B = 3$ are $B = 2$ are identical to CE loss (see the coincidence of the dashed green line and P-BIRADS = 3 in Fig. 2).

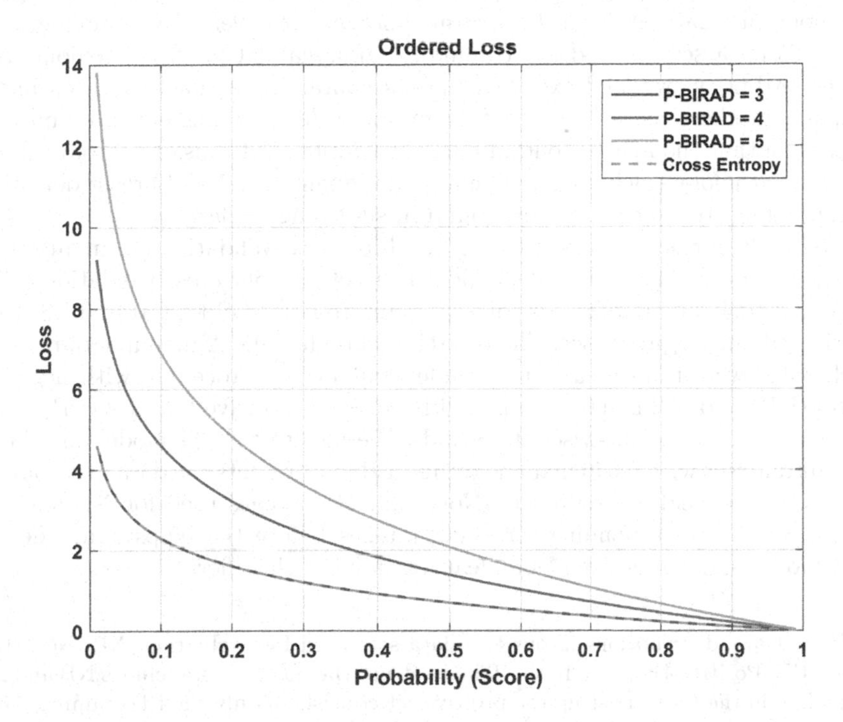

Fig. 2. An example of ordered loss for the positive set, with $\gamma_{FN} = 1$. The severity order is determined by pseudo-BIRADS (P-BIRADS). Note the increase in loss obtained by higher P-BIRADS. We associate true cancers (verified positive biopsy) with P-BIRADS = 5, and the negative biopsies with P-BIRADS = 3 (Color figure online).

Network Architecture. In order to validate our model we use a weakly supervised deep learning classifier. Our model is based on a customized Inception-ResNet V2 network architecture [9]. The network is composed of 14 Inception-ResNet blocks. This output is then fed into a global-max-pooling stage followed by a fully connected layer (256D) and a softmax classifier. The network was trained end-to-end. The input images were tight bounding boxes around the breast area, resized to 2048 × 1024. We trained the network using the Adam optimizer with a learning rate of 10^{-4}, dropout and batch-size of 2 images. In addition, we performed augmentations of flips, rotations, contrast, offsets, and zoom to add diversity in the training set and balance between the positive and the negative classes. For each fold we first ran the baseline training, in which we set $\gamma_{FN} = \gamma_{FP} = 0$. The network converged after 60 epochs, which took ~2.5 days on an IBM Power-AI machine with NVIDIA Tesla-V100-16G GPU.

3 Experimental Results

Dataset. To evaluate our method, we used a large multi-center full-field digital mammography data set from the screening stage, with over 16 K mammograms. From this data set we used the common cranial-caudal (CC) & mediolateral-oblique (MLO) views, and excluded images containing artifacts such as metal clips, skin markers, as well as large foreign bodies like pacemakers and implants. Other than that the images contain a wide variation, in terms of anatomical differences, pathology (including benign and malignant cases) and breast densities, corresponding to what is typically found in screening clinics.

We divide our data to two sets, one used for cross-validation (train and test) and the second as held-out (independent test set). In our cross validation (CV) set we had 10,424 mammograms corresponding to BIRADS 1,2 (B1 and B2) that constituted our **negative** set. The **positive** class had 4,412 mammograms from which 3,440 were from breasts that underwent biopsy procedure with negative results (NB) and 972 mammograms of breasts with positive biopsies (PB). This data was divided patient-wise into 3 folds. We also tested our model on a large held out data set with 6,046 mammograms. Table 1 presents the data distribution in our CV and held-out data sets. Note that the images used for NB and PB class, were the corresponding screening images before the biopsy, in order to avoid biopsy signs appearing as a "feature" for the classifier.

Table 1. Class distributions in our two data sets used for evaluation. NB: Negative-Biopsy, PB: Positive-Biopsy. * In the Triage set-up, the positive set includes NB and PB cases while in the Cancer set-up the positive set consists of only the PB mammograms (for more details see the Evaluation section).

	Negative set		Positive set*		
	BIRADS 1	BIRADS 2	NB	PB	Total
Cross-Validation	3,685	2,327	3,440	972	10,424
Held-Out	1,464	1,902	1,943	747	6,046

Evaluation. We consider two set-ups in our evaluation. First is the original *Triage* set-up with NB and PB used as positive class. The second set-up is the *Cancer* classifier where we measure the success of the trained triage classifier in true cancer detection. Overall performance of a triage classifier is defined by high values for these two measures.

In this section we present the results of our deep learning model for a binary classification task. We used $\mathcal{B}_N = \{1, 2\}$ and $\mathcal{B}_P = \{3, 5\}$ in our ordered loss configuration as described in Sect. 2. We compare the network trained with ordered loss to a baseline that uses standard cross entropy loss. Targeting the miss cases of true cancers (false negatives of type PB), we demonstrate the performance of the ordered loss using $\gamma_{FN} = 1$ and $\gamma_{FP} = 0$ (*i.e.*, false-positives are not penalized in an ordered manner here). We show the two ROC curves for one fold of our cross validation set in Fig. 3. The plots correspond to the triage set-up with the associated ROC for cancer detection (denoted by Cancer). The cancer detection ROC is computed from the triage model while considering only the PB mammograms in the positive set (removing the NB set). The plots show that in this case the ordered loss improves both the triage separation task along with cancer detection.

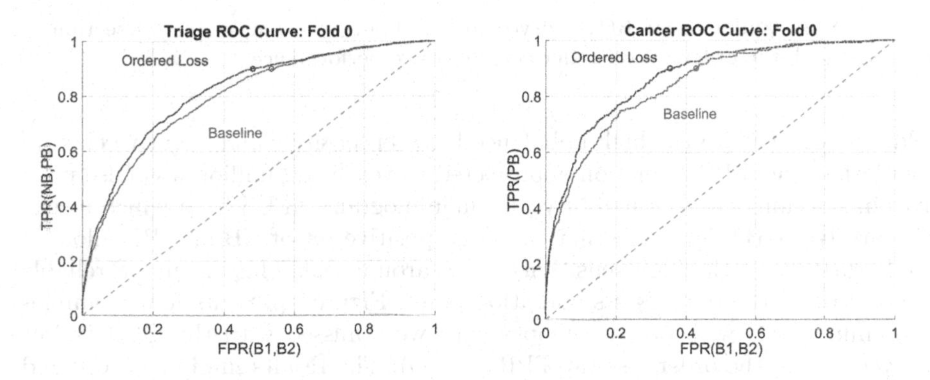

Fig. 3. Triage and Cancer ROC curves in a single fold of the cross-validation tests. Both triage and cancer detection performance are improved with the ordered loss.

In our second evaluation we measure the triage and cancer detection performance on the held-out set. To this end, we use an ensemble of trained models in the 3 folds of our cross-validation test set. The results depicted in Fig. 4 show a similar triage performance for both loss functions, while the ordered loss yields an improved cancer detection (note the higher blue ROC curve at the right plot).

We further report quantitative results by the following measures: ROC-AUC, false-negative rate (FNR) at operation point of false-positive rate FPR = 0.35, for triage and cancer detection curves (see Figs. 3 and 4). Note that FPR = 0.35 corresponds to sensitivity in the range [0.8,0.9], close to the average radiologists sensitivity in screening (0.87) [5]. Our last measure includes the FPR at true-positive rate TPR = 0.90 for true cancer detection (lower is better). Table 2 summarizes these results for our cross validation and held-out tests. The results show that under similar triage AUC, the operation point on triage is slightly improved indicated by lower FNR, although this was not targeted. Most importantly the ordered loss significantly improves the true cancer detection along with slight improvement in the triage performance. The cancer miss-rate is reduced by 19% in average in CV tests (from 12.97% to 10.50%) and by 10.5% for the held-out set (from 13.3% to 11.9%). While the AUC-Cancer is slightly raised by 1.2–2%, the FPR-Cancer is reduced by 9.5% in CV and 7.4% in the

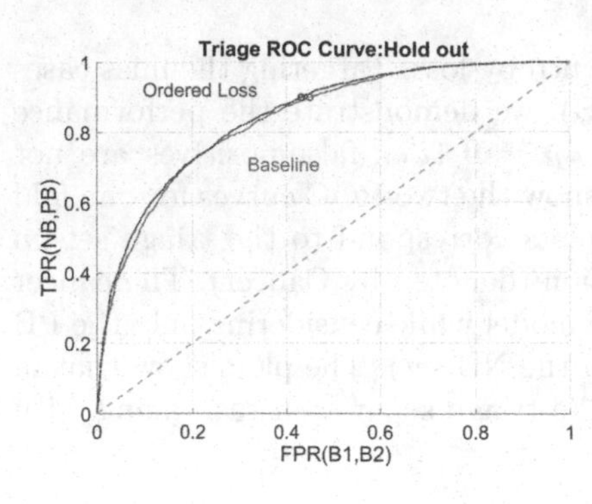

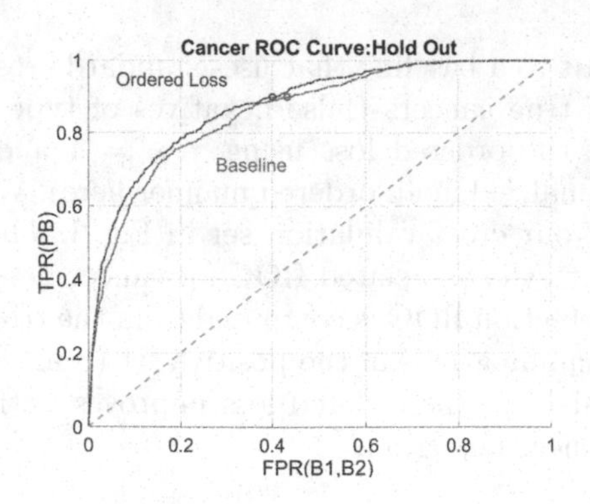

Fig. 4. Triage and Cancer ROC curves for held-out set. The cancer detection is improved with the ordered loss under similar triage performance.

held-out test. Due to the high imbalance between positive and negative class in population, this FPR reduction is projected to yearly 1.2 million less false positives in screening (based on 40M yearly mammograms in US from which nearly 90% are B1 or B2 [5]). Note that the large positive set of NB and PB allows a valid statistics at the high sensitivity rates around 0.9. This ensures a reliable comparison at the radiologists operation point. Figure 5 presents four examples of mammograms with positive biopsy that were missed with the CE loss but recovered with the order loss (at TPR = 0.90). The results imply that ordered loss better captures low signature cancers.

Table 2. Comparison of classification performance on cross-validation (CV) and held-out data sets. False-negative rates in triage task FNR(NB,PB) and the corresponding true cancer miss-rates, FNR(PB) are shown (lower is better). The positive class in each scenario is indicated, NB: Negative Biopsy, PB: Positive Biopsy. AUC values are for triage and cancer (higher is better). FPR-Cancer indicates the false-positive rate (FPR) at cancer detection rate of 0.90 (lower is better). For CV, mean values are shown. Best results are in bold.

Cross-Validation					
	FPR = 0.35 Corresponds to 0.8-0.9 TPR				TPR(**PB**) = **0.90**
Model	FNR(NB,PB) % triage	FNR(PB) % cancer	AUC triage	AUC cancer	FPR cancer
CE	16.93	12.97	0.832	0.868	0.392
OL	**16.47**	**10.50**	0.835	**0.879**	**0.358**
Held-Out					
CE	16.20	13.30	0.841	0.859	0.433
OL	**15.00**	**11.90**	0.848	**0.876**	**0.401**

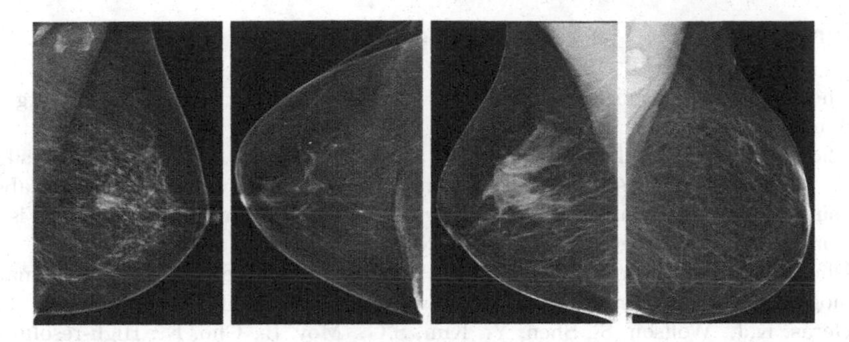

Fig. 5. Examples of mammograms with positive biopsy results (true cancers) that were missed by the standard CE loss but recovered by the ordered loss. Bright patterns in breast are mostly part of natural fibro-glandular tissues. Pathologies are often obscured and hardly visible.

Fine Tuning: In this experiment we first trained the network with CE loss to obtain the standard baseline approach. We then changed the loss to our ordered loss and continued training end-to-end. In this workflow we obtained convergence of the baseline model to the ordered loss optimum after only 20 epochs. This was verified by the classification results. Therefore, an existing model trained with CE loss can obtain ordered preference in a short extra training time, without the need to retrain the model from scratch.

4 Summary and Conclusions

Often in medical applications, radiologists are required for a binary decision. For example, whether to recall a patient or not. In such tasks, high sensitivity is expected while an emphasis is put on the subclass of more severe cases, such as missing true cancers. This clinical practice is often ignored in computational models, which aim to maximize AUC or an operation point. In this work, we propose a generalization of the common cross-entropy loss to an ordered loss to address this problem. In our formulation, the ordered term is reduced to the cross entropy loss by a simple parameter setting. We evaluated our new loss function with a large data set of over 16 K mammograms, including nearly 2,500 true cancers verified by biopsy. The results show a significant reduction in true cancer misses while preserving the overall classification performance in the original task. This outcome shows that the characteristics of severe cases are more emphasized in a model trained with our ordered loss function. We hope that this property will lead to a more reliable and trustworthy tool for radiologists. However, the current model is limited to binary classification and there is a need to further study the combination of augmentations (for different sub-classes) and ordered loss. While demonstrated on mammography, the method can be directly implemented to other medical domains. In our future work, we intend to implement the ordered loss in a detection scenario, to prioritize detection of lesions of more severe types.

References

1. The Digital Mammography DREAM Challenge. https://www.synapse.org/#!Synapse:syn4224222
2. Choukroun, Y., Bakalo, R., Ben-Ari, R., Akselrod-Ballin, A., Barkan, E., Kisilev, P.: Mammogram classification and abnormality detection from nonlocal labels using deep multiple instance neural network. In: Eurographics Workshop on Visual Computing for Biology and Medicine (2017)
3. Dhungle, N., Carnerio, G., Bradley, A.P.: Fully automated classification of mammograms using deep residual neural networks. In: ISBI (2017)
4. Geras, K.J., Wolfson, S., Shen, Y., Kim, S.G., Moy, L., Cho, K.: High-resolution breast cancer screening with multi-view deep convolutional neural networks. arXiv preprint: 1703.07047 (2017)
5. Lehman, C.D., et al.: National performance benchmarks for modern screening digital mammography: update from breast cancer surveillance consortium. Radiology **283**(1), 59–69 (2017)
6. Lotter, W., Sorensen, G., Cox, D.: A multi-scale CNN and curriculum learning strategy for mammogram classification. In: MICCAI-DLMIA (2017)
7. Niu, Z., Zhou, M., Wang, L., Gao, X., Hua, G.: Ordinal regression with multiple output CNN for age estimation. In: Proceedings of the IEEE Conference on Computer Vision and Pattern Recognition, pp. 4920–4928 (2016)
8. Rennie, J.D., Srebro, N.: Loss functions for preference levels: regression with discrete ordered labels. In: Proceedings of the IJCAI Multidisciplinary Workshop on Advances in Preference Handling, pp. 180–186. Kluwer, Norwell (2005)
9. Szegedy, C., Ioffe, S., Vanhoucke, V., Alemi, A.A.: Inception-v4, inception-resnet and the impact of residual connections on learning. In: AAAI, vol. 4, p. 12 (2017)
10. Zhu, W., Lou, Q., Vang, Y.S., Xie, X.: Deep multi-instance networks with sparse label assignment for whole mammogram classification. In: Descoteaux, M., Maier-Hein, L., Franz, A., Jannin, P., Collins, D.L., Duchesne, S. (eds.) MICCAI 2017. LNCS, vol. 10435, pp. 603–611. Springer, Cham (2017). https://doi.org/10.1007/978-3-319-66179-7_69

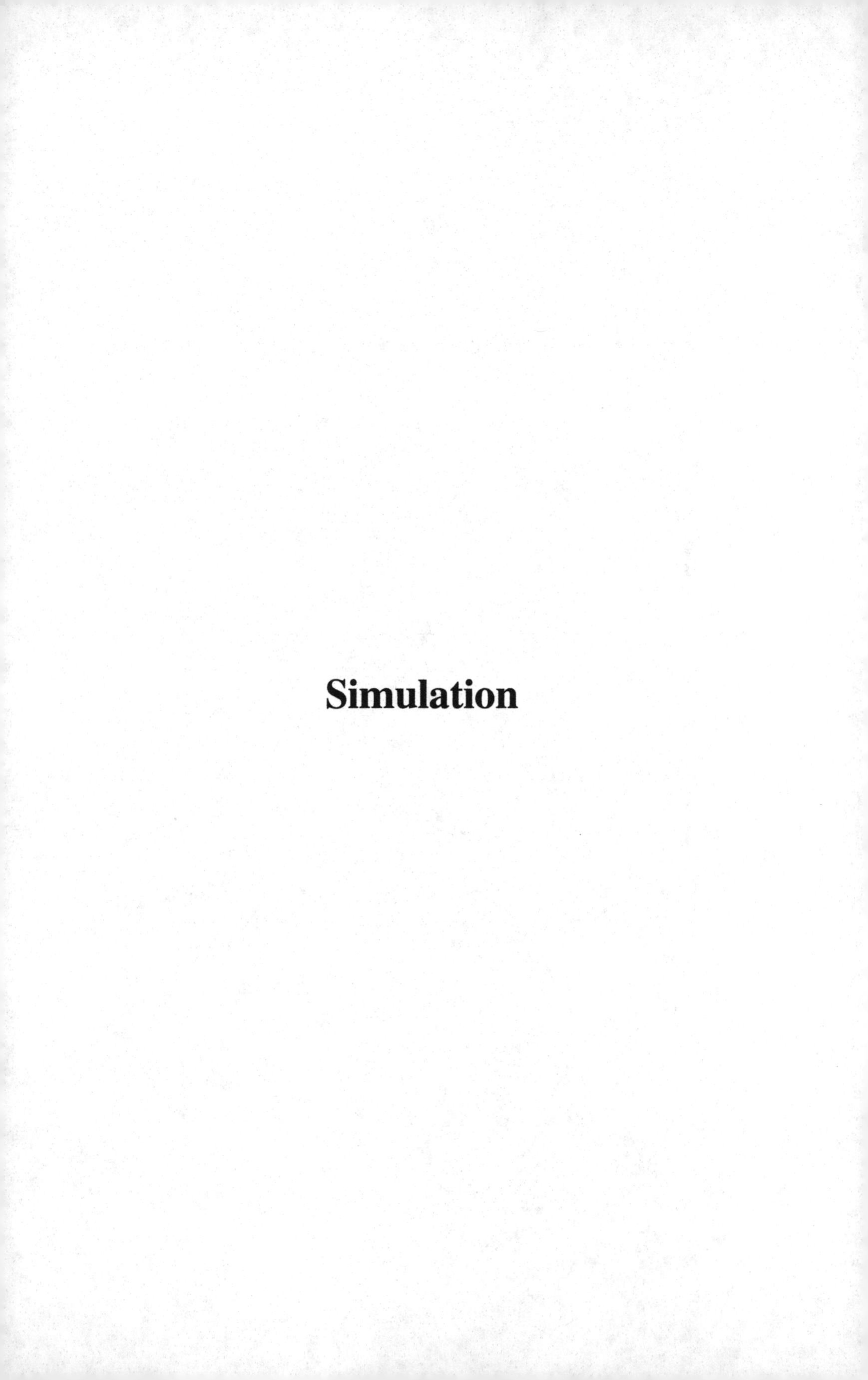

Simulation

Agent-Based Models and Spatial Enablement: A Simulation Tool to Improve Health and Wellbeing in Big Cities

Daniele Pala[1(✉)], John Holmes[2], José Pagàn[3], Enea Parimbelli[4], Marica Teresa Rocca[5], Vittorio Casella[5], and Riccardo Bellazzi[1]

[1] Department of Electrical, Computer and Biomedical Engineering, University of Pavia, Pavia, Italy
daniele.pala02@universitadipavia.it
[2] Center for Clinical Epidemiology and Biostatistics, University of Pennsylvania, Philadelphia, PA, USA
[3] Department of Public Health Policy and Management, College of Global Public Health, New York University, New York, NY, USA
[4] Telfer School of Management, University of Ottawa, Ottawa, ON, Canada
[5] Department of Civil Engineering and Architecture, University of Pavia, Pavia, Italy

Abstract. As the percentage of the population living in urban areas is constantly increasing throughout the world, big cities' municipalities and public health policy makers have to deal with raising socioeconomic disparities and need for environmental interventions to reduce pollution and improve wellbeing. The PULSE project, funded by the EU commission under the H2020 program, aims at providing an instrument that assesses health and wellbeing in cities through sensing technologies and data integration. The system has been deployed in 7 cities – Barcelona, Birmingham, Keelung, New York, Paris, Pavia and Singapore – and includes several state-of-the-art technologies, such as a smartphone App, a WebGIS, air quality sensors, a Decision Support System and dashboards. A crucial aspect of the project is the direct involvement of the citizens and the creation of Public Health Observatories (PHOs) that can help taking informed decisions and organize targeted interventions. To this end, PHOs are provided with *powerful visual analytics* to study different areas of the city, and with *simulation tools* that can be used to model the effect of interventions of public health authorities the city. In this paper, a first agent-based simulation model, based on the results of spatio-temporal data analytics, is presented. The model simulates the effect of traffic pollution, industrial land use and green areas on the probability of asthma hospitalizations in an area of East Harlem, one the neighborhoods with the highest asthma hospitalizations rate in New York City.

Keywords: Public health · Agent-based modeling · Simulation · Spatial analysis · Big data · Asthma

D. Riaño et al. (Eds.): AIME 2019, LNAI 11526, pp. 79–83, 2019.
https://doi.org/10.1007/978-3-030-21642-9_11

1 Introduction

The percentage of the world's population living in urban areas is projected to increase in the next decade [1]. In big cities social disparities are usually higher than in small towns and environmental factors are crucial to guarantee citizens' health and wellbeing. The EU project named PULSE (Participatory Urban Living for Sustainable Environments) aims at providing an instrument that assesses health and wellbeing in cities through sensing technologies and data integration. The project partners with 7 pilot cities – Barcelona, Birmingham, Keelung, New York, Paris, Pavia and Singapore – focusing on two main diseases: asthma, known for being related to air pollution [2], and type 2 diabetes, related to physical inactivity and unhealthy habits [3]. PULSE main innovation stands in the direct involvement of the citizens, that can send their own data and receive personalized feedbacks and communications that help them improve their health and lower their risk or diseases. PULSE culminates with the establishment of Public Health Observatories (PHOs), that, provided with visualization and simulation tools, can inspect the situation in the city and design proper interventions in the neighborhoods that need them. The data collection and integration in PULSE allows Spatial Enablement, i.e. the addition of geographic information to existing data, useful to study how health problems can change in the different areas of a city, depending on environmental and socioeconomic factors. In this paper, we briefly present the results of a spatial enablement study carried out within PULSE, where we model the relation between asthma hospitalizations and a number of factors in New York City, then, we present a first prototype of simulation tool of the interactions between several variables and asthma hospitalizations to be provided to the PHOs.

2 Methods

In this section, we briefly present the PULSE system architecture, the methods used in our spatial enablement analyses and the principles of agent-based modeling, with which we implemented the PHO simulation tool.

2.1 The PULSE System

Data integration is at the basis of the PULSE system. Within PULSE, both personal and public data are collected. Through an App, called PulseAir, each user can send his/her own data by (i) answering to health and well-being questionnaires, (ii) sharing his/her mobility patterns, measured with GPS and physical activity tracking and, finally, contributing to air quality monitoring by collecting air pollution data with low-cost measuring devices. These data are sent to a set of backend services, where their analysis allows defining citizen profiles that may include disease risk estimate and well-being scores, thus finally providing semi-automated feedbacks on life style to the citizen through the App. Besides personal data, also open data are acquired in the process, and used mainly to allow citywide analyses. This data is shown on the Pulse WebGIS (Geographical Information System) and on a dashboard available to the PHO. The PHOs will also be able to see aggregated anonymized user data, useful for decision making concerning public health interventions in the city.

2.2 Spatial Enablement Through Geographically Weighted Regression

One of the most important and innovative aspects of PULSE is its focus on each city, considering how population and environment can vary, sometimes widely, from a neighborhood to another. Using data collected within PULSE, we used Geographically Weighted Regression (GWR) [4] to model the relations between asthma hospitalizations and several factors that are known to be related to them according to the literature [5]. GWR can be thought as a generic regression model where the relation is multiplied by a weight that gives a geographic description to the results. In formula, for the simplest linear case:

$$\boldsymbol{Y} = \boldsymbol{X\beta}; \boldsymbol{\beta} := min \sum w_i \left(y_i - (\boldsymbol{X\beta})_i\right)^2; \beta_i' = (\boldsymbol{X}^t \boldsymbol{W}_i \boldsymbol{X}) - \boldsymbol{X}^t \boldsymbol{W}_i \boldsymbol{Y} \tag{1}$$

The introduction of a weight factor in the least squared estimation of the coefficients β can be noticed. For each β, a weight matrix $\boldsymbol{W}_i$ is defined specific to the i-th location, so that the nearest points have more weight than the ones further away. In our case, we overlapped a grid of points distant 1 km from each other to a polygon map of New York City. The weights are defined, following some examples found in literature, as the exponential of the negative squared ratio between the distance between the considered dot and the *i*-th centroid of the polygon and a threshold, in our case correspondent to 5 km.

2.3 Agent-Based Modeling

Agent-based models (ABM) are simulation tools used to study a lot of physical, social, epidemiological and economical phenomena. ABM are versatile tools that model interactions among agents in a certain environment, where agents can be people, animals, objects etc., and they interact with each other and with the environment changing their behavior or status according to underlying laws. In our study, we developed a first version of an ABM that exploits some of the GWR results and allows the observer to visualize how the trend of asthma hospitalizations would change if some specific variables were changed with targeted interventions.

3 Results and Implementation

3.1 GWR Results

We tested several covariates both in univariate and multivariate models, specifically average annual PM2.5, age, race, poverty rate, percentage of industrial land use, diffusion of Medicaid insured, obesity and recycling rate. The type and reliability of the correlations between each covariate and the independent variable change notably within the different neighborhoods of the city. Figure 1 shows an example of the GWR results concerning poverty rate, which demonstrates the extreme variability of the relations within neighborhoods. However, as a general rule, we observed that the areas with the highest hospitalization rates (the Bronx, East Harlem, Washington Heights and an area that includes Crown Heights, Brownsville and East Flatbush) are the same in which the average age is low, the concentration of Black and Hispanic is higher, the obesity rate, the poverty rate and Medicaid use are higher, there are more industrial sites and lower recycling rates. PM2.5, at least on an averaged measure, has generally a bland positive correlation with asthma, with the exception of most of Manhattan.

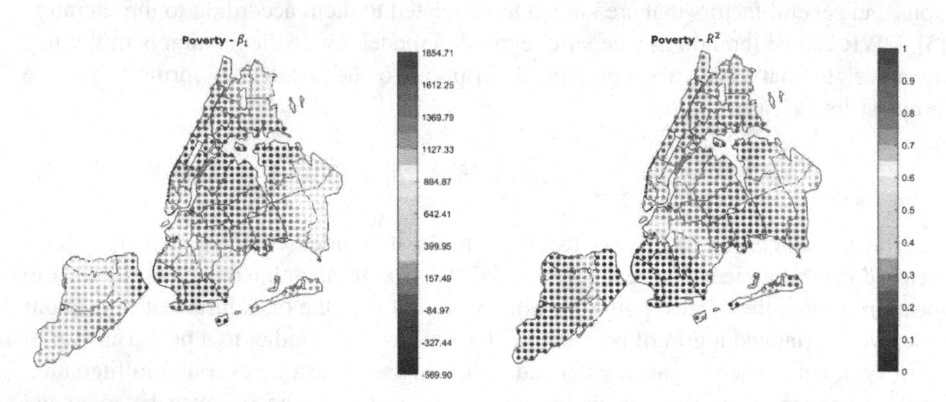

Fig. 1. Results of the GWR using poverty as covariate. On the left, values of the regression coefficient across the city. On the right, values of the coefficient of determination.

3.2 ABM Implementation

In line with the aim of PULSE of allowing the PHOs to plan interventions throughout the city, we developed a first example tool based on agent-based modeling, visually implementing the relations found with our GWR model. Our ABM (Fig. 2), developed in NetLogo, is based on real GIS data referred to part of East Harlem, with the following boundaries: Malcolm X Boulevard on the West, Tito Puente Way (E 100th Street) on the South, the FDR Drive on the East and E 126th Street on the North. The used shapefiles contain streets centerlines, sidewalks, buildings and parks. The observer can determine the initial population and the traffic density, in order to simulate how a variation of them could influence pollution and exposure to it. The interface features some sliders where the observer can increase or decrease the percentage of land used for industrial activities, the recycling rate and the obesity rate, in order to simulate the impact of interventions on land use, public services and food policies. The observer can also set the initial mean and standard deviation of the population's age, a specific age will be given to all people according to a normal distribution. The risk of hospitalizations changes with age (i.e. people under 18 and over 60 are more at risk), plus there's a probability of death that increases dramatically after 75 years of age. Each tick of the model corresponds to 6 months. Once the observer hits the "Go" button, cars are free to move on the streets and pollute the area, and people walk in the sidewalks and get exposed to pollution. A plot and some monitors show the current number of hospitalized people, based on the probability computed by the regression model and a reasonable discharge rate. The initialized quantities can be changed during the simulation to see the subsequent changes in the hospitalizations trend.

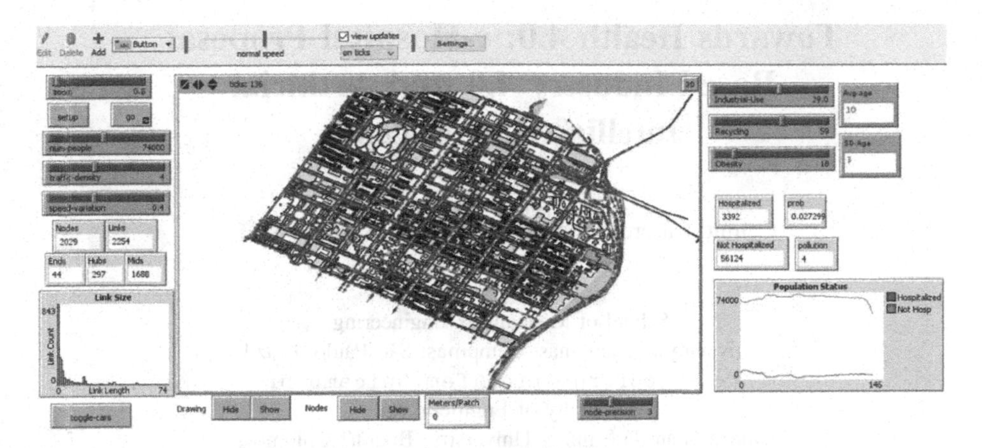

Fig. 2. Screenshot of the current ABM prototype interface.

4 Conclusions and Future Developments

Public health, wellbeing and disease risks are the result of a complex combination of factors, especially in big cities. In this paper we presented how the PULSE project is addressing these topics through the development of a collaborative system that allows both citizens and policy makers to take part of the health decision making, using data with high spatial and temporal resolution. As the project goes on, thanks to the higher availability of data, new risk models will be developed, and the PHOs will be provided with new ABM models for all the neighborhoods concerning also different health-related phenomena, such as the relations among environment, behavior and wellbeing.

References

1. United Nations Department of Economic and Social Affairs. (2018). https://www.un.org/development/desa/en/news/population/2018-revision-of-world-urbanization-prospects.html. Accessed 21 Jan 2019
2. Guarnieri, M., Balmes, J.R.: Outdoor air pollution and asthma. Lancet Lond. Engl. **383**(9928), 1581–1592 (2014)
3. Sigal, R.J., Kenny, G.P., Wasserman, D.H., Castaneda-Sceppa, C., White, R.D.: Physical activity/exercise and type 2 diabetes: a consensus statement from the American diabetes association. Diab. Care **29**(6), 1433–1438 (2006)
4. McMillen, D.P.: Geographically weighted regression: the analysis of spatially varying relationships. Am. J. Agric. Econ. **86**(2), 554–556 (2004)
5. Litonjua, A.A., Carey, V.J., Weiss, S.T., Gold, D.R.: Race, socioeconomic factors, and area of residence are associated with asthma prevalence. Pediatr. Pulmonol. **28**(6), 394–401 (1999)

Towards Health 4.0: e-Hospital Proposal Based Industry 4.0 and Artificial Intelligence Concepts

Camilo Cáceres[1(✉)], Joao Mauricio Rosário[1], and Dario Amaya[2]

[1] School of Mechanical Engineering, University of Campinas, Campinas, São Paulo, Brazil
{camilocf, rosario}@fem.unicamp.br
[2] Faculty of Engineering, Nueva Granada Military University, Bogotá, Colombia
dario.amaya@unimilitar.edu.co

Abstract. The implementation of the most recent technologies is a requirement in this fast-growing and competitive world, especially in the improvement and development of the healthcare area, that is fundamental in life quality enhancement. This article has as objective the utilization of AI and Industry 4.0 concepts oriented to the optimization of a hospital, using a case study an Emergency Department (ED). This proposal allows the development of a current proposal of e-Hospital based on Health 4.0 features and the use of computational ED models will allow the avoidance and detection of bottlenecks in the workflow. Those blockages are automatically removed using an improved shift management proposal based on control theory, AI, and telemedicine. The results show an optimization in the use of the resources and a reduction of the length of stay improving the service quality. The simulation tools allow the test and validation of novel proposals for e-health.

Keywords: Health 4.0 · Discrete Event Simulation · e-Hospital · Emergency Department · Evolutionary Algorithm · Optimization

1 Introduction

Health 4.0 was created as a response to the demographic and socio-economic changes in the last years. The main design principles are interoperability, virtualization, real-time capability decentralization, service orientation, modularity, safety, security and resilience [1, 2]. Its execution represents a complex challenge that can be solved by the adoption and implementation of concepts like Artificial Intelligence (AI), Cloud Computing, 5G, Internet of Things (IoT), Medical Internet of Things, Precision Medicine and Blockchain. That technology is complemented with healthcare simulation tools, like Discrete Event Simulation (DES), that is the main simulation tool to test and solve management issues.

A problem that can be solved by the implementation of Health 4.0 concepts and the use of simulation tools is the overcrowding in the Emergency Department (ED),

D. Riaño et al. (Eds.): AIME 2019, LNAI 11526, pp. 84–89, 2019.
https://doi.org/10.1007/978-3-030-21642-9_12

recognized as a world-class problem. The overcrowding is a situation that occurs when the number of patients is superior to the available resources. This is an unpredictable situation that occurs only 25% of the total operating time of an ED, it reduces the service quality and the staff productivity, creating agglomerations, increasing the patients waiting time, and others negative social factors [3]. The overcrowding problem is a current problem, making it a good case study to proof the design principles of Health 4.0 and the integration of up-to-date technologies. For its implementation is mandatory an analysis and understanding of the processes in a traditional Hospital, identifying the opportunities to upgrade the tools and improve the resources, for example with the use of teleconsultation, telepresence, and telesurgery, as a solution for remote services [4, 5].

This work has as objective to propose a solution for the overcrowding in a chosen ED, testing the solutions in a DES environment, measuring and performance comparison using the Key Performance Indicators (KPI). The comparison and development of simulation models for a traditional ED and an e-Hospital ED based on Health 4.0 were implemented in DES models, using Matlab-SimEvents®, based in [6, 7], Table 1 and Fig. 1.

Table 1. Service time distributions and staff number [6, 7].

Stage	Distribution (Minutes)	Staff	Abbrev	Current staff
Reception	Uniform (5,10)	Receptionist	R	2
Lab tests	Triangular (10,20,30)	Laboratory technician	T	3
Examination room	Uniform (10,20)	Physician	D	2
Reexamination room	Uniform (7,12)			
Treatment Room (TR)	Uniform (20,30)	TR nurse	TN	1
Emergency Room (ER)	Uniform (60,120)	ER nurse	EN	9

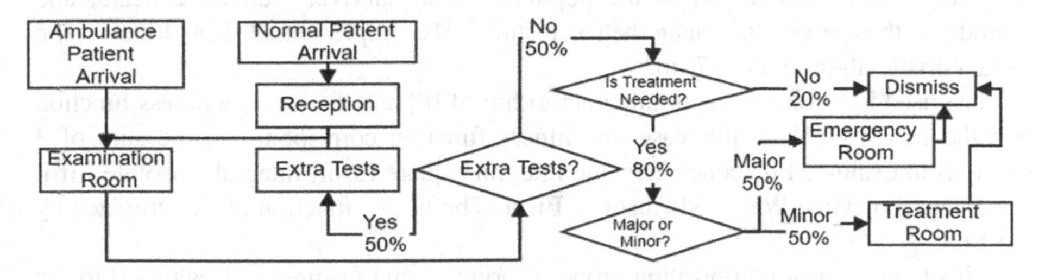

Fig. 1. ED high-level process view [6, 7].

2 Methodology

The proposed methodology for this work takes the study case of ED presented by [6], where a DES model was developed for a chosen ED to optimize the personnel, increasing the Hospital profits. By another hand, this work is the continuation of [7], where a DES model was developed with an automatic method to solve the ED overcrowding issues.

Initially was developed of a DES for the chosen the ED model, it is used for analysis of health services under a certain established condition. The DES ED follows the data presented in Fig. 1. The developed DES follows the implementation and patients arrival rate presented by [6, 7]. The improvement of the traditional ED makes of it an e-Hospital ED, that is achieved based on the analysis of the ED-DES. A bottleneck is detected in the Examination Room because of its high utilization (99 ± 1%.), it generates overcrowding. Also, other KPI in that stage are abnormal, the queue length and waiting time average of 18.51 ± 4.36 patients/h and 88.41 ± 20.41 min, respectively.

An e-Health ED is proposed based on the implementation of telemedicine, used technically in [5], in order to reduce the detected bottleneck. The proposed telediagnosis solution, based on automatic control, it is a controller that takes as reference the Examination Room queue length, in this case, 6 patients in the waiting line. That controller will automatically choose the moment when a tele-physician will be required.

The implemented controller corresponds to a PID controller [8], that following a mathematical model based on the error signal, it will be able to take some decisions, as it implemented by [7]. In order to set controller parameters (Proportional, Integral, and Derivative), traditional methods used for deterministic dynamic systems are useless, due because of the stochastic nature of the studied system. Instead, an AI technique was used to find the right parameters. An Evolutionary Algorithm was used to solve this problem, more specifically a Memetic Algorithm (MA) based on a Genetic Algorithm (GA).

The MA optimizes the PID controller parameters using metaheuristics, it uses a fitness function to determine how good or bad is a solution. The population size corresponds to 100 controllers with its constants, representing a set of controllers in the solution space. Each solution is graded by a fitness function, where the more fit individuals (better solutions) of the population can survive, mutate, replicate, and reproduce themselves to obtain better results. The implemented MA follows the metaheuristic algorithm of [7, 8].

The used MA follows the training algorithm of [8] and the use of a fitness function like the used by [7]. In this case, the fitness function corresponds to the sum of 3 methods to evaluate PID controllers, the Integral Square Error, Integral Absolute Error and Integral of Time-Weighted Absolute Error. The fitness function (J) is optimized by the MA.

This management optimization proposal brings some features of Health 4.0 to the ED like Virtualization, Real-Time Capability, and Modularity. The implemented solution of the e-Hospital DES model corresponds to an improvement of the Traditional Hospital ED, in Fig. 2. The implementation of the described telediagnosis room represents the design principle of a Hospital 4.0. This solution is based on the integration of concepts like AI, and Control theory, showing Health 4.0 principles like Virtualization, Real-Time Capability, and Modularity, improving some e-health characteristics like management, efficiency, and equity. Finally, the identification and improvement of the system allowed a KPI comparison of a traditional and an e-Hospital ED-based Health 4.0 concepts.

3 Results and Discussion

The simulation results of the developed DES of a traditional ED [6, 7] and the proposed e-Hospital, in Fig. 2. Table 2 shows the most remarkable obtained KPIs. The results summarize 1000 days of each DES model under different conditions (stochastic simulation). The main difference between the compared systems is the use of a telediagnosis room in the e-Hospital, in order to reduce the ER queue length bottleneck. The main KPI improvements are presented in the Examination Room, where the bottleneck was located. That improvement is reflected in the improvement of the e-Hospital ED performance, with the reduction of the Length of Stay (LOS) and the increased number of Discharged patients.

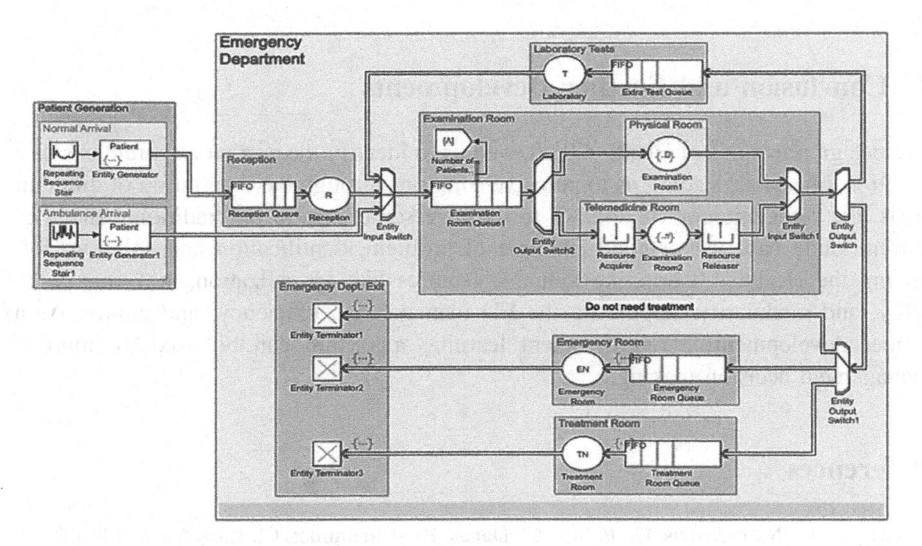

Fig. 2. DES simulator of an ED-based Health 4.0, following [7].

The main difference between the structures of the traditional and the e-Hospital ED proposal is the addition of *1.72* ± *0.96* tele-physicians were scheduled, using the based AI PID controller. The performance of the e-Hospital system showed that a minor improvement makes a difference in the service quality, reflected in the KPIs. Comparing the obtained results with the results of [6], the obtained results were better, since [6] do not consider dynamical shifts during the day. The current work can be also compared with [7], where following a similar methodology, they added an additional step in the analysis of bottlenecks after the first improvement. That second analysis was avoided in this work to respect the principle of Real-Time Capability of Health 4.0. The improvement in [7] for the LOS was around *23.86 ± 8.85%* adding 9.8 ± *0.94%* of personnel, while in the current work, the LOS improvement was around *13.65 ± 9.79%* adding *10.11 ± 5.64%* of personnel. This comparison showed the importance of secondary bottleneck analysis and improvement.

Table 2. Main KPI of the DES for traditional hospital and e-Health ED.

Key Performance Indicator (KPI)	Traditional ED		e-Hospital ED		Error
	Mean	Std	Mean	Std	
Length of stay (h)	3,88	0,56	3,35	0,38	0,53
Discharged patients (patient/h)	4,59	0,26	5,53	0,27	–0,94
Examination room queue length (patients/h)	18,51	4,36	6,40	0,39	12,11
Examination room queue wait (min)	88,41	20,29	29,34	3,05	59,06
Examination room queue (patients/h)	19,76	4,36	7,08	0,83	12,68
Examination room utilization (%)	0,99	0,01	0,98	0,01	0,01

4 Conclusion and Further Developments

The design principles of Health 4.0 allowed us to identify the current requirements of a traditional hospital, leading us to perceive solutions through the integration of different areas like automatic control and AI, to improve some gaps in the traditional ED. The current study case was a good example of problem identification and solution, following the Health 4.0 bringing valuable features like virtualization, real-time capability, and modularity, improving the ED management, efficiency, and equity. As a further development, a reinforcement learning algorithm can be used to improve management decision making.

References

1. Grigoriadis, N., Bakirtzis, C., Politis, C., Danas, K., Thuemmler, C., Lim, A.K.: A health 4.0 based approach towards the management of multiple sclerosis. In: Thuemmler, C., Bai, C. (eds.) Health 4.0: How Virtualization and Big Data are Revolutionizing Healthcare, pp. 205–218. Springer, Cham (2017). https://doi.org/10.1007/978-3-319-47617-9_10
2. Thuemmler, C.: The case for Health 4.0. In: Thuemmler, C., Bai, C. (eds.) Health 4.0: how virtualization and big data are revolutionizing healthcare, pp. 1–22. Springer, Cham (2017). https://doi.org/10.1007/978-3-319-47617-9_1

3. McHugh, M.: The Consequences of Emergency Department Crowding and Delays for Patients. In: Hall, R. (ed.) Patient Flow. International Series in Operations Research & Management Science, vol. 206, pp. 107–127. Springer, Boston (2013). https://doi.org/10.1007/978-1-4614-9512-3_5
4. Feussner, H., et al.: Surgery 4.0. In: Thuemmler, C., Bai, C. (eds.) Health 4.0: how virtualization and big data are revolutionizing healthcare, pp. 91–107. Springer, Cham (2017). https://doi.org/10.1007/978-3-319-47617-9_5
5. Marconi, G.P., Chang, T., Pham, P.K., Grajower, D.N., Nager, A.L.: Traditional nurse triage vs physician telepresence in a pediatric ED. Am. J. Emerg. Med. **32**, 325–329 (2014)
6. Ahmed, M.A., Alkhamis, T.M.: Simulation optimization for an emergency department healthcare unit in Kuwait. Eur. J. Oper. Res. **198**, 936–942 (2009)
7. Cáceres, C., Rosário, J.M., Amaya, D.: Proposal of a smart hospital based on Internet of Things (IoT) concept. In: Lepore, N., Brieva, J., Romero, E., Racoceanu, D., Joskowicz, L. (eds.) SaMBa 2018. LNCS, vol. 11379, pp. 93–104. Springer, Cham (2019). https://doi.org/10.1007/978-3-030-13835-6_11
8. Cáceres Flórez, C.A., Rosário, J.M., Amaya, D.: Control structure for a car-like robot using artificial neural networks and genetic algorithms. Neural Comput. Appl. 1–14 (2018). https://doi.org/10.1007/s00521-018-3514-1

Knowledge Representation

MitPlan: A Planning Approach to Mitigating Concurrently Applied Clinical Practice Guidelines

Martin Michalowski[1(✉)], Szymon Wilk[2], Wojtek Michalowski[3], and Marc Carrier[4]

[1] University of Minnesota, Minneapolis, MN 55455, USA
martinm@umn.edu
[2] Poznan University of Technology, Poznan, Poland
[3] University of Ottawa, Ottawa, Canada
[4] The Ottawa Hospital Research Institute, Ottawa, Canada

Abstract. As the overall population ages, patient complexity and the scope of their care is increasing. Over 60% of the population over 65 years of age suffers from multi-morbidity, which is associated with over two times as many patient-physician encounters. Yet clinical practice guidelines (CPGs) are developed to treat a single disease. To reconcile these two competing issues, we developed a framework for identifying and addressing adverse interactions in multi-morbid patients managed according to multiple CPGs. The framework relies on first-order logic (FOL) to represent CPGs and secondary medical knowledge and FOL theorem proving to establish valid patient management scenarios. In this work, we leverage the framework's representation capabilities to simplify its mitigation process and cast it as a planning problem represented using the Planning Domain Definition Language (PDDL). We demonstrate the framework's ability to identify and mitigate adverse interactions using planning actions, add support for durative clinical actions, and show the improved interpretability of management plan recommendations in the context of both proof-of-concept and clinical examples.

Keywords: Clinical practice guidelines · Multi-morbidity · Planning

1 Introduction

Clinical practice guidelines (CPGs) are statements developed systematically from available evidence to assist practitioners in appropriate management of a patient with a specific disease [1]. CPGs have demonstrated multiple benefits, including improved quality of care and patient outcomes [2]. Despite this, their practical adoption is limited and one of the major obstacles is their very limited support for complex patients, i.e., patients with discordant multi-morbidity [3]. On one hand, such patients are typically excluded from clinical trials used for CPG development [4]. On the other hand, population aging and the widening scope of care results in an increasing number of complex patients (over 60% of population over 65 suffers from multi-morbidity [5]). A straightforward application of disease-specific CPGs to such patients can lead to

D. Riaño et al. (Eds.): AIME 2019, LNAI 11526, pp. 93–103, 2019.
https://doi.org/10.1007/978-3-030-21642-9_13

adverse interactions between recommendations that were not considered when developing a CPG that significantly deteriorate the quality of provided care and may even be dangerous to a patient's health [6].

Thus, there is a need for methods and tools for identifying such adverse interactions and for addressing them by revising conflicting recommendations [4, 7] – we refer to this process as the mitigation of adverse interactions. Responding to this challenge, there is significant research on computer-interpretable CPGs (CIGs) and on mitigating adverse interactions between simultaneously applied CPGs (summarized in [3]). Two groups of approaches have been proposed: those aimed at merging treatments and those that are merging CIGs. Generally, the former take treatment recommendations constructed according to several CIGs, mitigate possible interactions, and construct a single management plan, while the latter combine several CIGs into a single patient-specific CIG that is later used to establish a safe (interaction-free) management plan.

In [8] we proposed a framework for mitigating adverse interactions that belongs to the first group of approaches. It relies on first-order logic (FOL) to represent CIGs and other medical knowledge (referred to as primary and secondary knowledge, respectively) and combines search techniques with FOL-based reasoning in order to find treatment recommendations from multiple CIGs into a safe management plan. When revising recommendations, it also considers patient preferences such that the most desired revisions are introduced. Although capable, our FOL-based framework is complex and in this paper we propose its simplification, where we use planning instead of a hybrid of search and logical reasoning. Specifically, we demonstrate how to translate our FOL-based clinical knowledge representation into the Planning Domain Definition Language (PDDL), and how using a PDDL-based representation with a planner eases mitigation and construction of management scenarios.

2 Background

In this section we conceptually define the FOL-based mitigation framework and introduce the planning paradigm to provide context for our new contributions.

2.1 FOL-Based Mitigation Framework

Our mitigation framework assumes that each disease-specific CPG capturing primary clinical knowledge is represented as an *actionable graph* (AG) [8, 9]. The AG is a directed graph with context, decision, action, and parallel nodes. A context node is the root of the graph, a decision node indicates a clinical decision, an action node indicates a clinical action, and a parallel node indicates the beginning or end of two or more sequences of clinical decisions or actions that are executed in parallel. Nodes are ascribed with additional properties, e.g., related to their temporal characteristics.

The AG relies on the task network-based representation used as a foundation in a number of representation languages (e.g., GLIF3, SAGE or PROforma) [3], however, it has been limited to the 4 types of nodes listed above as those elements are important from a mitigation perspective. AGs can be easily obtained from other task network-based languages [10] and we use these AGs as an intermediate representation to apply our mitigation framework to CIGs represented in different languages.

We formally represent the AGs using a FOL language. The two key components of this language are structural and temporal predicates. The structural predicates, shown in Fig. 1a, describe the structure of the AG, and the temporal predicates, shown in Fig. 1b, describe the temporal relationship of and between nodes in the AG. These predicates are used in combination to construct FOL theories (i.e., sets of logical sentences) describing specific patient encounters. Each theory represents all disease-specific CIGs applied to a particular patient and information (potentially incomplete) available about this patent. In order to generate a management scenario for an encounter, we apply model finding techniques to the corresponding FOL theory. These structural predicates constitute a starting point for MitPlan (described below).

Predicate	Description
node(x)	x is a node in an actionable graph
disease(x, d)	x is a context node indicating disease d
decision(x, t)	x is a decision node associated with decision t
action(x, a)	x is a action node associated with action a
parallel(x)	x is a parallel node
directPrec(x, y)	Node x directly precedes node y (there is an arc from x to y)
prec(x, y)	Node x precedes y (there is a path from x to y)
dosage(x, n)	Dosage of the drug administered in an action node x is n units
result(x, v)	Result of the decision made in a decision node x is v

Fig. 1a. Structural predicates [8].

Predicate	Description
timeOffset(x, to)	Node x occurs to time units after the preceding node
duration(x, dt)	Node x takes dt time units to complete
startTime(x, st)	Node x starts at time st
currentTime(ct)	Current patient time is ct
happensNowOrLater(x)	Activity (decision or action) from node x is happening now (given current time) or will happen in future
overlap(x, y)	Execution periods of nodes x and y overlap
overlapNowOrLater(x, y)	Execution periods of nodes x and y are overlapping now (given current time) or will overlap in the future

Fig. 1b. Temporal predicates [8].

To identify and address adverse interactions, the framework includes *revision operators* that encode two types of secondary medical knowledge: (1) knowledge required to mitigate adverse interactions due to discordant morbidities and, (2) knowledge about patient preferences that describe clinical circumstances (e.g., a sequence of actions) that are not consistent with a patient's preferences. All revision operators are defined as a logical sentence in the FOL language that describes the undesired circumstances for which the operator is applicable and a set of operations that need to be applied to address the applicability of the operator.

The mitigation of adverse interactions is then handled through a series of algorithms (see [8] for more detail). These algorithms operate on the FOL and its textual representation, applying generalized search and replace operations. This makes the process of generating a management scenario a complex operation. FOL was designed to express statements, propositions, and relations between objects. The goal is checking if a theory is consistent, determining if a certain formula holds in the context of a theory, and finding a model for the theory. Our framework uses FOL reasoning techniques to achieve these goals and then extracts certain parts of the model to form a management scenario. Thus, support for treatment planning is only indirect.

While our FOL-based mitigation framework is able to mitigate adverse interactions between concurrently applied CIGs, the FOL-based approach introduces extra complexity to the mitigation process. Numerical operations, including the calculation of time, theorem editing, and result interpretation to return management scenarios are challenging tasks that require significant domain engineering and complex processing. For example, numerical operations associated with revisions need to be performed at the textual representation level outside of FOL, while time-based calculations (establishing start and end times based on offsets and durations) are conducted inside FOL. Also, constructing a management scenario requires the parsing of logical sentences in a found model. To alleviate these drawbacks, we developed the planning-based approach Mit-Plan described in this paper. We note that since FOL is more expressive than standard planning formalisms, one can also use FOL to describe planning problems. However, due to the complete difference of the semantics of planning and FOL, one would have to "misuse" FOL for this purpose. FOL expresses statements, propositions, and relations between objects while planning is about the execution of actions and reasoning whether a sequence from initial state to the goal exists. These are two fundamentally different aims.

2.2 Planning

Planning, a field related to decision theory, finds a sequence of planning actions to realize a stated goal. Given an initial state of the world, a set of desired goals, and a set of planning actions, the planning problem is to identify a set of actions (ordered or non-ordered) that is guaranteed to generate a state from the initial one that contains the desired goal(s). Planning approaches fall into one of two categories: *state-space* and *plan-space*. State-space planning works at the level of the states and operators, where finding a plan is formulated as a search through state space, looking for a path from the start state to the goal state(s). This is most similar to constructive search. Plan-space planning operates at the level of plans, where finding a plan is formulated as a search through the space of plans. Planning starts with a partial, possibly incorrect plan, then applies changes to it to make it a full, correct plan. This approach is seen as an iterative improvement/repair process.

Planning (and hence problems described using PDDL) asks the question whether there exists a sequence of planning actions that transform the initial state (description of the world prior execution of an action) to some desired goal state. Thus, planning is about the execution of planning actions and reasoning whether such a sequence exists. Each planning action has a set of parameters (typed objects in the planning problem), preconditions that must be true for the action to be taken, and effects resulting from its execution. As we discuss later, planning actions can also be associated with durations, conditional effects, and costs. It is our hypothesis that planning is more naturally suited to the mitigation problem we are solving.

In this work, we use a state-space approach to interleave planning for specific paths through applied CIGs and applying revision and patient preference actions while trying to reach terminal states for each CIG. By defining the initial state to include all applicable CPGs and available patient information, our improved framework iteratively builds the plan and avoids or mitigates adverse interactions between the CPGs. We represent the planning problem using PDDL 2.1, which adds support for durative actions and both negative and conditional effects. PDDL 2.1 enables our framework to plan over parallel paths and actions with durations, when these are present in the AGs, within our mitigation framework. We use the Optic [11] planner, a forward-chaining partial-order state-space planner that supports PDDL 2.1, to find plans that can execute any node from the AGs using defined planning actions.

3 MitPlan

In this section we describe MitPlan, an updated component to our mitigation framework, that replaces the procedures, theorem proving, and model finding over FOL theories with a planning approach. MitPlan significantly reduces the complexity of our framework and supports durative clinical actions as first-class citizens in the planning process, rather than domain-driven manual additions as done using FOL. It also improves the interpretability of management scenario recommendations.

3.1 From FOL to PDDL

A planning problem is made up of two components: the domain and the problem instance. A domain contains the planning predicates, functions, and actions while the problem instance defines the objects, and the initial state and goal specification. The first step in transitioning from FOL to PDDL is to define the planning domain. Figures 1a and 1b show the predicates in our FOL-based mitigation framework that describe the structure of an AG and its nodes' temporal properties. We translate these into predicates in the planning domain and eliminate others by converting their relationships into planning functions. Table 1 shows only key planning predicates (due to space limitations). All of the temporal predicates present in our FOL-based approach are now encompassed in the semantics of durative actions in PDDL. We add new predicates *goal*: a goal (terminal) node in the AG; *interactionPresent*: an adverse interaction has been found; and *revisionOperator*: represents a revision operator.

Table 1. Predicate transition from FOL to PDDL.

FOL predicate	Planning predicate	Planning function
node	✓	–
disease	✓	–
decision	✓	–
action	✓	–
parallel	✓	–
directPrec	✓	–
prec[a]	–	–
dosage	–	✓
result	–	✓
–	*goal*	–
–	*interactionPresent*	–
–	*revisionOperator*	–

[a]prec is no longer needed in the planning domain

(action ?d – disease ?n - node) is a PDDL predicate that contains two typed objects (*d*: a disease, *n*: a node in an AG). The PDDL predicate *(action HTN A1)*, when true, indicates there is a node object *A1* in the AG for the disease object *HTN*. The PDDL

function *(patientValue ?v - patientData)* associates a numeric value with the patient data object *v*. The PDDL predicate *(decision ?d - disease ?n - node ?v - patientData)* associates patient data with a clinical decision such that object *n* is a clinical decision node associated with disease object *d* and patient data object *v*. *(revisionOperator ?n1 - node ?n2 - node ?r - node ?c - cost)* is a PDDL predicate that defines a revision operator that replaces nodes *n1* and *n2* with *r* with a cost of *c*. Future work expands this definition to support a wider range of replacement and insertion operations. All predicates also include a cost and duration.

To complete the planning domain, we include the set of planning actions described in Table 2. The terms listed for the preconditions and effects are all predicates and functions in the domain (+ denotes optional terms). Each action is defined by a set of parameters, a duration, preconditions, and effects. All preconditions and effects are required (precondition) or achieved (effect) at the start, over all, or at the end of the action. We do not list durations in Table 2 as they are action-instance dependent and action costs are numerical fluents added to an overall cost as part of an action's effect.

Table 2. Actions in the planning domain.

Action	Preconditions	Effects
takeAction	disease, action node, prec node executed, patient data value(+)	action taken, cost added(+)
makeDecision	disease, decision node, prec node executed, patient data value(+)	decision made, patient data value set, cost added(+)
startParallel	disease, parallel node, prec node executed, patient data value(+)	parallel path started
endParallel	disease, parallel node, parallel path started, all parallel nodes executed	parallel path completed
reachGoal	disease, goal node, prec node executed, patient data value(+)	goal node reached for disease
replaceNodes	disease(s), node(s), interaction present	existing nodes replaced with new nodes, precedence relationships set, cost added(+)

For each revision operator, its precondition is represented as the predicate *revisionOperator* and its operations as the planning action *replaceNodes*. In our previous work we supported both insertion and deletion operations. In this paper we only describe the replacement of actions with a new action, a combination of deletion and insertion. We will support finer grained revision operators in future work, by defining additional planning actions and updating the *revisionOperator* predicate.

The planning problem (referred to as the problem instance) inherits from the domain described above and contains the objects, initial state, and goal specification for a specific patient encounter. The objects in the instance are the diseases, nodes, and patient data (as defined in each AG), and the revision operators described in secondary knowledge sources or provided as patient preferences. The initial state represents the structure of each AG, the available patient information (patient data), and all known revision operators. The goal specification is to reach the goal (terminal) state of each AG with no adverse interactions present (optionally minimizing the overall cost).

4 Case Study

To show the feasibility of MitPlan within the context of our mitigation framework, we first describe proof-of-concept actionable graphs similar to those presented in our previous work [8]. We visually represent these AGs in Fig. 2. These AGs include action and decision nodes, a parallel path, share the action *A3*, and the planning instance contains two revision operators. The first operator states that if actions *A2* in disease *D1* and *A3* in disease *D2* are executed, then replace *A3* in *D2* with *newAction*. The second states that if action *C2* in *D1* and *A3* in *D2* are taken, then replace *A3* in *D2* with *newAction2*. MitPlan also supports more general revisions that replace all instances of *A3*, for example. We acknowledge that creating revision operators can be a time consuming task for a clinical expert, however interaction repositories (e.g., Cochrane) and ontologies are sources that can be used to (semi-)automatically generate these operators.

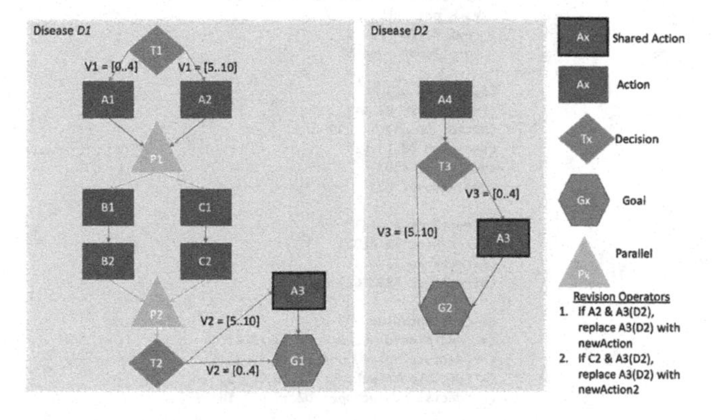

Fig. 2. Proof-of-concept AGs.

Figure 3 shows the PDDL problem instance for the AGs and revision operators in Fig. 2. Due to space considerations, we only show the description of the AG for *D2*, the revision operators, and goals. Patient data *V3* is set to 2, the costs of each revision operator are 10, and our goal is to minimize the total cost. The bounds for decision node *T3*'s branching points are set as [0..4] and [5..10]. Using the Optic planner to solve this problem instance, we get the solution presented in Fig. 4. We note that patient data values *V1 = 2* and *V2 = 4* are set in the initial state for this instance.

Each line shows the action taken, where the number at the start of each line is the time step in which the planning action is taken, and the number at the end the time duration for the action. These results show the feasibility of our approach in several ways. First, we see that the planner has successfully identified a plan to achieve the goal of both AGs while identifying and addressing an adverse interaction (*rev2*). Actions for each CPG are taken concurrently as they are independent of each other, as are actions within a parallel block for *D1* (*B1/C1*, *B2/C2*). In this problem, the planner checks for interactions at the end of reaching the goal states, although it is trivial to check them as a new node is executed (we discuss this below). While in this example all action durations and costs are 1, these can be varied by assigning values for duration and cost to each *action* and *decision* predicate in the initial state. Furthermore, we support revision operators that mitigated the same adverse interactions but at different costs. By minimizing the *total-cost*, the planner applies the revision operator(s) with the lowest cost. In our previous work all of this reasoning had to be encapsulated in the mitigation algorithm. Lastly, the management scenario is extracted from the plan by considering each line and mapping the action to its corresponding text and applying any revisions taken. This output generation makes integration into a CDSS much easier when compared to the manual interpretation, rearrangement, and processing of logical sentences in the found model of our FOL-based approach.

```
(define (problem CPG-mitigate)
   (:domain CPG-gen)
   (:objects D1 D2 - disease
             A4 T3 A3D2 G2 newAction newAction2 - node
             V3 - patientData
             rev1 rev2 - revID
             total-cost - cost)
   (:init
            (diagnosed D2)
            (initialNode D2 A4)
            (action D2 A4)
            (decision D2 T3 V3)
            (action D2 A3D2)
            (goalNode D2 G2)

            (directPrec A4 T3)
            (directPrec T3 A3D2)
            (directPrec T3 G2)
            (directPrec A3D2 G2)

            (= (patientValue V3) 2)
            (= (decisionValueLower D2 T3 A3D2) 0)
            (= (decisionValueUpper D2 T3 A3D2) 4)
            (= (decisionValueLower D2 T3 G2) 5)
            (= (decisionValueUpper D2 T3 G2) 10)


            (replaceRevision rev1 A2 A3D2 newAction)
            (replaceRevision rev2 C2 A3D2 newAction2)
            (= (interactionPresent rev1) 0)
            (= (interactionPresent rev2) 0)

            (= (total-cost) 0.0))
   )
   (:goal (and
                (reachedGoal D1)
                (reachedGoal D2)

                (noAdverseInteractions D1 D2 rev1)
                (noAdverseInteractions D1 D2 rev2)
            )
   )
   (:metric minimize (total-cost))
)
```

Fig. 3. Example problem instance PDDL.

Having conceptually demonstrated the feasibility of MitPlan, we applied it to the clinical case study in our previous work [8] by solving each patient case as a planning problem. In this study we combined the chronic kidney disease (CKD), atrial fibrillation (AFib), and hypertension (HTN) guidelines and used revision/patient preference operators for each case. Each patient case defined the problem instance and the planning domain was the same for all problem instances. We represented deletion revisions as replacements where the replacement action was an empty action and temporal revisions were encoded in the temporal aspects of durative actions. MitPlan was able to successfully find a plan for each patient case (successful identification of a management scenario was also the metric used in our previous work) with no additional computational costs (a cost that is insignificant overall).

```
0.000: (makedecision d1 t1 v1)  [0.001]
0.000: (takeaction d2 a4)  [0.001]
0.001: (makedecision d2 a4 t3 v3)  [0.001]
0.001: (takeaction d1 t1 a1 v1)  [0.001]
0.002: (takeparallel d1 a1 p1 b1 c1)  [0.001]
0.002: (takeaction d2 t3 a3d2 v3)  [0.001]
0.003: (reachgoal d2 a3d2 g2)  [0.001]
0.003: (takeaction d1 p1 c1)  [0.001]
0.003: (takeaction d1 p1 b1)  [0.001]
0.004: (takeaction d1 b1 b2)  [0.001]
0.004: (takeaction d1 c1 c2)  [0.001]
0.005: (endparallel d1 p2 p1 b2 c2)  [0.001]
0.006: (makedecision d1 p2 t2 v2)  [0.001]
0.007: (reachgoal d1 t2 g1 v2)  [0.001]
0.008: (norevisionneeded d1 d2 a2 a3d2 newaction rev1)  [0.001]
0.009: (replaceactions d1 d2 c2 a3d2 newaction2 rev2)  [0.001]
```

Fig. 4. Example problem instance resulting plan.

5 Discussion and Future Work

In this paper we presented MitPlan – a modification of our FOL-based mitigation framework where we replaced a hybrid approach combining search and FOL-based reasoning with a uniform planning approach employing PDDL. The revised framework significantly simplifies the mitigation process, as identification of interactions, revision of CIGs, and construction of management scenarios is handled by a planner and there is no need to switch between several representations and methods (e.g., FOL and text). Moreover, MitPlan provides support for additional criteria when developing a management scenario (e.g., the total cost of prescribed treatments and introduced revisions) and provides sound support for durative actions without the need for explicit specification of additional knowledge (e.g., logical rules for handling temporal action properties in FOL).

MitPlan shares some similarities with solutions described in [12, 13]. Similarly to GLARE-SSCPM [12], it takes into account temporal characteristics of CIG actions during mitigation and relies on knowledge-driven detection of interactions. However, it automatically derives a management scenario, while GLARE-SSCPM aims at planning the scenario through interactions with a physician. Automatic planning is employed in the multi-agent planning (MAP) framework [13] that handles temporal CIG characteristics and patient planning. The important difference between [13] and MitPlan lies in the representation of clinical knowledge. In MitPlan, secondary clinical knowledge on how to handle adverse interactions is captured by revision operators independent of CIGs, while MAP assumes the primary and secondary knowledge is combined and embedded in CIGs. Our approach facilitates knowledge management as adding new revision operators does not imply changes to CIGs.

As part of our future work we plan to expand our PDDL-based MitPlan to use the Action Description Language (ADL). ADL provides a richer and more compact representation that supports more of the PDDL formalism and enables a finer grained description of revisions introduced to CIGs as single insertions or deletions. We are also planning to implement MitPlan within the larger framework for clinical decision support presented in [8] and evaluate it practically in a clinical setting.

References

1. Rosenfeld, R.M., Shiffman, R.N.: Clinical practice guideline development manual: a quality-driven approach for translating evidence into action, otolaryngol. Head Neck Surg. **140**, S1–S43 (2009)
2. Goud, R., et al.: The effect of computerized decision support on barriers to guideline implementation: a qualitative study in outpatient cardiac rehabilitation. Int. J. Med. Inform. **79**, 430–437 (2010)
3. Peleg, M.: Computer-interpretable clinical guidelines: a methodological review. J. Biomed. Inform. **46**(4), 744–763 (2013)
4. Shekelle, P., Woolf, S., Grimshaw, J.M., Schunemann, H.J., Eccles, M.P.: Developing clinical practice guidelines: reviewing, reporting, and publishing guidelines; updating guidelines; and the emerging issues of enhancing guideline implementability and accounting for comorbid conditions in guideline development. Implement. Sci. **7**, 62 (2012)
5. Xu, J., Murphy, S.L., Kochanek, K.D., Arias, E.: Mortality in the United States, 2015. NCHS Data Brief **267**, 1–8 (2016)
6. Boyd, C.M., Darer, J., Boult, C., Fried, L.P., Boult, L., Wu, A.W.: Clinical practice guidelines and quality of care for older patients with multiple comorbid diseases: implications for pay for performance. JAMA **294**, 716–724 (2005)
7. Riaño, D., Ortega, W.: Computer technologies to integrate medical treatments to manage multimorbidity. J. Biomed. Inform. **75**, 1–13 (2017)
8. Wilk, Sz., Michalowski, M., Michalowski, W., Rosu, D., Carrier, M., Kezadri-Hamiaz, M.: Comprehensive mitigation framework for concurrent application of multiple clinical practice guidelines. J. Biomed. Inform. **66**(2), 52–71 (2017)
9. Wilk, Sz., Michalowski, M., Michalowski, W., Farion, K., Hing, M.M., Mohapatra, S.: Mitigation of adverse interactions in pairs of clinical practice guidelines using constraint logic programming. J. Biomed. Inform. **46**(2), 341–353 (2013)
10. Hing, M.M., Michalowski, M., Wilk, Sz., Michalowski, W., Farion, K.: Identifying inconsistencies in multiple clinical practice guidelines for a patient with co-morbidity. In: IEEE International Conference on Bioinformatics and Biomedicine Workshops (BIBMW), pp. 447–452. IEEE (2010)
11. Coles, A., Coles, A., Fox, M., Long, D.: Forward-chaining partial-order planning. In: Proceedings of the 20th International Conference on International Conference on Automated Planning and Scheduling ICAPS 2010, pp. 42–49. AAAI Press (2010)

12. Piovesan, L., Terenziani, P., Molino, G.: GLARE-SSCPM: an intelligent system to support the treatment of comorbid patients. IEEE Intell. Syst. **33**(6), 37–46 (2018)
13. Fdez-Olivares, J., Onaindia, E., Castillo, L., Jordán, J., Cózar, J.: Personalized conciliation of clinical guidelines for comorbid patients through multi-agent planning. Artif. Intel. Med. (2018, in press)

Automatic Alignment of Surgical Videos Using Kinematic Data

Hassan Ismail Fawaz[1(✉)], Germain Forestier[1,2], Jonathan Weber[1], François Petitjean[2], Lhassane Idoumghar[1], and Pierre-Alain Muller[1]

[1] IRIMAS, University of Haute-Alsace, Mulhouse, France
hassan.ismail-fawaz@uha.fr

[2] Faculty of Information Technology, Monash University, Melbourne, Australia

Abstract. Over the past one hundred years, the classic teaching methodology of "see one, do one, teach one" has governed the surgical education systems worldwide. With the advent of Operation Room 2.0, recording video, kinematic and many other types of data during the surgery became an easy task, thus allowing artificial intelligence systems to be deployed and used in surgical and medical practice. Recently, surgical videos has been shown to provide a structure for peer coaching enabling novice trainees to learn from experienced surgeons by replaying those videos. However, the high inter-operator variability in surgical gesture duration and execution renders learning from comparing novice to expert surgical videos a very difficult task. In this paper, we propose a novel technique to align multiple videos based on the alignment of their corresponding kinematic multivariate time series data. By leveraging the Dynamic Time Warping measure, our algorithm synchronizes a set of videos in order to show the same gesture being performed at different speed. We believe that the proposed approach is a valuable addition to the existing learning tools for surgery.

Keywords: Dynamic Time Warping · Multivariate time series · Video synchronization · Surgical education

1 Introduction

Educators have always searched for innovative ways of improving apprentices' learning rate. While classical lectures are still most commonly used, multimedia resources are becoming more and more adopted [22] especially in Massive Open Online Courses (MOOC) [13]. In this context, videos have been considered as especially interesting as they can combine images, text, graphics, audio and animation. The medical field is no exception, and the use of video-based resources is intensively adopted in medical curriculum [11] especially in the context of surgical training [9]. The advent of robotic surgery also simulates this trend as surgical robots, like the Da Vinci [7], generally record video feeds during the intervention. Consequently, a large amount of video data has been recorded in the last

D. Riaño et al. (Eds.): AIME 2019, LNAI 11526, pp. 104–113, 2019.
https://doi.org/10.1007/978-3-030-21642-9_14

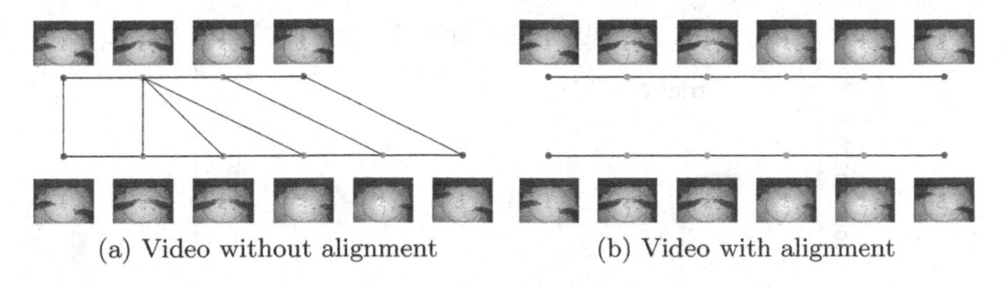

Fig. 1. Example on how a time series alignment is used to synchronize the videos by duplicating the gray-scale frames. Best viewed in color. (Color figure online)

ten years [19]. This new source of data represent an unprecedented opportunity for young surgeons to improve their knowledge and skills [5]. Furthermore, video can also be a tool for senior surgeons during teaching periods to assess the skills of the trainees. In fact, a recent study [14] showed that residents spend more time viewing videos than specialists, highlighting the need for young surgeons to fully benefit from the procedure. In [6], the authors showed that knot-tying scores and times for task completion improved significantly for the subjects that watched the videos of their own performance.

However, when the trainees are willing to asses their progress over several trials of the same surgical task by re-watching their recorded surgical videos simultaneously, the problem of videos being out-of-synch makes the comparison between different trials very difficult if not impossible. This problem is encountered in many real life case studies, since experts on average complete the surgical tasks in less time than novice surgeons [12]. Thus, when trainees do enhance their skills, providing them with a feedback that pinpoints the reason behind the surgical skill improvement becomes problematic since the recorded videos exhibit different duration and are not perfectly aligned.

Although synchronizing videos has been the center of interest for several computer vision research venues, contributions are generally focused on a special case where multiple simultaneously recorded videos (with different characteristics such as viewing angles and zoom factors) are being processed [15,25,26]. Another type of multiple video synchronization uses hand-engineered features (such as points of interest trajectories) from the videos [2,24], making the approach highly sensitive to the quality of the extracted features. This type of techniques was highly effective since the raw videos were the only source of information available, whereas in our case, the use of robotic surgical systems enables capturing an additional type of data: the kinematic variables such as the x, y, z Cartesian coordinates of the Da Vinci's end effectors [5].

In this paper, we propose to leverage the sequential aspect of the recorded kinematic data from the Da Vinci surgical system, in order to synchronize their corresponding video frames by aligning the time series data (see Fig. 1 for an example). When aligning two time series, the off-the-shelf algorithm is Dynamic Time Warping (DTW) [20] which we indeed used to align two videos. However, when aligning multiple sequences, the latter technique does not generalize in a

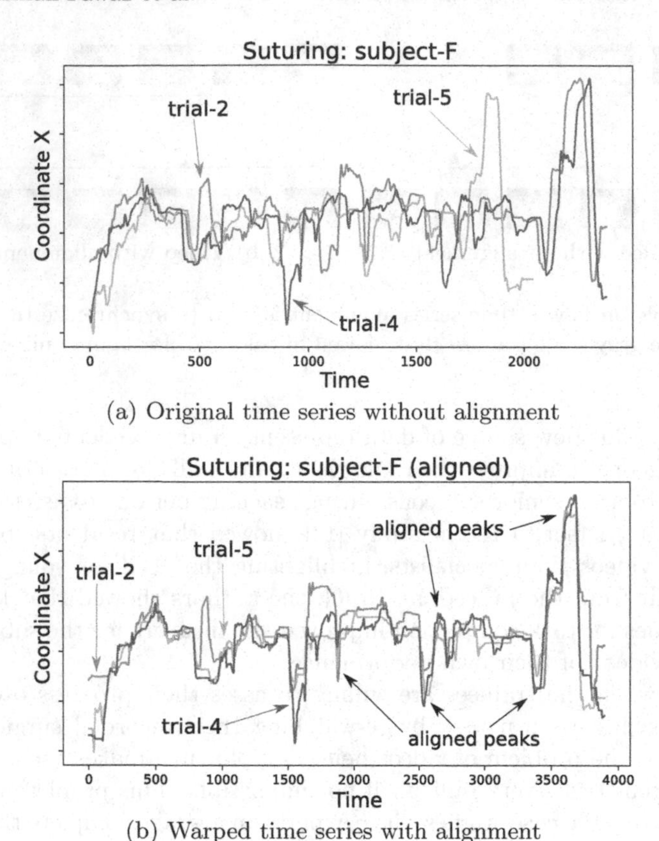

(a) Original time series without alignment

(b) Warped time series with alignment

Fig. 2. Example of aligning coordinate X's time series for subject F, when performing three trials of the suturing surgical task.

straightforward and computationally feasible manner [16]. Hence, for multiple video synchronization, we propose to align their corresponding time series to the average time series, computed using the DTW Barycenter Averaging (DBA) algorithm [16]. This process is called Non-Linear Temporal Scaling (NLTS) and has been proposed to find the multiple alignment of a set of discretized surgical gestures [3], which we extend in this work to continuous numerical kinematic data. Figure 2 depicts an example of stretching three different time series using the NLTS algorithm. Examples of the synchronized videos and the associated code can be found on our GitHub repository[1], where we used the JHU-ISI Gesture and Skill Assessment Working Set (JIGSAWS) [5] to validate our work.

The rest of the paper is organized as follows: in Sect. 2, we explain in details the algorithms we have used in order to synchronize the kinematic data and eventually their corresponding video frames. In Sect. 3, we present our experiments and finally conclude the paper and discuss our future work in Sect. 4.

---

[1] https://github.com/hfawaz/aime19.

2 Methods

In this section, we detail each step of our video synchronization approach. We start by describing the Dynamic Time Warping (DTW) algorithm which allows us to align two videos. Then, we describe how Non-Linear Temporal Scaling (NLTS) enables us to perform multiple video synchronization with respect to the reference average time series computed using the DTW Barycenter Averaging (DBA) algorithm.

2.1 Dynamic Time Warping

Dynamic Time Warping (DTW) was first proposed for speech recognition when aligning two audio signals [20]. Suppose we want to compute the dissimilarity between two time series, for example two different trials of the same surgical task, $A = (a_1, a_2, \ldots, a_m)$ and $B = (b_1, b_2, \ldots, b_n)$. The length of A and B are denoted respectively by m and n, which in our case correspond to the surgical trial's duration. Here, a_i is a vector that contains six real values, therefore A and B can be seen as two distinct Multivariate Time Series (MTS).

To compute the DTW dissimilarity between two MTS, several approaches were proposed by the time series data mining community [21], however in order to apply the subsequent algorithm NLTS, we adopted the "dependent" variant of DTW where the Euclidean distance is used to compute the difference between two instants i and j. Let $M(A, B)$ be the $m \times n$ point-wise dissimilarity matrix between A and B, where $M_{i,j} = ||a_i - b_j||^2$. A warping path $P = ((c_1, d_1), (c_2, d_2), \ldots, (c_s, d_s))$ is a series of points that define a crossing of M. The warping path must satisfy three conditions: (1) $(c_1, d_1) = (1, 1)$; (2) $(c_s, d_s) = (m, n)$; (3) $0 \leq c_{i+1} - c_i \leq 1$ and $0 \leq d_{j+1} - d_j \leq 1$ for all $i < m$ and $j < n$. The DTW measure between two series corresponds to the path through M that minimizes the total distance. In fact, the distance for any path P is equal to $D_P(A, B) = \sum_{i=1}^{s} P_i$. Hence if $\mathbf{P}$ is the space of all possible paths, the optimal one - whose cost is equal to $DTW(A, B)$ - is denoted by P^* and can be computed using: $\min_{P \in \mathbf{P}} D_P(A, B)$.

The optimal warping path can be obtained efficiently by applying a dynamic programming technique to fill the cost matrix M. Once we find this optimal warping path between A and B, we can deduce how each time series element in A is linked to the elements in B. We propose to exploit this link in order to identify which time stamp should be duplicated in order to align both time series, and by duplicating a time stamp, we are also duplicating its corresponding video frame. Concretely, if elements a_i, a_{i+1} and a_{i+2} are aligned with the element b_j when computing P^*, then by duplicating twice the video frame in B for the time stamp j, we are dilating the video of B to have a length that is equal to A's. Thus, re-aligning the video frames based on the aligned Cartesian coordinates: if subject S_1 completed "*inserting the needle*" gesture in 5 s, whereas subject S_2 performed the same gesture within 10 s, our algorithm finds the optimal warping path and duplicates the frames for subject S_1 in order to synchronize with subject S_2 the corresponding gesture. Figure 1 illustrates how the alignment computed by

DTW for two time series can be used in order to duplicate the corresponding frames and eventually synchronize the two videos.

2.2 Non-Linear Temporal Scaling

The previous DTW based algorithm works perfectly when synchronizing only two surgical videos. The problem arises when aligning three or more surgical trials simultaneously, which requires a multiple series alignment. The latter problem has been shown to be NP-Complete [23] with the exact solution requiring $O(L^N)$ operations for N sequences of length L. This is clearly not feasible in our case where L varies between 10^3 and 10^4 and $N \geq 3$, which is why we ought to leverage an approximation of the multiple sequence alignment solution provided by the DTW Barycenter Averaging (DBA) algorithm which we detail in the following paragraph.

DBA was originally proposed in [18] as a technique that averages a set of time series by leveraging an approximated multiple sequence alignment algorithm called Compact Multiple Alignment (CMA) [17]. DBA iteratively refines an average time series T and follows an expectation-maximization scheme by first considering T to be fixed and finding the best CMA between the set of sequences D (to be averaged) and the refined average sequence T. After computing the CMA, the alignment is now fixed and the average sequence T is updated in a way that minimizes the sum of DTW distances between T and D [16].

DBA requires an initial value for T. There exist many possible initializations for the average sequence [17], however, since our ultimate goal is to synchronize a set of sequences D by duplicating their elements (dilating the sequences), we initialize the average T to be equal to the longest instance in D. We then find precisely the exact optimal number of time series elements - and their associated video frames - to be duplicated in order to synchronize multiple videos, using the NLTS technique which we describe in details in the following paragraph.

Non-Linear Temporal Scaling (NLTS) was originally proposed for aligning discrete sequences of surgical gestures [3]. In this paper, we extend the technique for numerical continuous sequences (time series). The goal of this final step is to compute the approximated multiple alignment of a set of sequences D which will eventually contain the precise information on how much a certain frame from a certain series should be duplicated. We first start by computing the average sequence T (using DBA) for a set of time series D that we want to align simultaneously. Then, by recomputing the Compact Multiple Alignment (CMA) between the refined average T and the set of time series D, we can extract an alignment between T and each sequence in D. Thus, for each time series in D we will have the necessary information (extracted from CMA) in order to dilate the time series appropriately to have a length that is equal to T's, which also corresponds to the length of the longest time series in D. Figure 2, depicts an example of aligning three different time series using the NLTS algorithm.

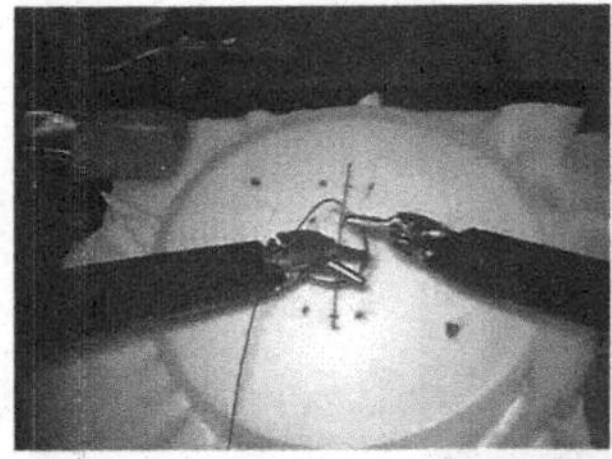

Fig. 3. Snapshots of the three surgical tasks in the JIGSAWS dataset (from left to right): suturing, knot-tying, needle-passing [5].

3 Experiments

We start by describing the JIGSAWS dataset we have used for evaluation, before presenting our experimental study.

3.1 Dataset

The JIGSAWS dataset [5] includes data for three basic surgical tasks performed by study subjects (surgeons). The three tasks (or their variants) are usually part of the surgical skills training program. Figure 3 shows a snapshot example for each one of the three surgical tasks (Suturing, Knot Tying and Needle Passing). The JIGSAWS dataset contains kinematic and video data from eight different subjects with varying surgical experience: two experts (E), two intermediates (I) and four novices (N) with each group having reported respectively more than 100 h, between 10 and 100 h and less than 10 h of training on the Da Vinci. All subjects were reportedly right-handed.

The subjects repeated each surgical task five times and for each trial the kinematic and video data were recorded. When performing the alignment, we used the kinematic data which are numeric variables of four manipulators: left and right masters (controlled directly by the subject) and left and right slaves (controlled indirectly by the subject via the master manipulators). These kinematic variables (76 in total) are captured at a frequency equal to 30 frames per second for each trial. Out of these 76 variables, we only consider the Cartesian coordinates (x, y, z) of the left and right slave manipulators, thus each trial will consist of an MTS with 6 temporal variables. We chose to work only with this subset of kinematic variables to make the alignment coherent with what is visible in the recorded scene: the robots' end-effectors which can be seen in Fig. 3. However other choices of kinematic variables are applicable, which we leave the exploration for our future work. Finally we should mention that in addition to the three self-proclaimed skill levels (N, I, E) JIGSAWS contains the modified Objective Structured Assessment of Technical Skill (OSATS) score [5], which corresponds to an expert surgeon observing the surgical trial and annotating the performance of the trainee.

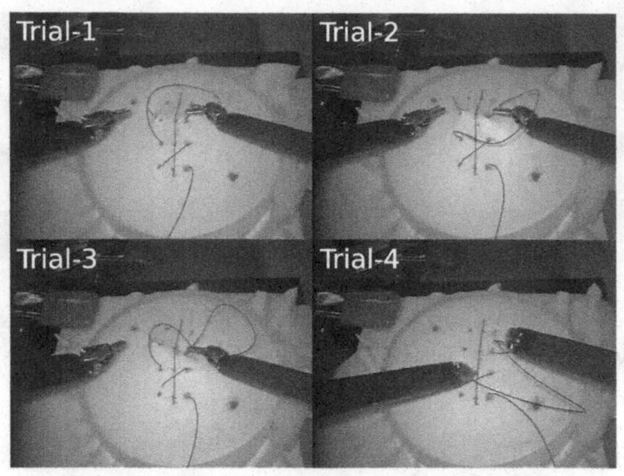

(a) Videos synchronization process

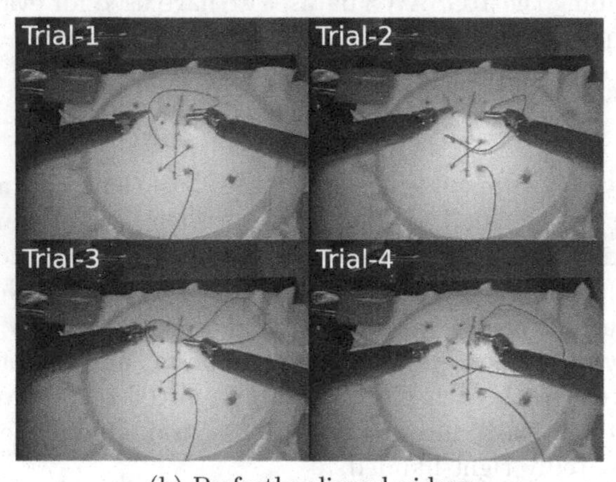

(b) Perfectly aligned videos

Fig. 4. Video alignment procedure with duplicated (gray-scale) frames.

3.2 Results

We have created a companion web page[2] to our paper where several examples of synchronized videos can be found. Figure 4 illustrates the multiple videos alignment procedure using our NLTS algorithm, where gray-scale images indicate duplicated frames (paused video) and colored images indicate a surgical motion (unpaused video). In Fig. 4a we can clearly see how the gray-scale surgical trials are perfectly aligned. Indeed, the frozen videos show the surgeon ready to perform "*pulling the needle*" gesture [5]. On the other hand, the colored trial (bottom right of Fig. 4a) shows a video that is being played, where the surgeon is performing "*inserting the needle*" gesture in order to catch up with the other paused trials in gray-scale. Finally, the result of aligning simultaneously these

[2] https://germain-forestier.info/src/aime2019/.

four surgical trials is depicted in Fig. 4b. By observing the four trials, one can clearly see that the surgeon is now performing the same surgical gesture "*pulling the needle*" simultaneously for the four trials. We believe that this type of observation will enable a novice surgeon to locate which surgical gestures still need some improvement in order to eventually become an expert surgeon.

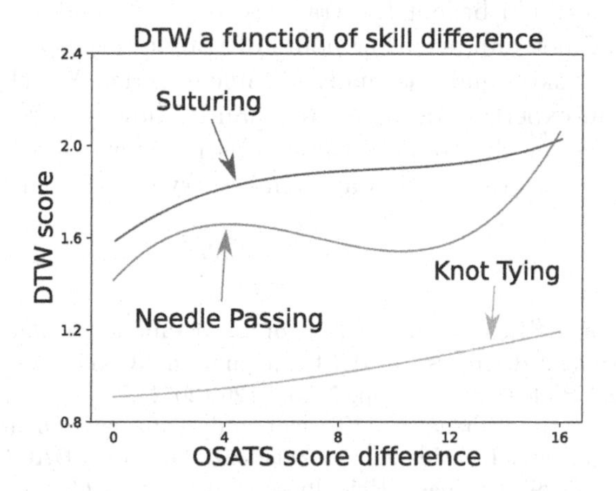

Fig. 5. A polynomial fit (degree 3) of DTW dissimilarity score (y-axis) as a function of the OSATS score difference between two surgeons (x-axis).

Furthermore, in order to validate our intuition that DTW is able to capture characteristics that are in relationship with the motor skill of a surgeon, we plotted the DTW distance as a function of the OSATS [5] score difference. For example, if two surgeons have both an OSATS score of 10 and 16 respectively, the corresponding difference is equal to $|10 - 16| = 6$. In Fig. 5, we can clearly see how the DTW score increases whenever the OSATS score difference increases. This observation suggests that the DTW score is low when both surgeons exhibit similar dexterity, and high whenever the trainees show different skill levels. Therefore, we conclude that the DTW score can serve as a heuristic for estimating the quality of the alignment (whenever annotated skill level is not available) - especially since we observed low quality alignments for surgeons with very distinct surgical skill levels.

Finally, we should note that this work is suitable for many research fields involving motion kinematic data with their corresponding video frames. Examples of such medical applications are assessing mental health from videos [27] where wearable sensor data can be seen as time series kinematic variables and leveraged in order to synchronize a patient's videos and compare how well the patient is responding to a certain treatment. Following the same line of thinking, this idea can be further applied to kinematic data from wearable sensors coupled with the corresponding video frames when evaluating the Parkinson's disease evolution [1] as well as infant grasp skills [10].

4 Conclusion

In this paper, we showed how kinematic time series data recorded from the Da Vinci's end effectors can be leveraged in order to synchronize the trainee's videos performing a surgical task. With personalized feedback during surgical training becoming a necessity [4,8], we believe that replaying *synchronized* and well *aligned* videos would benefit the trainees in understanding which surgical gestures did or did not improve after hours of training, thus enabling them to further reach higher skills and eventually become experts. We acknowledge that this work needs an experimental study to quantify how beneficial is replaying synchronized videos for the trainees versus observing non-synchronized trials. Therefore, we leave such exploration and clinical try outs to our future work.

References

1. Criss, K., McNames, J.: Video assessment of finger tapping for Parkinson's disease and other movement disorders. In: IEEE International Conference on Engineering in Medicine and Biology Society, pp. 7123–7126 (2011)
2. Evangelidis, G.D., Bauckhage, C.: Efficient and robust alignment of unsynchronized video sequences. In: Mester, R., Felsberg, M. (eds.) DAGM 2011. LNCS, vol. 6835, pp. 286–295. Springer, Heidelberg (2011). https://doi.org/10.1007/978-3-642-23123-0_29
3. Forestier, G., Petitjean, F., Riffaud, L., Jannin, P.: Non-linear temporal scaling of surgical processes. Artif. Intell. Med. **62**(3), 143–152 (2014)
4. Forestier, G., et al.: Surgical motion analysis using discriminative interpretable patterns. Artif. Intell. Med. **91**, 3–11 (2018)
5. Gao, Y., et al.: The JHU-ISI Gesture and Skill Assessment Working Set (JIGSAWS): a surgical activity dataset for human motion modeling. In: Modeling and Monitoring of Computer Assisted Interventions - MICCAI Workshop (2014)
6. Herrera-Almario, G.E., Kirk, K., Guerrero, V.T., Jeong, K., Kim, S., Hamad, G.G.: The effect of video review of resident laparoscopic surgical skills measured by self- and external assessment. Am. J. Surg. **211**(2), 315–320 (2016)
7. Intuitive Surgical Sunnyvale, C.A.: The Da Vinci surgical system. https://www.intuitive.com/en/products-and-services/da-vinci
8. Ismail Fawaz, H., Forestier, G., Weber, J., Idoumghar, L., Muller, P.-A.: Evaluating surgical skills from kinematic data using convolutional neural networks. In: Frangi, A.F., Schnabel, J.A., Davatzikos, C., Alberola-López, C., Fichtinger, G. (eds.) MICCAI 2018. LNCS, vol. 11073, pp. 214–221. Springer, Cham (2018). https://doi.org/10.1007/978-3-030-00937-3_25
9. Kneebone, R., Kidd, J., Nestel, D., Asvall, S., Paraskeva, P., Darzi, A.: An innovative model for teaching and learning clinical procedures. Med. Educ. **36**(7), 628–634 (2002)
10. Li, Z., Huang, Y., Cai, M., Sato, Y.: Manipulation-skill assessment from videos with spatial attention network. ArXiv (2019)
11. Masic, I.: E-learning as new method of medical education. Acta informatica medica **16**(2), 102 (2008)
12. McNatt, S., Smith, C.: A computer-based laparoscopic skills assessment device differentiates experienced from novice laparoscopic surgeons. Surg. Endosc. **15**(10), 1085–1089 (2001)

13. Means, B., Toyama, Y., Murphy, R., Bakia, M., Jones, K.: Evaluation of evidence-based practices in online learning: a meta-analysis and review of online learning studies (2009)
14. Mota, P., Carvalho, N., Carvalho-Dias, E., Costa, M.J., Correia-Pinto, J., Lima, E.: Video-based surgical learning: improving trainee education and preparation for surgery. J. Surg. Educ. **75**(3), 828–835 (2018)
15. Padua, F., Carceroni, R., Santos, G., Kutulakos, K.: Linear sequence-to-sequence alignment. IEEE Trans. Pattern Anal. Mach. Intell. **32**(2), 304–320 (2010)
16. Petitjean, F., Forestier, G., Webb, G.I., Nicholson, A.E., Chen, Y., Keogh, E.: Dynamic time warping averaging of time series allows faster and more accurate classification. In: IEEE International Conference on Data Mining, pp. 470–479 (2014)
17. Petitjean, F., Gançarski, P.: Summarizing a set of time series by averaging: from Steiner sequence to compact multiple alignment. Theoret. Comput. Sci. **414**(1), 76–91 (2012)
18. Petitjean, F., Ketterlin, A., Gançarski, P.: A global averaging method for dynamic time warping, with applications to clustering. Pattern Recognit. **44**(3), 678–693 (2011)
19. Rapp, A.K., Healy, M.G., Charlton, M.E., Keith, J.N., Rosenbaum, M.E., Kapadia, M.R.: Youtube is the most frequently used educational video source for surgical preparation. J. Surg. Educ. **73**(6), 1072–1076 (2016)
20. Sakoe, H., Chiba, S.: Dynamic programming algorithm optimization for spoken word recognition. IEEE Trans. Acoust. Speech Signal Process. **26**(1), 43–49 (1978)
21. Shokoohi-Yekta, M., Hu, B., Jin, H., Wang, J., Keogh, E.: Generalizing dtw to the multi-dimensional case requires an adaptive approach. Data Min. Knowl. Disc. **31**(1), 1–31 (2017)
22. Smith, T.L., Ransbottom, S.: Digital video in education. In: Distance Learning Technologies: Issues, Trends and Opportunities, pp. 124–142 (2000)
23. Wang, L., Jiang, T.: On the complexity of multiple sequence alignment. J. Comput. Biol. **1**(4), 337–348 (1994)
24. Wang, O., Schroers, C., Zimmer, H., Gross, M., Sorkine-Hornung, A.: Videosnapping: interactive synchronization of multiple videos. ACM Trans. Graph. **33**(4), 77 (2014)
25. Wedge, D., Kovesi, P., Huynh, D.: Trajectory based video sequence synchronization. In: Digital Image Computing: Techniques and Applications, p. 13 (2005)
26. Wolf, L., Zomet, A.: Sequence-to-sequence self calibration. In: Heyden, A., Sparr, G., Nielsen, M., Johansen, P. (eds.) ECCV 2002. LNCS, vol. 2351, pp. 370–382. Springer, Heidelberg (2002). https://doi.org/10.1007/3-540-47967-8_25
27. Yamada, Y., Kobayashi, M.: Detecting mental fatigue from eye-tracking data gathered while watching video. In: ten Teije, A., Popow, C., Holmes, J.H., Sacchi, L. (eds.) AIME 2017. LNCS (LNAI), vol. 10259, pp. 295–304. Springer, Cham (2017). https://doi.org/10.1007/978-3-319-59758-4_34

A Rule-Based Expert System for Automatic Implementation of Somatic Variant Clinical Interpretation Guidelines

Giovanna Nicora[1(✉)], Ivan Limongelli[2], Riccardo Cova[1], Matteo Giovanni Della Porta[3], Luca Malcovati[4], Mario Cazzola[4], and Riccardo Bellazzi[1]

[1] Department of Electrical, Computer and Biomedical Engineering, University of Pavia, Via Ferrata 1, 27100 Pavia, Italy
giovanna.nicora01@universitadipavia.it
[2] enGenome S.r.l, Via Ferrata 1, 27100 Pavia, Italy
[3] Cancer Center, Humanitas Research Hospital, Humanitas University, Via Alessandro Manzoni, 56, 20089 Rozzano, MI, Italy
[4] Department of Hematology and Oncology, IRCCS Policlinico San Matteo, V.le Golgi, 19, 27100 Pavia, Italy

Abstract. Precision oncology aims at integrating molecular data into clinical decision making, in order to provide the most suitable therapy and follow-up according to patient's specific characteristics. A critical step towards this goal is the interpretation of genomic variants, whose presence can be revealed by next generation sequencing. In particular, cancer variant interpretation defines whether the patient harbors genomic alterations that could be targeted by specific drugs, or that were observed as prognostic biomarkers. To standardize somatic interpretation, in 2017 guidelines have been proposed by a working group of associations, including the American Society of Clinical Oncology (ASCO). Automatic tools implementing such guidelines to ease their actual application in the clinical routine are needed.

We developed a Rule-based Expert System (ES) that automatically implements ASCO guidelines. ES is an Artificial Intelligence system able to reason over a set of rules and to perform classification, thus emulating human reasoning process. First, we developed automatic pipelines to extract information of over 1500 known diagnostic/prognostic/diagnostic biomarkers from six public databases, including COSMIC and CiVIC. The collected knowledge base is structured in an object-oriented model and the ES is implemented in a Python program through the PyKnow library.

Keywords: Somatic variant interpretation · Standard guidelines · Expert System

D. Riaño et al. (Eds.): AIME 2019, LNAI 11526, pp. 114–119, 2019.
https://doi.org/10.1007/978-3-030-21642-9_15

1 Introduction

1.1 Somatic Variant Interpretation: Principles and Current Status

The promise of Precision Medicine (PM) is to improve patient outcomes by identifying the best clinical strategy according to patient's individual characteristics. In this context, clinical decision making, i.e. the delineation of diagnosis, prevention and treatment approaches, could greatly benefit from genomic information extracted from DNA sequencing [1]. Oncology was one of the first fields of PM application, due to the intrinsic genetic nature of cancer: in fact, tumor originates from the accumulation in somatic cells of alterations in genes involved in apoptosis, cellular proliferation and cell cycle. Not all somatic alterations are able to driver cancer progression and, among those that confer an evolutive advantage, it is critical to identify potential biomarkers. Biomarkers are any molecular characteristics that have been correlated with response or resistance to a particular treatment (therapeutic biomarkers), influence disease prognosis (prognostic biomarker) or serve to establish a diagnosis (diagnostic biomarkers) [1]. As genomic screening is becoming routinely applied in clinical practice, standards and guidelines for the cataloging and interpretation of patients variants have been proposed [1, 2]. As somatic variant interpretation is concerned, the aim is to assess whether a somatic variant could have an impact on clinical care. In particular, a working group including the Association for Molecular Pathology (AMP), the American Society of Clinical Oncology (ASCO) and the College of American Pathologists (CAP) published in 2017 a set of guidelines to leverage information collected from different omics-resources in order to classify a somatic variant in 4 different tiers of clinical significance ("Strong Clinical Significance", "Potential Clinical Significance", "Unknown Clinical Significance" and "Benign") for three different categories ("Therapeutic", "Diagnosis" and "Prognosis"). Omics-resources include population allele frequency databases, in silico predictions of variant damaging impact, and repositories of previous interpretation. For instance, a variant observed at high frequency in population database is likely to be "Benign", while a very low frequency variant targeted by a FDA-approved therapy will fall into "Therapeutic Tier I" category. However, since these guidelines rely on the utilization of different and complex omics-resources, and they could be applied to potentially thousands of variants per patients, the demand for automatic tools implementing them is high.

2 Methods

We developed a Rule-based Expert System (ES) able to automatically interpret somatic variant according to AMP/ASCO/CAP guidelines. ESs are Artificial Intelligence systems that emulate expert human reasoning process over a set of rules and knowledge from a specific domain [3]. In our case, rules are represented by the AMP/ASCO/CAP guidelines, while the domain knowledge needs to be gathered from several public omics-resources. After knowledge base collection, the ES is implemented in a Python program thanks to the PyKnow library, which creates an environment to define Rules and fire them against Facts.

2.1 Preprocessing: Knowledge Base Collection

We collected information about known biomarkers from 6 different cancer-specific databases. These repositories provide evidence about variants clinical impact, public literature references, clinical trials and professional guidelines. Information about cancer-specific databases is listed in Table 1.

Table 1. Cancer-specific databases information.

Database	URL	Type of evidence
CGI [4]	https://www.cancergenomeinterpreter.org/biomarkers	Therapeutic
CiVIC [5]	https://civicdb.org/home	Therapeutic, Diagnostic, Prognostic
OncoKB [6]	http://oncokb.org/	Therapeutic
DEPO [7]	http://depo-dinglab.ddns.net/	Therapeutic
DOCM [8]	http://docm.info/	Diagnostic
COSMIC (Resistance mutation) [9]	https://cancer.sanger.ac.uk/cosmic/download	Therapeutic

We developed automatic pipelines that extract relevant information and standardize nomenclature from each resource. In fact, each database has different terminologies: for instance, OncoKB Therapeutic levels are "Resistance" or "Response", while in DEPO the same concept is represented by "Resistant" and "Sensitive". Moreover, we standardized cancer representation to Disease Ontology terms, and we select single nucleotide variations and indels.

2.2 ES Implementation

The ES is implemented in a Python program. Input files are the following: an annotation tab-delimited file with the lists of genomic coordinates of somatic variants that need to be classified and a tab-delimited file for each collected omics-resource, resulting from our preprocessing pipeline. Data are organized into an Object-oriented model. Rules, representing AMP/ASCO/CAP guidelines, are defined through PyKnow. For instance, the final rule for "Tier I" classification is composed by three "sub-rules": one is related to the allele population frequency, the second to in silico prediction of damaging impact, and the last one checks if the variant is actually reported as a biomarker. The final rules could is therefore the following: (IF variant allele frequency <=5% in DbSNP, ExAC and Esp population databases THEN variant has low allele frequency) AND (IF PaPI, Dann and dbscSNA prediction score >=0.8 THEN variant has damaging impact) AND (variant is reported in the knowledge base as "Therapeutic/Prognostic/Diagnostic" by FDA or professional guidelines) THEN variant is Tier I Therapeutic, Prognostic or Diagnostic. Rules could overlap since a variant could be interpreted as both Tier I Therapeutic and Tier I Prognostic, but it cannot be interpreted both as Tier I and Benign.

After classification process, the ES provides as output a tab-delimited file with final classification for each input variant and a JSON file for each variant, containing information about variant annotation and classification, following the minimal variant level data (MVLD), a recently proposed framework to standardize cancer variants data for clinical utility [2].

3 Results

3.1 Standardized Knowledge Base

The harmonization of cancer-databases contains 1277 prognostic assertions, 987 therapeutic and only 38 prognostic. Almost 200 types of different cancer where mapped to Disease Ontology terms. The majority of variants (1270) are found in DOCM (diagnostic biomarkers), while from OncoKB we collected only 40 therapeutic variants, among that only 31 were reported also in other cancer-databases. COSMIC contains about 190 therapeutic biomarkers, DEPO and CGI about 300. CIVIC final variant list has 38 prognostic biomarkers, 24 diagnostic and more than 330 therapeutic.

3.2 Case Study: Interpretation of Variants in Myelodysplastic Syndromes Patients

We interpreted 884 variants found in a cohort of 310 patients with myelodysplastic syndromes (MDS). MDS are heterogeneous hematopoietic disorders whose progression could lead to Acute Myeloid Leukemia. The ES took 6.15 s to interpret all 884 variants. Among these, 8 variants were classified as "Strong Clinical Significance": 5 variants were reported as Diagnostic biomarkers, 5 as Prognostic and 3 as Therapeutic (3 variants are reported as both Diagnostic and Prognostic, while a variant has been observed as Therapeutic, Diagnostic and Prognostic). 27 variants were interpreted as "Potential Clinical Significance" (34 as Diagnostic, 1 as Prognostic and 11 as Therapeutic). The remaining variants are interpreted as Uncertain. The 35 classified variants occurred in 115 different patients. We compared our classification of MDS variants with a previous study classifying mutations as oncogenic/possible oncogenic or uncertain, in 111 genes associated with MDS or closely related neoplasms [10]. We found that 225 variants in our cohort have been reported by this study as "oncogenic". Among that, we interpreted 7 as "Strong" and 34 as "Potential". Only one "Strong" variant is reported as uncertain by the previous study. Therefore, the 97% of variants interpreted to have a clinical impact are reported as oncogenic. It is important to underline that these guidelines are not supposed to predict the pathogenicity of a variant, but they provide a framework to evaluate the clinical impact of a variant according known studies.

4 Conclusions

Somatic variant interpretation is a complex process whose results could guide clinical decision making. However, the actual implementation of interpretation guidelines in clinical practice calls for tools able to reason over a heterogeneous and always growing knowledge base. We collected and standardized from 6 databases over 1500 mutations known to have a clinical impact in cancer. Within the annotation process, we associated each variants with further information. Developed ETL pipelines will allow future update of the knowledge base. We then implemented an Expert System that reasons over the collected standardized knowledge base and automatically interprets somatic variant according to standard AMP/ASCO/CAP guidelines. ES architecture will allow future updates of the Rules, avoiding complex alteration of the application code. The ES receives as input a list of genomic variants, it performs inference, and then provides as output a JSON file for each variant, reporting variant annotation and AMP/ASCO/CAP interpretation, according to MVLD. Thanks to output files, the ES allows user to follow the reasoning process that lead to the final classification. We interpreted more than 800 variants in patients with myelodysplastic syndromes, suggesting that almost half of the cohort carried variants of strong or potential clinical significance. This information could therefore help clinicians in clinical decision-making process. Future improvements will be the possibility to interpret also complex alteration and the development of a web tool where user could query the ES interpretations. Moreover, other databases could be included in the knowledge base.

References

1. Li, M.M., Datto, M., Duncavage, E.J., et al.: Standards and guidelines for the interpretation and reporting of sequence variants in cancer: a joint consensus recommendation of the Association for Molecular Pathology, American Society of Clinical Oncology, and College of American Pathologists. J. Mol. Diagn. **19**, 4–23 (2017). https://doi.org/10.1016/j.jmoldx.2016.10.002
2. Ritter, D.I., Roychowdhury, S., Roy, A., et al.: ClinGen Somatic Cancer Working Group: Somatic cancer variant curation and harmonization through consensus minimum variant level data. Genome Med. **8**, 117 (2016). https://doi.org/10.1186/s13073-016-0367-z
3. Liao, S.-H.: Expert system methodologies and applications—a decade review from 1995 to 2004. Expert Syst. Appl. **28**, 93–103 (2005). https://doi.org/10.1016/j.eswa.2004.08.003
4. Tamborero, D., Rubio-Perez, C., Deu-Pons, J., et al.: Cancer Genome Interpreter annotates the biological and clinical relevance of tumor alterations. Genome Med. **10**, 25 (2018). https://doi.org/10.1186/s13073-018-0531-8
5. Griffith, M., Spies, N.C., Krysiak, K., et al.: CIViC is a community knowledgebase for expert crowdsourcing the clinical interpretation of variants in cancer. Nat. Genet. **49**, 170–174 (2017). https://doi.org/10.1038/ng.3774
6. Chakravarty, D., Gao, J., Phillips, S., et al.: OncoKB: a precision oncology knowledge base. JCO Precis. Oncol. **1**, 1–16 (2017). https://doi.org/10.1200/PO.17.00011
7. Sun, S.Q., Mashl, R.J., Sengupta, S., et al.: Database of evidence for precision oncology portal. Bioinformatics **34**, 4315–4317 (2018). https://doi.org/10.1093/bioinformatics/bty531

8. Ainscough, B.J., Griffith, M., Coffman, A.C., et al.: DoCM: a database of curated mutations in cancer. Nat. Methods **13**, 806–807 (2016). https://doi.org/10.1038/nmeth.4000
9. Forbes, S.A., Beare, D., Boutselakis, H., et al.: COSMIC: somatic cancer genetics at high-resolution. Nucleic Acids Res. **45**, D777–D783 (2017). https://doi.org/10.1093/nar/gkw1121
10. Papaemmanuil, E., Gerstung, M., Malcovati, L., et al.: Clinical and biological implications of driver mutations in myelodysplastic syndromes. Blood **122**, 3616–3627 (2013). https://doi.org/10.1182/blood-2013-08-518886. Quiz 3699

Considering Temporal Preferences and Probabilities in Guideline Interaction Analysis

Paolo Terenziani[1(✉)] and Antonella Andolina[2]

[1] DISIT, Institute of Computer Science, Università del Piemonte Orientale, Alessandria, Italy
paolo.terenziani@uniupo.it
[2] ITCS Sommeiller, Corso Duca degli Abruzzi 20, 10129 Turin, Italy
antoando@libero.it

Abstract. The treatment of patients affected by multiple diseases (comorbid patients) is one of the main challenges of the modern healthcare, involving the analysis of the interactions of the guidelines for the specific diseases. Practically speaking, such interactions occur in time. The GLARE project explicitly provides temporal representation and temporal reasoning methodologies to cope with such a fundamental issue. In this paper, we propose a further improvement, to take into account that, often, (i) the actions in the guidelines can be executed by physicians at different times with different *preferences*, and that (ii) the effects of such actions have a *probabilistic* distribution in time. In our approach, physicians may investigate what are the preferences of their choices on the execution-time of guideline actions, and the probabilities that their effects temporally intersect (interactions may occur only in case effects intersect in time).

Keywords: Comorbidities · CIG interactions · Temporal reasoning · Probabilities · Preferences

1 Introduction

Clinical practice guidelines are the major tool that has been introduced to grant both the quality and the standardization of healthcare services, on the basis of evidence-based recommendations. The adoption of computerized approaches to acquire, represent, execute and reason with Computer–Interpretable Guidelines (CIGs) provides crucial additional advantages so that, in the last twenty years, many different approaches and projects have been developed to manage CIGs (consider, e.g., the book [1] and the survey [2]). One of such approaches is GLARE (Guideline Acquisition, Representation and Execution) [3]. By definition, clinical guidelines address specific pathologies. However, comorbid patients are affected by more than one pathology. The problem is that, in comorbid patients, the treatments of single pathologies may interact with each other, and the approach of proposing an ad-hoc "combined" treatment to cope with each possible comorbidity does not scale up.

D. Riaño et al. (Eds.): AIME 2019, LNAI 11526, pp. 120–124, 2019.
https://doi.org/10.1007/978-3-030-21642-9_16

In the last years, many approaches in the Medical Informatics literature have faced different aspects of the treatment of comorbid patients (see the survey in [4]). Some of them have focused on the *knowledge-based automatic detection* of possible interactions between CIGs [5, 6], considering (i) the actions in the CIGs and (ii) their effects. In particular, in GLARE-SSCPM [7], specific attention has been devoted to temporal data [8–10] and to the temporal analysis of interactions [11], taking into account complex forms of temporal reasoning and query answering about temporal constraints between events. However, the approach in [11] only considers "crisp" temporal constraints, while a more flexible analysis may be needed, considering the fact that:

(1) CIG may contain guideline preferences among the execution-time of CIG actions [12] and
(2) the effects of CIG actions may have a probabilistic distribution in time (derivable, e.g., from pharmacokinetic and a pharmacodynamic studies).

Consider, e.g., Example 1, about the interaction between calcium carbonate administration (CCA) and nalidixic acid administration (NAA), concerning gastric absorption.

Example 1. A patient is affected by gastroesophageal reflux (GR) and by urinary tract infection (UTI). The CIG for GR may recommend CCA, to be administered as soon as possible, and within three hours. Considering as granularity units of 15 min, and assuming preferences in a scale from 0 (minimum preference) to 1 (maximum preference) the administration can be in the first two units (first 30 min) with preference 1, in units 3 and 4 with preference 0.75, in units 5, 6, 7 and 8 with preference 0.5, units 9, 10, 11 and 12 with preference 0.25. CCA has the effect of decreasing gastric absorption (DGA). DGA can start after 1 unit with probability 0.4, after 2 with probability 0.4, and after 3, with probability 0.2. Additionally, the duration of DGA may be 4 units (probability 0.1), 5 (0.3), 6 (0.4), 7 (0.1), or 8 (0.1). The CIG for UTI may recommend NAA, to be administered within two hours, with decreasing preferences (preference 1 for the units 1 and 2, 0.75 for units 3 and 4, 0.5 for 5 and 6, and 0.25 for 7 and 8). NAA has as effect nalidixic acid gastric absorption (NAGA), starting after 1 unit (probability 0.4) or 2 (probability (0.6). The duration of NAGA may be 1 (probability 0.05), 2 (0.05), 3 (0.15), 4 (0.15), 5 (0.25), 6 (0.25), 7 (0.05), 8 (0.05). ■

In order to support physicians in the study of the interaction between CCA and NAA, one must take into account not only the temporal constraints, but also their preferences and probabilities. This is essential to answer physician's queries such as:

(Q1) *If I perform on the patient CCA in unit 1 or 2 (i.e., in the following 30* min*), and NAA in units 1 or 2, what is the guideline preference of my choices and what is the probability that the effects of such two actions intersect in time (i.e., what is the probability of the interaction between CCA and NAA)?*

The approach in [11] does not consider preferences nor probabilities on temporal constraints. Such an approach has been extended in [13] to consider *probabilities*. However, no work in the literature has proposed a comprehensive approach coping with both preferences and probabilities.

2 Managing Temporal Constraints with Preferences and Probabilities

Temporal Formalism. The first step of our approach is the definition of an extended temporal formalism, in which temporal constraints are paired with preferences and/or probabilities. We base our approach on STP (Simple Temporal Problem [14]). In STP, temporal constraints have the form $P_i[l, u]P_j$, where P_i and P_j denote time points, and l and u (l <= u) are integer numbers, stating that the *temporal distance between P_i and P_j ranges between l and u*. Notably, in STP, pairs of time points may represent time intervals, to cope with durative facts/actions.

As discussed in the introduction, certain constraints are "purely" preferential, and others are "purely" probabilistic. Additionally, while performing the propagation of temporal constraints, "mixed" probabilistic+preferential constraints, which model probabilities and preferences along paths of events, can arise. To simplify the technical treatment, we choose to represent all types of constraints (preferential, probabilistic, and "mixed") in an homogeneous way. Thus, in our approach, each constraint has both a preference and a probability, and we use the special symbols "%" and "#" to denote undefined probability and preference, respectively.

Definition. Probabilistic+Preferential Quantitative Temporal Constraint (P+PQTC). Let

- let $t_i, t_j \in R$ be time points
- let $p_1, \ldots, p_n \in R$ be probabilities; $0 < p_1 \leq 1, \ldots, 0 < p_n \leq 1$ or $p_1 = \ldots = p_n = \%$
- let $P_1, \ldots, P_n \in R$ be preferences; $0 \leq P_1 \leq 1, \ldots, 0 \leq P_n \leq 1$ or $P_1 = \ldots = P_n = \#$
- let $d_1, \ldots, d_n \in Z$ be distances (between points)

A **Probabilistic+Preferential Quantitative Temporal Constraint (P+PQTC)** is a constraint of the form $t_i < (d_1, p_1, P_1), \ldots, (d_n, p_n, P_n) > t_j$ where $t_i, t_j \in R$ are time points and where either $p_1 = \ldots = p_n = \%$ or $p_1, \ldots, p_n \in [0, 1]$ and conforms a probability distribution. ■

The intended *meaning* of a constraint $t_i < (d_1, p_1, P_1), \ldots, (d_n, p_n, P_n) > t_j$ is that the *distance* $t_j - t_i$ between t_j and t_i can be d_1 with probability p_1 and preference P_1, or … or d_n with probability p_n and preference P_n (where preferences or probabilities may also be undefined, when they are denoted by # or %).

Example 2. The constraint between calcium carbonate administration (CCA) and the beginning of decreasing gastric absorption (DGA_S) in Example 1 can be represented by the following **P+PQTC**:

CCA <(1,0.4,#),(2,0.4,#),(3,0.2,#)> DGA_S

and the constraint in Example 1 about the calcium carbonate administration (CCA), relating it to a time point X_0 representing the starting time in the execution of the guideline, can be represented by the following **P+PQTC**:

X_0 <(1,%,1),(2,%,1),(3,%,0.75)(4,%,0.75),(5,%,0.5),(6,%,0.5),(7,%,0.5),(8,%,0.5),(9,%,0.25),(10,%,0.25),(11,%,0.25),(12,%,0.25)> CCA ■

Temporal reasoning. In STP, as well as in most AI approaches, temporal reasoning is based on two operations on temporal constraints: ***intersection*** and ***composition***. Given two constraints C1 and C2 between two temporal entities A and B, temporal intersection (henceforth ∩) determines the most constraining relation between A and B (e.g., A[20,40]B ∩ A[30,50]B → A[30,40]B). On the other hand, given a constraint C1 between A and B and a constraint C2 between B and C, composition (@) gives the resulting constraint between A and C (e.g., A[20,40]B @ B [10,20]C → A[30,60]C).

In STP, constraint propagation can be performed applying Floyd-Warshall's *all-pairs shortest path* algorithm, to repeatedly apply intersection and composition of temporal constraints. Floyd-Warshall's algorithm is *correct* and *complete* on STP [14], operates in cubic time, and provides as output the ***minimal network*** of the input constraints, i.e., the tightest equivalent STP, or an inconsistency.

In our approach, we extend such an approach to operate on the **P+PQTC** constraints. We propose a version of the general Floyd-Warshall's algorithm in which the operations of intersection and composition used for STP are extended to operate also on preferences and probabilities (our formal definition of intersection and composition is quite technical and long, and is omitted for the sake of brevity). The application of Floyd-Warshall algorithm (considering our new definition of intersection and composition) provides as output the minimal network of our **P+PQTC** constraints, i.e., the possible distances between each pair of time points, and the preference and probability of each distance.

Query Answering. To facilitate the interaction with physicians, we provide users with facilities to query such a minimal network, to ask for (i) the extraction (from the minimal network) of the temporal constraints between actions (or their endpoints), and their preferences and probabilities; (ii) Boolean queries, concerning whether a set of P +PQTC temporal constraints holds; (iii) Temporal interaction queries, devoted to the check of whether two events (effects of CIG actions) can interact in time, and what is the probability of such an interaction; (iv) Hypothetical queries, in which queries of types (i)–(iii) above are asked while assuming a set of temporal constraints. For example, query Q1 can be expressed in our approach as

IF{X_0<1,2>CAA, X_0<1,2>NAA}THEN Pref&Prob(INTERSECT(DGA, NAGA))?

The answer is <pref: **1**, prob: **0,9486**> (i.e., the choice of the execution time has an high preference, but there is a strong probability of interactions between the effects). Notably, after Q1, physicians might ask a query like Q2 (to check the probability of interaction in case NAA is executed in the first 30 min, and CAA between two and three hours from the current time, and the guideline preference of such a choice of the execution-time of the CIG actions):

(Q2) IF {X_0<9,10,11,12>CAA,X_0<1,2>NAA} THEN Pref&Prob(INTERSECT (DGA,NAGA))?

The answer is <pref: **0,625**, prob: **0,02455**>, suggesting to physicians that, delaying CAA, they can still comply with the CIG constraints (though obtaining a lower preference with respect to the choice in Q1), but sharply decrease the probability of interactions. Notably, considering our running example, using a "standard" temporal reasoner (i.e., not considering probabilities and preferences), physicians could only infer that an interaction may occur, both in case CAA is executed within the first 30 min, and in case it is executed after two or three hours.

3 Conclusions

We propose the first temporal reasoning approach in the AI literature coping with *both preferences and probabilities*. Our approach provides significant advantages to support physicians in interaction detection for comorbid patient. Future work regards the development of user-friendly interfaces, and an experimental evaluation.

References

1. Ten Teije, A., Miksch, S., Lucas, P. (eds.): Computer-Based Medical Guidelines and Protocols: A Primer and Current Trends. IOS Press, Amsterdam (2008)
2. Peleg, M.: Computer-interpretable clinical guidelines: a methodological review. J. Biomed. Inform. **46**, 744–763 (2013)
3. Bottrighi, A., Terenziani, P.: META-GLARE: a meta-system for defining your own computer interpretable guideline system - architecture and acquisition. Artif. Intell. Med. **72**, 22–41 (2016)
4. Riaño, D., Ortega, W.: Computer technologies to integrate medical treatments to manage multimorbidity. J. Biomed. Inform. **75**, 1–13 (2017)
5. Piovesan, L., Molino, G., Terenziani, P.: Supporting multi-level user-driven detection of guideline interactions. In: Proceedings of HEALTHINF, pp. 413–422 (2015)
6. Zamborlini, V., da Silveira, M., Pruski, C., ten Teije, A., van Harmelen, F.: Towards a conceptual model for enhancing reasoning about clinical guidelines. In: Miksch, S., Riaño, D., ten Teije, A. (eds.) KR4HC 2014. LNCS (LNAI), vol. 8903, pp. 29–44. Springer, Cham (2014). https://doi.org/10.1007/978-3-319-13281-5_3
7. Piovesan, L., Terenziani, P., Molino, G.: GLARE-SSCPM: an intelligent system to support the treatment of comorbid patients. IEEE Intell. Syst. **33**(6), 37–46 (2018)
8. Terenziani, P.: Irregular indeterminate repeated facts in temporal relational databases. IEEE Trans. Knowl. Data Eng. **28**(4), 1075–1079 (2016)
9. Anselma, L., Piovesan, L., Sattar, A., Stantic, B., Terenziani, P.: A comprehensive approach to 'Now' in temporal relational databases: semantics and representation. IEEE Trans. Knowl. Data Eng. **28**(10), 2538–2551 (2016)
10. Terenziani, P.: Nearly periodic facts in temporal relational databases. IEEE Trans. Knowl. Data Eng. **28**(10), 2822–2826 (2016)
11. Anselma, L., Piovesan, L., Terenziani, P.: Temporal detection and analysis of guideline interactions. Artif. Intell. Med. **76**, 40–62 (2017)
12. Terenziani, P., Andolina, A., Piovesan, L.: Managing temporal constraints with preferences: representation, reasoning, and querying. IEEE Trans. Knowl. Data Eng. **29**(9), 2067–2071 (2017)
13. Andolina, A., Anselma, L., Piovesan, L., Terenziani, P.: Querying probabilistic temporal constraints for guideline interaction analysis: GLARE's approach. In: Simari, G.R., Fermé, E., Gutiérrez Segura, F., Rodríguez Melquiades, J.A. (eds.) IBERAMIA 2018. LNCS (LNAI), vol. 11238, pp. 3–15. Springer, Cham (2018). https://doi.org/10.1007/978-3-030-03928-8_1
14. Dechter, R., Meiri, I., Pearl, J.: Temporal constraint networks. Artif. Intell. **49**, 61–95 (1991)

Predicting Patient's Diagnoses and Diagnostic Categories from Clinical-Events in EHR Data

Seyedsalim Malakouti(✉) and Milos Hauskrecht

University of Pittsburgh, Pittsburgh, PA 15260, USA
salimm@cs.pitt.edu

Abstract. In this paper we develop and study machine learning based models based on latent semantic indexing capable of automatically assigning diagnoses and diagnostic categories to patients based on structured clinical data in their Electronic Health record (EHR). These models can be either used for automatic coding of patient's diagnoses from structured EHR data at the time of discharge, or for supporting dynamic diagnosis and summarization of the patient condition. We study the performance of our diagnostic models on MIMIC-III EHR data.

Keywords: Lower dimensional representation · Singular value decomposition · Electronic health records · Machine learning · ICD-9 diagnosis

1 Introduction

Healthcare is one of the most promising areas for applications of data mining and machine learning methodologies. Since the adoption of electronic health records (EHRs), there has been an explosion in digital clinical data available for learning and analysis. However, the development of models that are derived from such data and that can solve important clinical problems still lags the advances in data collection. One important that can use such data is the problem of automated assignment of diagnoses to EHRs. Motivation behind solving this problem can be summarized as follows. First, automated diagnostic assignments can be used as a utility that informs clinician about the diagnoses associated with the current patient. Second, it can be used as a patient condition summarization tool to define proper context for analysis of patient management steps or to support improved prediction of future outcomes.

Going from structured EHR data to automated diagnoses is not easy. First, structured EHRs consist of a large number of time series that represent variety of labs, physiological measurements, symptoms, treatments, procedures, etc. Hence it is not easy to automatically associate the signals in these time series with specific diagnoses, especially when the diagnoses are defined by a combination of these signals or the same diagnosis can be confirmed by multiple

D. Riaño et al. (Eds.): AIME 2019, LNAI 11526, pp. 125–130, 2019.
https://doi.org/10.1007/978-3-030-21642-9_17

alternative signals. This problem is even more challenging when data are sparse (data are collected at irregular times) and many time series for the patient cases are unknown or missing. Second, the assignment of diagnoses to patient case is typically done at the time of the discharge, which means it is not only unclear what the signals related to the specific diagnosis are but also when they occurred in time. Finally, some diagnoses are very rare and even with moderate to large EHR repositories the number of patients suffering from the specific disease is very small, so learning of diagnostic models for such diseases is not feasible.

In this work we study this important problem by investigating methods from text mining, natural language processing (NLP) and information retrieval, but apply them to structured EHR data. Briefly we consider each patient's EHR to be equivalent to a document, and clinical events of different kinds recorded in EHR as words or terms in the document. To represent different events describing the patient case we consider the bag-of-word (BoW) representation that uses individual event counts and transform it using a lower-dimensional projection, based on Latent Semantic Indexing [5] that aims to better reflect semantic relations between events. The advantage of such a representation is that it permits us to consider a large number of events of different types typically found in the EHR data, and is also robust in handling missing and unknown data sources very common in EHRs. Additionally, it helps us to define the meaningful similarity among the patients as well as similarities among the words (clinical events). We use this new patient case representation to build models for individual diagnoses, as well as, diagnostic categories we define with the help of icd-9 hierarchy. Through experiments we demonstrate our new representation is able define accurate diagnostic models at different levels of abstractions.

2 Related Work

Majority of existing work modelling patient diagnostic process fall into one of two categories. The first group tackles prediction of future patient visit diagnosis. Lipton et al. proposed a Recurrent Neural Network (RNN) architecture based on Long Short Term Memory units to predict future patient visit diagnosis from a collection of 13 clinical variables [8]. GRAM [3] is an attention based RNN network that uses a BoW representation of patient's previous diagnosis as their input and take advantage of diagnosis hierarchies to extend low level diagnosis to categories. The second category of existing work studies the problem of automatic diagnosis assignment at the end of hospitalization and it is mainly motivated by improving hospital billing process. Other solutions were also proposed based on Autoencoder and LSTM neural network architectures [10,12].

The data models and SVD-based lower dimensional projections we propose in our work are typically used for analysis of text data. For example, SVD has been applied in addition to information retrieval and document analysis [2]. In terms of clinical applications, SVD and other lower dimensional representation methods including non-negative matrix factorization have been used on EHR data for missing value imputation [1], future visit diagnosis prediction [9] and medical phenotyping [11]. Despite numerous studies in diagnoses prediction and assignment, the existing work has not attempted to take advantage of the entire span of structured clinical data in EHR nor they have studied the advantage of looking at diagnostic categories as target variables.

3 Methodology

Let V_i denote a patient visit i and let $D = \{V_1, V_2, ..., V_{|D|}\}$ be a set of all patient visits in our data. A visit can be defined as $V_i = \{x_i, y_i\}$ where x_i and y_i are respectively a set of clinical events and diagnoses assigned to the patient during the visit. Clinical events are formed by a discrete representation of clinical information derived from Electronic Health Records (see below for details). Additionally, we adopt a bag-of-word (BoW) representation of a patient's EHR, therefore, $x_i \in \mathbb{N}^E$ reflects the number of occurrences of each clinical event during a patient's stay where E is the total number of event types.

Low dimensional representation of patient's clinical information is a key step in summarizing the information important for learning of diagnostic models. We define a low dimensional embedding as a mapping $E \mapsto \mathbb{R}^k : x_i : u_i$ that maps a patient's visit's data to a new lower dimension dense vector $u_i \in \mathbb{R}^k$ while $k << |E|$. Automatic learning of a low dimensional representation of complex data vectors is one of the most actively studied topics in machine learning research [4]. Our goal in this work is to show that these methods are capable supporting our problem - automatic assignment of diagnoses to patient's clinical data. Briefly, Electronic Health Records contain tens of thousands of different information including medications, procedures and surgeries, lab results, vital signs, pain scores and etc. However, often this data contain missing values. Additionally, much of this information is interrelated, conveying interchangeable or opposite information regarding patient condition. For example, various medications are used to treat blood pressure related conditions including Diuretics, Beta blockers and Alpha-1-Agonist medications. However, the first two are prescribed to patients with high blood pressure and the third group is ordered for patients with low blood pressure. Therefore, with the help of lower dimensional representation methods one can learn compact representations of patient data that a simple bag-of-word model fails to do.

Latent semantic indexing is a statistical method for analyzing the relationship between a set of documents and terms used in information retrieval by finding underlying concepts [5]. This is done by finding a Singular Value Decomposition (SVD) of original term-document matrix A. We consider each patient's EHR to be equivalent to a document, and clinical events of different kinds recorded in EHR as words or terms in the document. The underlying concepts are in fact eigenvectors of symmetric matrix $X^T X$ and are represented in the left singular vector matrix in $A = U \Sigma V^T$. Therefore, rank k Singular value decomposition of patient matrix $X_{|D|,|E|}$ can be obtained as:

$$X_{|D|,|E|} = U_{|D|,k} \Sigma_{kk} V^T{}_{k,|E|} \quad (1)$$

The lower dimensional representation of u_i can be obtained as $u_i = x_i V \Sigma^T$.

Learning diagnostic models includes learning one model per y_i (diagnosis or diagnostic category) using logistic regression with L2 regularization to capture the input-output relations. All models use low dimensional vectors as their inputs. If the lower dimensional representation is successful in capturing all important information about the patient visit in a compact form, we expect it to be sufficient. We note that this approach is not optimized to capture the relations among different diagnoses and their categories. We leave the study of these models to our future work.

4 Experiments

We experiment with our models on MetaVision part of MIMIC-III [7], an open access EHR dataset obtained over a 12-year time span that covers 22K patient visits or hospitalization to ICU. MIMIC-III encodes patients' diagnoses using standard ICD-9 codes. We enrich the ICD-9 codes with diagnostic categories defined by ICD-9 hierarchy. We limited our experiments to ICD-9 codes with at least 0.02 for prior probability of positive examples chosen to guarantee enough positive examples for learning and cross validation. This results in 421 diagnoses and diagnosis categories. We evaluate the performance of our models on the post-discharge diagnostic assignments expressed in terms of icd-9 diagnoses and their categories using the area under receiver operating characteristics curve (AUROC) and area under precision recall curve (AUPRC). The latter statistics is known to be more appropriate in the presence of imbalanced data [6].

Data processing is needed before creating a bag-of-word representation of patient data. We convert patient information in EHR to a set of meaningful binary events. We used medication and procedure orders by converting them to occurrence indicators. Laboratory results and physiological measurements with

Table 1. Performance of models for diagnoses on different ICD-9 hierarchy levels

Task name	Prior	AUROC	AUCPRC	Task name	Prior	AUROC	AUPRC
Root ICD9 codes average	0.434	0.74	0.647	Forms of heart failure	0.509	0.822	0.822
All ICD-9 codes average	0.096	0.771	0.262	Heart failure	0.249	0.852	0.681
Diseases of Genitourinary	0.495	0.862	0.869	Systolic heart failure	0.096	0.808	0.324
Nephritis related diseases	0.371	0.931	0.891	Chr systolic hrt failure	0.035	0.732	0.091
Acute renal failure	0.269	0.878	0.718	Diastolic heart failure	0.103	0.81	0.331
Ac kidney fail, tubr necr	0.056	0.897	0.376	Cardiac dysrhythmias	0.352	0.793	0.674
Ac kidney failure NOS	0.213	0.832	0.527	Cardiac arrest	0.026	0.849	0.211
Chronic kidney disease	0.198	0.917	0.742	Atr fibrillation & flutter	0.269	0.831	0.642
Chr kidney dis stage III	0.029	0.863	0.158	Atrial fibrillation	0.261	0.829	0.63
End stage renal disease	0.053	0.971	0.79	Atrial flutter	0.032	0.751	0.095
Chronic kidney dis NOS	0.097	0.838	0.323	Liver disease & cirrhosis	0.073	0.876	0.586
Diseases of Urinary Sys.	0.183	0.732	0.351	Alcohol cirrhosis liver	0.028	0.924	0.419
Disease of male Genital.	0.061	0.657	0.102	Cirrhosis of liver NOS	0.03	0.885	0.273

numerical values were converted to Abnormal Low, Normal or Abnormal High events based on their standard normal ranges, discrete valued measurements were converted to events matching these values. Finally, pain assessments were converted to special events reflecting the different pain levels. After the conversion our new events data covered 4826 clinical events including 2420 for medication orders, 116 for procedure orders, 2012 for laboratory results and 278 for physiological and pain assessment measurements.

Results in Table 1 show that more accurate models can be learned by using higher level (more general) diagnoses from disease hierarchy by taking advantage of their higher priors. However, it is important to mention that moving up the hierarchy may not always improve the models as generic categories might be harder to learn. An example of this case is the category "Diseases of Genitourinary" in Table 1 that has lower AUPRC and AUROC from its immediate sub-category. Additionally, generic categories may not be as informative.

Acknowledgement. This work was supported by NIH grant R01GM088224. The content of the paper is solely the responsibility of the authors and does not necessarily represent the official views of NIH.

References

1. Beaulieu-Jones, B.K., Moore, J.H.: Missing data imputation in the electronic health record using deeply learned autoencoders. In: Pacific Symposium on Biocomputing 2017, pp. 207–218. World Scientific (2017)
2. Berry, M.W., Drmac, Z., Jessup, E.R.: Matrices, vector spaces, and information retrieval. SIAM Rev. **41**(2), 335–362 (1999)
3. Choi, E., Bahadori, M.T., Song, L., Stewart, W.F., Sun, J.: Gram: graph-based attention model for healthcare representation learning. In: Proceedings of the 23rd ACM SIGKDD International Conference on Knowledge Discovery and Data Mining, pp. 787–795. ACM (2017)
4. Collobert, R., Weston, J., Bottou, L., Karlen, M., Kavukcuoglu, K., Kuksa, P.: Natural language processing (almost) from scratch. J. Mach. Learn. Res. **12**, 2493–2537 (2011)
5. Deerwester, S., Dumais, S.T., Furnas, G.W., Landauer, T.K., Harshman, R.: Indexing by latent semantic analysis. J. Am. Soc. Inform. Sci. Technol. **41**(6), 391–407 (1990)
6. He, H., Garcia, E.A.: Learning from imbalanced data. IEEE Trans. Knowl. Data Eng. **9**, 1263–1284 (2008)
7. Johnson, A.E., et al.: MIMIC-III, a freely accessible critical care database. Sci. Data **3**, 160035 (2016)
8. Lipton, Z.C., Kale, D.C., Elkan, C., Wetzel, R.: Learning to diagnose with LSTM recurrent neural networks. arXiv preprint arXiv:1511.03677 (2015)
9. Miotto, R., Li, L., Kidd, B.A., Dudley, J.T.: Deep patient: an unsupervised representation to predict the future of patients from the electronic health records. Sci. Rep. **6**, 26094 (2016)
10. Pakhomov, S.V., Buntrock, J.D., Chute, C.G.: Automating the assignment of diagnosis codes to patient encounters using example-based and machine learning techniques. J. Am. Med. Inform. Assoc. **13**(5), 516–525 (2006)

11. Wang, Y., et al.: Rubik: knowledge guided tensor factorization and completion for health data analytics. In: Proceedings of the 21th ACM SIGKDD International Conference on Knowledge Discovery and Data Mining, pp. 1265–1274. ACM (2015)
12. Xie, P., Xing, E.: A neural architecture for automated ICD coding. In: Proceedings of the 56th Annual Meeting of the Association for Computational Linguistics (Volume 1: Long Papers), vol. 1, pp. 1066–1076 (2018)

Assessing the Effectiveness of Sequences of Treatments Using Sequential Patterns

Maciej Piernik(✉), Joanna Solomiewicz, and Arkadiusz Jachnik

Institute of Computing Science, Poznan University of Technology,
ul. Piotrowo 2, 60-965 Poznan, Poland
maciej.piernik@cs.put.poznan.pl

Abstract. In this paper, we tackle the issue of assessing the effectiveness of sequences of treatments by introducing the concept of state-changing sequential patterns. Our proposal aims at identifying sequential patterns in an environment where certain actions are taken for patients (medical procedures, administration of pharmaceuticals, etc.) while simultaneously measuring some indicator of their health (e.g., blood pressure). We propose to combine the information about the events with the information about the states of the patients targeted by these events when mining for sequential patterns. To be able to properly interpret the changes in states as outcomes of sequences of events, we rely on the concept of a control group known from clinical trials. We illustrate the usefulness of our proposal with a proof-of-concept experiment.

Keywords: Sequential data · Frequent patterns · Modeling change

1 Introduction

Sequential patterns are an extension of frequent patterns (or frequent itemsets, known from association rule mining) to sequential data. They find many applications in domains such as customer transaction analysis, web mining, software bug analysis, chemical and biological analysis [1]. Just like with traditional frequent patterns, there are many versions of sequential patterns, depending on the structure of the sequences. In scenarios such as classification or regression, target attribute can be added to each element in each sequence. This results in a setting where a dataset contains sequences of pairs ⟨event, target⟩. In many real-world scenarios, however, such a setting is impossible to achieve, as the value of the target attribute may be provided with a delay or even completely asynchronously from the analyzed events. Consider a sequence of treatments prescribed to a given patient for a certain disease measured by some indicator (e.g., blood pressure). After a series of events (e.g., administered pharmaceuticals, medical procedures, dietary regulations) the indicator may either improve, worsen, or stay unchanged. However, this result does not necessarily coincide with any of the events nor need it be a result of one, all, or any of the preceding events. This scenario is universal when modeling people's behavior, opinion,

D. Riaño et al. (Eds.): AIME 2019, LNAI 11526, pp. 131–135, 2019.
https://doi.org/10.1007/978-3-030-21642-9_18

or—more generally speaking—*state*. As illustrated by the examples above, this problem is no longer described by a single sequence of events (like in classical sequential pattern mining), but rather by two connected sequences—one with the events and the other with target values. To the best of our knowledge, processing of sequential data of such composition has not yet been considered and is the focus of this research.

In this paper, we introduce the concept of state-changing sequential patterns along with a method to find them. Unlike in regular clinical trials, where we try to nullify the impact of all other factors, state-changing sequential patterns focus on discovering potentially hidden dependencies between medical events. We showcase the applicability of the presented concept in practical situations by performing a proof-of-concept experiment.

2 Related Work

Sequential pattern mining has first been introduced by Agrawal and Srikant [2] through a market basket analysis model. Since mining of such patterns is very costly, many optimisation algorithms have been created to improve sequential pattern mining. Giannotti et al. [3] propose an annotation solution to a problem of distinction between patterns with the same sequence but different transition times. Gebser et al. [4] propose to use knowledge-based sequence mining which takes into account expert knowledge in order to extract fewer patterns but of greater relevance.

Associating data with additional information not only can help in pattern distinction or evaluation of relevance but also in classifying it into categories. This was suggested by Pinto et al. [5]. Their algorithm focuses on multi-dimensional data and describes how certain patterns might apply to certain categories of data. Multi-dimensional data has also been examined by Plantevit et al. [6]. Their framework concentrates on relevant frequent sequences in multi-dimensional and multi-level data. It is a solution to mining relevant patterns in data of various dimensions, but there are other proposals for standard sequential data. One of such papers [7] proposes an algorithm for mining the most relevant sequential patterns and also provides a ranking according to their interestingness. Another paper [8] about mining interesting sequential patterns uses leverage (difference between observed and expected frequencies of a pattern) as a measure of interest.

A solution to mining patterns with a user-centric approach has been described by Guidotti et al. [9]. In their market basket prediction model the focus is on single users history by using four characteristics: co-occurrence (items often bought together), sequentiality (set of items often bought after another one), periodicity (sequential purchases in specific periods), recurrence (frequency of sequential purchases in a given period).

The described papers aim at finding more meaningful sequential patterns, however, none of them studies patterns with an impact on certain objects' state. To the best of our knowledge, such a problem has not yet been considered.

3 State-Changing Sequential Patterns

Assume we have a history of medical events (procedures, pharmaceuticals, dietary regulations, etc.) of a given patient. Additionally, between these events the health of the patient was being recorded in a form of some indicators (e.g., blood test results). Given a database of such records for many patients, we can look for patterns of events which increase the chances of improving patients' health. Typically, one would analyze each medical event in isolation from others to assess its sole impact on patients' health (e.g., in clinical trials). However, given historical data of the above-described composition, we can look for patterns of different events appearing in a certain order, i.e., sequential patterns.

Given the above, the problem of state-changing sequential patterns can be formulated as follows. Is it possible to find a sequence of events which will have a high probability of influencing the patients' state in a desired manner.

Formally, the concept of state-changing sequential patterns can be defined as follows. By a *sequence* $s =< s_1, s_2, ..., s_n >$ we understand an ordered multi-set of *elements*, where each element s_i is drawn from the same set. We distinguish two types of sequences: sequences of events and sequences of states. Each events sequence has a corresponding states sequence. The corresponding sequences can be combined into a single sequence of events and states and there exists a total order between the elements of the combined sequences such that the order of the elements from each sequence is preserved.

A sequence s' which elements form a subset of elements of another sequence s is called a *subsequence* of s and is denoted as $s' \subseteq s$. Given a set of sequences $\mathcal{S}$, a sequence p is called a *sequential pattern* (or *pattern*), if it is a subsequence of at least *minsup* sequences in $\mathcal{S}$: $|\{s \in \mathcal{S} : p \subseteq s\}| \geq minsup$, where *minsup* is a user-defined minimal support parameter. We denote that a sequence s *contains* a pattern p if $p \subseteq s$. Given a sequence $s =< s_1, s_2, ..., s_n >$, its subsequence $s' =< s_{i_1}, s_{i2}, ..., s_{i_m} >$, $1 \leq m \leq n$, and an element $s_x \in s$, we say that s_x appears in s after s' if $i_m < x \leq n$, and before s' if $1 \leq x < i_1$, denoted respectively as $s' \prec^s s_x$ and $s' \succ^s s_x$. Given the above, a *state-changing sequential pattern* can be generally defined as a pattern p, for which the probability of a certain change in state (positive or negative) appearing in any given sequence s after this pattern is higher than the probability of this change appearing without this pattern by at least *minchange*:

$$P(\underset{s_i,s_j \in s}{\exists} s_i < s_j | s_i \prec^s p \succ^s s_j) - P(\underset{s_i,s_j \in s}{\exists} s_i < s_j | \neg(s_i \prec^s p \succ^s s_j)) > minchange$$

where s_i and s_j indicate states, $i < j$ for positive change, $j < i$ for negative change, and *minchange* is a user-defined threshold.

Ideally, to calculate the second probability, i.e., the change happening without the pattern, we would use a separate control group. However, unfortunately such data in historical patients' records are very rare. Therefore, to make this definition usable on any given dataset, let us split each sequence into smaller sequences based on the following principle. For any three consecutive states s_i, s_j, s_k, if $sign(s_k - s_j) \neq sign(s_j - s_i)$ then s_j marks the end of one sequence of events an the beginning of another. For every such sequence of events we calculate the difference between the state which marks the beginning and the end of this sequence. The sequences with positive and negative differences fall into separate datasets: $\mathcal{S}^+$ and $\mathcal{S}^-$, respectively. Given the above datasets, we can mine for sequential patterns in each of these sets separately and select the state-changing sequential patterns as those p for which:

$$\frac{|sup(p,\mathcal{S}^+) - sup(p,\mathcal{S}^-)|}{\text{total number of sequences}} > minchange \tag{1}$$

where $sup(p,\mathcal{S}) = |\{s \in \mathcal{S} : p \subseteq s\}|$.

4 Application

Let us now illustrate the usefulness of state-changing sequential patterns with a simple proof-of-concept experiment. In the experiment we use the *diabetes* dataset, which is publicly available through the UCI Machine Learning Repository [10]. It includes medical events performed on patients suffering from diabetes along with their blood sugar level measurements. The dataset consists of 3883 sequences composed of 20 different elements with an average length of 7.6 elements per sequence. The code for the experiment was written in Python programming language and is available at https://github.com/joanna-solomiewicz/state-changing-sequential-patterns. The experiment was carried out using the procedure described at the end of Sect. 3.

Table 1. Top 5 patterns: left—in order of their support, right—in order of their change (calculated using Eq. 1) [R = Regular insulin dose, N = NPH insulin dose].

Pattern p	$sup(p,\mathcal{S}^-)$	$sup(p,\mathcal{S}^+)$	Change	Pattern p	$sup(p,\mathcal{S}^-)$	$sup(p,\mathcal{S}^+)$	Change
R	982	785	0.096	N, R, N	460	185	0.134
N	771	603	0.082	N, N	503	243	0.127
R, N	688	454	0.114	R, R, N	499	243	0.125
R, R	677	555	0.059	N, R	550	305	0.119
N, R	550	305	0.119	R, N	688	454	0.114

In Table 1 we present the top 5 patterns found according to their support in $\mathcal{S}^-$ and contrast them with the top 5 state-changing sequential patterns. As the support of the patterns is already calculated based on the dataset transformed according to our method, it is difficult to objectively compare the measurements. Still, we can clearly observe that the ranking produced by support is significantly different from the one produced by the change indicator. This suggests that the concept of state-changing sequential patterns can be potentially used to discover new causal relationships between sequences of events and changes in state which could otherwise be omitted.

5 Conclusions

In this paper, we introduced the concept of state-changing sequential patterns along with a simple way of discovering them. The concept allows for finding patterns of events which have high probability of causing a certain change of state. We define and formalize the concept and illustrate its applicability in medical scenarios with an empirical example. As this paper reports a work-in-progress research, there is still much theoretical and experimental work to be done. After exploring the theoretical properties and thoroughly experimenting with the introduced concept, we plan on including time constraints in the analysis as some previous studies suggest they can add important information from the pattern mining perspective [11]. The constraints could concern both, events (e.g., restricting time gaps between events) and states (e.g., the certainty of a given object's state can decay over time until new state appears). We also intend to quantify the magnitude of change in state caused by the discovered patterns, as currently we solely focus on the direction of change. Moreover, we plan to create new sequential pattern evaluation measures dedicated for this problem as well as an efficient algorithm which would cut the unpromising patterns at an earlier stage to enhance efficiency.

Acknowledgments. This research is partly funded by the Polish National Science Center under Grant No. DEC-2015/19/B/ST6/02637.

References

1. Aggarwal, C.C., Han, J. (eds.): Frequent Pattern Mining. Springer, Cham (2014). https://doi.org/10.1007/978-3-319-07821-2
2. Agrawal, R., Srikant, R.: Mining sequential patterns. In: Proceedings of the 11th ICDE, pp. 3–14 (1995)
3. Giannotti, F., Nanni, M., Pedreschi, D.: Efficient mining of temporally annotated sequences. In: SIAM International Conference on Data Mining, pp. 348–359 (2006)
4. Gebser, M., Guyet, T., Quiniou, R., Romero, J., Schaub, T.: Knowledge-based sequence mining with ASP. In: Proceedings of the 25th IJCAI, pp. 1497–1504 (2016)
5. Pinto, H., Han, J., Pei, J., Wang, K., Chen, Q., Dayal, U.: Multi-dimensional sequential pattern mining. In: Proceedings of the 10th CIKM, pp. 81–88 (2001)
6. Plantevit, M., Laurent, A., Laurent, D., Teisseire, M., Choong, Y.W.: Mining multidimensional and multilevel sequential patterns. ACM Trans. Knowl. Discov. Data (TKDD) **4**, 1–37 (2010)
7. Fowkes, J., Sutton, C.: A subsequence interleaving model for sequential pattern mining. In: Proceedings of the 22nd ACM SIGKDD, pp. 835–844 (2016)
8. Li, T., Webb, G.I., Petitjean, F.: Exact discovery of the most interesting sequential patterns. CoRR abs/1506.08009 (2015)
9. Guidotti, R., Rossetti, G., Pappalardo, L., Giannotti, F., Pedreschi, D.: Market basket prediction using user-centric temporal annotated recurring sequences. In: Proceedings of the 33rd ICDM, vol. 00, pp. 895–900 (2018)
10. Kahn, M.: UCI Machine Learning Repository (1994)
11. Gay, P., López, B., Meléndez, J.: Learning complex events from sequences with informed gaps. In: ICMLA, pp. 1089–1094. IEEE (2015)

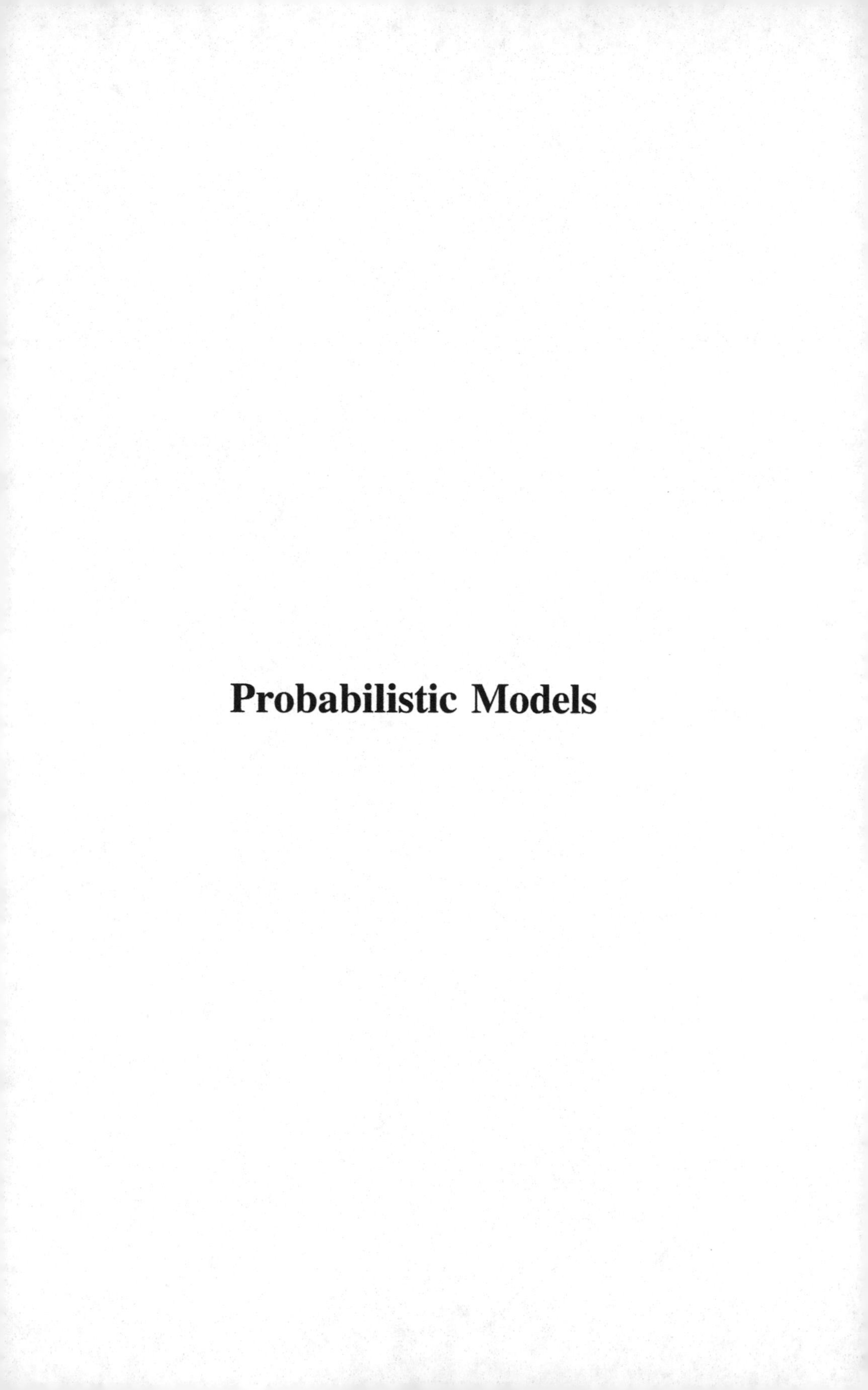

Probabilistic Models

Bayesian Network vs. Cox's Proportional Hazard Model of PAH Risk: A Comparison

Jidapa Kraisangka[1(✉)], Marek J. Druzdzel[1,2], Lisa C. Lohmueller[3], Manreet K. Kanwar[4], James F. Antaki[5], and Raymond L. Benza[4]

[1] University of Pittsburgh, Pittsburgh, PA 15213, USA
jik41@pitt.edu
[2] Białystok University of Technology, Wiejska 45A, 15-351 Białystok, Poland
[3] Carnegie Mellon University, Pittsburgh, PA 15213, USA
[4] Cardiovascular Institute, Allegheny General Hospital, Pittsburgh, PA 15212, USA,
[5] Cornell University, Ithaca, NY 14850, USA

Abstract. Pulmonary arterial hypertension (PAH) is a severe and often deadly disease, originating from an increase in pulmonary vascular resistance. The REVEAL risk score calculator [3] has been widely used and extensively validated by health-care professionals to predict PAH risks. The calculator is based on the Cox's Proportional Hazard (CPH) model, a popular statistical technique used in risk estimation and survival analysis. In this study, we explore an alternative approach to the PAH patient risk assessment based on a Bayesian network (BN) model using the same variables and discretization cut points as the REVEAL risk score calculator. We applied a Tree Augmented Naïve Bayes algorithm for structure and parameter learning from a data set of 2,456 adult patients from the REVEAL registry. We compared our BN model against the original CPH-based calculator quantitatively and qualitatively. Our BN model relaxes some of the CPH model assumptions, which seems to lead to a higher accuracy (AUC = 0.77) than that of the original calculator (AUC = 0.71). We show that hazard ratios, expressing strength of influence in the CPH model, are static and insensitive to changes in context, which limits applicability of the CPH model to personalized medical care.

Keywords: Bayesian networks · Risk assessment · Cox's proportional hazard model · Hazard ratios · Pulmonary arterial hypertension

1 Introduction

Pulmonary arterial hypertension (PAH) is a chronic and life-changing disease, originating from an increase in pulmonary vascular resistance, and leading to high blood pressure in the lung. One of the most widely used tools in prognosis and management of PAH is the REVEAL risk score calculator [3], which assesses the risk

D. Riaño et al. (Eds.): AIME 2019, LNAI 11526, pp. 139–149, 2019.
https://doi.org/10.1007/978-3-030-21642-9_19

of death of a PAH patient based on various risk factors. The REVEAL calculator is based on the Cox's Proportional Hazard (CPH) [4] model, a popular statistical technique used in risk estimation and survival analysis. One weakness of this approach is that CPH models can be only learned from data and are not readily amenable to refinement based on expert knowledge. Another limitation is that the CPH models rest on several assumptions simplifying the interactions between the risk factors and the disease. While the CPH-based risk assessment models have been successfully used for decades, it is interesting to study whether recent developments at the intersection of statistic and artificial intelligence, such as Bayesian networks (BNs), offer more modeling flexibility and possibly superior performance.

In this paper, we describe our effort to replace the CPH model underlying the existing REVEAL risk score calculator by a BN. We created a BN model with the same variables and variable states as the REVEAL risk score calculator using data from the REVEAL database [3]. In an earlier paper [10], we described a BN model that mimicked the REVEAL CPH model and, hence, offered precisely the same accuracy [9]. In this paper, we are presenting a BN model that is learned directly from the REVEAL data set and compare it to the original CPH model. We applied a Tree Augmented Naïve Bayes (TAN) algorithm for structure and parameter learning and 10-fold cross-validation to measure the BN performance, which we report as the area under the Receiver Operating Characteristic curve (AUC). The BN model has the AUC of 0.77, compared to the original REVEAL calculator's AUC of 0.71. We attribute this difference to relaxing some of the modeling assumptions that are not satisfied by the data set. We also show restrictive assumption of the CPH model, notably static and context-invariant character of influence of individual risk factors, as expressed by CPH hazard ratios.

The remainder of our paper is structured as follows. Sections 2 and 3 provide background knowledge on the REVEAL risk score calculator and BNs respectively. Section 4 describes our approach to building and validation of the BN model. Finally, Sects. 5 and 6 focus on both qualitative and quantitative comparison of the two modeling techniques.

2 The REVEAL Risk Score Calculator

The REVEAL risk score calculator was developed to predict PAH disease progression and guide physician's therapeutic decision making. The calculator is comprised of 19 demographic, functional, laboratory, and hemodynamic parameters and is based on a multivariate Cox's proportional hazard (CPH) model. The CPH model was developed based on data from 2,529 newly and previously diagnosed PAH patients. The model was simplified into a risk score calculator, which was later validated on 504 newly diagnosed patients. The original AUC of the risk score calculator validated on the validation cohort of 504 adult patients was reported to be 0.724 [2].

The REVEAL risk score calculator has been used for almost two decades. Although it offers good quality of predictions, it has some limitations. The model does not take into account interactions among its variables and is not robust to missing key variables.

3 Bayesian Networks

Bayesian networks [11] are probabilistic graphical models capable of modeling the joint probability distribution over a finite set of random variables. The structure of a BN is an acyclic directed graph in which each node corresponds to a single variable and directed arcs denote direct dependencies between pairs of variables. A conditional probability table (CPT) of a variable X contains probability distributions over the states of X for all combinations of states of X's parents. The joint probability distribution over all variables of the network can be calculated by taking the product of all prior and conditional probability distributions, i.e.,

$$\Pr(\mathbf{X}) = \Pr(X_1, \ldots, X_n) = \prod_{i=1}^{n} \Pr(X_i | Pa(X_i)) . \tag{1}$$

BNs have been used in numerous practical applications and because they are capable of deriving the posterior marginal probability distribution over any variables of interest, given values of other variables in the model, it is quite natural to apply them to risk assessment. BNs are compact and intuitive, while also being theoretically sound [7]. They can be based purely on literature or expert knowledge, can be learned from data, or a combination of the two. Calculation in BNs, which ks worst case NP-hard, is very efficient for most practical models. Prior Bayesian networks have been previously applied to risk assessment (e.g. a BN model for predicting cardiovascular risk [1] and a BN risk assessment model for patients with the left ventricular assist devices [8]).

4 Application of a Bayesian Network to the REVEAL Risk Score Calculator

We used a total of 2,456 patient records from the REVEAL registry to develop a BN model. The data did not include censored patients. We preserved the list of variables from the REVEAL risk score calculator along with their discretization levels (see Table 1). It is clear that some of the variables in the table have been artificially created for the purpose of CPH modeling. For example, the three *WHO Group I Subgroup* variables are mutually exclusive states of a single variable. The same holds for the *NYHA functional class*, *6 min walking distance*, *BNP* and *% DLCO* variables. The CPH model required them to be risk factors, modeled as states of binary variables. We combined these states back into single variables, as the laws of probability require. For all numerical variables, which we had to discretize in order to include them into the Bayesian network model, we applied the cut points used by the REVEAL risk score calculator. We also added a baseline state (shown in **bold**), wherever needed but not explicitly defined in the calculator.

Table 1. A list of 19 binary risk factors from the REVEAL risk score calculator [3] along with their counterparts in the Bayesian network. Baseline states are shown in **bold**.

Risk factors	HR	Random variable	States
APAH-CTD	1.59	WHO group I subgroup	APAH-CTD
FPAH	3.60		FPAH
APAH-PoPH	2.17		APAH-PoPH
			Other
Renal insufficiency	1.90	Renal insufficiency	Yes
			No
Male >60 years age	2.18	Male >60 years	Yes
			No
NYHA/WHO FC I	0.42	NYHA/WHO FC	I
			II
NYHA/WHO FC III	1.41		III
NYHA/WHO FC IV	3.13		IV
SBP <110 mmHg	1.67	Systolic BP	<110
			≥110
Heart Rate >92 bpm	1.39	Heart rate	>92
			≤92
6MWD <165 m	1.68	Six minute walking distance	<165
			165-<440
6MWD ≥440 m	0.58		≥440
BNP <50 pg/ML	0.50	BNP	<50
			50–180
BNP >180 pg/ML	1.68		>180
Pericardial effusion	1.35	Pericardial effusion	Yes
			No
% DLCO ≤32%	1.46	% pred. DLCO	≤32
			>32-<80
% DLCO ≥80%	0.59		≥80
Mean RAP >20 mmHg	1.79	Mean RAP	>20
			≤20
PVR >32 WU	4.08	PVR	>32
			≤32

Note: HR: hazard ratio; APAH-CTD: PAH associated with connective tissue disease; FPAH: familial PAH; APAH-PoPH; PAH associated with portal hypertension; WHO: World Health Organization; NYHA: New York Heart Association; FC: functional class; 6MWD: 6-min walk distance; SBP: systolic blood pressure; BNP: brain natriuretic protein; % pred. DLCO: % predicted diffusing capacity of the lung for carbon monoxide; RAP: right atrial pressure; PVR: pulmonary vascular resistance; WU: Wood unit

We used the REVEAL registry data to learn a Tree Augmented Naïve (TAN) BN model [5] that predicts 1-year survival. The TAN learning algorithm is one of the most popular learning methods for BN classification. TAN extends the Naïve Bayes structure by adding most important interdependencies among feature variables. At the same time, the algorithm constraints the maximum number of incoming arcs to two and, by this, keeps the conditional probability tables (CPTs) in individual nodes small. Small CPTs mean a small number of parameters, which can be learned reliably even when the learning data set is small. Effectively, when the learning data set is small, the quality of the parameters remains high and the entire TAN model typically matches well the joint probability distribution that generated the data.

Figure 1 shows the TAN Bayesian network learned for the purpose of our study.

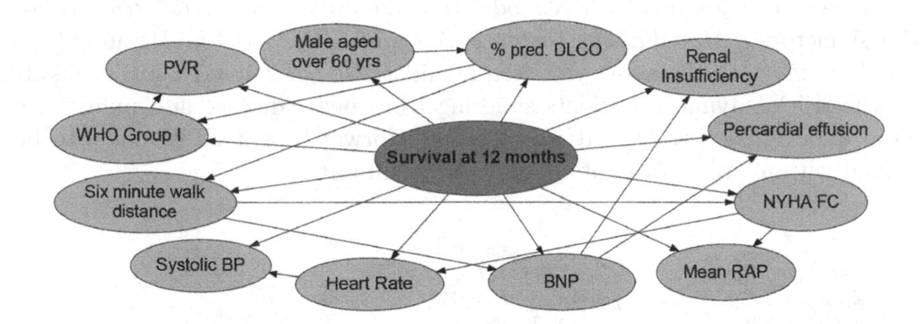

Fig. 1. The TAN Bayesian network learned from the REVEAL registry data.

We applied 10-fold cross-validation on the 2,456 adult patients records from the REVEAL registry to validate this BN model. We validated the REVEAL risk score calculator on the same data set. The BN model demonstrated an improved AUC of 0.77 to the AUC of 0.71 for the REVEAL calculator.

5 Bayesian Networks vs. CPH Model in Risk Assessment

CPH models are prevalent in medical literature. They allow for estimating the effect of multiple risk factors on survival. Impact of each individual risk factor is expressed by a number called *hazard ratio* (HR). The HR is defined as a ratio of the hazard in the corresponding risk group to the hazard in the baseline group (i.e., a hypothetical group in which none of the risk factors is present). For example, Table 1 reports the HR for *Renal insufficiency* as 1.90. This means that patients with renal insufficiency have a 90% higher risk of dying from PAH than patients without renal insufficiency. This ratio is, by one of the assumptions of the CPH model, constant over time and the same regardless of what other risk factors are present [4]. It is clear that in practice HRs may potentially change over time [6] and there exists an extension of the CPH model that relaxes this assumption [14]. A problem that has to our knowledge not received attention is that HRs are sensitive to context and change with presence or absence of other risk factors.

BNs, on the other hands, do not require restrictive modeling assumptions outside of expressing independencies. The structure of a BN and all its numerical probabilities can be obtained from experts or learned from data. While it is conceivable that CPH model parameters can be elicited from experts, most clinical CPH models are derived directly from data. In practice, data sets used for model construction usually contain missing values. The standard CPH model is not capable of handling missing values and require statisticians to perform proper imputation. BNs naturally allow for estimating their parameters from data sets with missing values.

BNs also allow researchers to combine multiple risk models. For example, Fig. 2 shows an example of a BN model that combines (*Heart-Related Deaths*, with risk factors *6 Min Walking Distance*, *Age* and *SBP* > *110* mmHg and *PAH-Related Deaths* with the above risk factors and *PVR* > *32 Wood Unit*) to determine the risk of dying of patients suffering from heart disease and pulmonary arterial hypertension (PAH). It is not straightforward for a CPH model to be extended without re-learning its parameters from data.

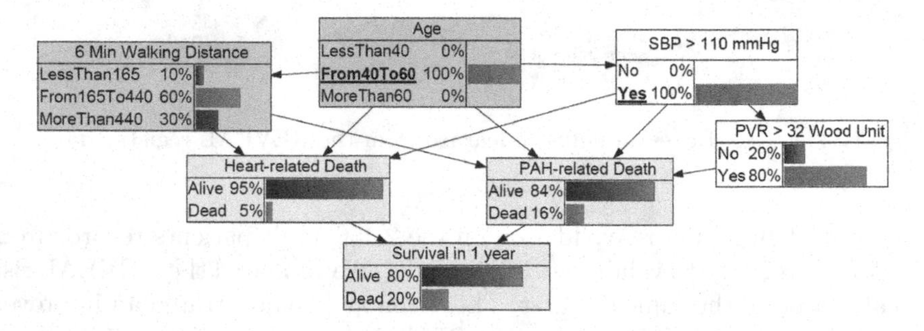

Fig. 2. An example Bayesian network predicting survival based on two risk models of patients with partial observations (only *Age* and *SBP* are observed)

When performing prediction, BNs allow for estimating the outcome probability based on partial observations, while the CPH model is not designed for that, even though one could extend it along the lines of BN inference. For example, if we know that a given patient is subjected to only one or two risk factors, we can make a prediction of survival without knowledge of presence or absence of the remaining risk factors. CPH models require that we know for each risk factor whether it is present or absent.

6 Context Sensitivity of CPH Hazard Ratios

We mentioned above that one of the assumptions of the CPH model is that the individual HRs do not change with presence or absence of other risk factors. This assumption did not seem realistic to us, so we decided to test it on the REVEAL registry data. The REVEAL registry data set is far too small to test this assumption, so we decided to use an artificial data set of 30,000 records generated from our TAN network. As we argued above, TAN networks match the joint probability distributions from which they were learned reasonably well and, effectively, statistical properties of a data set generated from the TAN model will not depart too far from the statistical properties of the original data set. Given a 30,000 record data set, we were able to simulate situations in which some of the risk factors have been observed (this amounted to selecting a subset of the data) and to learn a new CPH model from the resulting data. Our goal was to check whether the HRs for those variables that have not been observed yet are indeed constant, i.e., the same in the selected subset of records. Figure 3 shows the result of this experiment. Figure 3a shows the HRs calculated for subsets in which a single risk factor (listed in the header of the table) has been observed. All columns differ from the first column, which contains the original CPH parameters. Figure 3b shows differences between the hazard ratios calculated for each of the cases relative to the original parameters and expressed as a percentage of change. Colors give a visual indication of where the largest differences are. Some of the hazard ratios in the table have changed as much as 700%!

Figure 4 shows NYHA-I column of Fig. 3b in graphical format. We can see that these risk factors, e.g., *SIXMWD_165*, *MRAP*, become very important once we observe that the patient belongs to *NYHA Functional Class I*. However, in theory HRs are static and are not capturing this context-induced change.

Modeling with BNs does not require us to make such assumptions. In fact, varying degree of influence of risk factors is a natural consequence of varying context. As some of the risk factors are observed, the role of other risks, expressed by their potential to impact of the survival variable, changes. Figure 5 shows a scatter plot of HR and entropy (used typically in BNs as an expression at the amount of information) for the *NYHA Functional Class I* case. The plot shows the baseline situation, i.e., when no risk factors are observed (triangle marks) and a change in context, when *NYHA-I* is observed (circles). The two measures are correlated with each other at the baseline. However, the entropy changes with context, while the hazard ratios stay the same by definition.

CPH	HR	APAH_CTD	APAH_PoPH	FPAH	NYHA_I	NYHA_III	NYHA_IV	SIXMWD_165	SIXMWD_440	BNP_50	BNP_180	DLCO_32	DLCO_80	SYSBP	HR	MRAP	PVR	RI	MALE_60YR	PERI_EFFU
APAH_CTD	1.52				1.30	1.68	1.20	1.62	1.09	1.54	1.52	1.65	0.27	1.55	1.27	1.39	4.59	1.20	1.71	1.51
APAH_PoPH	2.42				5.87	2.58	1.67	1.89	2.90	3.62	2.12	2.63	3.40	2.51	2.15	2.72	3.22	1.96	1.57	2.33
FPAH	1.82				0.70	1.56	1.63	1.55	2.69	2.91	1.53	1.02	2.71	1.83	1.39	1.88	5.32	1.32	0.54	1.77
NYHA_I	0.35	0.40	0.51	0.09				1.09	0.28	0.36	0.37	0.44	0.19	0.37	0.52	2.67	1.62	0.20	0.36	0.41
NYHA_III	1.46	1.56	1.44	1.02				1.04	1.48	1.52	1.43	1.53	1.29	1.42	1.58	0.93	2.14	1.44	1.38	1.53
NYHA_IV	4.20	3.76	3.39	3.91				3.03	1.40	5.13	3.93	3.83	4.33	3.77	3.55	4.57	4.36	3.65	2.76	4.02
SIXMWD_165	1.46	1.53	1.23	1.83	11.80	1.36	1.48			1.91	1.51	1.43	1.31	1.52	1.45	1.52	1.28	1.48	1.34	1.43
SIXMWD_440	0.58	0.46	0.55	0.74	0.77	0.62	0.16			1.41	0.48	0.97	0.76	0.55	0.73	0.67	0.61	0.39	0.52	0.53
BNP_50	0.41	0.40	0.46	0.49	0.68	0.37	0.38	0.52	0.89			0.52	0.53	0.39	0.39	0.38	0.49	0.25	0.49	0.45
BNP_180	1.67	1.72	1.36	1.25	1.94	1.65	1.51	1.91	1.79			1.73	1.60	1.64	1.56	1.58	1.97	1.18	1.81	1.76
DLCO_32	1.39	1.38	1.56	0.98	1.08	1.42	1.25	1.23	3.12	2.41	1.32			1.44	1.41	1.18	1.33	1.61	1.44	1.39
DLCO_80	0.84	0.14	1.13	1.30	0.59	0.79	0.75	0.56	1.14	1.03	0.79			0.99	0.84	0.86	0.88	0.75	0.17	0.89
SYSBP	1.72	1.71	1.79	1.81	2.13	1.73	1.54	1.78	1.68	1.64	1.70	1.78	2.17		1.96	1.56	1.88	1.41	1.73	1.66
HR	1.33	1.16	1.35	1.23	2.55	1.46	0.93	1.21	1.92	1.40	1.28	1.24	1.61	1.47		1.30	1.61	1.41	1.11	1.34
MRAP	1.46	1.40	1.48	1.85	10.87	1.11	1.81	1.50	1.60	1.44	1.47	1.42	1.67	1.40	1.50		2.98	1.95	1.69	1.36
PVR	1.99	3.56	2.15	4.10	9.00	2.50	1.44	2.13	2.24	2.71	2.02	1.53	2.25	2.00	1.95	2.19		1.98	1.59	1.54
RI	1.60	1.43	1.51	1.32	0.54	1.66	1.51	1.52	1.24	1.38	1.54	1.73	1.43	1.42	1.74	2.17	0.98		1.42	1.56
MALE_60YR	1.83	2.01	1.24	0.79	1.46	1.91	1.45	1.65	1.75	2.10	1.80	1.98	0.42	1.83	1.58	2.41	1.40	1.68		1.65
PERI_EFFU	1.56	1.54	1.54	1.65	1.62	1.64	1.42	1.48	1.47	1.70	1.59	1.54	1.73	1.51	1.60	1.34	0.87	1.50	1.32	

(a) Hazard ratios of each observed group

CPH	HR	APAH_CTD	APAH_PoPH	FPAH	NYHA_I	NYHA_III	NYHA_IV	SIXMWD_165	SIXMWD_440	BNP_50	BNP_180	DLCO_32	DLCO_80	SYSBP	HR	MRAP	PVR	RI	MALE_60YR	PERI_EFFU
APAH_CTD	1.52				-15%	10%	-21%	6%	-29%	1%	0%	8%	-82%	2%	-17%	-9%	201%	-21%	12%	-1%
APAH_PoPH	2.42				142%	7%	-31%	-22%	20%	49%	-12%	9%	40%	4%	-11%	12%	33%	-19%	-35%	-4%
FPAH	1.82				-61%	-14%	-10%	-15%	48%	60%	-16%	-44%	49%	1%	-23%	3%	193%	-27%	-70%	-3%
NYHA_I	0.35	17%	49%	-74%				216%	-19%	3%	6%	28%	-44%	7%	50%	673%	369%	-42%	5%	18%
NYHA_III	1.46	6%	-2%	-30%				-29%	1%	4%	-2%	5%	-12%	-3%	8%	-37%	46%	-2%	-6%	4%
NYHA_IV	4.20	-10%	-19%	-7%				-28%	-67%	22%	-6%	-9%	3%	-10%	-15%	9%	4%	-13%	-34%	-4%
SIXMWD_165	1.46	4%	-16%	25%	706%	-7%	1%			31%	3%	-2%	-11%	4%	-1%	4%	-13%	1%	-9%	-3%
SIXMWD_440	0.58	-22%	-5%	27%	31%	7%	-72%			142%	-17%	66%	31%	-5%	25%	15%	5%	-33%	-12%	-10%
BNP_50	0.41	-2%	12%	19%	67%	-9%	-6%	28%	117%			28%	29%	-4%	-5%	-6%	20%	-38%	21%	9%
BNP_180	1.67	3%	-19%	-25%	17%	-1%	-10%	15%	7%			4%	-4%	-1%	-6%	-5%	18%	-29%	9%	5%
DLCO_32	1.39	-1%	12%	-30%	-23%	2%	-10%	-11%	125%	74%	-5%			4%	1%	-15%	-4%	15%	4%	0%
DLCO_80	0.84	-83%	34%	55%	-30%	-5%	-11%	-33%	36%	23%	-6%			17%	0%	2%	5%	-11%	-80%	6%
SYSBP	1.72	-1%	4%	5%	24%	1%	-10%	4%	-2%	-5%	-1%	4%	26%		14%	-9%	9%	-18%	1%	-3%
HR	1.33	-13%	1%	-7%	92%	10%	-30%	-9%	45%	5%	-3%	-7%	21%	11%		-2%	21%	6%	-17%	1%
MRAP	1.46	-4%	2%	27%	646%	-24%	24%	3%	10%	-1%	1%	-2%	15%	-4%	3%		105%	34%	16%	-7%
PVR	1.99	79%	8%	106%	353%	26%	-27%	7%	13%	36%	2%	-23%	13%	1%	-2%	10%		-1%	-20%	-23%
RI	1.60	-11%	-6%	-18%	-66%	4%	-6%	-5%	-23%	-14%	-4%	8%	-11%	-11%	8%	35%	-39%		-11%	-3%
MALE_60YR	1.83	10%	-32%	-57%	-20%	4%	-21%	-10%	-4%	14%	-2%	8%	-77%	0%	-14%	31%	-24%	-8%		-10%
PERI_EFFU	1.56	-2%	-2%	5%	4%	5%	-9%	-5%	-6%	9%	1%	-2%	11%	-3%	2%	-14%	-44%	-4%	-15%	

(b) Percent relative change of the hazard ratio from the baseline

Fig. 3. Effect of observing one of the risk factors on the hazard ratios of the remaining variables

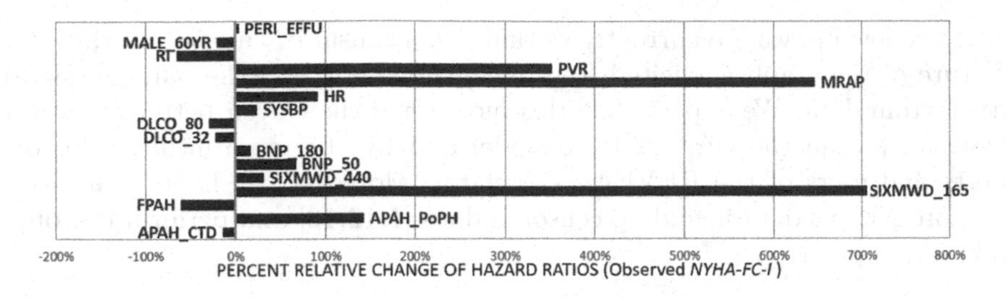

Fig. 4. Percent relative change of hazard ratios when we observed *NYHA-FC-I*

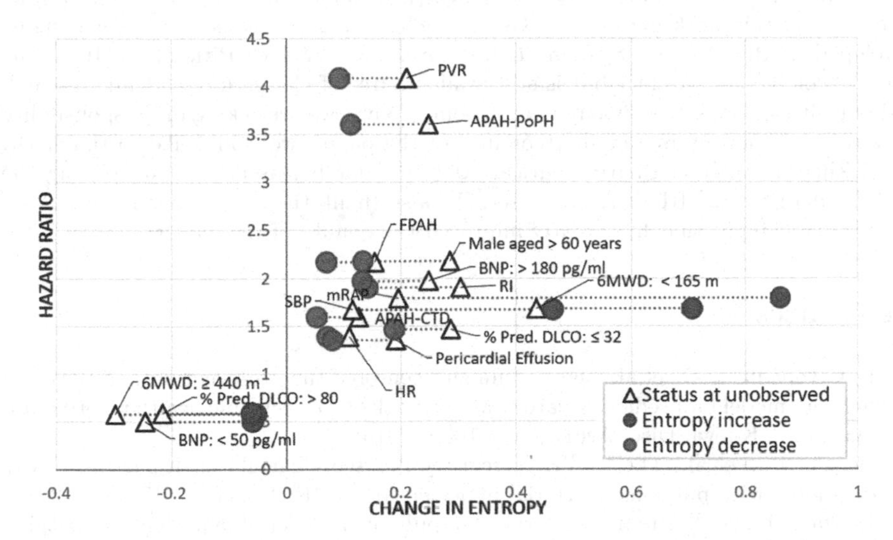

Fig. 5. An example of the movement of the entropy when we observed *NYHA-I*. The entropy change or the influence of the risk factors is context-sensitive.

7 Conclusion

In this paper, we proposed a simple TAN BN model to replace the CPH model underlying the REVEAL risk calculator. The TAN model is a compromise between our desire to relax the CPH assumptions and what can be extracted from a limited size data set. While the REVEAL data set is not small, it is not large enough to learn the general structure of a BN reliably. While the TAN model fits the joint probability distribution over its variables reasonably well, it does not mimic the causal structure of interactions among the model variables. Still, with this important disadvantage, it offers an improved numerical accuracy, which we believe stems from relaxing the CPH model assumptions. As we demonstrate in this paper, the assumptions of the CPH model may be unrealistic in practice. BNs model naturally varying magnitude of influence of risk factors as other factors are observed.

As a follow up, we are currently working on a causal BN model, in which the structure of the graph is elicited from medical experts and the parameters are learned from data. We expect that this model will show even better numerical performance than the current TAN model and will be more intuitive for our experts and users of the REVEAL calculator. Moreover, we hope to incorporate more patient data including censored data [1,12,13] and perform thorough validation in practice.

Acknowledgments. We acknowledge the support of the National Institute of Health (1R01HL134673-01), Department of Defence (W81XWH-17-1-0556), and the Faculty of Information and Communication Technology, Mahidol University, Thailand. Implementation of this work is based on GeNIe and SMILE, a Bayesian inference engine developed at the Decision Systems Laboratory, University of Pittsburgh. It is currently a commercial product but is still available free of charge for academic research and teaching at https://www.bayesfusion.com/. While we are taking full responsibility for any remaining errors and shortcomings of the paper, we would like to thank Dr. Carol Zhao of Actelion Pharmaceuticals US, Inc., for her assistance in learning the TAN model from the REVEAL data set. We also thank the anonymous reviewers for their valuable input that has greatly improved the quality of this paper.

References

1. Bandyopadhyay, S., et al.: Data mining for censored time-to-event data: a Bayesian network model for predicting cardiovascular risk from electronic health record data. Data Min. Knowl. Disc. **29**(4), 1033–1069 (2015)
2. Benza, R.L., et al.: The REVEAL registry risk score calculator in patients newly diagnosed with pulmonary arterial hypertension. CHEST **141**(2), 354–362 (2012)
3. Benza, R.L., et al.: Predicting survival in pulmonary arterial hypertension: insights from the Registry to Evaluate Early and Long-Term Pulmonary Arterial Hypertension Disease Management (REVEAL). Circulation **122**(2), 164–172 (2010)
4. Cox, D.R.: Regression models and life-tables. J. Roy. Stat. Soc. Ser. B (Methodol.) **34**(2), 187–220 (1972)
5. Friedman, N., Geiger, D., Goldszmidt, M.: Bayesian network classifiers. Mach. Learn. **29**(2–3), 131–163 (1997)
6. Hernán, M.A.: The hazards of hazard ratios. Epidemiology (Cambridge, Mass.) **21**(1), 13 (2010)
7. Husmeier, D., Dybowski, R., Roberts, S.: Probabilistic Modeling in Bioinformatics and Medical Informatics. Springer, London (2005)
8. Kanwar, M.K., et al.: A Bayesian model to predict survival after left ventricular assist device implantation. JACC Heart Fail. **6**(9), 771–779 (2018)
9. Kraisangka, J., Druzdzel, M.J.: A Bayesian network interpretation of the Cox's proportional hazard model. Int. J. Approximate Reasoning **103**, 195–211 (2018)
10. Kraisangka, J., Druzdzel, M.J., Benza, R.L.: A risk calculator for the pulmonary arterial hypertension based on a Bayesian network. In: Proceedings of the 13th UAI Bayesian Modeling Applications Workshop, pp. 1–6 (2016)
11. Pearl, J.: Probabilistic Reasoning in Intelligent Systems: Networks of Plausible Inference. Morgan Kaufmann Publishers Inc., San Francisco (1988)

12. Štajduhar, I., Dalbelo-Bašić, B.: Learning Bayesian networks from survival data using weighting censored instances. J. Biomed. Inform. **43**(4), 613–622 (2010)
13. Štajduhar, I., Dalbelo-Bašić, B.: Uncensoring censored data for machine learning: a likelihood-based approach. Expert Syst. Appl. **39**(8), 7226–7234 (2012)
14. Zhang, Z., Reinikainen, J., Adeleke, K.A., Pieterse, M.E., Groothuis-Oudshoorn, C.G.: Time-varying covariates and coefficients in Cox regression models. Ann. Transl. Med. **6**(7), 121 (2018)

Pursuing Optimal Prediction of Discharge Time in ICUs with Machine Learning Methods

David Cuadrado[1], David Riaño[1(✉)], Josep Gómez[1,2], María Bodí[2], Gonzalo Sirgo[2], Federico Esteban[2], Rafael García[2], and Alejandro Rodríguez[2]

[1] Universitat Rovira i Virgili, Tarragona, Spain
david.riano@urv.cat
[2] Intensive Care Unit, University Hospital Joan XXIII, Tarragona, Spain
mbodi.hj23.ics@gencat.cat

Abstract. In hospital intensive care units (ICU), patients are under continuous evaluation. One of the purposes of this evaluation is to determine the expected number of days to discharge. This value is important to manage ICUs. Some studies show that health care professionals are good at predicting short-term discharge times, but not as good at long-term predictions. Machine learning methods can achieve 1.79-day average prediction error. We performed a study on 3,787 patient-days in the ICU of the Hospital Joan XXIII (Spain) to obtain a data-driven model to predict the discharge time of ICU patients, in a daily basis. Our model, which is based on random forest technology, obtained an error of 1.34 days. We studied the progression of the model as more data are available and predicted that the number of instances required to reduce the error below one day is 4,745. When we trained the model with all the available data, we obtained a mean error of less than half a day with a coefficient of determination (R2) above 97% in their predictions on either ICU survivors and not survivors. Similar results were obtained differentiating by patients' gender and age, confirming our approach as a good means to achieve optimal performance when more data will be available.

Keywords: Intelligent data analysis · Intensive care units · Discharge time prediction · Data-driven models

1 Introduction

Intensive Care Units (ICUs) are hospital services with an intense and complex activity, where heterogeneous critical patients are admitted to receive continuous care from an organized team of health care professionals (mainly physicians and nurses) till the patients' clinical condition and parameters reach an acceptable state for them to be moved to other hospital services or discharged.

The proper functioning of an ICU depends on the quality of care and the correct management and planning of resources. In order to face these challenges, a series of descriptive parameters and indicators of the ICUs are defined, among which there are the length of stay (LOS) and the time to discharge (TTD). While LOS is a static

D. Riaño et al. (Eds.): AIME 2019, LNAI 11526, pp. 150–154, 2019.
https://doi.org/10.1007/978-3-030-21642-9_20

indication of the time that a patient remains in the ICU between the admission and the discharge, TTD is dynamic and it informs about the time a patient will remain in the ICU from a given moment.

According to these definitions, LOS prediction is based on the patient's condition at the time of admission (or in the first 24 h after admission), it is useful during admission, but it loses interest as the patient spends more days in the ICU and his/her clinical condition evolves. On the contrary, the prediction of TTD takes into account the patient's condition (and optionally its evolution) at any moment when the prediction is required. In this sense, TTD can be calculated at any time (e.g., in a daily routine) and adapts to the patient's evolution over time.

The task of correctly predicting the LOS, or the TTD, is difficult for physicians [1], particularly for predictions of five or more days [2], but even one-day predictions are not free of errors [3].

For many years, statistical and machine learning methods have been recurrently applied to predict LOS. Only one study on TTD prediction was found with good results for short term TTD [4], but less effective for long term TTD. For LOS, datasets on ICU patients have been used to produce predictive models [5] of various types and qualities. Four of the most used methods are artificial neural networks [6–8], regression [9–11], random forest [11, 12], and support vector machines [11, 13]. Some comparative studies showed that random forest outperforms artificial neural networks and support vector machines [12, 13], while regression models are not always recommended [9]. Random forest is, therefore, recommended if we want to predict TTD.

The quality of these predictive models uses to be measured in terms of the mean average error (MAE), the root mean square error (RMSE), or the coefficient of determination (R^2) between the time of the real stay and the predicted value. Our bibliographic search found that some of the best current predictors are [12] with $R^2 = 0.81$ and [11] with MAE = 1.79. Errors of 1–2 days are not always acceptable. This introduces the challenge of producing models with prediction errors of less than one day. Moreover, these models should be robust at the time of predicting stays of both ICU patients who survive and those who do not survive. Optimally, they should also be stable with regard to other parameters such as the gender and the age of the patient.

In this paper, we present the result of a process of searching a robust and precise data-driven model to predict TTD in ICUs. In particular, we propose the use of random forests. When applied on our data, we obtained an MAE below 1.34 days, after a 10-fold cross validation. We also studied the progression of error reduction as new data is added and we concluded that, by the end of 2019, the UCI of the hospital will have treated so many patients as to be able to obtain models with prediction errors of less than one day. To analyze the potential of our approach, we also trained a model with all the available data. This model showed MAE and RMSE values below one day and R^2 above 97%, and it was robust when considering surviving and not surviving patients, gender, and age group. These results do not only represent a significant improvement with respect to previous predictive models, but they are also an important step towards the construction of optimal models for TTD prediction and their incorporation in ICUs.

2 Methods

The University Hospital Joan XXIII is a tertiary hospital in Tarragona (Spain). In the ICU of this hospital, approximately 900 patients are attended every year. For these patients, data about monitoring systems, laboratory results, diagnoses by the image information, assessments and scales (APACHE 2, NAS, SOFA, etc.), adverse events, administrative data, and treatments are electronically registered in the hospital records through a Clinical Information System (CIS). In the ICU, the computer tool ICU-DaMa [14] allows the extraction of data from the CIS and the definition of clinical indicators that are used by physicians and UCI managers to provide a better service.

In our retrospective study, we used ICU-DaMa to extract all the patients admitted in the UCI between January 2016 and November 2017. We removed patients who stayed more than 14 days in the ICU (i.e., outliers). In order to predict TTD rather than LOS, each day of stay of each patient was considered a clinical case. The data was anonymized. From the CIS, we considered 49 clinically relevant variables related to demographic (e.g., age, gender, origin, etc.), laboratory (e.g., creatinine, max-min glycaemia values, platelets, etc.), clinical (e.g., O_2 therapy, primary diagnosis, heat rate, etc.), and pharmacological (e.g., sedatives and analgesics, vasopressors, insulin, antibiotics, etc.) information. For continuous values (e.g., heart rate) the mean value within each day was taken. The data also included daily values of fifteen scales such as APACHE II, CaM-ICU, CHE, EMINA, GCS, NAS, six specific SOFA values, and others.

This dataset was used to learn a TTD predictive model with the random forest machine learning method. A 10-fold cross validation approach was followed. The quality of the approach was measured in terms of the MAE, RMSE, and R^2. Later on, we studied the progression of the RMSE as new training data are added to the learning algorithm. For this purpose, 10% of the available data was reserved for testing, and 10%, 20%, …, 100% of the remaining data progressively used to train the model. The process was repeated 10 times with different training and testing sets, using random selection of the data. The average MAE for each percentage group was calculated. With the help of linear extrapolation, we calculated the expected amount of data required by our method to construct a model that predicts TTD with a RMSE below one day.

To complement the study and in order to determine the quality of the approach, we trained a random forest with all the available data and obtained the MAE, RMSE, and R^2 values, when tested on all the data in 2016–17.

3 Results

The database contained 3,787 cases concerning 62% of males and an average (±std. deviation) age of 61.4 (±17.2) years, arriving at the ICU with cardiovascular problems (15%) or postoperative (13%), from the emergency department (54%) or after surgery (20%), with admission SOFA value 3.74 (±3.26), who stayed 4.7 (±3.04) days in the ICU. 90.51% of them were discharged alive. See Fig. 1(a) for a detailed distribution of the patients across the TTD values. Most of the cases survived, and 85% of them have a TTD between 2 and 7 days. When 10-fold cross validation was applied, we obtained the values MAE = 1.34, RMSE = 1.73 and R^2 = 0.61. These results are significantly better than previous published results [6–12].

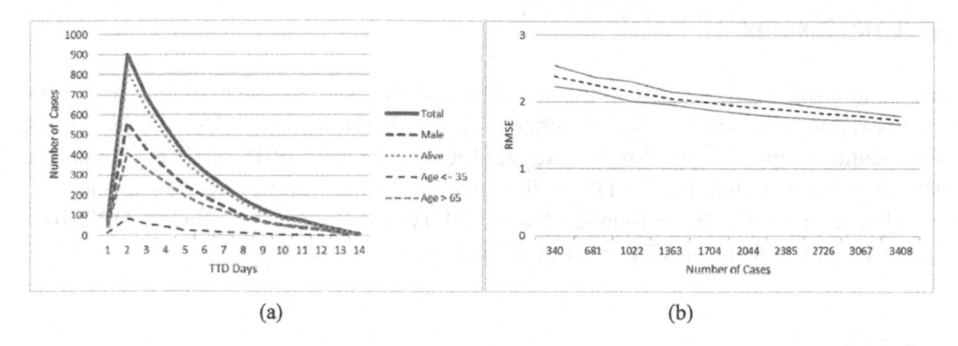

(a) (b)

Fig. 1. (a) Prevalence of cases for different TTDs and (b) Progression of RMSE mean (dotted line) and RMSE mean ± st.dev (straight lines).

Figure 1(b) shows the progression of RMSE as new ICU data is considered to train the TTD predicting model. When we extend the curve with a linear extrapolation, we obtain that, with 4,745 data (to be achieved by 2019), a model could be made with an RMSE below one day. Also, the model generated with the 3,787 data available showed an average MAE below 0.5 days for all the data, regardless the gender and the age. Only patients that did not survive had a mean value slightly higher. See Table 1. RMSE showed a similar behavior. R^2 is 97% or above, except for younger patients.

Table 1. MEA, RMSE, and R^2 values for DT prediction within the groups studied.

	All	Age <=35	Age 35–65	Age >65	Male	Female	Alive	Dead
MEA	0.4802	0.4821	0.4760	0.4837	0.4799	0.4806	0.4726	0.5508
RMSE	0.6204	0.6944	0.6900	0.6955	0.6928	0.6933	0.6875	0.7422
R^2	0.9723	0.9632	0.9724	0.9732	0.9722	0.9725	0.9733	0.9689

When we studied the MAE and RMSE values for the different TTDs (Fig. 2), we observed a homogeneous shape for all the groups of patients, except for young patients, whose prediction errors were lower. In general, the model predicts below one-day error for all the stays shorter than 10–11 days and below 2 days for stays of 11–13 days.

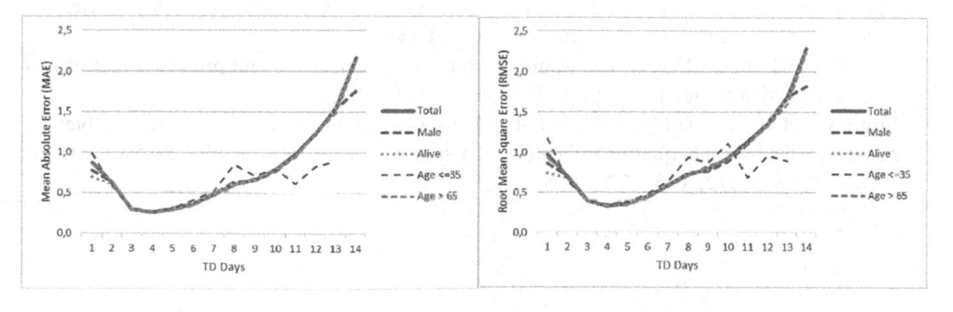

Fig. 2. TTD predictive errors for patients discharged in 1–14 days.

4 Conclusions

Anticipating the time of discharge of patients in ICUs is of great importance. The incorporation of a reliable and accurate T2D predictor into the ICU routine could significantly help improve clinical care and the management of ICUs. We proposed a data-driven approach to build TTD predictive models with average error below one day.

This research has been founded by the RETOS P-BreasTreat project (DPI2016-77415-R) of the Spanish Ministerio de Economia y Competitividad.

References

1. Nassar, A.P., Caruso, P.: ICU physicians are unable to accurately predict length of stay at admission: a prospective study. Int. J. Qual. Heal. care **28**(1), 99–103 (2016)
2. Vicente, F.G., et al.: Can the experienced ICU physician predict ICU length of stay and outcome better than less experienced colleagues? Int. Care Med. **30**(4), 655–659 (2004)
3. van Walraven, C., Forster, A.J.: The TEND (Tomorrow's Expected Number of Discharges) model accurately predicted the number of patients who were discharged from the hospital the next day. J. Hosp. Med. **13**(3), 158–163 (2018)
4. Temple, M.W., Lehnmann, C.U., Fabbri, D.: Predicting discharge dates from the NICU using progress note data. Pediatrics **136**(2), e395-405 (2015)
5. Awad, A., Bader-El-Den, M., McNicholas, J.: Patient length of stay and mortality prediction: a survey. Heal. Serv. Manag. Res. **30**(2), 105–120 (2017)
6. Gholipour, C., et al.: Using an Artificial Neural Networks (ANNs) model for prediction of Intensive Care Unit (ICU) outcome and length of stay at Hospital in Traumatic Patients. J. Clin. Diagn. Res. **9**(4), OC19-23 (2015)
7. Rowan, M., Ryan, T., Hegarty, F., O'Hare, N.: The use of artificial neural networks to stratify the length of stay of cardiac patients based on preoperative and initial postoperative factors. Artif. Intell. Med. **40**(3), 211–221 (2007)
8. LaFaro, R.J., et al.: Neural network prediction of ICU length of stay following cardiac surgery based on pre-incision variables. PLoS ONE **10**(12), e0145395 (2015)
9. Verburg, I.W.M., et al.: Comparison of regression methods for modeling intensive care length of stay. PLoS ONE **9**(10), e109684 (2014)
10. Van Houdenhoven, M., et al.: Optimizing intensive care capacity using individual length-of-stay prediction models. Crit. Care **11**(2), R42 (2007)
11. Houthooft, R., et al.: Predictive modelling of survival and length of stay in critically ill patients using sequential organ failure scores. Artif. Intell. Med. **63**(3), 191–207 (2015)
12. Caetano, N., Laureano, R.M.S., Cortez, P.: A data-driven approach to predict hospital length of stay - a Portuguese case study. In: Proceedings of the 16th International Conference on Enterprise Information Systems, pp. 407–414 (2014)
13. Hachesu, P.R., et al.: Use of data mining techniques to determine and predict length of stay of cardiac patients. Healthc. Inform. Res. **19**(2), 121–129 (2013)
14. Sirgo, G., et al.: Validation of the ICU-DaMa tool for automatically extracting variables for minimum dataset and quality indicators: the importance of data quality assessment. Int. J. Med. Inform. **112**, 166–172 (2018)

Towards the Economic Evaluation of Two Mini-invasive Surgical Techniques for Head&Neck Cancer: A Customizable Model for Different Populations

Elisa Salvi[1(✉)], Enea Parimbelli[2], Lucia Sacchi[1], Silvana Quaglini[1], Erika Maggi[1], Lorry Duchoud[3], Gian Luca Armas[3], John De Almeida[4], and Christian Simon[3]

[1] University of Pavia, Pavia, Italy
elisa.salvi01@universitadipavia.it
[2] University of Ottawa, Ottawa, ON, Canada
[3] Service d'Oto-rhino-laryngologie - Chirurgie cervico-faciale, Centre Hospitalier Universitaire Vaudois (CHUV), Université de Lausanne (UNIL), Lausanne, Switzerland
[4] Otolaryngology-Head and Neck Surgery-Surgical Oncology, Princess Margaret Cancer Centre, University of Toronto, Toronto, Canada

Abstract. Sometimes representing a decision process involves building multi-stage models, where some outputs of $stage_i$ become the inputs of $stage_{i+1}$. This was the case for an economic evaluation study of two mini-invasive techniques for head&neck cancer, namely Transoral Robotic Surgery and Transoral Laser Microsurgery. We built a two-stage model, composed by a first decision tree accounting for the surgical complications and need for additional treatments, which in turn are used as initial conditions for a second decision tree that models long-term outcomes through a Markov process. To allow the automatic concatenation of decision trees, we developed a Java extension to the TreeAgePro software, a well-known tool for decision analysis. Moreover, we integrated the resulting model into UceWeb, a framework we developed over the last two years, which allows personalizing costs and preferences for different target populations. In this way the same model may be re-used to perform cost/utility and cost/effectiveness analysis in different settings.

Keywords: Decision analysis · Economic evaluation · Head&neck cancer

1 Introduction

Oro-pharyngeal carcinomas represent a significant health burden with 400,000 new cases a year in the world. Three types of treatment can be offered to patients, namely surgery, radiotherapy (RT) and chemotherapy (CT). Some of them may be combined to improve treatment efficacy. For example, radiotherapy or chemo-radiation may be added after the surgery, and in this case they are referred to as "adjuvant". In order to improve functional recovery after surgery, new, mini-invasive, techniques have been

D. Riaño et al. (Eds.): AIME 2019, LNAI 11526, pp. 155–159, 2019.
https://doi.org/10.1007/978-3-030-21642-9_21

developed. They consist of fully endoscopic approaches to the tumor through the mouth, avoiding access through the neck and thus reducing access related morbidity. The two major techniques practiced nowadays are Trans-Oral Robotic Surgery (TORS) and Trans-oral Laser Microsurgery (TLM). In both strategies, a retractor is used to open the mouth to create room for intervention. In TORS, the surgeon uses a surgical robot to view and access structures in the pharynx, while in TLM he uses a microscope and a laser. A limitation of TLM is that the visual field of the microscope is quite small, so the tumor has to be removed in pieces. This may affect surgical margins and trigger further adjuvant treatments. However, the precision of TLM is exceptionally high, since the resection is done under high magnification. TORS on the contrary allows a resection in one piece, leading to more precise margins, but less precise dissection [1]. Also, costs are different, with TORS showing a higher economic burden for the intervention. Since the two techniques show pros and cons, with different outcomes and different costs, a decision analytic approach is reasonable to compare them.

2 Methods

2.1 Decision Trees and Markov Models

For performing decision analysis, we used the decision tree formalism. A decision tree is a graphical model including decision nodes, i.e. the options to be compared, followed by the consequences of those options, represented as a series of probabilistic nodes. Each path of the tree ends with a so-called "value node", which is valued with one or more payoffs, for example survival, quality-adjusted survival, and costs. Solving the tree provides the option that maximizes/minimizes a given payoff. In economical evaluations, solving the tree allows to find incremental cost-effectiveness and cost-utility ratios, which may inform decision makers about the convenience of adopting a new intervention. A decision tree may embed a Markov process [2] that represents patients through their transitions among mutually exclusive health states, each one lasting a fixed length of time (Markov cycle). Each transition is given a 'transition probability'. Each health state is then assigned a quality of life coefficient (utility coefficient) and costs for the resources used to manage it. By running the model over a number of cycles, life-years, quality-adjusted life years and costs associated with the different options can be estimated. As a computational tool, we employed TreeAgePro, which is a well-known commercial tool for representing and running decision trees.

2.2 Concatenation of Decision Trees

The TreeAgePro standard user interface only allows running one tree at a time. However, the tool provides the Object Interface API that allows to access and modify tree components. The extension developed in this work allows to automatically propagate values from a tree to another tree. To this aim, we developed a Java application exploiting the UceWeb repository of decision trees. UceWeb [3] is a framework we developed in the last years to elicit utility coefficients, use them to quantify decision trees developed with TreeAgePro, run them, and present analysis results. In particular, the

UceWeb repository contains a description of all the decision trees that we developed, and all the JSON (Javascript Object Notation) files with the mapping information for those decision trees that need to be concatenated. Once the user runs a decision tree (e.g., DT1) from UceWeb, the application checks if the JSON file reports other trees that are concatenated with DT1. In that case, it runs the trees in the proper sequence to obtain and present the final result.

3 The Model

Starting from a model already published by one of the authors [4] (comparing TORS with chemo-radiotherapy), additional literature [5], experts' opinions, and data provided by the Lausanne University Hospital CHUV, we built and quantified a two-stage model. As a matter of fact, the clinical path of patients may be separated in two conceptually different phases: the short/mid-term consequences of the surgery, and the long-term remission-relapse process. The first-stage model is summarized in Fig. 1, and accounts for the events that may happen during the first months after surgery. More precisely, for the two interventions under comparison, the first variable to consider is the need for adjuvant therapy (Fig. 1a). The second one is represented by complications (Fig. 1b) that are related both to the intervention and to possible adjuvant treatment. The third one (Fig. 1c) is the possible need for re-interventions, which again may lead to complications (with higher probability with respect to the first intervention).

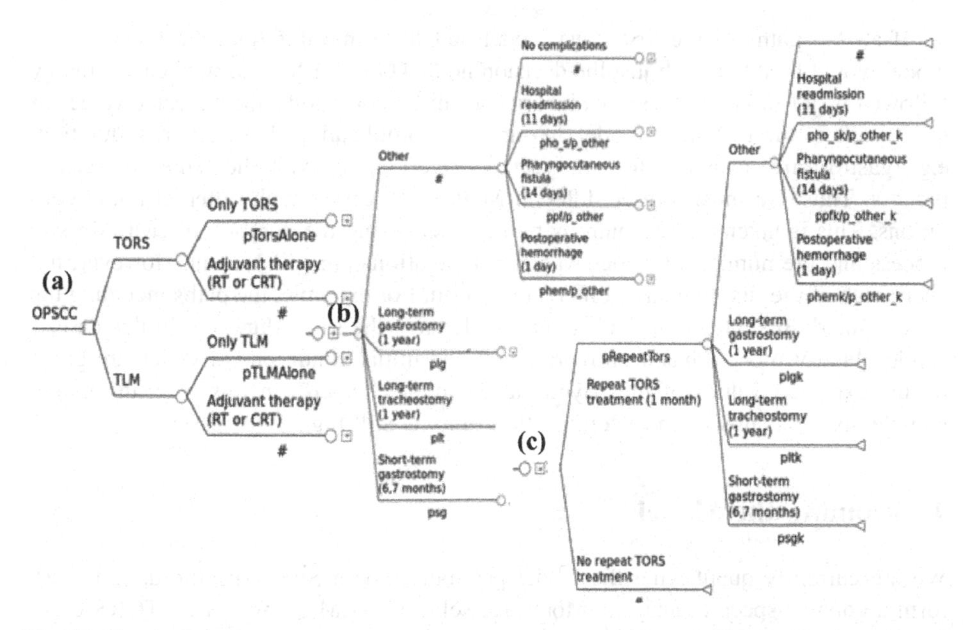

Fig. 1. Sketch of the first-stage model. Each path in tree (a) continues with subtree (b), and each path in subtree (b) continues with paths in subtree (c). Variables under the branches indicate the event probability and the symbol # indicates the complement to 1. The + symbol after a node indicates that the branches originating from that node are not shown in this figure.

The second-stage model must embed the Markov process that starts at the end of each first model path, depicting the patient's transition among different health states, as shown in Fig. 2. For the study, we chose a time horizon of 10 years and a Markov cycle of 3 months, corresponding to the typical rate of the patients' follow-up visits.

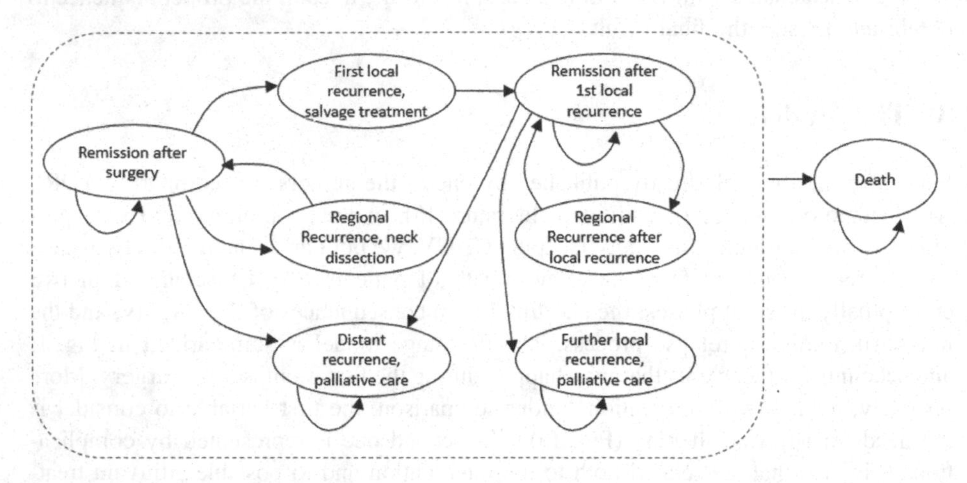

Fig. 2. Sketch of the Markov model to represent the remission-relapse process following the first treatment phase. From every state there is a transition probability to the state "Death".

If all the paths of the first-stage model had the same duration, the second-stage model would be a tree with just the decision node TORS/TLM, and with each strategy followed by a Markov process that inherits, as initial conditions, the expected values of the payoffs of the first-stage-model. In our case, complications have different duration, e.g., gastrostomy may last for one year, so overlapping with the remission-relapse process. Thus, we must foresee different Markov processes with different initial conditions. This is taken into account by properly assessing the duration of each Markov process and the number of cycles where complications are still present. However, not every path of the first tree has a different duration. For example, the paths including the node "Other" in subtrees (b) and (c) in Fig. 1, have the same duration. In this case, a single Markov process has been represented, the initial conditions of which are given by the expected values of the payoffs at the node "Other". In this way, the model complexity, and subsequently computation time, is still highly decreased.

4 Running the Model

We are currently quantifying the model parameters with Swiss-specific data, to perform a country-specific analysis. More precisely, 45 TLM cases and 62 TORS cases have been collected and analyzed to estimate the model parameters. Since we still lack some data about costs and quality of life (see next section), we ran the model using literature estimates for some of those parameters. Thus, our results are not final. On the other hand, the main purpose of this paper was to describe the technical solutions for tree concatenation and tree quantification with different populations. That said, we report some figures to provide the preliminary results of our analysis. Using the current parameters, the expected survival is almost the same for the two strategies, being 110.30 months for TORS and 110.25 for TLM. After adjusting for quality of life, the difference is still low, but in favor of TLM, being 92.79 quality-adjusted life months for TORS and 92.91 for TLM. Costs have been estimated as 45547.96 CHF for TORS and 32484.25 for TLM.

5 Conclusion and Future Developments

The work presented in this paper has two values. From the medical standpoint, despite mini-invasive techniques have been proven to be cost-effective in head&neck cancer, in our knowledge there is no literature yet comparing TORS and TLM. From the technical standpoint, motivated by the two-stage nature of our model, we developed a generic approach allowing to automatically concatenate multiple decision trees. The ex-ample in Sect. 3 shows only one of the potential advantages that tree concatenation provides. In fact, our application may be used to (i) decrease a tree complexity avoiding duplication of identical subtrees, (ii) represent separate models for conceptually separated phases of clinical histories, facilitating model development and interpretability, and (iii) re-use the same model, when it may be used in conjunction with more than one possible model for subsequent stages. Since our models run within UceWeb, which also manages a utility coefficient repository, the next steps are to run the TORS/TLM model with utilities collected from a Swiss population and analyze the differences with respect to different settings. Also costs are different in different settings, and sensitivity analyses will be used to figure out if results are more sensible to variations in costs or utilities, or both.

References

1. Cracchiolo, J.R., Roman, B.R., Kutler, D.I., Kuhel, W.I., Cohen, M.A.: Adoption of transoral robotic surgery compared with other surgical modalities for treatment of oropharyngeal squamous cell carcinoma. J. Surg. Oncol. **114**, 405–411 (2016)
2. Sonnenberg, F.A., Beck, J.R.: Markov models in medical decision making: a practical guide. Med. Decis. Mak. Int. J. Soc. Med. Decis. Mak. **13**, 322–338 (1993)
3. Parimbelli, E., Sacchi, L., Rubrichi, S., Mazzanti, A., Quaglini, S.: UceWeb: a web-based collaborative tool for collecting and sharing quality of life data. Methods Inf. Med. **54**, 156–163 (2015)
4. de Almeida, J.R., et al.: Cost-effectiveness of transoral robotic surgery versus (chemo) radiotherapy for early T classification oropharyngeal carcinoma: A cost-utility analysis. Head Neck **38**, 589–600 (2016)
5. Li, H., et al.: Clinical value of transoral robotic surgery: Nationwide results from the first 5 years of adoption. Laryngoscope (2018)

Gated Hidden Markov Models for Early Prediction of Outcome of Internet-Based Cognitive Behavioral Therapy

Negar Safinianaini[1(✉)], Henrik Boström[1(✉)], and Viktor Kaldo[2,3(✉)]

[1] School of Electrical Engineering and Computer Science, KTH Royal Institute of Technology, Stockholm, Sweden
{negars,bostromh}@kth.se

[2] Department of Psychology, Faculty of Health and Life Sciences, Linnaeus University, Växjö, Sweden

[3] Centre for Psychiatry Research, Department of Clinical Neuroscience, Karolinska Institutet, and Stockholm Health Care Services, Stockholm County Council, Stockholm, Sweden
viktor.kaldo@ki.se

Abstract. Depression is a major threat to public health and its mitigation is considered to be of utmost importance. Internet-based Cognitive Behavioral Therapy (ICBT) is one of the employed treatments for depression. However, for the approach to be effective, it is crucial that the outcome of the treatment is accurately predicted as early as possible, to allow for its adaptation to the individual patient. Hidden Markov models (HMMs) have been commonly applied to characterize systematic changes in multivariate time series within health care. However, they have limited capabilities in capturing long-range interactions between emitted symbols. For the task of analyzing ICBT data, one such long-range interaction concerns the dependence of state transition on fractional change of emitted symbols. Gated Hidden Markov Models (GHMMs) are proposed as a solution to this problem. They extend standard HMMs by modifying the Expectation Maximization algorithm; for each observation sequence, the new algorithm regulates the transition probability update based on the fractional change, as specified by domain knowledge. GHMMs are compared to standard HMMs and a recently proposed approach, Inertial Hidden Markov Models, on the task of early prediction of ICBT outcome for treating depression; the algorithms are evaluated on outcome prediction, up to 7 weeks before ICBT ends. GHMMs are shown to outperform both alternative models, with an improvement of AUC ranging from 12 to 23%. These promising results indicate that considering fractional change of the observation sequence when updating state transition probabilities may indeed have a positive effect on early prediction of ICBT outcome.

Keywords: Hidden Markov Models · Expectation Maximization · Depression · Internet-based Cognitive Behavioral Therapy

D. Riaño et al. (Eds.): AIME 2019, LNAI 11526, pp. 160–169, 2019.
https://doi.org/10.1007/978-3-030-21642-9_22

1 Introduction

Depression affects about a hundred million people worldwide and it is estimated to reach second place in the ranking of Disability Adjusted Life Years for all ages in 2020 [15]. It is vital to consider depression an issue of public health importance, thereby prompting effective treatment of the patients and minimizing the disease burden [15]. Internet-based Cognitive Behavioral Therapy (ICBT) is an effective treatment for depression [22]. Machine learning can be used to solve different computational challenges in the analysis of ICBT, such as predicting patient adherence to depression treatment [20] and outcome prediction for obsessive-compulsive disorder [11].

A goal in the analysis of treatment outcome in ICBT for depression is to perform early predictions; thus, the patients with unsuccessful treatment can early on be detected and receive better care by the therapists. However, the accuracy of the predictions are also affected by at what time they are made; there is hence a trade-off for the psychologist to decide when to perform the early prediction. For example, if waiting one extra week gives better accuracy in predicting the final outcome, it may be preferred over deciding on a treatment earlier, based on a less accurate prediction. At the same time, waiting too long means less time to step in and adjust the treatment to better suit the patient [16].

The main motivation of this work is to explore a suitable machine learning method which improves the performance of early predictions on ICBT outcome for patients suffering from depression. We focus on graphical models, as they are interpretable, often easy to customize and allow for probabilistic modeling [1,23]. In particular, they allow for incorporating prior knowledge and handling missing data without imputation, through marginalization [2]. The latter is of particular importance as there are several, often unknown, reasons for why data is missing, and imputation may often not be appropriate in healthcare applications [9,21].

ICBT involves changes in human behavior; these have stochastic properties resulting in health state transitions. In this particular study, we have categorical observations (self-rated scores established by questionnaires) and a latent variable (treatment outcome) over time. As the state transitions can be modeled as Markov chains, Hidden Markov Models (HMMs) is a natural choice. However, one limitation of HMMs is the lack of context [24], which becomes a challenge when a state transition is dependent on the fractional change (defined as the difference between two values in time divided by the first value) of the observation sequence. We propose Gated Hidden Markov Models (GHMMs) as a potential solution to the problem. GHMMs extend standard HMMs by modifying the Expectation Maximization (EM) algorithm; for each observation sequence, the new algorithm regulates the transition probability update based on the fractional change, as specified by domain knowledge.

In the next section, we provide some notation and background on HMMs. In Sect. 3, we introduce the GHMMs. In Sect. 4, we evaluate and compare this approach to standard HMMs and a recently proposed approach, Inertial Hidden Markov Models (IHMMs) [13], on the task of early prediction of the outcome of ICBT for treatment of depression. In Sect. 5, we discuss related work, and finally, in Sect. 6, we summarize the main findings and point out directions for future research.

2 Preliminaries

An HMM [2] is a statistical Markov model in which one observes a sequence of emitted symbols (observation sequence), but does not know the sequence of states the model went through to generate the observation sequence. The Markov property implies that the next state only depends on the current state. We define an HMM with N time steps, an observation sequence denoted as $X = \{x_1, \ldots, x_N\}$ containing N emitted symbols, and hidden states defined as $Z = \{z_1, \ldots, z_N\}$. An HMM has a parameter set, θ, which contains: initial probabilities, $p(z_1)$; transition probabilities, $p(z_n|z_{n-1})$; and emission probabilities, $p(x_n|z_n)$ where $n \in [1, N]$. The learning of the parameters of an HMM can be done by maximizing likelihood, using EM, which comprises two steps: the *E-step*, calculating the expected values; and the *M-step*, maximizing likelihood based on the expected values. Baum-Welch [2], shown in Algorithm 1, is an instance of EM suitable for HMMs. The E-step is done by calculating the marginal posterior distribution of a latent variable z_n, denoted as $\gamma(z_n)$, and the joint posterior distribution of two successive latent variables, $\varepsilon(z_{n-1}, z_n)$. In the M-step, θ is updated using $\gamma(z_n)$ and $\varepsilon(z_{n-1}, z_n)$. Forward and backward probabilities, $\alpha(z_n)$ and $\beta(z_n)$ [2], are used in the calculations of $\gamma(z_n)$ and $\varepsilon(z_{n-1}, z_n)$ as below. For details we refer to [2].

$$\gamma(z_n) = \frac{\alpha(z_n)\beta(z_n)}{p(X)} \qquad \varepsilon(z_{n-1}, z_n) = \frac{\alpha(z_{n-1})p(x_n|z_n)p(z_n|z_{n-1})\beta(z_n)}{p(X)} \tag{1}$$

Algorithm 1. Baum-Welch

1: **procedure** LEARN(*trainingData*):
2: Initialise θ
3:     **repeat**
4:         **for each** $X \in trainingData$ **do**
5:             E-step: calculate ε, γ in Equation (1)
6:             M-step: update θ using ε, γ
7:     **until** convergence
8: **return** θ

3 Gated Hidden Markov Models

Although HMMs are quite powerful, as demonstrated by their wide variety of applications, they have limitations in capturing long-range interactions between emitted symbols in the observation sequence; e.g. *Palindrome Language* [24].

Rather than considering more powerful (and less explored) models, we will in this proposal instead consider modifying the learning algorithm, i.e. Baum-Welch, to incorporate information regarding such long-range interactions through regulating the transition probabilities. In particular, we will consider global properties of the observation sequences, and when certain conditions are met, the algorithm, in the E-step, will be forced to set certain transition probabilities to zero. The latter can be thought of as gates being closed; hence the name Gated Hidden Markov Models (GHMMs).

Our algorithm, as presented in Algorithm 2, modifies Algorithm 1 by adding lines 6 through 8. Moreover, three additional input arguments (*policy*, *threshold*, *label*) and one new local variable (*change*) are added:

- *policy*: the rule defining how to calculate the fractional change
- *change*: the fractional change within an observation sequence, X, as calculated by *policy*.
- *threshold*: the specified threshold to compare with *change* (as determined by domain knowledge)
- *label*: the hidden state of the GHMM, which the algorithm regulates.

Algorithm 2. Modified Baum-Welch

```
 1: procedure LEARN(trainingData, policy, threshold, label):
 2: Initialise θ
 3:     repeat
 4:         for each X ∈ trainingData do
 5:             Calculate γ in Equation (1)
 6:             Calculate change by X and policy
 7:      Gate { if change <threshold then
 8:                 p(z_n = label|z_{n-1}) = 0
 9:             Calculate ε in Equation (1) by applying p(z_n = label|z_{n-1}) = 0
10:             M-step: update θ using ε, γ
11:     until convergence
12: return θ
```

Conceptually, the if-clause (line 7, Algorithm 2) represents the *Gate* concept. When the transition probability is set to zero, it means that the *Gate* is closed. Whenever this occurs, the update of θ in the M-step of EM is affected. The semantics of Baum-Welch is retained because the regulation only concerns the value of a transition probability and does not change any formulas calculated in the E-step or M-step. The algorithm can be viewed as updating the transition probabilities not only based on EM, but also based on the domain knowledge. Notice that the algorithm targets cases where the state transition is dependent on the fractional change of the observation sequence. The parameters *threshold*, *policy* and *label* may be customized for other situations, with similar types of data.

The worst-case time complexity of the modified algorithm is the same as for the original Baum-Welch algorithm; the worst case scenario of calculating the fractional change requires parsing the whole length of sequence, which results in that the original complexity is multiplied with a constant.

4 Empirical Investigation

4.1 Experimental Setup

Dataset. The data, based on the depression rating scale MADRS-S (Montgomery Åsberg Depression Rating Scale) [4], contain self-score replies to treatment questionnaires filled in by 2076 patients with depression and which have been assessed as suitable to, and willing to try, ICBT. The project, in which the data has been collected, has been approved by the regional ethical board in Stockholm (ref. no. 2011/2091-31/3, 2016/21-32 and 2017/2320-32).

The data points consist of ordinal values, ranging from 0 to 6, reflecting the severity of the mental state, as assessed by the patients themselves. The highest score represents the worst mental situation a patient can experience. The data is for each patient collected over 13 weeks, where for each week, the patient is requested to answer the same set of nine standardized questions. The data for the first week, week 0, is based on screening, before introducing the patient to ICBT, which contains the same questions. Week 0 is used when there is missing data regarding week 1 (week 1 corresponds to the pre-measurement week in [16]). Only patients that answered the questionnaires for the final week are included in the dataset (required for supervised learning). Let $q_i w_j$ denote the answer (a score from 0 to 6) to question i at week j. The observation sequence of $q_i w_j$s for each patient is assumed to be a merge of time-based (e.g. the step from $q_9 w_0$ to $q_1 w_1$) and event-based steps (e.g. the step from $q_1 w_0$ to $q_2 w_0$). For increasing the sequence size, which improves the learning of HMMs, we consider each step as a generic step in an HMM (regardless of whether it is event-based or time-based). The final observation sequence then becomes: $q_1 w_0, q_2 w_0, .., q_9 w_0, q_1 w_1, q_2 w_1, .., q_9 w_1, ..., q_1 w_{12}, q_2 w_{12}, .., q_9 w_{12}$.

The labels representing treatment outcome are "success" and "failure". Below, we show the rule concerning the class "success" based on clinical expertise [4] using the data from week 1 and week 12; the "failure" class does not satisfy the rule. The left inequality in Eq. 2 concerns the fractional change—called symptom reduction—being compared to the threshold of 50%; the right inequality in Eq. 2 defines the cut-off for the healthy score at the end of the treatment:

$$\frac{\sum_{i=1}^{9} q_i w_1 - \sum_{i=1}^{9} q_i w_{12}}{\sum_{i=1}^{9} q_i w_1} >= 0.50 \quad \vee \quad \sum_{i=1}^{9} q_i w_{12} <= 10 \tag{2}$$

The average symptom reduction over time and the frequency of the missing scores are shown in Fig. 1.

Experimental Protocol. This section explains the technical configurations of our experiments. For each of the considered algorithms, the same underlying structure is considered, consisting of the observation sequence and two hidden states, each corresponding to one of the two possible labels ("success" and "failure"). Here, we assume to know which hidden state corresponds to "success". For learning of the HMM parameters, we make the last latent variable observable by assigning the label to it, for each sequence X; we incorporate these changes into Baum-Welch.

For handling missing observations, marginalization is used based on [14]. We set the initial emission probabilities inspired by the prior knowledge; a patient's score for the "success" class has higher probabilities for the lower scores while for the "failure" class has higher probabilities for the higher scores.

The input parameters of Algorithm 2 are set to meet the requirements of the specific application of depression treatment using ICBT. The parameter *change* is set to be fractional change as defined in Eq. (2); *threshold* is set to 0.50 as in Eq. (2), and finally, *label* is set to "success". This means that if a patient's fractional change is less than 0.50, the transition probability of the outcome becoming "success" is set to zero (it is known which state transition probability to set to zero since the hidden state corresponding to "success" is known). The *Gate* here disallows EM to independently decide over the probability of treatment success if patients have insufficient fractional change, symptom reductions, according to the psychological measures. By this we have a hypothesis of reducing false negatives—the patients incorrectly predicted to belong to the "success" class—, which is critical for detecting that treatment is not successful.

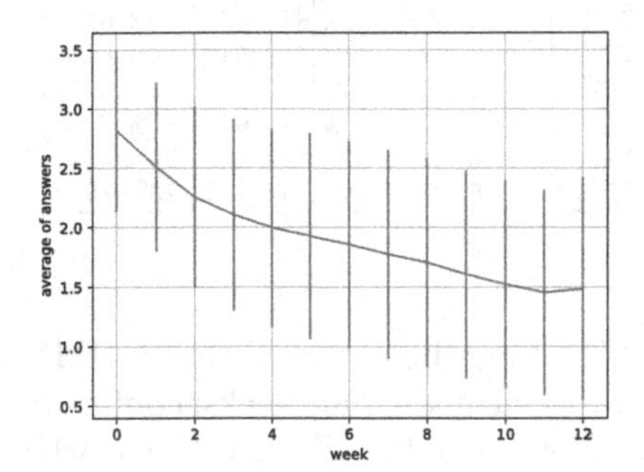

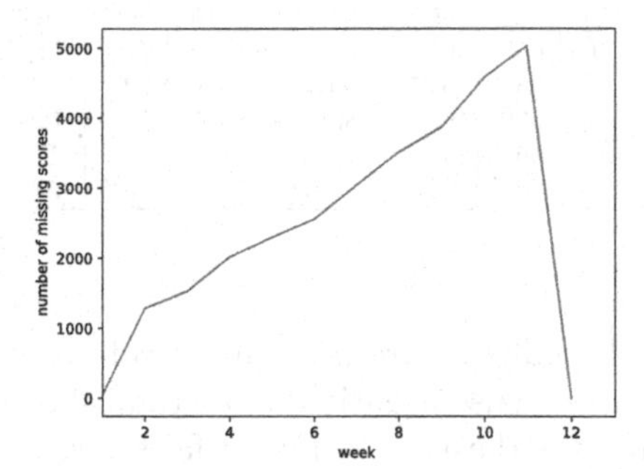

Fig. 1. On the left side, the average score for all patients through 13 weeks is shown, presenting the trend of symptom reduction. The vertical bars represent the standard deviation. To the right, the trend of missing scores is illustrated.

We compare the novel algorithm with HMMs and IHMMs. The latter regularizes the update of the transition matrix so that self-transitions, i.e., transitions to the same state as previous, have a higher probability than non-self-transitions. It is relevant to compare our algorithm with IHMMs since they satisfy the slow state transition property concerning a patient's behavior. We perform the comparison on a separate test dataset and for each early prediction, data corresponding to the later weeks is withheld. The IHMM is trained with a set of values for the regularization parameter and the value resulting in the highest AUC in the validation set is chosen to be the regularization value. AUC, accuracy, precision and recall are used to evaluate the performance of the algorithms.

For the implementation of GHMMs, we refer to GHMMs.

4.2 Experimental Results

In Table 1, results are presented for GHMMs, HMMs and IHMMs regarding AUC, accuracy, precision and recall. The comparison is done for different early predictions with the earliest prediction taking place at week 5. This week, which corresponds to week 4 in [16], has shown to be the best week for measuring early change for ICBT [16]. GHMMs outperform HMMs and IHMMs with respect to AUC by between 12 to 23% and with respect to accuracy, with a probability threshold of 0.5, by between 2 to 8%. In Fig. 2, the performance comparison with respect to AUC is plotted. Evidently, GHMMs outperform the other models regarding all predictions which are up to 7 weeks before the final week.

Table 1. AUC, accuracy, precision and recall are compared for early predictions among three algorithms: HMMs; IHMMs, GHMMs.

% AUC

Week	HMM	IHMM	GHMM
12	67%	68%	91%
11	65%	65%	85%
10	65%	65%	85%
9	66%	67%	83%
8	63%	64%	80%
7	63%	63%	80%
6	66%	66%	78%
5	64%	64%	77%

% Accuracy (threshold 0.5)

Week	HMM	IHMM	GHMM
12	77%	77%	79%
11	69%	69%	77%
10	70%	70%	77%
9	71%	71%	77%
8	70%	70%	76%
7	72%	72%	74%
6	71%	71%	74%
5	70%	70%	73%

% Precision % Recall (threshold 0.5)

Week	HMM	IHMM	GHMM
12	99% 56%	99% 56%	98% 60%
11	90% 44%	90% 45%	90% 60%
10	88% 47%	88% 47%	88% 62%
9	84% 53%	84% 53%	86% 66%
8	78% 56%	79% 56%	81% 69%
7	78% 65%	78% 65%	77% 71%
6	75% 65%	75% 65%	76% 73%
5	73% 68%	73% 68%	73% 75%

Looking at precision and recall, in Table 1, it can be observed that GHMMs decrease false negatives more than the other algorithms for all weeks; confirming our hypothesis of reducing false negatives. Note that for all algorithms, when using probability threshold 0.50, precision gets higher but recall gets lower for later weeks; as shown in Fig. 2, however, week 12 dominates week 5, hence choosing a different threshold can lead to higher values for both precision and recall at later predictions.

5 Related Work

In medical applications, Markov models have been used to capture disease patterns regarding discrete mutually exclusive health states and the transitions between them over time. Markov models are useful in particular when the pattern involves clinical changes across the states; one clinical example being the

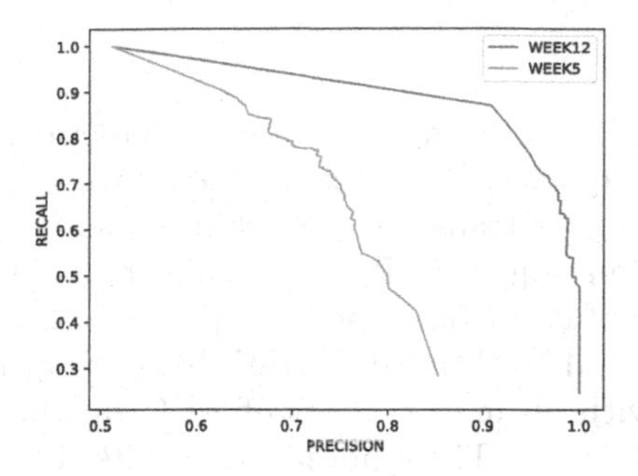

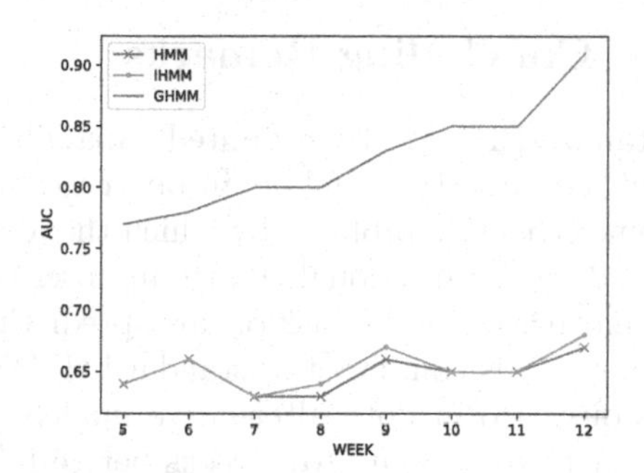

Fig. 2. On the left side, latest and earliest predictions by GHMMs, are compared concerning precision and recall. To the right, AUC of GHMMs, HMMs and IHMMs is compared for different early predictions.

progression of Alzheimer's Disease (AD) over time [6]. Shirley *et al.* [17] apply HMMs in alcoholism treatment analysis, by which different drinking behaviors are recognized. Assessment of preterm babies' health is another application of HMMs where the measurements are linked to state of health [10]. Capturing the quality of healthcare has been studied using HMMs for geriatric patient data by modelling quality as hidden states [12]. The clinical state of patients have also been estimated using infinite-HMM (an HMM with an unknown number of latent variables) [7].

Similar ideas to what have been proposed here, have also been used in integrating domain knowledge into machine learning; e.g. [3, 8], where domain knowledge is applied in form of a framework or new components in the learning model. In contrast, GHMMs do not add any extra components to the model, as these may be expensive and complex. GHMMs instead apply domain knowledge through modifying the learning algorithm. Fung *et al.* [5] improve a binary classifier by incorporating two linear inequalities—so called knowledge sets—, corresponding to the classes, into the error minimization term of the classifier. We similarly use the linear inequalities between fractional change and the defined threshold as a constraint bundled in the optimization algorithm, EM.

Concerning context-sensitive HMMs for handling long-ranged interactions between symbols, in [24] an approach is proposed which stores symbols, emitted at certain states, in an auxiliary memory; the stored data serves as the context that affects the emission and the transition probabilities of the model. GHMMs also considers a symbol-based context, although without introducing extra components in HMMs. The early detection of neonatal sepsis has been studied using Autoregressive HMMs [18]; this work tackles HMMs' context limitation by introducing direct dependencies only between consecutive symbols. Similarly, GHMMs consider symbols dependencies but in a longer range.

6 Concluding Remarks

Standard HMMs have limited capabilities in capturing long-range interactions between emitted symbols in observation sequence. We introduce GHMMs as a remedy to this problem by which the learning of transition probabilities is regulated by the fractional change in observation sequence. This particular problem is motivated by the task of early prediction of ICBT outcome for depression. The approach is compared to standard HMMs and IHMMs, and GHMMs are shown to outperform both alternative models, with an improvement of AUC ranging from 12 to 23%, up to 7 weeks before ICBT ends. These promising results show that considering fractional change of observation sequence when updating state transition probabilities may have a positive effect on early prediction of ICBT outcome. These results, obtained through a collaboration project led by the Internet Psychiatry Clinic in Stockholm, indicate that GHMM may be a potentially effective tool in practice to improve predictions regarding ICBT [19].

The proposed approach can be applied and further tested in contexts of other psychological disorders and similar data types where the fractional change of an observation sequence should be allowed to affect state transitions. This work opens up for several different research paths, as there are still room for improvement, such as incorporating other forms of domain knowledge, considering additional data types, modelling missing values in the graphical model and combining the GHMMs with other machine learning and time series methods. Finally, regarding GHMMs, directions for future research include investigating soft thresholds and more complex gate mechanisms as well as techniques to avoid over-training of GHMMs.

References

1. Belgrave, D.: Machine learning for personalized health (ICML 2018). https://mlhealthtutorial.files.wordpress.com/2018/07/tutorial-ml-for-health1.pdf. Accessed 14 Aug 2018
2. Bishop, C.: Pattern Recognition and Machine Learning. Information Science and Statistics. Springer, New York (2006)
3. Constantinou, A., Fenton, N., Neil, M.: Integrating expert knowledge with data in Bayesian networks: preserving data-driven expectations when the expert variables remain unobserved. Expert. Syst. Appl. **56**, 197–208 (2016)
4. Fantino, B., Moore, N.: The self-reported montgomery-asberg depression rating scale is a useful evaluative tool in major depressive disorder. BMC Psychiatry **9**(1), 26 (2009)
5. Fung, G., Mangasarian, O., Shavlik, J.: Knowledge-based support vector machine classifiers. In: NIPS (2003)
6. Green, C.: Modelling disease progression in Alzheimer's disease. Pharm. Econ. **25**(9), 735–750 (2007)
7. Hoiles, W., Van Der Schaar, M.: A non-parametric learning method for confidently estimating patient's clinical state and dynamics. In: NIPS (2016)
8. Kuusisto, F., Dutra, I., Elezaby, M., Mendonca, E., Shavlik, J., Burnside, E.: Leveraging expert knowledge to improve machine-learned decision support systems. AMIA Jt. Summits Transl. Sci. Proc., 87–91 (2015)

9. Kwak, Y., Yang, Y., Park, S.: Missing data analysis in drug-naïve Alzheimer's disease with behavioral and psychological symptoms. Yonsei Med. J. **54**(4), 825–831 (2013)
10. Lee, D., Roscoe, J., Russell, G.: Developing Hidden Markov Models for aiding the assessment of preterm babies- health. In: International Conference on Biomedical and Pharmaceutical Engineering, pp. 104–109 (2006)
11. Lenhard, F., et al.: Prediction of outcome in internet delivered cognitive behaviour therapy for paediatric obsessive compulsive disorder: a machine learning approach. Int. J. Methods Psychiatr. Res. **27**(1), e1576 (2018)
12. Mitchell, H., Marshall, A., Zenga, M.: Using the Hidden Markov Model to capture quality of care in lombardy geriatric wards. Oper. Res. Health Care **7**, 103–110 (2015)
13. Montanez, G., Amizadeh, S., Laptev, N.: Inertial Hidden Markov Models: modeling change in multivariate time series. In: AAAI Conference on Artificial Intelligence (2015)
14. Popov, A., Gultyaeva, T., Uvarov, V.: A comparison of some methods for training Hidden Markov Models on sequences with missing observations. In: 2016 11th International Forum on Strategic Technology (IFOST), pp. 431–435 (2016)
15. Reddy, M.: Depression: the disorder and the burden. Indian J. Psychol. Med. **32**(1), 1–2 (2010)
16. Schibbye, P., et al.: Using early change to predict outcome in cognitive behaviour therapy: exploring timeframe, calculation method, and differences of disorder-specific versus general measures. PLoS ONE **9**(6), e100614 (2014)
17. Shirley, K., Small, D., Lynch, K., Maisto, S., Oslin, D.: Hidden Markov Models for alcoholism treatment trial data. Ann. Appl. Stat. **4**(1), 366–395 (2010)
18. Stanculescu, I., Williams, C., Freer, Y.: Autoregressive Hidden Markov Models for the early detection of neonatal sepsis. IEEE J. Biomed. Health Inform. **18**(5), 1560–1570 (2014)
19. Titov, N., et al.: ICBT in routine care: a descriptive analysis of successful clinics in five countries. Internet Interv. **13**, 108–115 (2018)
20. Wallert, J., et al.: Predicting adherence to internet-delivered psychotherapy for symptoms of depression and anxiety after myocardial infarction: machine learning insights from the u-care heart randomized controlled trial. J. Med. Internet Res. **20**(10), e10754 (2018)
21. Wartella, E.A.: The Prevention and Treatment of Missing Data in Clinical Trials. National Academies Press, Washington, D.C. (2010)
22. Williams, A., Andrews, G.: The effectiveness of internet cognitive behavioural therapy (ICBT) for depression in primary care: a quality assurance study. PLoS ONE **8**(2), E57447 (2013)
23. Xing, Z., Pei, J., Keogh, E.: A brief survey on sequence classification. ACM SIGKDD Explor. Newsl. **12**(1), 40–48 (2010)
24. Yoon, B., Vaidyanathan, P.: Context-sensitive Hidden Markov Models for modeling long-range dependencies in symbol sequences. IEEE Trans. Signal Process. **54**(11), 4169–4184 (2006)

A Data-Driven Exploration of Hypotheses on Disease Dynamics

Marcos L. P. Bueno[1,2](✉), Arjen Hommersom[1,3], Peter J. F. Lucas[1,4], and Joost Janzing[5]

[1] iCIS, Radboud University Nijmegen, Nijmegen, The Netherlands
{mbueno,arjenh,peterl}@cs.ru.nl

[2] Department of Computer Science, Federal University of Uberlândia, Uberlândia, Brazil

[3] Faculty of Management, Science and Technology, Open University, Heerlen, The Netherlands

[4] LIACS, Leiden University, Leiden, The Netherlands

[5] Department of Psychiatry, Radboud UMC, Nijmegen, The Netherlands
Joost.Janzing@radboudumc.nl

Abstract. Unsupervised learning is often used to obtain insight into the underlying structure of medical data. In this paper, we show that unsupervised methods, in particular hidden Markov models, can go beyond this by guiding the generation of clinical outcome measures and hypotheses, which play a crucial role in medical research. The usage of the data-driven approach facilitates selecting which hypotheses to further investigate. We demonstrate this by using clinical trial data for psychotic depression treatment as a case study. The discovered latent structure and proposed outcome are shown to provide new insight into the heterogeneity of psychotic depression in terms of predictive symptoms.

Keywords: Machine learning · Psychiatry · Depression · Latent variable · Hidden Markov model · Unsupervised learning · Outcome measure

1 Introduction

Much about disease processes is unknown, as often the only available information about a disease are the patient's symptoms and signs. This results in an incomplete understanding of a medical disorder, which can be overcome by latent variable modeling. Latent variables can enhance our understanding of the problem domain by capturing unmeasured quantities (e.g. related to the underlying physiology) and their relationship to observed quantities [14], and might provide better fitted models [15]. Hence, by using latent variables, one can try to reconstruct the underlying structure of the process at hand by using observed data.

Unsupervised learning is the machine learning task that aims to generate representations of the underlying structure of the data. Applications of unsupervised

D. Riaño et al. (Eds.): AIME 2019, LNAI 11526, pp. 170–179, 2019.
https://doi.org/10.1007/978-3-030-21642-9_23

learning to medical data include, e.g., the discovery of underlying patient groups using clustering methods [7,8], which might help improve diagnosis and provide new insight into more effective treatment selection [1]. Yet, when applied to medical data, unsupervised techniques generate output that often makes experts confront themselves with questions like *what else can we do with this structure?*. We show in this paper that unsupervised learning methods, in particular hidden Markov models (HMMs) [9], can be used not only to describe the underlying structure but also to support the formulation of meaningful medical outcomes. Previous research suggested that the formulation of clinical outcomes might be guided by latent-variable models [5], with the advantage of reducing the hypothesis space to be explored by inspecting model properties. By using HMMs, we claim that one can explore hypotheses on disease dynamics by inspecting model characteristics such as transition dynamics, latent states, etc.

In order to illustrate the usage of HMMs on disease dynamics, we make use of data from a clinical trial originally designed to compare pharmacological treatments to psychotic depression (PD) [13]. PD is a severe medical condition that is associated with a high burden of disease and relatively low remission rates following pharmacological treatment [10]. Although recent research has considered PD as a homogeneous subtype of major depressive disorder [12], the possibility that this subtype itself is heterogeneous should also be considered, which would stimulate the development of subgroup adjusted prognostics and treatment modifications. In this work, we apply HMMs to one of the largest pharmacological trials of patients with PD conducted so far [13], aiming to explore potential differences in course characteristics.

The contributions of this paper are as follows. We present a procedure to guide the exploration of hypotheses on disease dynamics by means of HMMs. We then apply this methodology to yield insight into the dynamics of PD treatments by exploring clinically meaningful outcomes. The results are then assessed based on standard clinical response and remission in PD. To the best of our knowledge, this is the first effort into a more systematic approach for exploring hypotheses on disease dynamics based on probabilistic graphical models.

The remainder of this paper is organized as follows. In Sect. 2, the relevant work related to this paper is discussed. In Sect. 3, a method for exploring hypotheses on latent disease dynamics is proposed. In Sect. 4, the PD data used as case study is described. The experimental results are presented in Sect. 5. Section 6 summarizes the paper and suggests future work.

2 Related Work

Hidden Markov models have been extensively used in medicine in general, as well as in psychiatry. It is often the case that the number of latent states in HMMs is determined in advance, as researchers might be interested in a specific subset among all possible models. Previous research [4] used a two-state HMM to investigate the hypothesis that patients switch between two stable states (symptom-free versus depressed) in major depressive disorder. In the context of Alzheimer's disease, a four-state continuous-time HMM was developed to investigate the relationship between cognition and psychotic symptoms [11]. However, one might argue that by not imposing an *a priori* number of or already known latent states, a larger set of possible models is considered, which can lead to more insight into disease dynamics, at a potential cost of having an increased difficulty interpreting the models.

The typical usage of HMMs is in prediction or as a model to describe the underlying structure of the data. While prediction is self-explanatory, the underlying structure is usually be seen as a set of clusters, thus it is a more abstract and more difficult to be used representation. A much more specialized usage of latent variables lies in the development of data-driven outcome measures [5], which was based on models other than HMMs (namely, the item response theory). Such data-driven approach to generating outcomes has the advantage that latent states might provide a more natural, compact and empirically-oriented way to measure multiple relationships between symptoms and other observables.

More recently, HMMs have been applied to electronic health records (see e.g. [6]). Such datasets are often large and heterogeneous, requiring models such as HMMs for gaining relevant insights.

3 Capturing Latent Disease Dynamics

In this section we discuss models suitable for capturing latent disease dynamics and propose a data-driven method for exploring medical outcomes.

3.1 Bayesian Networks and Hidden Markov Models

Hidden Markov models are models based on latent variables that are able to cope with uncertainty and sequential phenomena, which makes HMMs suitable for many biomedical problems [4,6,11]. In HMMs, the observable variables typically interact only via the latent (or state) variable [9], which is known as the naive-structure HMM. In this work we opt for modeling the observation space as a Bayesian network (BN), thus allowing for more general representations of symptom interaction. By such modeling, more insight into the problem can be obtained by a more concise latent-state representation [2].

3.2 State Trajectories

Before we describe how to use HMMs to obtain insight into disease processes, we introduce the notation. Let us denote by S the random variable representing the latent states to be modeled, where S takes values on the set $\text{dom}(S) = \{s_1, \ldots, s_k\}$. The remaining variables $\{X_1, \ldots, X_m\}$ are observable variables with associated domains $\text{dom}(X_i)$. In medical domains, each X_i will often refer to measured data such as symptoms, lab exams, medication, etc., while the latent variable S will refer to some state of the underlying disease (e.g. a disease remitting situation). The disease process of interest is assumed discrete over the time points $\{0, \ldots, T\}$, where the value of the latent variable and the observables that hold at time t will be denoted by $S^{(t)}$ and $X_i^{(t)}$ respectively. For a discrete time interval $[t_1, t_2]$, the notation $S^{(t_1:t_2)}$ will be used.

HMMs can be used to predict the hidden states that better explain the observations [9]. Prediction is achieved by first computing the distribution of latent states at each week t conditional on the complete patient's symptom data (i.e. his or her data over all the weeks):

$$\gamma_t(s) = P(S^{(t)} = s \mid X_1^{(0:T)}, \ldots, X_m^{(0:T)}) \tag{1}$$

where $\gamma_t(s)$ is the notation used in standard forward-backward algorithms for HMMs [9]. After this has been done, the sequence of states for a given patient is obtained by selecting the most likely state at each time t:

$$\gamma_t^* = \arg\max_{s \in \text{dom}(S)} \gamma_t(s) \tag{2}$$

for all $t \in \{0, \ldots, T\}$. This can be interpreted as "placing" patients in states. Note that the predicted states are the individually most likely states, obtained by maximizing Eq. 2 for each time point independently, in contrast to the so-called Viterbi path, where the maximization is applied over the probability of state trajectories over $\{0, \ldots, T\}$.

3.3 Exploring Medical Outcomes

Understanding disease dynamics in a multi-variate setting is challenging because of potential complex interactions between diseases, symptoms, and findings. Therefore, we propose to investigate the transition dynamics between latent states. This is convenient because each latent state can take into account multiple symptom dimensions at once, which makes reasoning over patient trajectory very feasible. Once the states are discovered, a detailed outcome measure that provide insight into treatment dynamics can be formulated.

We propose a procedure to build outcome measures in Fig. 1. The procedure selects a set of *baseline states* S_b based on a selection criterion. From the remaining states, a set of *target states* S_e are to be selected based on its own criterion. Once S_b and S_e are obtained, *state reachabilities* from S_b states to S_e states are calculated. By varying the time interval between two given states of S_b and S_e, the resulting probabilities $reach(i, j, t_1, t_2)$ indicate the temporal influence of a baseline state over a target state. Such state reachabilities can then be used to compose a rich outcome measure, e.g., by making $t_1 = 0$ and $t_2 \in \{1, \ldots, T\}$, which will result in a reachability trend as indicated in Fig. 1.

In this paper, a state is classified as a baseline state if one or more patients are predicted to be in this state at the process start. These states can be computed by first determining the state trajectories $\gamma_t(s)$ for each patient (see Eq. 1). Then we define $s \in S_b$ if and only if $\gamma_0^* = s$ holds for at least one patient. To determine the target states, we look at the model parameters, such that $P(s \to s) \geq \rho$, where $0 \leq \rho \leq 1$. In particular, we chose $\rho = 0.95$, resulting in states that are not in S_b and have a high self-transition probability.

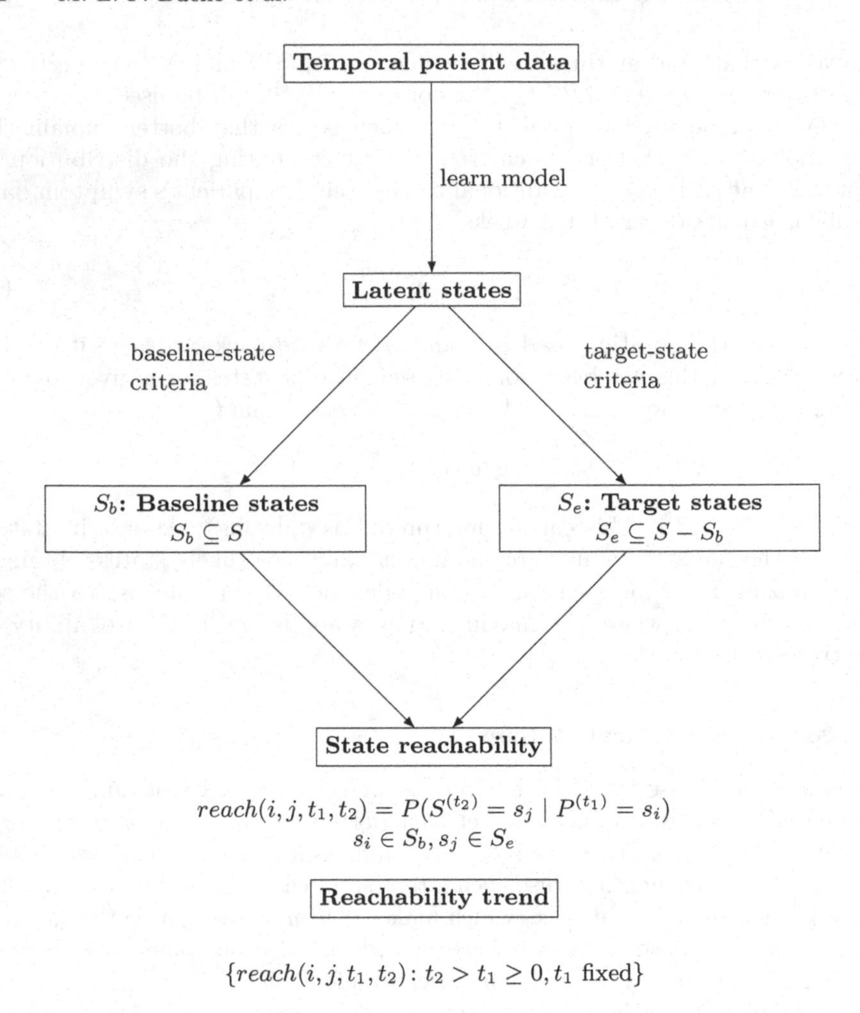

Fig. 1. Procedure to guide the generation of outcome measures based on latent states.

4 Data

4.1 Patients and Variables

All patients had participated in the DUDG (Dutch University Depression Group) study [13], a 7 week double-blind randomized clinical trial originally designed for comparing the effectiveness of venlafaxine, imipramine and venlafaxine plus quetiapine (V+Q, for brevity) in psychotic depression. The dataset originally included 122 participants aged 18–65 who met DSM-IV-TR criteria for a unipolar major depressive episode with psychotic symptoms and a 17-item Hamilton Depression Rating Scale (HAM-D [3]) score of at least 18 (both at the screening visit and at baseline). The 17-item HAM-D indicates severity of depression as follows:

normal (0–7), mild depression (8–13), moderate depression(14–18), severe depression (19–22), and very severe depression (greater than or equal to 23).

Because of insufficient information about the specific nature of psychotic symptoms, three patients were not included in the current study resulting in a dataset with 119 patients. From the total group, 59 (49,6%) were females; the mean age was 51.1 (SD 10.9) years. Forty patients were randomized to treatment with imipramine, 38 to venlafaxine and 41 to V+Q.

Severity of depression (HAM-D, continuous) and the presence of psychotic symptoms (dichotomized) were measured at baseline (i.e. before treatment starts) and weekly thereafter. A total of 98 patients completed the trial (34 in imipramine, 30 in venlafaxine, and 34 in V+Q). Data on patients who dropped out was imputed by the last-observation-carried-forward approach [13].

4.2 Depression Assessment

At the end of medical treatment, patients were assessed according to conventional criteria for response and remission of depression [13]. Response was defined as a reduction of at least 50% on the HAM-D score compared to baseline and a score of 14 or below, and remission as a score of 7 or below.

5 Experimental Results

5.1 Model of Observations

The observable variables in the HMM used in this work are modeled according to the BN shown in Fig. 2, which allows for a more expressive representation than the naive-Bayes structure by connecting Hal and Del via HAM-D. By doing so, we impose less independence assumptions than the naive solution, thus the model becomes more flexible in that more dependences can be induced from data. Hence, once in a state the observables are parameterized as follows: the psychotic symptoms are encoded as binary random variables, while the depressive symptom (the HAM-D score) is a conditional Gaussian distribution (conditioned on the state and on both psychotic symptoms, as shown in Fig. 2).

At any time point, the parameterization of each symptom is given by the factorization entailed by the BN structure of Fig. 2. This modeling dictates that HAM-D will be given by a mixture of four Gaussians, one for each configuration of Del and Hal (assuming the state is fixed). For a given state $s \in S$, the distribution of HAM-D can be obtained by marginalizing out Del and Hal and by applying the Bayesian network factorization as follows (we omit the time index as it is equal to t):

$$p(\text{HAM-D} \mid s) = \sum_{\text{Del},\text{Hal}} p(\text{HAM-D}, \text{Hal}, \text{Del} \mid s) \tag{3}$$

$$= \sum_{\text{Del},\text{Hal}} P(\text{Del} \mid s) P(\text{Hal} \mid s) p(\text{HAM-D} \mid \text{Del}, \text{Hal}, s) \tag{4}$$

Thus, the distribution of HAM-D conditional on state s is Gaussian as it is a linear combination of the Gaussians associated to the possible parent values.

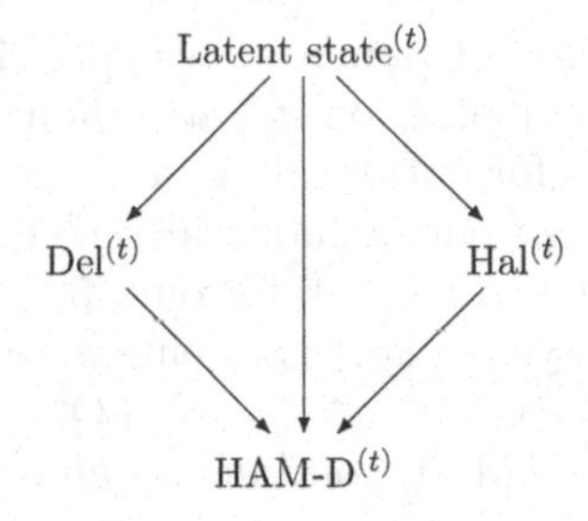

Fig. 2. BN structure of the space of observations. The domain of the state variable depends on the experiments.

5.2 Model Dimension

The number of latent states was obtained by balancing model fit and interpretability. Log-likelihoods were obtained from a 10-fold cross validation procedure, where models can have from two states up to the number of states obtained prior to model overfitting. Suppose we denote by $\mathcal{L}(k)$ [95% CI] the mean log-likelihood obtained by the model with k states. The obtained means are as follows: $\mathcal{L}(2) = -431[-441; -421]$, $\mathcal{L}(3) = -410[-422; -398]$, $\mathcal{L}(4) = -416[-436; -396]$, $\mathcal{L}(5) = -410[-431; -389]$, $\mathcal{L}(6) = -405[-423; -387]$, $\mathcal{L}(7) = -399[-418; -381]$, $\mathcal{L}(8) = -400[-417; -383]$. We do not show $\mathcal{L}$ values for $k > 8$ states as $\mathcal{L}$ approximately saturates at that point. As the values of $\mathcal{L}$ are to be maximized and the 95% CIs highly overlap for $k \geq 3$, then adding more than 3 states is not likely to lead to a significant improvement to $\mathcal{L}$. Hence, we choose $k = 3$ as the number of states. The selected number of states also takes into account that simpler models are preferred for the formulation of outcomes.

5.3 Identified States

The learned model has 3 latent states, as shown in Fig. 3 (top row), where in each latent state there is one distribution for each symptom measurement (i.e., Del, Hal and HAM-D). The states can be interpreted as follows:

- The **state Hallucinations (abbreviated as state H)** is associated with patients with high prevalence of hallucinations and moderate prevalence of delusions, with the highest mean HAM-D score and low variance.
- The **state Delusions (abbreviated as state D)** is associated with patients with high prevalence of delusions and low prevalence of hallucinations. Its mean HAM-D score is moderate and has wide tails.
- The **state No Psychosis (abbreviated as state NP)** is associated with patients with low prevalence of psychotic symptoms and moderate HAM-D score (though with high variance).

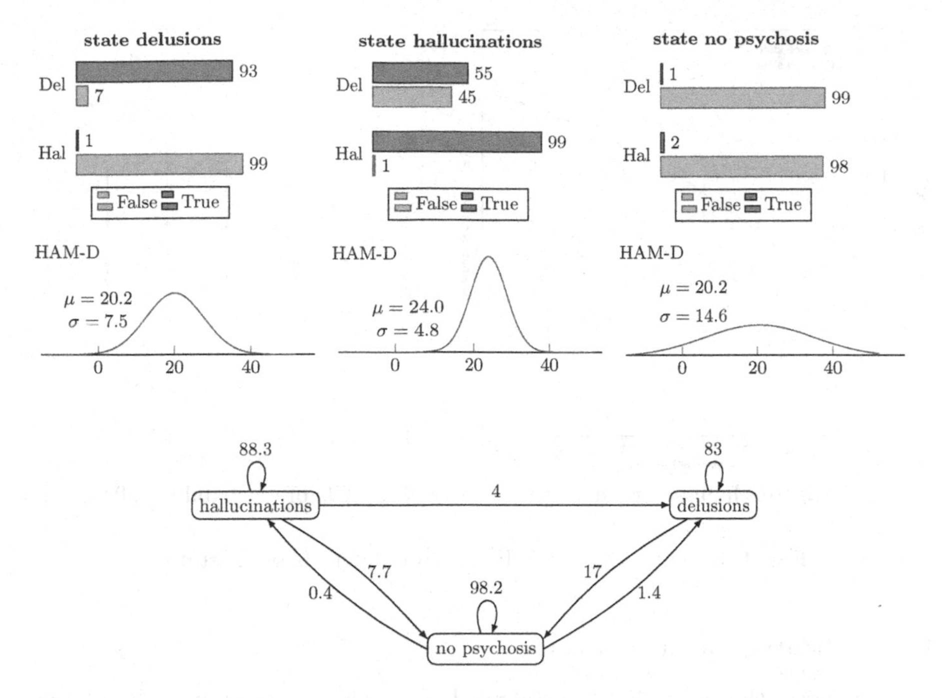

Fig. 3. Top: marginal distributions of symptoms in the latent states of the learned model (Del and Hal stand for symptom measurements). Bottom: state transitions.

5.4 Dynamics

Figure 3 (bottom row) shows the transition behavior of the learned HMM. The arcs indicate transition probabilities between latent states, e.g. the looping probability of 88.3% in state H represents the chance for remaining in such state over two adjacent weeks. Based on Fig. 3 (top row) and on the previous characterization of the states, D and H can be seen as starting states that are primarily distinguished based on the prevalence of hallucinations in patient. Later on, depending on their response to treatment, the patient will potentially move to state NP. The state NP can be seen as a healthier state due to the absence of psychotic symptoms, but the state does not imply depression remission or response due to its moderate mean HAM-D. In fact, the state NP characterizes a wide range of no-psychosis patients in terms of HAM-D score.

Figure 4-a shows the reachability trends given the baseline states, while Fig. 4-b shows the 95% bootstrap confidence intervals (BCIs) for the difference between the trends. In these cases, positive values indicate a stronger trend in favor of state D. The difference between the area under the curve of each trend was also computed, resulting in a 95% BCI equal to [0.17; 2.29]. The 95% BCI for the slope difference was [0.02; 0.17]. These results suggest that the initial state of the patient is relevant, i.e. starting in state D allows for a significantly stronger reachability to state NP than the reachability when starting in state H.

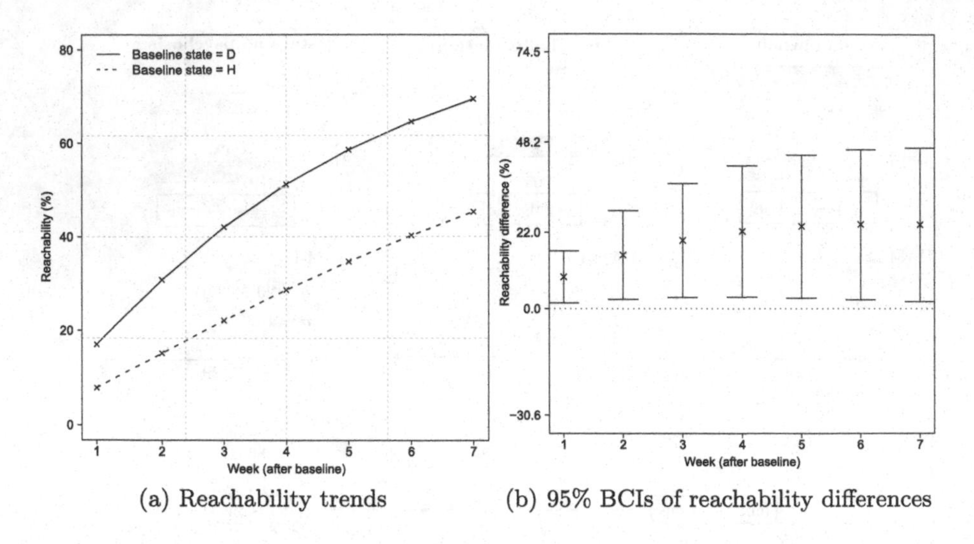

(a) Reachability trends (b) 95% BCIs of reachability differences

Fig. 4. Reachability trends based on different baseline states.

5.5 Validation of the Outcome

We now assess the claim that the state at baseline leads to significantly different state reachability. To this end, two distinct groups of patients were considered: patients with hallucinations at baseline (29 patients), and patients with no hallucinations at baseline (90 patients). The HAM-D scores of these groups at treatment endpoint were compared using a Mann-Whitney test for independent samples, which resulted in a p-value = 0.0007, suggesting that these two groups differ significantly (under a 95% confidence level). As a result, the psychotic symptom at baseline is predictive to depression recovery of patients in general.

6 Conclusions

This paper demonstrated that probabilistic graphical models can reveal insight into disease dynamics by considering not only the underlying structure, but also using meaningful outcome measures built from such structure. We illustrated the proposed methodology by applying hidden Markov models to psychotic depression treatment data, which were learned in a fully data-driven way.

The identified underlying symptom structure revealed two clinically significant results. First, the remission of psychotic symptoms preceded the decrease of depressive symptoms in PD treatment, which is in accordance with clinical observation. Second, it was shown that patients differed in their prognosis depending on the type of psychotic symptoms they exhibited at baseline (hallucinations versus delusions). Hence, our methodology allowed to shed light on the heterogeneity of psychotic depression. As future work, we will further investigate the clinical significance of the results, and will investigate the sensitivity of different treatment groups to treatment. The combination of graphical models and a data-driven approach can be easily integrated into the investigation of other disorders as well.

Acknowledgments. This work was partially funded by project "NORTE-01-0145-FEDER-000016" (NanoSTIMA). NanoSTIMA is financed by the North Portugal Regional Operational Programme (NORTE2020), under the PORTUGAL 2020 Partnership Agreement, and through the European Regional Development Fund (ERDF).

References

1. Ahlqvist, E., et al.: Novel subgroups of adult-onset diabetes and their association with outcomes: a data-driven cluster analysis of six variables. Lancet Diab. Endocrinol. **6**(5), 361–369 (2018)
2. Bueno, M.L.P., Hommersom, A., Lucas, P.J.F., Linard, A.: Asymmetric hidden Markov models. Intl. J. Approximate Reasoning **88**, 169–191 (2017)
3. Hamilton, M.: A rating scale for depression. J. Neurol. Neurosurg. Psychiatry **23**(1), 56–62 (1960)
4. Hosenfeld, B., et al.: Major depressive disorder as a nonlinear dynamic system: bimodality in the frequency distribution of depressive symptoms over time. BMC Psychiatry **15**(1), 222 (2015)
5. IsHak, W.W., et al.: The recovery index: a novel approach to measuring recovery and predicting remission in major depressive disorder. J. Affect Disord. **208**, 369–374 (2017)
6. Meier, J., Dietz, A., Boehm, A., Neumuth, T.: Predicting treatment process steps from events. J Biomed. Inform. **53**, 308–319 (2015)
7. Paoletti, M., et al.: Explorative data analysis techniques and unsupervised clustering methods to support clinical assessment of chronic obstructive pulmonary disease (COPD) phenotypes. J. Biomed. Inform. **42**(6), 1013–1021 (2009)
8. Paulus, M.P., et al.: Latent variable analysis of positive and negative valence processing focused on symptom and behavioral units of analysis in mood and anxiety disorders. J. Affect. Disord. **216**, 17–29 (2017)
9. Rabiner, L.R.: A tutorial on hidden Markov models and selected applications in speech recognition. Proc. IEEE **77**(2), 257–286 (1989)
10. Rothschild, A.J.: Challenges in the treatment of major depressive disorder with psychotic features. Schizophrenia Bull. **39**(4), 787–796 (2013)
11. Seltman, H.J., Mitchell, S., Sweet, R.A.: A Bayesian model of psychosis symptom trajectory in Alzheimer's disease. Int. J. Geriatr. Psychiatry **31**(2), 204–210 (2016)
12. Wijkstra, J., Lijmer, J., Burger, H., Cipriani, A., Geddes, J., Nolen, W.: Pharmacological treatment for psychotic depression. Cochrane Database Syst. Rev. **7** (2015)
13. Wijkstra, J., et al.: Treatment of unipolar psychotic depression: a randomized, double-blind study comparing imipramine, venlafaxine, and venlafaxine plus quetiapine. Acta Psychiatr. Scand. **121**(3), 190–200 (2010)
14. Yet, B., Perkins, Z., Fenton, N., Tai, N., Marsh, W.: Not just data: a method for improving prediction with knowledge. J. Biomed. Inform. **48**, 28–37 (2014)
15. Zhang, N.L., Nielsen, T.D., Jensen, F.V.: Latent variable discovery in classification models. Artif. Intell. Med. **30**(3), 283–299 (2004)

On Predicting the Outcomes of Chemotherapy Treatments in Breast Cancer

Agastya Silvina[1], Juliana Bowles[1(✉)], and Peter Hall[2]

[1] School of Computer Science, University of St Andrews, St Andrews KY16 9SX, UK
{as362,jkfb}@st-andrews.ac.uk
[2] Edinburgh Cancer Research Centre, University of Edinburgh, Western General Hospital, Crewe Road South, Edinburgh EH4 2XR, UK
p.s.hall@ed.ac.uk

Abstract. Chemotherapy is the main treatment commonly used for treating cancer patients. However, chemotherapy usually causes side effects some of which can be severe. The effects depend on a variety of factors including the type of drugs used, dosage, length of treatment and patient characteristics. In this paper, we use a data extraction from an oncology department in Scotland with information on treatment cycles, recorded toxicity level, and various observations concerning breast cancer patients for three years. The objective of our paper is to compare several different techniques applied to the same data set to predict the toxicity outcome of the treatment. We use a Markov model, Hidden Markov model, Random Forest and Recurrent Neural Network in our comparison. Through analysis and evaluation of the performance of these techniques, we can determine which method is more suitable in different situations to assist the medical oncologist in real-time clinical practice. We discuss the context of our work more generally and further work.

Keywords: Breast cancer data · Toxicity prediction · Modelling · Machine learning

1 Introduction

Cancer is a vast medical problem and a major cause of mortality in the UK and worldwide. Each year, one in every 250 men and one in every 300 women get diagnosed with cancer [12]. Cancer itself includes more than 200 different diseases which are characterised by the uncontrolled proliferation of cells. The rapid and abnormal reproduction of the cells can happen in several different organs and tissues within the human body (e.g., breast, lungs, bone, etc.) [12]. In this paper, we focus on chemeotherapy-based treatments for patients with breast cancer.

This work is supported by the DataLab and EU H2020 SU-TDS-02-2018 project Serums (No. 826278).

D. Riaño et al. (Eds.): AIME 2019, LNAI 11526, pp. 180–190, 2019.
https://doi.org/10.1007/978-3-030-21642-9_24

Chemotherapy in breast cancer is considered one of the major therapeutic treatments. Although introduced only fairly recently, it has gained increasing use both in primary management (also known as adjuvant therapy) and for patients with metastatic disease (for palliative care). Since the treatments are toxic and expensive, it is important to gain further insight into the consequences of their use when treating patients with cancer. One methodology to obtain knowledge about chemotherapy is by using a digital system (e.g., trained model or simulation). This system can evaluate the treatments applied to patients throughout several cycles.

Today, machine learning enables us to create a system which can be used to observe the outcome of the chemotherapy by feeding the data into several different learning algorithms [3]. With the right combination of data and techniques, we can improve the performance of the system and gain new insights that can guide and improve patient treatment in the future.

In this paper, we compare several different techniques, including Markov model (MM), Hidden Markov model (HMM), Random Forest (RF) and Recurrent Neural Network (RNN), to predict the outcome (e.g., toxicity) of chemotherapy treatments for breast cancer. The toxicity level is a scale obtained by measuring the condition of a patient based on several side effects of chemotherapy treatments (e.g., vomiting, diarrhoea, constipation, hand/foot and skin conditions). By comparing the result of several different techniques, we can find the connection between the treatment and its side effect. Finding this correlation among the recorded patient data can help guide clinicians and patients to decide which treatment is the most suitable for them when treating breast cancer.

This paper is structured as follows. We present related work in Sect. 2, describe the data and its features in Sect. 3, and our models in Sect. 4. We discuss our results in Sect. 5, and conclude with suggestions for further work in Sect. 6.

2 Related Work

In the past decade, many multivariate programs have been used to help diagnose and stage cancers, such as prostate cancer, as well as forecast the prognoses of patients [5]. As more facts about cancer are known, some cancer experts argue that every patient cancer is unique which explains why treating cancer is so difficult.

Motivated by this issue, there has been a lot of ongoing research to develop a multivariate system for personalised cancer treatment, e.g., IBM Watson [7], Microsoft Research [11], NHS [13]. Most of these approaches treat cancer as a data problem and should only be used for guidance.

Chen et al. [2] used the Breast Cancer Wisconsin (Diagnostic) Data Set, which describes characteristics of the cell nuclei in an image of a fine needle aspirate (FNA) of a breast mass [1], to train a support vector machine classifier for breast cancer diagnosis. Other studies by Nguyen et al. [10] used random forest to predict breast cancer diagnosis and prognostic. By using another machine learning technique, namely Bayesian logistic regression, Mani et al. [8] investigated the application of machine learning techniques to imaging data for predicting the eventual therapeutic response of breast cancer patients after a single cycle of neoadjuvant chemotherapy.

In our case, our data extraction consists of sequence data, and that makes it possible to explore other techniques commonly used in Natural Language Processing (NLP) such as Hidden Markov Model (HMM) [6] and Recurrent Neural Network (RNN) [4].

HMM is a sequence model for part-of-speech tagging. A sequence model, aka sequence classification-sequence model, is one whose job is to assign a label or class to each unit in a sequence, thus mapping a sequence of observations to a sequence of labels. Given a sequence of units (words, letters, morphemes, sentences, and so on), a HMM computes the probability distribution over possible sequences of labels and chooses the best label sequence [6].

RNN is an enhancement to a neural network. There is a known limitation with artificial neural networks (ANNs) and convolution neural networks (CNNs) that constrain their API. Both CNN and ANN only accept a fixed size of input or output (one sequence) [3]. RNN instead consists of several layers of ANNs, which allows us to process sequence data for which the input can be longer than one sequence [4].

In this paper, we adjust our data extraction, which is time series data, to create models using HMM and RNN and then we compare the result with the other machine learning classifiers to predict the toxicity level of a patient.

3 Data Analysis

3.1 Data Characteristics

In this paper, we use a data extraction from an oncology department in Scotland with information on treatment cycles, recorded side effects (here, toxicity level), and various observations concerning breast cancer patients for three years (from 2014 to 2016).

The extraction has data for 51661 treatments of which 13030 are of breast cancer treatments. There are 933 unique patients, and some patients have two or three different treatments/regimes during the time period. Each regime has several cycles ranging from one to more than 50 cycles (e.g., 85). Table 1 shows the number of patients for different intentions. We exclude the Curative regime because we do not have enough data for training our model.

Along with an extraction of general patient characteristics, we received the toxicity level and measurement of the patients in separate flat files. We combine the data by connecting the treatment appointment date with the date when the toxicity and other measurements (i.e., weight, height, surface area) were

Table 1. The treatment's Intentions

Intention	Total patients
Adjuvant	620
Neo-Adjuvant	427
Palliative	483
Curative	17

obtained. In this paper, we ignore patient data with no toxicity information. After we performed data cleansing, we are left with 2752 instances (i.e., 213 patients) for the palliative treatment, 1855 instances (i.e., 382 patients) for the adjuvant treatment, and 1209 instances (i.e., 205 patients) for the neo-adjuvant treatment.

3.2 Feature Analysis

Before we feed the data into the model, we analyse our datasets. First, we order the data by the cycles to make sequences. We then determine the target answer (i.e., toxicity) and predictors. After we categorised the fields, we check the relation between each predictor in the dataset to the toxicity outcome. Figure 1(a) shows that at the beginning of the treatment, most of the patients have low toxicity which is to be expected.

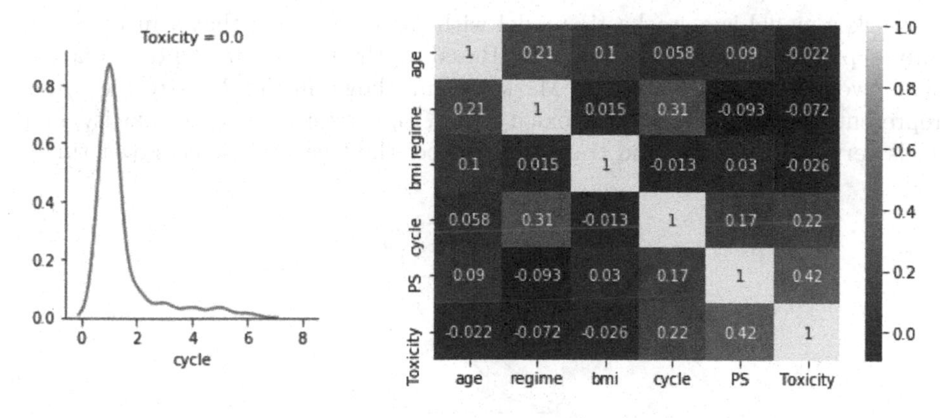

Fig. 1. Features analysis and correlation: (a) Patients' proportion against low toxicity (b) Adjuvant therapies fields' correlation map

Next, we calculate the correlation between all the predictors to the target answers as shown in Fig. 1(b). High correlation implies that there is a relationship between the variable and the target class. We want to include variables with high correlation because they are the ones with higher predictive power (signal), and leave out variables with low correlation because they are likely less relevant [9]. Even though including more relevant features during the training helps to improve the prediction power, we still include all features in the model training and then gradually exclude the irrelevant features as it is not always possible to know the features that have high predictive influence in advance.

Finally, we clean the data by replacing the missing/invalid data in our predictors. We use the mean average for fields like age or body mass index (BMI) while we use regression for the performance status (PS). To avoid the class imbalance problem, when some regime has more data than the others, we create a new dataset by duplicating some of the data. We perform this only for the RF model training because, unlike for the other models used (in our case HMM and RNN), our RF model is not dependent on the previous observation. For example, we have 141 patients treated with *FEC (D)* while only 80 patients treated with *PACLITAX*. Here, we duplicate some of the data from the *PACLITAX* to match the number of patients treated with *FEC (D)*.

4 Model Creation

As usual after analysis, we split the data into training and evaluation subset. The split ratio is 90% for training and 10% for evaluation. Hence, we randomly choose 20 patients as the test data for both adjuvant and neo-adjuvant treatments and 30 patients for the palliative treatments. All others are used to train the models.

4.1 Markov Model (MM)

A Markov model is a stochastic model with the assumption that a future state only depends on the current state [6]. Based on the toxicity in the data extractions, we created a discrete time Markov chain shown in Fig. 2 where the states represent the different levels of toxicity (e.g., T_0 corresponds to no toxicity, and T_3 is very high toxicity) and transitions reflect the treatment effects over time.

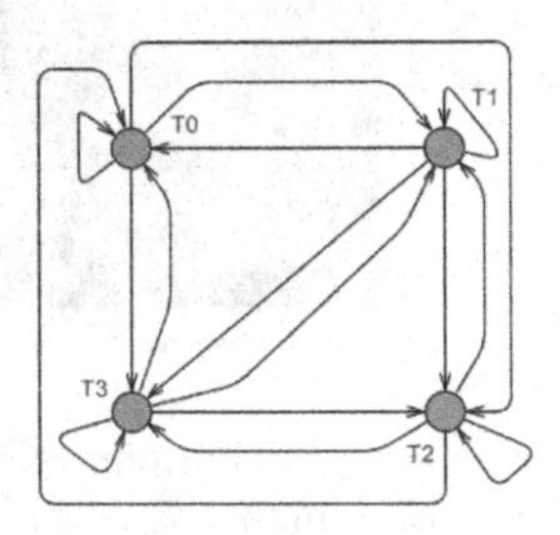

Fig. 2. The diagram representing the Markov chain for patients' toxicity outcome

Table 2 denotes the transition probability matrix for Fig. 2. From our data extraction, we calculate both the transition probability matrix and the initial probability distribution.

Table 2. The transition probability for all adjuvant treatment regimes

	T0	T1	T2	T3
T0	0.06177606	0.4980695	0.40926641	0.03088803
T1	0.03555556	0.63407407	0.30962963	0.02074074
T2	0.00524934	0.44356955	0.51968504	0.03149606
T3	0	0.32142857	0.64285714	0.03571429

We have three different Markov models for each Intention (i.e., adjuvant, neo-adjuvant, palliative). We have the model for all regimes, individual regimes, and the patient's body mass index (BMI).

4.2 Hidden Markov Model (HMM)

A HMM is based on augmenting a Markov chain to observe the hidden states of events. In our case, we want to infer/predict the toxicity level based on the patient's characteristics. Table 3 specifies the components of our HMM.

Table 3. The HMM components for predicting the toxicity outcome

Component	Description
States	The toxicity level of the patients (i.e., T0, T1, T2, T3)
Transition probabilities	The transition from one toxicity level to another toxicity level (e.g., from T0 to T1, from T1 to T3, etc)
Observations	The observed events obtained from the data extraction (i.e., cycle, age, BMI, regime). We categorise the value of each observation to simplify the process of training our HMM. For example, 1-2-3-1 denotes the observation for an overweight patient who gets the *FEC-D (D)* in their first cycle and is aged less than 50 years old
Emission probabilities	Each member represents the probability of the observations generated from the toxicity state

To predict the toxicity from the sequence of the patients' events, and as is usual for HMM, we use the *Viterbi* algorithm. The *Viterbi* algorithm is a dynamic programming algorithm used for finding the most likely sequence of hidden states (aka path) [6].

Table 4 shows an example of using HMM to predict the toxicity outcome for patients.

4.3 Random Forest (RF)

Random forest (RF) is an ensemble of decision trees for solving classification problems. The random forest classifier uses several features to predict the out-

Table 4. The HMM classification result example

Observed events	Toxicity outcome
1-2-3-2/1-2-3-2/2-2-3-2	T0/T1/T1
1-2-3-4/1-2-3-4/2-2-3-4	T3/T3/T2
1-2-3-4/1-2-3-4/2-2-3-4/2-2-3-4	T3/T3/T2/T2

come [3]. For our RF model, we use the following features: *age*, *BMI*, *cycle*, *Regime*, *previous performance status*, *previous toxicity level* to predict the toxicity outcome of the treatment. We created three RF models for each treatment intention (i.e., adjuvant, neo-adjuvant, palliative), and categorised most of the features (except age) for training our model. After we created our first RF model, we manipulate the hyperparameters to get a better prediction result. Those hyperparameters are *number of estimators*, *minimum sample leafs*, *minimum sample splits*, and *the maximum depth of each tree*.

Lastly, we observe the feature importance of each field. We get an estimate of the importance of a feature by computing the average depth at which it appears across all trees in the forest [3]. The RF libraries we used for this work allowed us to compute the feature importance automatically for every feature after training. Figure 3 shows the graph of the feature importance for the fields used to predict the toxicity outcome.

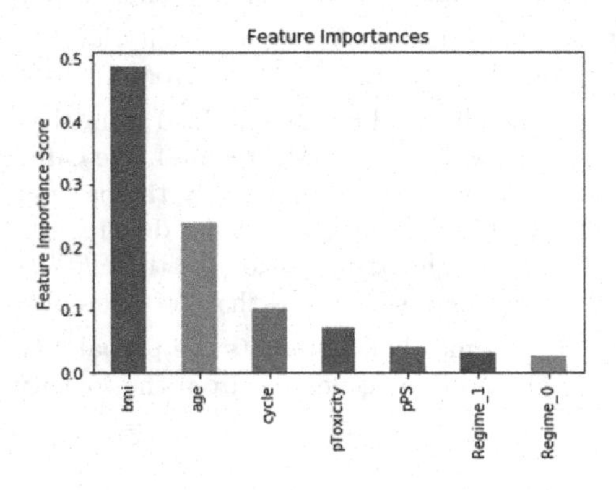

Fig. 3. The feature importance in neo-adjuvant treatments

4.4 Recurrent Neural Network (RNN)

The RNN models we created take several inputs and produce one output for each input based on the treatment cycle. During the training, we used similar features as for our RF model. However, we do not use the previous performance status and previous toxicity fields because an RNN model preserves states across time steps (in other words, has memory cells) [3]. For both models, we use the Long short-term memory (LSTM) [4] units.

5 Model Evaluation

For the MM, we observe the general pattern of the treatment outcome for each cycle and then compare it with the outcome distribution obtained from the data extraction. Figure 4 shows both datasets plotted together. The dashed line represents the value obtained from the Markov chain. From that we get the steady-state probability after 5/6 cycles. The distribution obtained from the MM resembles the distribution obtained from the real data.

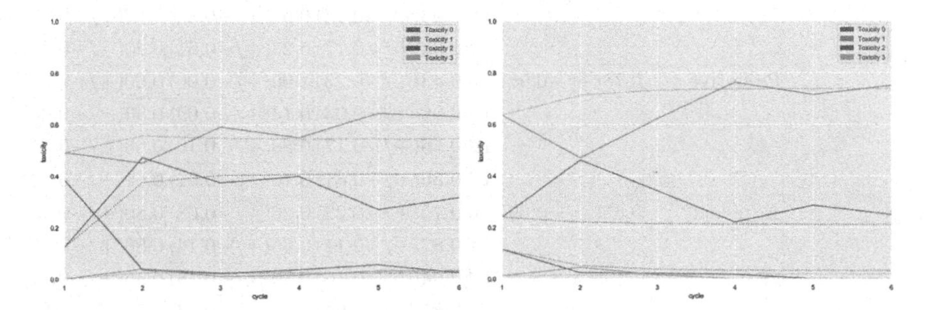

Fig. 4. The Distributions for Chemotherapy treatments: (a) Adjuvant treatments (b) Palliative treatments

We measure the performance of our classifier models by using several metrics (i.e., precision, recall, accuracy, and f1-score) after performing the cross-validation test [3]. We choose 5-fold cross validation (instead of 10-fold CV) to get more records for the validation (i.e., around 20% of total records). By duplicating the data for tackling the class imbalance issues, our model, especially the Random Forest model, is susceptible to overfitting. Here, we need more sample in the validation set to evaluate our models confidently. We do not perform the cross-validation for the HMM because the performance measured with the splitting train-test method is much lower compared to the RF or RNN models. Table 5 shows the result of the evaluation for all classifier models.

We need more data to train the corner cases (i.e., initial and end of the treatments) for the HMM models. The accuracy for the corner cases is significantly lower than the middle/transition case because the dataset has more transition cases than the initial (cycle 1) or end cases. Similarly, we can see the same characteristic for the F1-score for each treatment outcome. The T1 and T2 have higher F1-score compared to the extreme case, T0 and T3 because our datasets have more data with T1/T2 as its outcome. The RNN models outperform the RF models because, unlike RF, the RNN has LSTM units which allow the model to consider all the observations since the first treatment. Since our datasets are given as a time series, the previous treatments may affect the result of the current treatment. Hence, the RNN has an advantage compared to the RF that only considers the current state and limited information about the previous treatment result.

Table 5. Model test result (mean-std)

Model	Regime	Accuracy	Precision	Recall	F1-score
RF	Adjuvant	0.81(+/−0.11)	T0:0.55(+/−0.50)	0.92(+/−0.32)	0.65(+/−0.46)
			T1:0.85(+/−0.15)	0.83(+/−0.09)	0.84(+/− 0.09)
			T2:0.82(+/−0.09)	0.78(+/−0.20)	0.80(+/−0.13)
			T3:0.57(+/−0.52)	0.83(+/−0.67)	0.67(+/−0.55)
	Neo-Adjuvant	0.72(+/−0.09)	T0:0.53(+/−0.80)	0.60(+/−0.88)	0.52(+/−0.74)
			T1:0.77(+/−0.11)	0.80(+/−0.08)	0.79(+/−0.07)
			T2:0.63(+/−0.16)	0.61(+/−0.22)	0.62(+/−0.17)
			T3:0.33(+/−0.77)	0.23(+/−0.61)	0.23(+/−0.49)
	Palliative	0.78(+/−0.08)	T0:0.43(+/−0.23)	1.00(+/−0.00)	0.60(+/−0.24)
			T1:0.96(+/−0.03)	0.73(+/−0.09)	0.83(+/−0.06)
			T2:0.56(+/−0.17)	0.88(+/−0.10)	0.68(+/−0.15)
			T3:0.55(+/−0.81)	0.70(+/−0.92)	0.61(+/−0.83)
RNN	Adjuvant	0.85(+/−0.09)	T0:0.70(+/−0.22)	0.96(+/−0.08)	0.80(+/−0.15)
			T1:0.87(+/−0.11)	0.86(+/−0.14)	0.86(+/−0.10)
			T2:0.94(+/−0.10)	0.79(+/−0.16)	0.85(+/−0.11)
			T3:0.85(+/−0.64)	0.72(+/−0.63)	0.77(+/−0.61)
	Neo-Adjuvant	0.81(+/−0.09)	T0:0.58(+/−0.35)	0.82(+/−0.25)	0.67(+/−0.31)
			T1:0.84(+/−0.10)	0.84(+/−0.11)	0.84(+/−0.09)
			T2:0.85(+/−0.17)	0.77(+/−0.12)	0.81(+/−0.13)
			T3:0.95(+/−0.30)	0.78(+/−0.47)	0.82(+/−0.34)
	Palliative	0.85(+/−0.09)	T0:0.67(+/−0.94)	0.24(+/−0.44)	0.33(+/−0.57)
			T1:0.85(+/−0.12)	0.94(+/−0.05)	0.89(+/−0.07)
			T2:0.83(+/−0.17)	0.75(+/−0.20)	0.79(+/−0.15)
			T3:0.53(+/−0.96)	0.56(+/−0.99)	0.54(+/−0.97)
HMM (corner)	Adjuvant	0.53(+/−0.00)	NA	NA	NA
HMM (middle)		0.70(+/−0.00)	NA	NA	NA
HMM (corner)	Neo-Adjuvant	0.62(+/−0.00)	NA	NA	NA
HMM (middle)		0.70(+/−0.00)	NA	NA	NA
HMM (corner)	Palliative	0.4(+/−0.00)	NA	NA	NA
HMM (middle)		0.72(+/−0.00)	NA	NA	NA

6 Conclusion

The real value of predicting outcome/toxicity for individual patients in real-time is to help the patient and clinician understand the potential consequences of the treatment, where the patient needs to make a decision on whether to undergo treatment or not. Whereas attempts have been made to predict mortality from cancer, prediction of toxicity is much less common in the literature and where it has taken place has used simple logistic regression. The novelty of our approach is to explore the use of machine learning for these purposes.

With our classifiers, we can predict the toxicity outcome of the chemotherapy with around 0.8/0.85 accuracy. The RNN model performed better overall, because it considers all patient's treatments. Both RF and HMM only have limited observations (one previous state). However, RF has advantages because it does not differentiate between corner cases (first/last treatment) and the middle cases. Furthermore, the datasets we use for our RF models have a less class-imbalance problem than HMM. In comparison to the MM, the classifiers are more tailored for an individual patient. The MM shows the general pattern of the treatment while the classifiers can help predict the toxicity outcome of the patient. Both the MM and the classifiers complement each other.

We can improve the accuracy of our models further with more data regarding cancer characteristics or comorbidities. In our datasets, the information regarding the cancer stage is limited. We presently lack crucial information (e.g., TNM, ER/HER2 status [13]), which makes it difficult to reliably recommend suitable regimes for different patients, as we need both the toxicity outcome and cancer TNM to evaluate the treatment efficacy. For instance, some treatments might more effectively inhibit cancer growth but give higher toxicity in the short term. We are currently retraining our models with richer data extractions for more informed results on the suitability of different regimes for individual patients.

References

1. Breast cancer wisconsin (diagnostic) data set. https://data.world/health/breast-cancer-wisconsin. Accessed 15 Jan 2019
2. Chen, H., Yang, B., Liu, J., Liu, D.: A support vector machine classifier with rough set-based feature selection for breast cancer diagnosis. Expert Syst. Appl. **38**, 9014–9022 (2011)
3. Geron, A.: Hands-On Machine Learning with Scikit-Learn and TensorFlow, 1st edn. O'Reilly Media, Stanford (2017)
4. Graves, A.: Supervised Sequence Labelling with Recurrent Neural Networks. Studies in Computational Intelligence. Springer, Heidelberg (2012). https://doi.org/10.1007/978-3-642-24797-2
5. Hu, X., Cammann, H., Meyer, H., Miller, K., Jung, K., Stephan, C.: Artificial neural networks and prostate cancer tools for diagnosis and management. Nat. Rev. Urol.y **10**, 174–182 (2013)
6. Jurafsky, D., Martin, J.: Speech and Language Processing. Pearson Education, Upper Saddle River (2009)
7. Malin, J.: Envisioning Watson as a rapid-learning system for oncology. J. Oncol. Pract. **9**, 155–157 (2013)
8. Mani, S., et al.: Machine learning for predicting the response of breast cancer to neoadjuvant chemotherapy. J. Am. Med. Inform. Assoc. **20**, 688–695 (2013)
9. Machine learning concepts. http://docs.aws.amazon.com/machine-learning/latest/dg/machine-learning-concepts.html. Accessed 12 Oct 2017

10. Nguyen, C., Wang, Y., Nguyen, H.: Random forest classifier combined with feature selection for breast cancer diagnosis and prognostic. J. Biomed. Sci. Eng. **6**, 551–560 (2013)
11. Project hanover. https://hanover.azurewebsites.net/. Accessed 15 Jan 2019
12. Souhami, R., Tobias, J.: Cancer and its Management, 5th edn. Blackwell Publishing, Hoboken (2005)
13. Wishart, G., et al.: PREDICT: a new UK prognostic model that predicts survival following surgery for invasive breast cancer. Breast Cancer Res. **12** (2010)

External Validation of a "Black-Box" Clinical Predictive Model in Nephrology: Can Interpretability Methods Help Illuminate Performance Differences?

Harry F. da Cruz[1](✉), Boris Pfahringer[2], Frederic Schneider[1], Alexander Meyer[2], and Matthieu-P. Schapranow[1]

[1] Digital Health Center, Hasso Plattner Institute, Rudolf-Breitscheid-Straße 187, 14482 Potsdam, Germany
{harry.freitasdacruz,frederic.schneider,matthieu.schapranow}@hpi.de
[2] Department of Cardiothoracic and Vascular Surgery, German Heart Center Berlin, Augustenburger Platz 1, 13353 Berlin, Germany
{meyera,pfahringerb}@dhzb.de

Abstract. The number of machine learning clinical prediction models being published is rising, especially as new fields of application are being explored in medicine. Notwithstanding these advances, only few of such models are actually deployed in clinical contexts for a lack of validation studies. In this paper, we present and discuss the validation results of a machine learning model for the prediction of acute kidney injury in cardiac surgery patients when applied to an external cohort of a German research hospital. To help account for the performance differences observed, we utilized interpretability methods which allowed experts to scrutinize model behavior both at the global and local level, making it possible to gain further insights into why it did not behave as expected on the validation cohort. We argue that such methods should be considered by practitioners as a further tool to help explain performance differences and inform model update in validation studies.

Keywords: Clinical predictive modeling · Nephrology · Validation · Interpretability methods

1 Introduction

Clinical Prediction Models (CPM), more specifically prognostic models, or simply *models* are "tools for helping decision making that combine two or more items of patient data to predict clinical outcomes" [25]. In this context, validation studies are crucial, since for a model to be utilized, physicians must trust that it generalizes well to unseen patients. Therefore, to overcome a lack of trust, a prediction model should not only be internally but also externally validated

A. Meyer and M-P. Schapranow—Share the senior authorship.

D. Riaño et al. (Eds.): AIME 2019, LNAI 11526, pp. 191–201, 2019.
https://doi.org/10.1007/978-3-030-21642-9_25

to ensure the results achieved upon derivation hold true for a diverse patient population. In particular, Machine Learning (ML) models are specially prone to'learning' dataset-specific characteristics which might fail to generalize on a wide range of cohorts, further compounding the issue of lack of trust [9]. Previously, we derived a CPM to predict incidence of post-surgery Acute Kidney Injury (AKI) before the surgical procedure [3]. The model was based on the MIMIC-III database and utilized Gradient Boosting Decision Trees (GBDT) as modelling algorithm, achieving promising discrimination results (C-statistic of 0.9). In this paper, we validate the model developed on an external cohort of the German Heart Center Berlin (DHZB), evaluating its performance in terms of discrimination and calibration. As a rule, model performance obtained on the validation cohort is usually poorer when compared to that of the derivation cohort, with models often showing "disappointing accuracy" on new cohorts [17]. When this happens, a procedure called 'model updating' can be applied which consists in adapting model parameters or the model itself to the characteristics of the external validation cohort [24]. This is a straight-forward process when it comes to models such as logistic regression, in which it suffices to update regression weights or adjust decision thresholds. This is a less trivial task with respect to "black box" approaches such as GBDT or Random Forest (RF), since "the reasoning behind the function is not understandable by humans and the outcome returned does not provide any clue for its choice" [8].

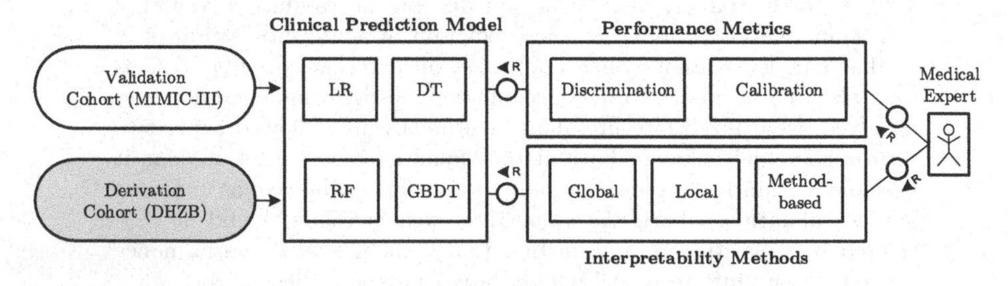

Fig. 1. Graphical abstract using Fundamental Modeling Concepts language: The Clinical Prediction Model developed was applied to two different cohorts, derivation (MIMIC-III) and validation (DHZB). Abbreviations: LR = Logistic Regression, DT = Decision Tree, RF = Random Forest and GBDT = Gradient-Boosted Decision Tree.

In addition to evaluating discrimination and calibration for both cohorts, we apply three interpretability methods, global and local surrogate, along with method-based feature importance, to help shed light on how the ML model works and possibly inform future model updating. Figure 1 provides a graphical abstract of this paper using a Fundamental Modeling Concepts (FMC) block diagram [12]. The remainder of this article proceeds as follows. Section 2 provides an overview of related work on the development and validation of CPM in Nephrology, especially for AKidney Injury. In Sect. 3 we lay out the methodology e pursued in this research, including interpretability methods. Section 4 summarizes the results obtained in both cohorts for discrimination and calibration, as well interpretability. Finally, in Sect. 5 we analyze the results obtained in light of expert feedback, especially how the interpretability methods can help inform model updating.

2 Related Work

A wealth of research has been conducted concerning the development of predictive models for cardiac-surgery related AKI. While the majority of published models rely on traditional statistical methods, such as the Cleveland score [22], more and more ML techniques are being utilized for AKI prediction. Thottakkara et al. utilized a number of ML techniques, with Generalized Additive Model as the best-performing for AKI prediction in a large patient cohort (N = 50,318) with AUC = 0.858 [23]. With a substantially smaller cohort (N = 212), Legrand et al. achieved AUC = 0.760 using a super learner estimator [15]. In a similarly sized cohort of heart surgical patients (N = 212), Eyck et al. were able to achieve AUC = 0.8339 [5]. An ensemble of learners achieved the best result in a cohort of North American patients f with AUC = 0.760 [11]. Flechet et al. achieved AUC = 0.84 using random forests on a general surgery multicenter cohort (N = 50,318) for AKI stage 2–3 [6]. More recently, in cohort of Korean patients (N = 2,010), Lee et al. 2010 [14] reported AUC = 0.78 with gradient boosting machine (XGBoost). In contrast, our model based on GBDT derived previously achieved AUC = 0.9 (N = 6,782). It is worth noting that excepting the work of Flechet et al. none of the other works offered an external validation of their ML models. Further, Doshi-Velez and Been define interpretability as "the ability to explain or to present in understandable terms to a human", which can be assessed globally, i.e., for the model as whole, or locally, for specific instances [4]. Taken together, these strategies rely on deriving surrogate models based on the original model to be explained. They are therefore termed global and local surrogates, respectively. Additionally, if one is dealing with a tree-based model, is it possible to derive rank of feature importance, which also provides some degree of insight into the inner-workings of the model at the global level. Given that multiple interpretability methods exist, we follow Hall and Gill's recommendation, and combine both global and local interpretability methods, along with method-based feature importance [9]. To the best of the authors' knowledge, this is the first work comparing three different interpretability methods applied to the validation of a CPM.

3 Methodology

The existing model relied on a cohort (N = 6,782) of surgical heart patients extracted from the MIMIC-III critical database [10]. From an initial set of 103 features encompassing demographics, laboratory values and comorbidities, 53 were selected via mutual information criteria [13]. We implemented Logistic Regression (LR), Decision Tree (DT), RF, and GBDT, the latter two being the best-performing algorithms. The model thus developed achieved an AUC = 0.9 in the prediction of post-surgical AKI with GBDT, comprising an ensemble of 126 tree stumps of depth = 3 and learning rate of 0.1 [3]. The algorithms' hyperparameters were optimized via gridsearch. Missing data was handled with multiple imputation using kNN (k Nearest Neighbors) with $k = 3$.

For development, we utilized Python version 3.6.1 [21]. The code used in the experiments can be accessed on-line[1]. Because of data privacy issues, we were not able to have physical access to the German hospital's data. Therefore, using the language-specific module 'pickle', we were able to exchange the trained models in binary form with the medical institution, thereby foregoing the need for data exchange. Model exchange took place via an encrypted channel using git-crypt [1]. After exchange of models between the institutions, we utilized two experiment set-ups. The first consisted in running the original model without any modification, in order to ascertain its generalizability. The second set-up consisted in updating the original model. In this work, model update consisted in re-training the original classifiers exclusively on the validation dataset – therefore not including the derivation cohort – while also optimizing the respective algorithms' hyperparameters using gridsearch with 5-fold cross-validation. The train/test split chosen was 80:20. As such, the metrics reported refer to the performance on the held-out test set. To measure model performance, we relied on both discrimination as measured by Area under the Receiver Operating Characteristic (AUROC) and calibration using Brier score and calibration plots. Furthermore, we included Diagnostic Odds Ratio (DOR), a metric commonly applied in medicine [7], alongside precision and recall.

3.1 Validation Cohort

Data for external validation was drawn from 54,958 admissions in the period of 2013–2018 at the German Heart Center Berlin (DHZB), a hospital specialized in the care of cardiac patients. Exclusion criteria entailed admissions of non-adult patients (5,853) and those that had no surgery or only minor surgery (31,635), with a final cohort of N = 14,191 admissions. In this cohort, AKI incidence was approximately 38.4% (5,449 out of 14,191) and therefore somewhat higher than usually reported in the literature [19], also differing from the AKI incidence in the derivation cohort (9.83%). The initial set of feature attributes were derived from literature and expert consultation and comprised demographics, comorbidities and laboratory data up to three days leading up to surgery. The complete list of attributes for both the derivation and validation cohort is provided as supplementary material[2].

3.2 Interpretability Methods

For our purposes, we define interpretability methods as "tools which quantify or visualize feature effects or feature importance", describing how features con-

[1] https://github.com/hpi-dhc/akilearner.

[2] https://goo.gl/aV8YLv.

tribute to the predictions of the model globally or locally [18]. This task is typically achieved by means of surrogate models or method-based approaches.

Global Surrogate. Global surrogates seek to distill the knowledge captured by a black-box ML model into a more interpretable model. In this method, also termed mimic learning, a simple more interpretable or *student* model is trained on the outputs of the original or *teacher* model. Instead of being trained on the outputs of interest, the student model is trained on the predicted probabilities of the teacher model while retaining the same original input features [2]. Formally, given a prediction model defined by $f(x, y) = y'$, where x is the model input and y' its output for a true label y, we train a mimic model $g(x, y') = y'_*$ where $g \in G$, i.e., a class of interpretable models. As such, the mimic model is obtained by minimizing $\sum_{i=1}^{N} ||y'_i - y'_{i*}||^2$ for N training samples. It is worth noting that the student model is only as accurate as its teacher. In this paper, we utilized Bayesian Ridge Regression (BRR) as mimic model.

Local Surrogate. The Locally Interpretable Model-agnostic Explanations (LIME) method makes use of a more interpretable model, e.g. linear regression, to explain the behavior of a black-box algorithm when applied to a given sample, i.e., a specific patient instance [20]. Given an instance x, in our case a surgical patient, LIME generates a number of 'perturbed' samples weighted by their distance to x and fits an interpretable model to these new samples. As per Eq. 1 the explanations are given by minimizing a loss function $\mathcal{L}$ that measures how well the local surrogate g belonging to a class of interpretable models G approximates our model f in the vicinity of the instance of interest defined by π_x. The loss function is further penalized by model complexity $\Omega(x)$. As such, the explanations are given by the regression coefficients of the surrogate model, which are deemed to be locally but not globally faithful. To obtain a global understanding of the model's behavior, this method provides the so-called submodular pick, in which explanations are chosen that have the highest explanation coverage, thereby offering some insight into the model's global behavior. We chose the number of explanations to be 25% of the dataset size. We then computed the mean absolute value of each feature's contribution across all explanations. Given that LIME is prone to unstable explanations because of the randomness of the perturbed samples, we excluded features that appeared in less than 10% of the explanations.

$$\xi(x) = argmin_{g \in G} \mathcal{L}(f, g, \pi_x) + \Omega(x) \quad (1)$$

Method-Based Feature Importance. In addition to the two interpretability methods discussed above, we also include feature importance provided by the algorithm's library implementation. For tree-based methods such as the ones we used, scikit-learn computes the fraction of samples to which a given feature contributes along with the associated decrease in impurity using it as split to estimate the relative feature importance. In other words, nodes closer to the top of the tree will be considered more important. For ensemble methods, decreases in impurity are averaged over all constituent trees, i.e., mean decrease in impurity [16]. In the case of random forests, one can calculate the importance of a variable X_m over all N_T trees as defined in Eq. 2, where $p(t)\Delta i(s_t, t)$ is the weighted decrease in impurity over all nodes t which include X_m. In Eq. 2, $v(s_t)$ is the variable used in split s_t and $p(t) = N_t/N$, i.e., the proportion of samples reaching t. A disadvantage of the mean decrease in impurity method is that is algorithm-dependent, i.e., can only be used with tree-based approaches.

$$Imp(X_m) = \frac{1}{N_T} \sum_{T} \sum_{t \in T: v(s_t) = X_m} p(t) \Delta i(s_t, t) \tag{2}$$

4 Results

When it comes to discriminative performance in the derivation cohort, we can observe from Table 1 that the ensemble methods employed RF and GBDT performed substantially better than LR and DT in all considered metrics except recall, where DT outperformed the GBDT, i.e., 0.66 × 0.48. Similarly, DOR of the ensemble models is significantly larger (>10-fold increase). Although not always guaranteed, ensemble learners have been empirically shown to provide better discrimination than single learners in a wide variety of applications. Nevertheless, all algorithms consistently presented relatively low precision and recall, with LR presenting lowest precision, possibly resulting from poor model calibration. In effect, as depicted in Fig. 2, we can visualize the degree of model miscalibration for both ensemble methods. After applying Platt's method (sigmoid calibration), model calibration could be improved, even though the model upon derivation still showed considerably higher miscalibration (cf. Fig. 2-A and Fig. 2-B).

Table 1. Precision, recall, diagnostic odds ratio (DOR), and area under the curve (AUC) for AKI=yes achieved in the different cohorts (derivation and validation) employing logistic regression (LR), decision tree (DT), random forest (RF) and gradient-boosted decision trees (GBDT).

Metrics	Derivation Cohort				Validation (w/o Update)				Validation (w/ Update)			
	Prec.	Rec.	DOR	AUC	Prec.	Rec.	DOR	AUC	Prec.	Rec.	DOR	AUC
LR	0.63	0.25	19.14	0.84	0.00	0.00	n/a	0.56	0.68	0.33	4.27	0.69
DT	0.35	0.66	11.22	0.80	0.68	0.16	3.78	0.52	0.75	0.30	6.08	0.71
GBDT	0.90	0.48	149.92	0.90	0.58	0.22	2.54	0.62	0.72	0.42	5.94	0.75
RF	0.92	0.41	169.58	0.90	0.90	0.02	14.44	0.70	0.75	0.42	6.90	0.76

With regards to performance in the validation cohort, we analyzed two variants, one considering the derivation model "as-is" and another by training the algorithms with the local dataset, i.e., model updating. As expected, a sharp deterioration in most metrics can be observed when applying the models without any changes on the validation cohort, with the AUROC of the ensemble methods being reduced in 30%, e.g., with respect to GBDT. Exception to this is the precision from the RF classifier, which remained similar to that of the original model (≈0.9). After model update, including hyperparameter tuning with gridsearch, a modest increase in performance could be achieved with AUROC = 0.76 for RF. Nevertheless, the updated model still performed significantly worse than the original model (derivation cohort). Most strikingly, a difference of about 12% in AUROC was accompanied by 20 times lower DOR, i.e., 169.58 vs. 6.90 for RF.

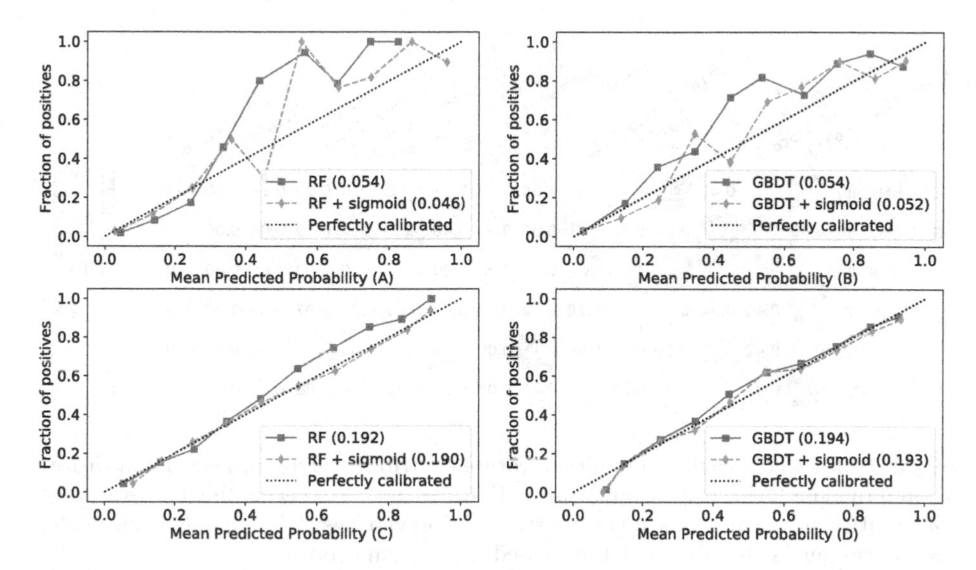

Fig. 2. Calibration plots for both the gradient boosting classifier (GDBT) and random forest classifier (RF) for the derivation (A and B) and validation cohorts (C and D). Graphs show model output without calibration and after applying Platt's method along with Brier scores.

We chose the best performing method as per DOR metric as the target for the interpretability methods (LIME, mimic learning and model-based feature importance). Figure 3 displays the feature importances provided by each of the methods employed for both cohorts (derivation and validation). The heatmap thus generated is colored according to the intensity of the given feature's importance after normalization (0–1) to allow comparison between the different methods. For plotting, features have been removed for which the maximal normalized importance for any method was below 0.3. When compared side by side, the outputs of the different interpretability methods provide some insight into the relevance of model features. The heatmap in Fig. 3 shows that blood urea values shortly before the procedure, a clinically recognized biomarker of kidney function, as well as Elixhauser score, a measure of combined comorbidity, were considered important for all three methods. In contrast, pre-surgical creatinine was deemed to be important by both LIME (local surrogate) and feature importance methods, while absent from mimic learning (global surrogate). Furthermore, creatinine and urea values in the days leading up to the procedure, e.g., two or three days before, are likewise relevant, albeit with a somewhat smaller magnitude. It is worth noting, however, that these values are absent from the explanations provided by the mimic learning method. Nevertheless, mimic learning was able to capture the importance of features such as pre-existing chronic kidney disease and fluid electrolyte imbalance, which were assigned a substantially lower importance.

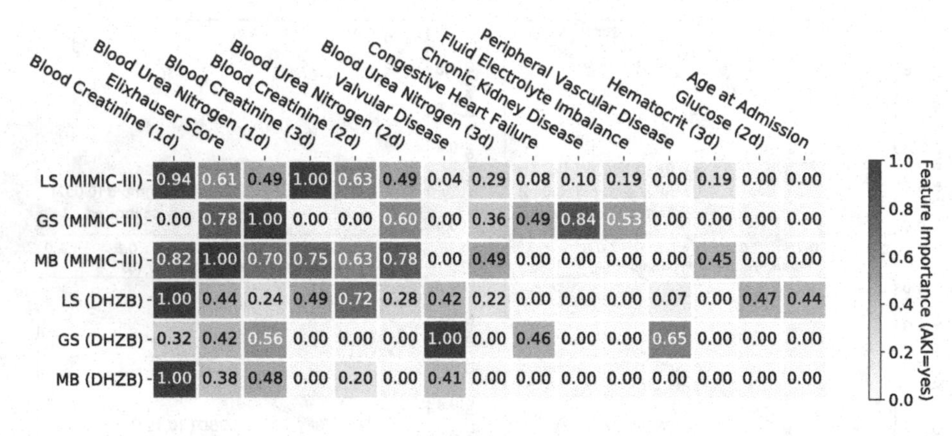

Blood Creatinine (1d)
Elixhauser Score
Blood Urea Nitrogen (1d)
Blood Creatinine (3d)
Blood Creatinine (2d)
Blood Urea Nitrogen (2d)
Valvular Disease
Blood Urea Nitrogen (3d)
Congestive Heart Failure
Chronic Kidney Disease
Fluid Electrolyte Imbalance
Peripheral Vascular Disease
Hematocrit (3d)
Glucose (2d)
Age at Admission
LS (MIMIC-III) 0.94 0.61 0.49 1.00 0.63 0.49 0.04 0.29 0.08 0.10 0.19 0.00 0.19 0.00 0.00
GS (MIMIC-III) 0.00 0.78 1.00 0.00 0.00 0.60 0.00 0.36 0.49 0.84 0.53 0.00 0.00 0.00 0.00
MB (MIMIC-III) 0.82 1.00 0.70 0.75 0.63 0.78 0.00 0.49 0.00 0.00 0.00 0.00 0.45 0.00 0.00
LS (DHZB) 1.00 0.44 0.24 0.49 0.72 0.28 0.42 0.22 0.00 0.00 0.00 0.07 0.00 0.47 0.44
GS (DHZB) 0.32 0.42 0.56 0.00 0.00 0.00 1.00 0.00 0.46 0.00 0.00 0.65 0.00 0.00 0.00
MB (DHZB) 1.00 0.38 0.48 0.00 0.20 0.00 0.41 0.00 0.00 0.00 0.00 0.00 0.00 0.00 0.00
1.0
0.8
0.6
0.4
0.2
0.0
Feature Importance (AKI=yes)

Fig. 3. Heatmap displaying normalized feature contributions from LIME, mimic learning and random forest feature importance. Features are sorted according to the average contribution of all methods taken together. Abbreviations: GS = Global Surrogate, LS= Local Surrogate, MB = Method-based feature importance.

5 Discussion

In the derivation cohort, while a high AUROC could be achieved, an imbalance with respect to precision and recall could be observed, which was also present in the validation cohort. In our context, higher precision than recall means that the model is particularly sure that patients it classifies as under risk of AKI do in fact develop it, though it was very selective when doing so. As a result, it means that the model will likely 'miss out' on patients under risk (false negatives). This tendency of the model was further exacerbated in the validation cohort as illustrated by the large difference in terms of DOR. While precision and recall differed upon validation, they did so to a considerable smaller extent when compared to DOR. This difference can be explained by a much larger ratio of false negatives to true negatives in the validation cohort. For AKI management, the costs of misclassifying a high-risk patient as not under risk are higher than enabling protective measures for patients who are not under risk. Therefore, thorough model calibration would be necessary before clinical deployment with other techniques beyond Platt's method. The differences in performance upon validation, particularly in the set-up *without* model update, can be traced back in part to a difference in the prevalence of the outcome of interest (9.83% vs. 38.4%).

The interpretability algorithms indicated the Elixhauser score among the most predictive features in the validation cohort. The exact same pattern could not be verified upon validation, where Elixhauser score did not play as significant role, i.e., much lower feature importance. Upon closer inspection of the distribution of this variable, one can observe that its mean value in the derivation cohort is substantially higher than in the validation cohort. This might suggest that either patients in the derivation cohort are on average more severely ill or that coding for this variable in the validation cohort might be inconsistent. In a similar fashion, creatinine and blood urea values played an important role for the predictions upon derivation, a result not observed to the same extent in the validation cohort. These lab values present a similar distribution and standard deviation in both cohorts, with a somewhat higher rate of missing values in the validation cohort, specially as days before surgery increase (cf. Supplementary Material). Appraisal with the medical expert revealed that in this German hospital surgical patients are usually admitted only shortly before the procedure takes place. As such, laboratory values days before surgery are not available as a rule. This observation agrees with the suggestion that in the derivation cohort patients are being treated who are in a more critical condition and likely had been admitted to the hospital and therefore under monitoring before surgery took place. Building on these insights, model update should, e.g., exclude Elixhauser score in favor of its constituent comordidities and limit itself to lab values available upon admission. A mere analysis of the distribution of different variables while informative, offers little towards understanding underlying causes for performance differences from the perspective of the model itself.

Limitations. The interpretability algorithms themselves are also not entirely without downsides and pitfalls. For one, while global surrogates such as mimic learning are flexible, the conclusions drawn concern the model, not the data, since the surrogate model does not have access to the actual data. As such, explanations tend to be only as good as the original model. Furthermore, since the choice of surrogate impacts how well it can mimic the original model, i.e., its R^2 values, the surrogate itself will bring along a host of potential issues. Second, local surrogates such as LIME tend to exhibit a considerable degree of instability for their explanations. This happens because of the random neighborhood sampling inherent to this method. In other words, if the sampling process is repeated, one might obtain different explanations for the same instance, calling into question its robustness. We sought to mitigate this effect by applying submodular pick and averaging out the contributions across many different explanations. For these reasons, not only more methods should be included, but also hypotheses drawn from such interpretability algorithms must be validated by experiment for claims to be considered valid. Third, considering the precipitous drop in performance upon validation, the original model might have suffered from overfitting, therefore warranting the use of regularization techniques such as elastic net or class balancing. Finally, we were not able to include the insights gleaned with the interpretability approaches into a new model. This and the other issues pointed out shall be addressed in future work.

6 Conclusion

In this paper, we reported the validation results of applying a ML-based CPM on a external cohort, observing a substantial deterioration of its performance upon validation. While a performance difference was expected to some degree, the issue is compounded by the use of "black-box" algorithms. We applied three interpretability methods in order to illuminate possible reasons that could account for performance differences. The methods employed highlighted particular characteristics of the CPM developed, which, as it turned out, relied considerably upon longitudinal lab values and an aggregated comorbidity index, the Elixhauser score. Even though the insights obtained can potentially inform model update for external validation, hypotheses drawn from these methods must be validated by experiment, i.e., via iteratively refining the model. Potentially, this might lead to more generalizable models that are easier to understand for practitioners. Open questions remain, however, regarding the robustness of interpretability methods, an issue that warrants further investigation.

Acknowledgments. Parts of the given work were generously supported by the European Union's Horizon 2020 research and innovation programme under grant agreement No. 780495.

References

1. Andrew, A.: Git-crypt (2013). https://github.com/AGWA/git-crypt
2. Che, Z., Purushotham, S., Khemani, R., Liu, Y.: Interpretable deep models for ICU outcome prediction. In: AMIA Symposium 2016, pp. 371–380 (2016)
3. Freitas da Cruz, H., Schneider, F., Schapranow, M.P.: Prediction of acute kidney injury in cardiac surgery patients. In: Proceedings of the 12th International Conference on Biomedical Engineering Systems and Technologies, vol. 5, pp. 380–387 (2019)
4. Doshi-Velez, F., Kim, B.: Towards A Rigorous Science of Interpretable Machine Learning. arXiv e-prints arXiv:1702.08608, February 2017
5. Eyck, J.V., et al.: Data mining techniques for predicting acute kidney injury after elective cardiac surgery. Crit. Care **16**(Suppl 1), P344 (2012)
6. Flechet, M., et al.: AKIpredictor, an on-line prognostic calculator for acute kidney injury in adult critically ill patients. Intensive Care Med. **43**(6), 764–773 (2017)
7. Glas, A.S., et al.: The diagnostic odds ratio: a single indicator of test performance. J. Clin. Epidemiol. **56**(11), 1129–1135 (2003)
8. Guidotti, R. et al.: A Survey of Methods for Explaining Black Box Models. arXiv e-prints arXiv:1802.01933, February 2018

9. Hall, P., Gill, N.: An Introduction to Machine Learning Interpretability. O'Reilly, Boca Raton (2018)
10. Johnson, A., et al.: MIMIC-III, a freely accessible critical care database. Sci. Data **3**, 160035 (2016)
11. Kate, R.J., et al.: Prediction and detection models for acute kidney injury in hospitalized older adults. BMC Med. Inform. Decis. Mak. **16**(1), 39 (2016)
12. Knöpfel, A., Gröne, B., Tabeling, P.: Fundamental Modeling Concepts: Effective Communication of IT Systems. Wiley, Hoboken (2005)
13. Kraskov, A., Stögbauer, H., Grassberger, P.: Estimating mutual information. Phys. Rev. **69**(6), 16 (2004)
14. Lee, H.C., et al.: Derivation and validation of machine learning approaches to predict acute kidney injury after cardiac surgery. J. Clin. Med. **7**(10), 322 (2018)
15. Legrand, M., et al.: Incidence, risk factors and prediction of post-operative acute kidney injury following cardiac surgery for active inefective endocarditis: an observational study. Crit. Care **17**(5), R220 (2013)
16. Louppe, G., Wehenkel, L., Sutera, A., Geurts, P.: Understanding variable importances in forests of randomized trees. In: Neural Information Processing Systems, pp. 1–9 (2013)
17. Moons, K.G.M., Altman, D.G., Vergouwe, Y., Royston, P.: Prognosis and prognostic research: application and impact of prognostic models in clinical practice. Brit. Med. J. **338**, b606 (2009)
18. Murdoch, W.J., et al.: Interpretable Machine Learning: Definitions, Methods, and Applications. arXiv e-prints arXiv:1901.04592, January 2019
19. O'Neal, J.B., et al.: Acute kidney injury following cardiac surgery: current understanding and future directions. Crit. Care **20**(1), 187 (2016)
20. Ribeiro, M., Singh, S., Guestrin, C.: "Why should i trust you?": explaining the predictions of any classifier. In: Proceedings of 22nd ACM SIGKDD, pp. 1135–1144, NY, USA (2016)
21. Rossum, G.V., Drake, F.L.: Python tutorial. History **42**(4), 1–122 (2010)
22. Thakar, C.V., et al.: A clinical score to predict acute renal failure after cardiac surgery. J. Am. Soc. Nephrol. **14**(8), 2176–7 (2004)
23. Thottakkara, P., et al.: Application of machine learning techniques to high-dimensional clinical data to forecast postoperative complications. PLoS ONE **11**(5), 1–19 (2016)
24. Toll, D., Janssen, K., Vergouwe, Y., Moons, K.: Validation, updating and impact of clinical prediction rules: a review. J. Clin. Epidemiol. **61**, 1085–1094 (2008)
25. Wyatt, J.C., Altman, D.G.: Commentary: prognostic models: clinically useful or quickly forgotten? Brit. Med. J. **311**(7019), 1539–1541 (1995)

Behavior Monitoring

Classifying Small Volumes of Tissue for Real-Time Monitoring Radiofrequency Ablation

Emre Besler[1], Yearnchee Curtis Wang[1], Terence Chan[1], and Alan Varteres Sahakian[1,2](✉)

[1] Department of Electrical and Computer Engineering, Northwestern University, Evanston, USA
{emrebesler2020,ycwang,terencechan2016}@u.northwestern.edu, a-sahakian@northwestern.edu

[2] Department of Biomedical Engineering, Northwestern University, Evanston, USA

Abstract. An increasingly-popular treatment for ablation of cancerous and non-cancerous masses is thermal ablation by radiofrequency joule heating. Real-time monitoring of the thermal tissue ablation process is essential in order to maintain the reliability of the treatment technique. Common methods for monitoring the extent of ablation have proven to be accurate, though they are time-consuming and often require powerful computers to run on, which makes the clinical ablation process more cumbersome and expensive due to the time-dependent nature of the clinical procedure. In this study, a Machine Learning (ML) approach is presented to reduce the time to calculate the progress of ablation while keeping the accuracy of the conventional methods. Different setups were used to perform the ablation and collect impedance data at the same time and different ML algorithms were tested to predict the ablation depth in three dimensions, based on the collected data. In the end, it is shown that an optimal pair of hardware setup and ML algorithm were able to control the ablation by estimating the lesion depth within an average of micrometer-magnitude error range while keeping the estimation time within 5.5 s on conventional x86-64 computing hardware.

Keywords: Radiofrequency · Ablation · Monitoring · Machine learning · Data · Ensemble · Lesion · Artificial intelligence

1 Introduction

Radiofrequency ablation (RFA) is a minimally invasive medical procedure that involves thermally destroying undesired tissue by delivering high-frequency alternating current via needle-sized electrodes or catheter [13]. When the undesired tissue is heated enough, coagulative necrosis begins to occur. It is possible to

Supported by National Science Foundation grants 1622842 and 1738541.

D. Riaño et al. (Eds.): AIME 2019, LNAI 11526, pp. 205–215, 2019.
https://doi.org/10.1007/978-3-030-21642-9_26

make a probabilistic estimation on the death of the target cells, depending on the tissue temperature and the exposure time to heat [2]. 43 °C+ for 10 min, 50 °C+ for 5 min, and 57 °C+ for 2 s are commonly-accepted scenarios for certain mammalian cell death [4,5].

A major drawback of RFA has been its uncontrollable nature that may result in undesired tissue left unablated and/or destruction of non-targeted tissue [15]. The need for real-time monitoring of RFA arises due to this uncontrollable nature as well as the opacity of human tissue, that hinders the visual control of the medical personnel during the treatment. There is quite a significant amount of work done by the scientific community to develop monitoring techniques that are more automated and data-based. The majority of these techniques utilize changes in tissue properties undergoing thermal ablation, including electrical, acoustic, and optical behaviors.

The requirements of the RFA monitoring in our case are that: (1) monitoring should be performed in real-time to avoid the aforementioned complications, (2) monitoring should be performed using equipment that could be low-cost for purchase by community hospitals for public health, and (3) monitoring should be performed with as little external equipment as possible. Due to the large capital equipment required, external imaging modalities such as Magnetic Resonance Imaging (MRI) and Contrast-enhanced Computed Tomography (CT) are not easily accessible, even though they construct very accurate depth maps [7,9].

A very common method for real-time monitoring is checking the temperature of the ablated volume via a temperature probe that is inserted closely to the ablation electrode [4]. A similar approach is applied with a thermographic camera that captures the radiation within the long-infrared range [11]. These methods are faster than an MRI or CT scan, however they make the whole treatment more invasive, defeating the main purpose of ablation therapy. For thermographic imaging, the emitted infrared waves cannot leave the tissue, so the camera should be inside the body, close to the targeted tissue. Moreover, they are both manual techniques, meaning that there should be a staff or the physician who is handling the monitoring equipment. Such manual methods tend to be cumbersome, especially in comparison with automated methods.

Three local methods that do not require large external imaging machines have been explored for RFA monitoring: acoustic, optoacoustic, and electrical. Ultrasound imaging (primarily Nakagami imaging) is real-time, but cannot image muscular tissue [17]. Optoacoustic imaging is very accurate (95%+), but can take up to 400+ s to compute a depth map, preventing true real-time use until processors are fast enough to reduce this time [10]. Electrical methods, primarily using EIT, are very accurate (90%+), but require $\geq$ 100 s to compute [6,8]. Thus, there is a need for a faster method to compute a depth map in order to provide real-time actuation. We present a machine learning (ML) based approach in this paper, using tree-based classifiers, to compute depth maps from tissue electrical impedance for real-time control of RFA. The results in this study are built on and compared to [14] that uses an Artificial Neural Network (ANN) classifier for depth estimation. The contribution of this study is a new hardware setup for data collection, two tree-based ML models that are predicted to outperform the

ANN and lastly, a simple linear search algorithm to interpret the classification performances both in this study and in [14] as actual depth predictions.

2 Materials and Methods

2.1 The Hardware Configuration and Data Collection

The tissue model that simulates breast tissue and the spherical design of the ablation device were the same as used in [14]. The electrodes on the device were connected to a system that activates or deactivates all electrodes on one side at a time. The control system was connected to a RF generator to deliver the alternating current (AC) and to an impedance analyzer for impedance measurements. The impedance data was collected by the same RFA device that performs the ablation, removing the need for any additional equipment that would affect the comfort of a patient in real life.

The thermal ablation was created by the AC delivered through the electrodes. A constant frequency of 100 kHz was used for the entire ablation process. After the ablation cycle was complete for all 6 sides, the impedance data was collected for all the sides through the same steel electrodes and measured at the impedance analyzer connected to them. The side of the activated electrodes was also added to the impedance data as another feature in the data logger. After the data was recorded, another ablation cycle starts and the whole process was repeated. The entire workflow is shown in Fig. 1a and the timing diagram of the ablation, the measurements and the data logging are shown in Fig. 1c.

The data for the models were collected with two different hardware setups. The first dataset used off-the-shelf equipment, based on a mid-tier LCR (inductance (L), capacitance (C), impedance (R)) meter (Hameg Rohde and Schwarz HM8118, Munich, DE) and matrix switch (National Instruments PXIe-2529, Austin, TX, USA). This setup is not feasible within clinical environments, however, as it is not low-cost and is in separate units that make setup difficult. Thus, a new low-cost embedded-system-based hardware setup was created, utilizing the AD5933 (Analog Devices, Norwood, MA, USA) in the setup for low-impedance measurement, as provided by the CN-0217 reference design from Analog Devices. The total cost for this embedded hardware setup is less than $200. The embedded system connects directly inline between the RFA generator and the RFA device. The second dataset was collected using the embedded hardware setup, shown in Fig. 1b. It was predicted that there would be a difference in the level of noise between the two datasets. Both this noise profile and the embedded-system based hardware setup were presented in more detail in [1].

In order to obtain the ground truth for ML, a resistance thermometer detector (RTD) input module was inserted through the clearances on each side. The temperature data for each side after each ablation cycle was recorded using temperature probes and platinum 100 Ω resistance temperature detectors on this module. These detectors were placed at 0 mm, 5 mm, 10 mm and 15 mm depths from the side of the device. The temperature values for the depths in between were linearly interpolated. After the temperature was recorded for all depth values from 0.0 mm to 15.0 mm with a stepsize of 0.1 mm, the temperature and exposure time thresholds introduced in Sect. 1 were used to determine whether the volume at every depth was ablated or not.

After the data was collected, the prediction of the lesion depth was posed as a classification task. The four numerical features from the impedance data were the initial and final magnitude and phase of tissue impedance. The side the activated electrodes were on was added as a categorical feature (1 to 6) in order to create a 3D depth map. Lastly, the depth value was added as the fifth numerical feature. The categorical feature is one-hot-encoded into 6 binary features, so the dataset has 11 features in total. The targets were binary labels; 0 for an unablated volume and 1 for an ablated one.

The first hardware setup collected 1,872,000 data points, which was referred as the first dataset in this study and the second one collected 1,561,944 data points, which was referred as the second dataset. Each sample ablation was comprised of 20–50 ablate/measure cycles. Each data measurement generates 6 data groups per cycle (for each side). Each data group has one sample per each depth value between 0.0 and 15.0 mm, with a resolution of 0.1 mm, containing 151 samples in total. Since it took many cycles for the tissue to be completely ablated, the classes in these datasets were not balanced, having only one third

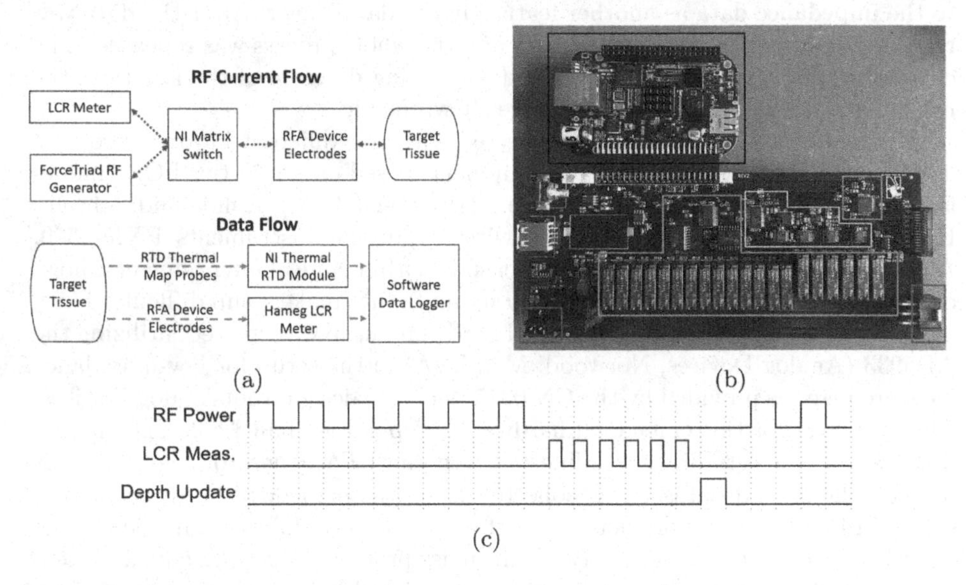

Fig. 1. Hardware setups. (a) shows system flow of off-the-shelf setup, (b) shows the embedded hardware setup, and (c) shows the timing diagram for the ablation, measurement and computation. The colored labels on the embedded hardware are as follows: red is the AC input, green is the relay network, yellow is the impedance measurement subsystem, purple is the connection to the RFA device, orange is the temperature measurement subsystem, and black is the microcontroller that controls the whole process. (Color figure online)

of their total number of samples labeled as the positive class. This will be taken into consideration when the ML metrics are presented.

2.2 Machine Learning Models

Aside from the ANN that was trained in [14], two different tree-based ensemble classifiers are introduced in this study. Tree-based models are known for their simplicity when compared to other ML models just as complex. They have fewer hyperparameters to tune and take considerably less time to train. On the other hand, while ANNs function well with unstructured and high-dimensional datasets, such as images and sequential data, they can be overly complex for structured datasets that are not very high-dimensional, like the data at hand for this study. Therefore, the tree-based models in this study were predicted to outperform the ANN in [14] for both datasets. Before making a comparison, it was verified that the ANN in [14] indeed has the optimum performance by tuning its hyperparameters and architecture.

The first ensemble classifier tested on the RFA data was a Random Forest, a bagging ensemble that is further randomized with a simple tweak, making it more robust to changes in data than just a number of trees trained in parallel on the same data [16]. The Random Forest algorithm was based on a Decision Tree that was trained with the CART algorithm [3]. All the trees in a Random Forest were trained in parallel, on different subsets of data. The slight modification that creates more randomness and gives the model its name is that the trees are given only a random subset of the features. This avoids the trees having high structural similarity and correlation in their predictions. After the model was trained, the final class decision was given with majority voting among the trees.

The second ensemble classifier was an Adaptive Boosting classifier, that was based on Decision Trees as well. For Adaptive Boosting, trees are trained sequentially instead of in parallel, each correcting the mistakes of the previous one, all of them adding up to a complex ensemble model [12]. For Adaptive Boosting, each tree was trained on the entire dataset instead of just a subset of it.

Since decision trees can keep branching out until all the training data was separated into pure leaf nodes, the models based on them are prone to high overfitting. To avoid this, the models were always regulated by putting restrictions on some of their hyperparameters. In this study, the maximum number of leaf nodes was picked as the regularization hyperparameter. The number of trees in the ensembles was another hyperparameter to be tuned.

2.3 Final Interpretation

Aside from some classification metrics to evaluate the models in pure ML sense, the classification results should also be interpreted according to how they reflect on the real-life ablation task this study is about.

This is made possible by using a linear search to obtain numerical depth estimations (between 0.0 and 15.0 mm) from the predicted labels and comparing them to their corresponding true depth values, which are obtained from the true

Algorithm 1. Ablation depth estimation from the classification results

Require: 151 instances that pertain to the same target depth, trained classifier
1: **for** Current Depth = [0.0:0.1:15.0] **do**
2: **if** Predicted class for the instance with depth as Current Depth = 0 **then**
3: **return** Current Depth
4: **end if**
5: **end for**

labels. In order to implement this conversion, the samples that belong to the same data group of 151 samples are used together. For each group, the search starts from the sample with 0.0 mm and went through the samples in an increasing order of their depth value as long as their predicted label is 1 (ablated). The algorithm stops when it sees a 0 (unablated) label and returns the last ablated depth. This linear search is summarized in Algorithm 1.

After the conversion with the linear search was done, the true and predicted ablation depths were compared in a way that is similar to a regression task. The metrics for this evaluation were Root Mean Squared Error (RMSE) and a residual map. These metrics were also considered for a comparison between datasets and ML models.

3 Results

This study has two datasets from two different hardware configurations and three different ML models that were tested on these sets. The ANN architecture from [14] was kept the same but the hyperparameters of the ensemble models were tuned with a two-dimensional grid search.

The training-test split was 70%–30% for both datasets. A 10-fold cross-validation on training data was performed to tune the hyperparameters with a grid search and the test data was held out to evaluate how well the models generalize after training.

The results of all models trained with the first dataset can be seen in Table 1. As mentioned in Sect. 2.1, the classes were unbalanced, so an f1-score was added as a second metric to eliminate any possible bias this unbalance might add to the accuracy results. The Random Forest was trained with 20 trees and 3000 maximum leaf nodes. The Adaptive Boosting model was trained with 40 trees and 2000 leaf nodes. (These optimum hyperparameter values were obtained with the grid search.)

Table 1. Performances of all the ML models on the first dataset

ML model	Classification metrics			
	Test Acc.	Test F1 Score	CV-Average Acc.	CV-Average F1 Score
ANN	92.72%	0.886	92.93%	0.888
Random forest	93.58%	0.890	92.65%	0.891
Adaptive boosting	93.74%	0.891	93.78%	0.891

The results with the second dataset are in Table 2. The Random Forest was trained with 20 trees and 7000 maximum leaf nodes. The Adaptive Boosting model was trained with 30 trees and 3000 maximum leaf nodes.

Table 2. Performances of all the ML models on the second dataset

ML model	Classification metrics			
	Test Acc.	Test F1 Score	CV-Average Acc.	CV-Average F1 Score
ANN	97.12%	0.932	97.10%	0.931
Random forest	99.83%	0.994	99.85%	0.995
Adaptive boosting	99.97%	0.999	99.96%	0.999

After getting the predicted and true ablation depths from predicted and true labels for the entire datasets, respectively, the depth prediction performances and their residual maps for all ML models are shown in Table 3 and Fig. 1.

Table 3. Depth prediction results of all the ML models on both datasets

ML model	RMSEs (mm)	
	First dataset	Second dataset
ANN	2.24	0.68
Random forest	2.18	0.19
Adaptive boosting	2.20	0.13

4 Discussion

Both classification metrics as well as the depth estimation results indicate that the second dataset was much less noisy and therefore, easier to predict. This makes sense because the second dataset was collected in a consistent manner by the embedded hardware setup, not affected by any noise introduced by the wiring of multiple pieces of external, off-the-shelf equipment, which was the setup for the first dataset. For both tree-based ML models on the first dataset, the classification performance reached a 94% test accuracy and 0.89 F1 score. Both of these models were outperformed by the models trained on the second dataset. The Random Forest had a 99.8% accuracy with both the test and the validation data and moreover, the Adaptive Boosting model reached an almost perfect test performance with 99.96% accuracy. The impact of data quality on ML results became much more obvious when the classification results were converted into depth estimations. The RMSEs and residues in Fig. 2 decreased dramatically for all models, reaching the lowest value of 0.13 mm for Adaptive Boosting, which corresponds to only one single misprediction. Especially this almost perfect prediction performance of Adaptive Boosting shows how effectively a ML model

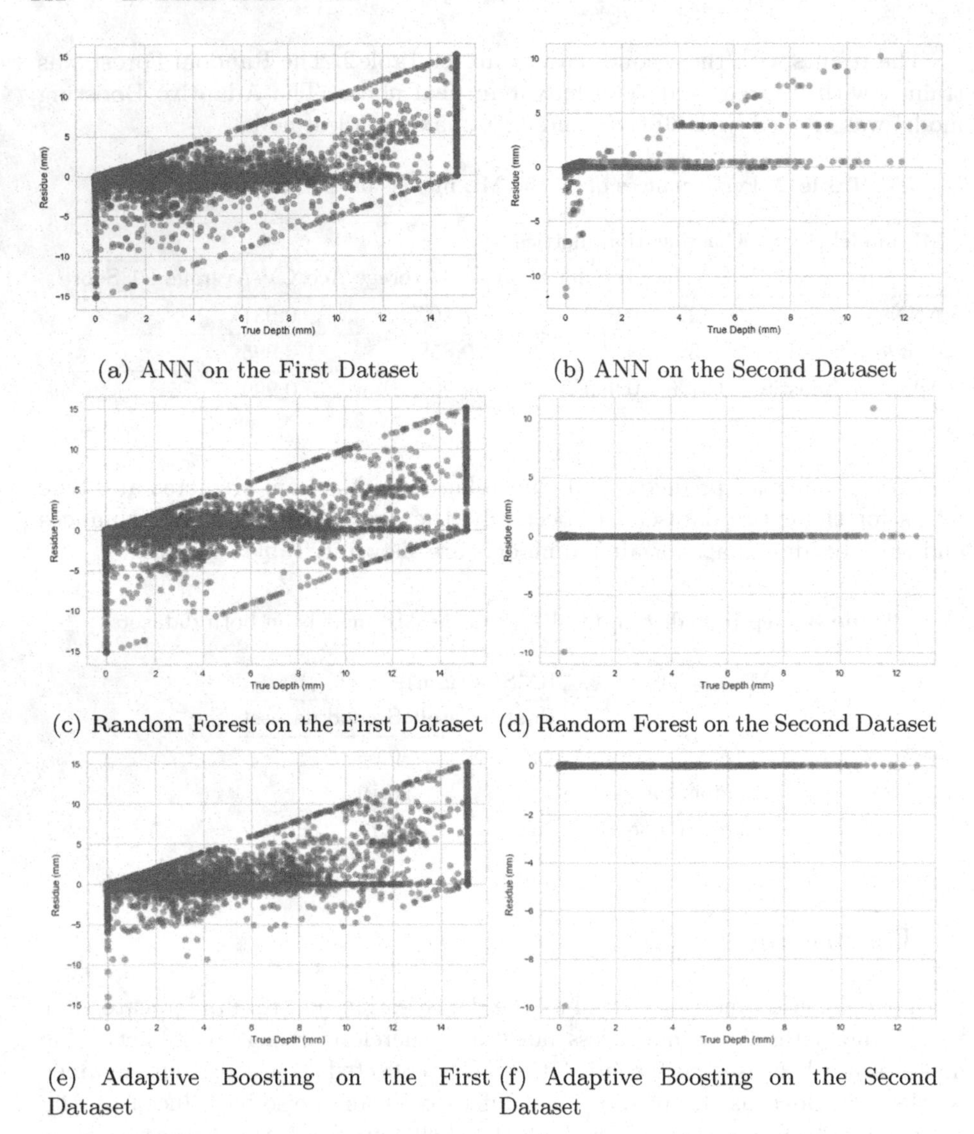

(a) ANN on the First Dataset (b) ANN on the Second Dataset

(c) Random Forest on the First Dataset (d) Random Forest on the Second Dataset

(e) Adaptive Boosting on the First Dataset (f) Adaptive Boosting on the Second Dataset

Fig. 2. Residual (prediction error) maps of all the models on both datasets after converting the class predictions to depth estimations. For the maps of the second dataset, the horizontal patterns at 0 mm residue for all true depth values indicate a high prediction performance without much error. The vertical and diagonal patterns on the maps of the first dataset correspond to inaccurate depth predictions for impedance measurements that actually belong to 0 or 15 mm depth and predicting 0 or 15 mm depth when the thermal lesion was not there, respectively. Both of these cases would cause serious medical issues in real-life tumor ablation such as a recurrent cancer from unablated tumor volume or ablated healthy tissue volume that can lead to body deformation or the collapse of an organ.

can function as the depth estimator in an ablation scheme, given that the noise was eliminated and enough data was collected. Since the ML approach to the RFA monitoring problem is a novelty, a direct comparison with any other study is not possible. However, it is safe to say that the monitoring precision is on par with the conventional local methods introduced in Sect. 1.

Among the ML models, the ANN from [14] should be compared to the tree-based ensemble models first. For both datasets and with all metrics, the ensembles outperformed the ANN. As mentioned in Sect. 2.2, it was made certain that the optimum architecture was used for the ANN, so it is safe to say that the ensemble models were a better fit for both datasets in this study, as predicted. As for the performance comparison between the ensembles, they performed similarly when trained with the first dataset, the reason for this being the noise that prevented both models from performing above a certain level. When trained with the second dataset though, the Adaptive Boosting model outperformed the Random Forest, with better test accuracy and a lower RMSE for depth prediction. The 10-fold CV results agree with the test results to show that these performances are indeed statistically significant and not due to random chance. The performance of Adaptive Boosting, which was the best in this study, came with a computational cost. Training the Adaptive Boosting model took 81.5 s which was more than four times how long it took to train the Random Forest. (18.4 s) This cost was expected because the Adaptive Boosting was a sequential ensemble, in which each tree goes through the entire training set one by one, whereas all trees in a Random Forest were trained in parallel, with random subsets of the training set. Lastly, prediction times were 5.5 s for the Adaptive Boosting model and 3.6 s for the Random Forest. These prediction times showed that the ML approach proposed in this study proved to be considerably faster than conventional methods while retaining their accuracy.

5 Conclusion

The results of this study show that different machine learning models can be used to predict the lesion depth for RFA successfully. The data from the embedded system proved that data collection setups with minimal noise are essential even before choosing the ML model. Adaptive Boosting seemed to have the best performance, for both classification and depth estimation. All these results, along with the short prediction times, indicate that an efficient real-time monitoring scheme for RFA can be successfully implemented with a noise-minimized hardware setup and appropriate ML classifier that is able to capture the nonlinear complexities in collected data.

References

1. Besler, E., Wang, Y., Chan, T., Sahakian, A.: Real-time monitoring radiofrequency ablation using tree-based ensemble learning models. Int. J. Hyperth. (2019). https://doi.org/10.1080/02656736.2019.1587008
2. Chu, K.F., Dupuy, D.E.: Thermal ablation of tumours: biological mechanisms and advances in therapy. Nat. Rev. Cancer **14**(3), 199 (2014)
3. Dietterich, Thomas G.: Ensemble methods in machine learning. In: Kittler, Josef, Roli, Fabio (eds.) MCS 2000. LNCS, vol. 1857, pp. 1–15. Springer, Heidelberg (2000). https://doi.org/10.1007/3-540-45014-9_1
4. Goldberg, S.N.: Radiofrequency tumor ablation: principles and techniques. In: Habib, N.A. (ed.) Multi-Treatment Modalities of Liver Tumours, pp. 87–118. Springer, Boston (2002). https://doi.org/10.1007/978-1-4615-0547-1_9
5. Goldberg, S.N., Gazelle, G.S., Dawson, S.L., Rittman, W.J., Mueller, P.R., Rosenthal, D.I.: Tissue ablation with radiofrequency: effect of probe size, gauge, duration, and temperature on lesion volume. Acad. Radiol. **2**(5), 399–404 (1995)
6. Javaherian, A., Soleimani, M., Moeller, K.: A fast time-difference inverse solver for 3D EIT with application to lung imaging. Med. Biol. Eng. Comput. **54**, 1243–1255 (2016)
7. Lardo, A.C., et al.: Visualization and temporal/spatial characterization of cardiac radiofrequency ablation lesions using magnetic resonance imaging. Circulation **102**(6), 698–705 (2000)
8. Martin, S., Choi, C.T.: A post-processing method for three-dimensional electrical impedance tomography. Sci. Rep. **7**, 7212 (2017)
9. Minami, Y., Nishida, N., Kudo, M.: Therapeutic response assessment of RFA for HCC: contrast-enhanced US, CT and MRI. World J. Gastroenterol. WJG **20**(15), 4160 (2014)
10. Pang, G.A., Bay, E., Deán-Ben, X.L., Razansky, D.: Three-dimensional optoacoustic monitoring of lesion formation in real time during radiofrequency catheter ablation. J. Cardiovasc. Electrophysiol. **26**(3), 339–345 (2015)
11. Primavesi, F., et al.: Thermographic real-time-monitoring of surgical radiofrequency and microwave ablation in a perfused porcine liver model. Oncol. Lett. **15**(3), 2913–2920 (2018)
12. Schapire, R.E.: The boosting approach to machine learning: an overview. In: Denison, D.D., Hansen, M.H., Holmes, C.C., Mallick, B., Yu, B. (eds.) Nonlinear Estimation and Classification, pp. 149–171. Springer, New York (2003). https://doi.org/10.1007/978-0-387-21579-2_9
13. Tateishi, R., et al.: Percutaneous radiofrequency ablation for hepatocellular carcinoma: an analysis of 1000 cases. Cancer Interdisc. Int. J. Am. Cancer Soc. **103**(6), 1201–1209 (2005)
14. Wang, Y.C., Chan, T.C., Sahakian, A.V.: Real-time estimation of lesion depth and control of radiofrequency ablation within ex vivo animal tissues using a neural network. Int. J. Hyperth. **34**, 1104–1113 (2018)
15. Wi, H., McEwan, A.L., Lam, V., Kim, H.J., Woo, E.J., Oh, T.I.: Real-time conductivity imaging of temperature and tissue property changes during radiofrequency ablation: an ex vivo model using weighted frequency difference. Bioelectromagnetics **36**(4), 277–286 (2015)

16. Zhang, C., Ma, Y.: Ensemble Machine Learning: Methods and Applications. Springer, New York (2012). https://doi.org/10.1007/978-1-4419-9326-7
17. Zhou, Z., Wu, S., Wang, C.Y., Ma, H.Y., Lin, C.C., Tsui, P.H.: Monitoring radiofrequency ablation using real-time ultrasound Nakagami imaging combined with frequency and temporal compounding techniques. PLoS ONE **10**(2), e0118030 (2015). https://doi.org/10.1371/journal.pone.0118030

Mobile Indoor Localization with Bluetooth Beacons in a Pediatric Emergency Department Using Clustering, Rule-Based Classification and High-Level Heuristics

Patrice C. Roy[1(✉)], William Van Woensel[1], Andrew Wilcox[2], and Syed Sibte Raza Abidi[1]

[1] NICHE Research Group, Dalhousie University, Halifax, Canada
{patrice.roy, william.van.woensel, ssrabidi}@dal.ca
[2] EverAge Consulting, Bedford, Canada
awilcox@everage.ca

Abstract. To mitigate anxiety, pain and dehydration in Pediatric Emergency Departments (PED), it is paramount to tailor educational, motivational and self-help content towards the current location inside the PED, since this reflects the current stage in their PED visit. However, accurately identifying the patient's indoor location in a real-world complex environment, such as a hospital, is still a challenging problem, with interference and attenuation from patients, staff, walls and various electromagnetic sources (e.g., imaging devices). We present an indoor localization methodology that achieve a *best-effort* localization accuracy given the available sensors, (low-quality) motion data and computational platforms. First, we utilize machine learning methods to find a suitable accuracy/granularity balance and then proceed by training a localization model. Then, we apply a set of heuristics based on motion data to eliminate false location estimates. We validated of our approach in a real-life busy and noisy PED with a 92% accuracy.

Keywords: Indoor localization · Machine learning · Mobile health · Bluetooth low energy beacons

1 Introduction

In Pediatric Emergency Departments (PED), children often experience pain, dehydration and anxiety, which can be mitigated by digital solutions that engage patients, and their families, to learn about their condition and care process. *iCare Adventure* is a mobile e-therapeutic app that seeks a patient's information through a series of questions embedded within games [1]. Based on their responses, the app invokes a variety of therapeutic protocols, such as the self-administration of Pedialyte for vomiting or Ventolin for asthma, and presents educational videos tailored to their condition.

PED are spread across a large area, with rooms designated for different assessments and interventions. Patients move through different rooms during a visit, with the patient's triaging determining the order and types of rooms. Throughout patients' visits,

D. Riaño et al. (Eds.): AIME 2019, LNAI 11526, pp. 216–226, 2019.
https://doi.org/10.1007/978-3-030-21642-9_27

it is essential to provide information that is contextually-salient—i.e., associated with the type and function of their current room—which raises the challenge of localizing the patient in the PED. Localization is useful for other healthcare-related purposes as well, such as tagging medical observations with the exam room, and enhancing the efficacy of self-management programs (e.g., medication adherence) [2]. Still, accurate indoor localization is a challenging problem, especially in a complex building such as a hospital, with attenuation due to walls, humans, and radio wave generators [3].

Our case study took place at the PED in the IWK Hospital in Halifax, Nova Scotia (Canada). The PED had iPad Minis (1st generation) [4] for deploying health-related content, while a set of 14 iBeacons [5] provided a reasonable coverage of the PED's rooms. Beacons are small devices that, among others, allow other devices to determine their relative location to the beacon. Our approach involves the standard process of *fingerprinting* [6], i.e., collecting Received Signal Strength (RSS) measures per discrete location (e.g., waiting, exam rooms). At runtime, this allows estimating distances between transmitters (i.e., iBeacon) and receivers (i.e., an iPad) and thus triangulating the user's location. However, our initial evaluation showed that many of the aforesaid factors lead to a loss of accuracy. To lessen the issue, we tried to increase sensor density. However, this merely led to signal strength differences between neighboring beacons dropping below the signal noise. We observe that, in our case, where localization is required in *discrete* locations (e.g., rooms) instead of continuous coordinates, an opportunity exists to offer a *best-effort* localization accuracy. In particular, by merging discrete locations into cohesive regions, we improve localization accuracy at a coarser granularity, which may nonetheless still be in line with application needs. Secondly, there are some opportunities for *sensor fusion*. Although the platform (1st gen iPad Mini) supplies only low-quality motion data, a set of heuristics can leverage the motion data, combined with a semantic location model, to rule out false location estimates.

We present an indoor localization methodology that applies (**a**) hierarchical clustering to obtain a suitable accuracy/granularity balance; (**b**) a decision tree algorithm to train a localization model that targets discrete locations; (**c**) a sliding window method to smooth beacon data; and (**d**) a set of heuristics to rule out false locations. We developed a mobile library, called *iLocate*, which implements our approach. We validated our approach in a real-life busy and noisy PED, and achieved a 92% location accuracy.

2 Background and Related Work

Received Signal Strength (RSS) is one of the simplest and widely used measuring metrics for indoor positioning based on radio waves (RF) [6], although it is quite susceptible to attenuation from walls and other obstacles [6]. By capturing the RSS and transmission power at the receiver, one can estimate the distance between a transmitter and a receiver. Several RF technologies are available on mobile devices, such as WiFi and Bluetooth Low Energy (BLE). While virtually ubiquitous, WiFi networks are mostly optimized for data throughput and coverage rather than positioning [6], and protocol differences lead to lower positioning accuracy than BLE [7]. In general, RF-based positioning is based on establishing proximity between an individual and several beacons based on an RF metric (e.g., RSS). Fingerprinting is the most utilized localization method [6]: it involves an offline phase, where a representative set of metric values are collected, and an online phase, where real-time values are compared with the fingerprint set to estimate the user's location. Fingerprinting thus allows coping with the idiosyncrasies of an indoor setting, where attenuations (e.g., due to walls and equipment) will interfere with positioning. Given a fingerprint set, k-Nearest Neighbor (kNN) is often utilized to estimate the user's location, i.e., where the average of k nearest fingerprint locations (e.g., using Root Mean Square Error) is taken as the user's location [6]. For identifying *discrete* locations, this method requires a highly detailed indoor map to connect absolute coordinates to discrete locations. Instead, we utilize a machine learning (ML) method to estimate an user's discrete location. E.g., the LoCo system [8] relies on supervised ensemble learning (boosting) to build a set of *room-specific* classifiers. Zhang et al. [9] compared the accuracy of Neural Networks and Support Vector Machines (SVM). To the best of our knowledge, no prior work has utilized Decision Trees for indoor localization, or applied clustering for balancing accuracy with granularity.

3 Indoor Localization Methodology

Our baseline localization approach involves a standard fingerprinting method, where RSS measures are collected together with their discrete location in an offline phase [6]. Then, we train indoor localization models using ML and the fingerprint dataset. A trained localization model can estimate the user's discrete location by correlating RSS values at runtime. Figure 1 shows an overview of the indoor localization process.

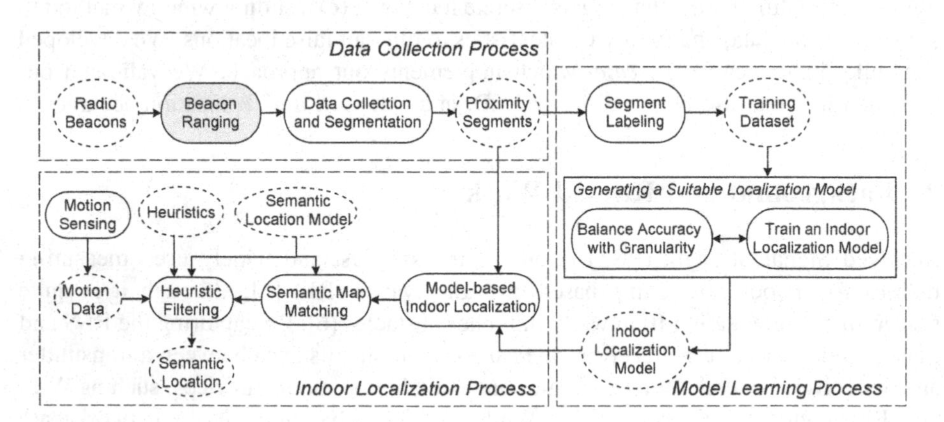

Fig. 1. Overview of the different processes involved in indoor localization.

During the *Data Collection Process*, the *Beacon Ranging* step continuously detects beacon proximity data (i.e., *RSS*, derived *proximity*, and *proximity accuracy*). The *Data Collection and Segmentation* step collects and segments this data stream using a sliding window, coping with beacon data fluctuations by extracting discriminative features (i.e., medians) per segment. These *Proximity Segments* are then utilized to either collect datasets for training localization models (*Model Learning Process*) or track patient locations at runtime (*Indoor Localization Process*). In the *Model Learning Process*, a supervisor performs *Segment Labeling* by labeling each segment with its location (*Training Dataset*). Our approach then proceeds by *Generating a Suitable Localization Model*, which involves ML methods to (a) *Balance Accuracy with Granularity*, where discrete locations are merged into cohesive regions; and (b) *Train an Indoor Localization Model*, where the labeled segments are utilized to train an *Indoor Localization Model*. These two steps may be executed iteratively and in any order (Sect. 3.2).

In the *Indoor Localization Process*, the *Model-Based Indoor Localization* step continuously estimates the user's location, based on detected *Proximity Segments* and a trained *Indoor Localization Model*. The *Semantic Location Model* represents a semantic indoor location model, highlighting the meaning, purpose and connectivity of each location. To locate the user in discrete locations, *Semantic Map Matching* utilizes this model to map estimated positions to *Semantic Locations* (e.g., waiting, exam room). The *Heuristic Filtering* step employs a set of *Heuristics*, together with *Motion Data* and the *Semantic Location Model*, to rule out false location estimates. The *iLocate* mobile library incorporates the *Data Collection* and the *Indoor Localization* processes, and deploys both a data-driven *Indoor Localization Model* and knowledge-driven *Semantic Location Model*. The *Model Learning Process* is carried out on a server, which will create indoor localization models to be run locally on mobile devices.

3.1 Data Collection Process

Within the IWK PED environment, 14 iBeacons are situated at different locations; 1 beacon per exam room (12 rooms, R1 to R13), 1 beacon in the waiting room (R101), and 1 beacon at the nurse's station. Each iBeacon is assigned an UUID (Universally Unique IDentifier). To estimate the proximity of iBeacons, we perform a beacon ranging process where each sample supplies the RSS and the derived *proximity value*, i.e., relative beacon distance (*immediate* = 1, *near* = 2, *far* = 3, *unknown* = 4). We utilize the iOS Core Location framework [10] although any framework providing UUID and RSS values could be utilized. Proximity values are considered more robust as they estimate the source/receiver distance from both the measured RSS as well as the transmission power. Nevertheless, it is still subject to variation due to antenna angles, noisy environments [11] and low sampling rates (we chose 1 Hz to preserve battery).

To further reduce the impact of noisy sensor data, we apply a sliding window approach to separate proximity data streams (i.e., time series) into discrete segments, and then extract *median proximity values* per segment, since medians are more robust to outliers and skewed distributions [12]. A window w is moved by an amount s over the proximity data stream, with each window delineating a segment. The sizes of w and s present a compromise between localization accuracy and delay: a larger segment leads to less fluctuations but there is a longer delay until a coherent segment is collected. In our case, we found that $|w| = 4$ and $|s| = 2$ s presents the best compromise.

3.2 Model Learning Process

After a training dataset is created, we utilize ML to (**a**) balance accuracy with granularity by merging discrete locations into cohesive regions; and (**b**) train an indoor localization model (proximity segment as input, discrete location as output) based on the (merged) training dataset. These sub-processes may be executed iteratively and in any order, e.g., when the model from (**b**) lacks accuracy, step (**a**) may be (re-)executed until the desirable accuracy/granularity balance is attained.

Balancing Accuracy with Granularity. Our initial goal was to locate patients, with reasonable accuracy (ca. 90%), to a room-level granularity inside the IWK PED. However, our initial evaluation only achieved ca. 84% accuracy (see Results). Due to our focus on discrete indoor locations, an opportunity exists to create a *best-effort* localization model where accuracy is better balanced with granularity. By merging discrete locations into geographically cohesive regions, accuracy can be improved at a coarser-granularity that is nevertheless still in line with application needs. For instance, the *iCare Adventure* app [1] personalizes educational content based on whether the patient is in the waiting room or the PED's "operational" part (i.e., exam rooms)—by merging all locations into three regions, i.e., waiting room, exam rooms, and "outside" (i.e., hallway), we may thus improve localization accuracy at no loss of functionality.

To that end, we apply a clustering method to group discrete locations with similar beacon measurements. We use hierarchical agglomerative clustering with the Ward criterion [13] using a set of dissimilarities computed using the Manhattan distance between proximity data segments. To select the best clustering, we rely on the average silhouette width [14], where the silhouette width of an observation (i.e. segment) indicates how similar an observation is to its own cluster (cohesion) compared to other clusters (separation). A large, small and negative width respectively mean that an observation is clustered very well, lying between 2 clusters, or likely in the wrong cluster. Thus, we kept the clusterings with the highest widths. For each clustering, we assign a room to a cluster in case that cluster has the highest proportion of the room's segments. As a second criteria, if the highest proportion of a room's segments is lower than 60%, we reject the clustering—in that case, a room belongs with similar likelihood to 2 regions, which will reduce the performance of the model. Based on the clustering, the relevant proximity segments are re-labeled with the corresponding region. An important but reasonable assumption here is that grouping locations with similar proximity data improves localization accuracy. We revisit this assumption in our discussion.

Training an Indoor Localization Model. We utilize the C5.0 algorithm [15] to learn ruleset-based localization models, where an initial decision tree is grown and collapsed into rules, which are further simplified (pruning, reducing ruleset size). This choice was made because, at least currently, deploying a ML structure on the mobile platform requires a rule-based format (see *Indoor Localization Process* section). Despite their relative simplicity, C5.0 models have been shown to achieve high accuracy [16]. We used 10-repeated 10-fold cross-validation to tune the learning parameters.

3.3 Indoor Localization Process

To perform indoor localization, the *iLocate* library is loaded with a trained *indoor localization model* and *semantic location model* for a specific indoor location. Further, the library is loaded with a set of heuristics to improve localization accuracy.

Semantic Location Model. To create a semantic location model, we utilize *SpatialModeler*, a custom Java application, to delineate discrete locations based on the indoor schematic, and indicate entrances to these locations (see Fig. 2, entrances are orange). By choosing a suitable concept from the hierarchy, we associate the discrete location with a concrete meaning (e.g., exam room). The application stores the location model as a set of First Order Logic (FOL) facts using the *Flora-2* [17] system. A set of FOL rules then determines the connectivity between locations, based on their associated entrances, and estimates the walking time between the connected locations.

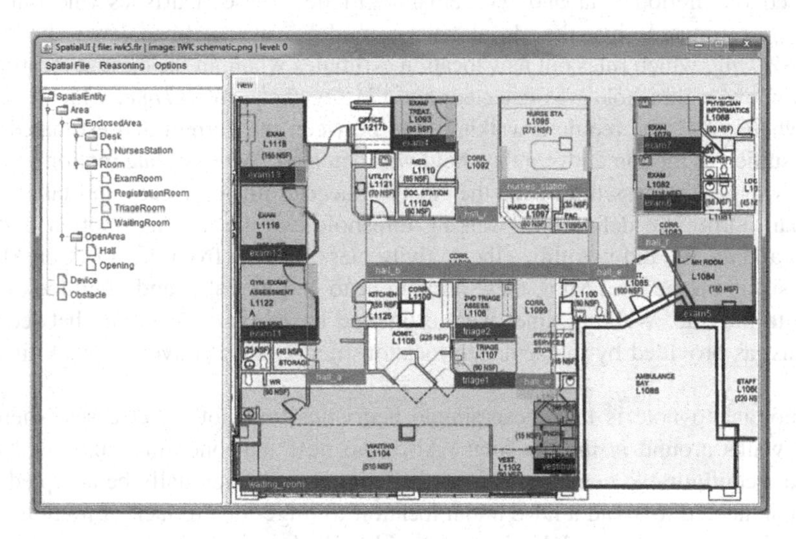

Fig. 2. Screenshot of the SpatialModeler application. (Color figure online)

Location Reasoner. *iLocate* currently relies on a custom probabilistic rule engine (iPad Mini 1st gen is restricted to iOS v.9.3.5 and lacks the Core ML framework for ML support) for executing the localization model, which requires it to be available in rule-based form. Since we rely on the C5.0 algorithm, models can be easily learnt as a set of rules with associated confidence values. In this setup, an inferred fact with the highest cumulative confidence value (i.e., multiple rules may infer the same fact) will be chosen as the user's most likely location. The output of the *indoor localization model*, i.e., the label of an indoor location (i.e., room or region), is used to obtain the patient's semantic location (e.g., waiting or exam room) from the *semantic location model*. This high-level location, together with the semantic location model, allows a third-party app to tailor its behavior and content to the patient's current location.

Heuristic Filter to Eliminate False Location Estimates. A well-known solution to optimizing localization accuracy is *sensor fusion*, where data from other sensors, typically motion-related, are correlated with the RSS readings using probabilistic models [18] (e.g. hidden Markov models, particle filters and conditional random fields). Other types of sensing modalities have known issues, such as acoustic and ultrasound signals (busy hospital settings, atmospheric disturbances), inherent magnetic fields (multitudes of magnetic interference in hospitals), computer vision (obvious privacy issues and budgetary implications), and WiFi (see related work) [6]. In our IWK data collection step, we found that mobile platforms (1st gen iPad Mini from 2012) only provided low-quality motion data and had underpowered hardware (1 GHz dual-core ARM, 512 MB), while they also had to concurrently run heavyweight apps (i.e., iCare Adventure app). Hence, instead of using the aforesaid resource-intensive methods, we opted for a set of low-computation heuristics that rely on motion data and a semantic location model.

Based on motion data and the semantic model, these heuristics rule out false locations estimated by the localization model. We defined two heuristics: (1) *NoWalking*, which rules out new location estimates when, in the current location, no motion above a threshold was detected; and (2) *InsufficientWalkTime*, which does the same when, given the required walking time between the current and estimated location, insufficient motion above a threshold was detected. We calculate motion intensity by applying a low-pass filter over the X, Y, Z accelerometer values and taking their Euclidian norm. We define the "walking threshold" as the average motion intensity while walking. We did not utilize the Activity classification from the iOS CoreMotion library since, on the iPad Mini, these turned out to be unreliable and often took over a minute to provide. Walking times were estimated based on connectivity between two locations, as provided by the semantic location model, and an average walk time per meter.

Important to note is that these simple heuristics will not reject cases where the patient walks around *inside* the room. Also, we note that once the same location is rejected a configurable number of times, the location will eventually be accepted. This was implemented to avoid a false initial location to linger due to lack of motion—else, when one e.g., enters room 7 that is wrongly identified as room 6, any location estimate *correctly* putting them in room 6 would be rejected in case they do not walk around.

4 IWK Case Study Results

Proximity data was collected during several sessions on different days at the IWK PED. The data from the first session was used as the training dataset (2012 segments), while data from the second session was used for testing (841 segments). First, we used the training dataset to identify suitable clusterings (i.e., coherent regions). Then, we trained a localization model for each clustering. Next, we loaded the *iLocate* test app with each localization model, the test dataset, semantic location model, and the two heuristics for evaluation on an iPad Mini (1st gen), i.e., the target platform in our case study. The average duration for the indoor localization process is 28.54 ms per segment.

According to the silhouette criterion, the best clustering involves $2 \leq K \leq 7$ (K = # of clusters or cohesive regions). According to the second criterion, we rejected clusterings with $K > 7$ since those involved cases where, for one room, less than 60% of the segments are inside the same cluster. In our results, we denote a particular clustering as M[*# regions + outside*]. Figure 3 illustrates the regions in the IWK for M8 and M3. We trained an indoor localization model for each clustering, where each room/region is a class, and an *outside* class includes all locations not inside a room/region. Each trained localization model was converted to a list of decision rules (*if-then rules*), where the proximity values of iBeacons are conditions and the output are the patient location.

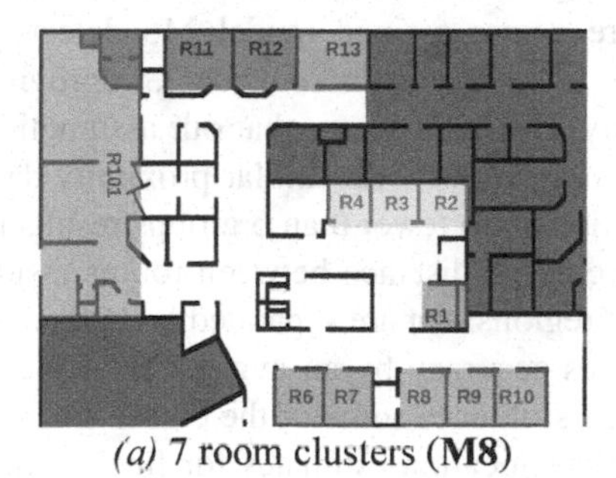

(a) 7 room clusters (**M8**)

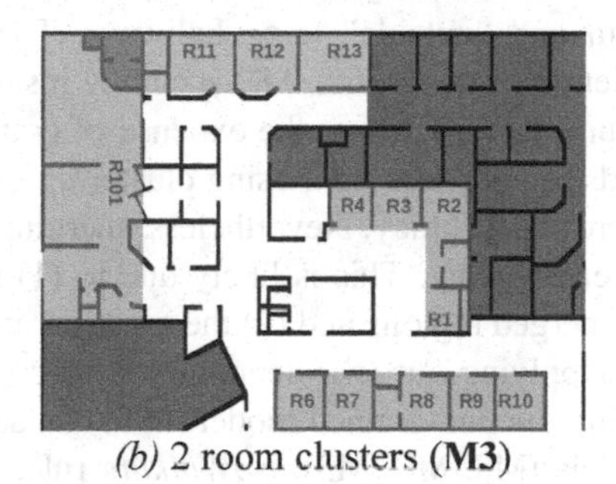

(b) 2 room clusters (**M3**)

Fig. 3. Indoor localization models **M8** and **M3**.

For instance, the indoor localization model M6 includes the following rules:

```
- b_1 <= 1 AND b_8 <= 1 AND b_9 > 1 -> @outside ~ 0.714
- b_7 <= 2 AND b_10 <= 2 AND b_100 > 3 -> @C1 ~ 0.994
- b_3 <= 2 AND b_12 > 2 AND b_12 <= 3 -> @C2 ~ 0.988
- b_11 <= 1 AND b_12 <= 2 -> @C3 ~ 0.984
- b_7 <= 1 -> @C4 ~ 0.942
- b_12 > 2 AND b_13 > 3 AND b_101 <= 2 -> @C5 ~ 0.998
```

E.g., in case the beacons b_1 and b_8 proximity value is ≤ 1 and the beacon b_9 proximity value is >1, then the confidence value of the patient being *outside* is 0.714. Table 1 presents the localization accuracy of each test configuration on unseen proximity data (test dataset). Columns correspond to the number of regions (clusters), and rows indicate whether or not heuristics were used.

Table 1. Evaluation results (accuracy) for different test configurations.

	M14	M8	M7	M6	M5	M4	M3
Default	0.83	0.87	0.87	0.89	0.85	0.85	0.87
2 heuristics	0.90	0.90	0.90	0.92	0.91	0.92	0.90

When utilizing heuristics, *NoWalking* ruled out between 102 (M3) and 170 location estimates (M14), whereas *InsufficientWalkTime* only ruled out 2 locations (M14).

5 Discussion of Evaluation Results

Our baseline approach, i.e., the fine-grained model (M14), detects a patient's location with 0.83 accuracy across 14 locations, including 12 exam rooms, the waiting/triage room and "outside" (e.g., hallway). The more coarse-grained model M6 detects a patient's location with 0.89 accuracy inside one of 5 regions (or "outside"), improving accuracy by ca. 6% at the expense of granularity. Hence, it seems that our assumption holds to an extent, i.e., using clustering to group locations with similar proximity data improves accuracy. Nevertheless, merging locations into fewer than 6 groups results in lower accuracy. This is likely due to (1) the increased distance between rooms inside the merged regions and (2) the creation of large regions that are separated by hallways.

Applying our two heuristics further improves accuracy by an avg. of ca. 4,5%—giving the fine-grained model (M14) an accuracy similar to some of the coarse-grained models (M8, M7, M3). *NoWalking* rules out more location estimates for fine-grained (e.g., 170 for M14) than for coarse-grained models (e.g., 102 for M3)—merging nearby locations into coherent regions reduces the likelihood of false locations due to beacon proximity. *InsufficientWalkTime* ruled out only 2 locations, both for M14. We found that the vast majority of false locations were already rejected by *NoWalking*. Also, once starting to merge locations into larger regions, to avoid false negatives, one must assume the most optimistic scenario—i.e., walking from the current region's edge to the nearest edge of the other region. This greatly reduces the "required" walking times between locations, and hence the effectiveness of this heuristic. Note that the number of rejected locations is not necessarily linear w.r.t. the increase in accuracy, since rejecting a false location "early" on while inside a room will have a larger impact than later.

6 Conclusions and Future Work

The ability to identify the location of movable resources, patients and care providers in a healthcare facility allows us to optimize resource utilization, as well as the tailored delivery of educational material and care services. In an assisted living setting, it can enhance the efficacy of patient self-management programs [2]. We presented a novel application of ML methods to improve indoor localization accuracy by merging rooms into cohesive regions. We further evaluated the effectiveness of two motion-based heuristics in improving accuracy. Our approach trains indoor localization models using the C5.0 ML algorithm, for different clusterings of locations, which estimate the user's position by correlating beacon proximity values. Together with a semantic location model, these localization models are used by the *iLocate* app library to identify the patient's semantic location in the indoor environment. We validated our localization models in a real-life busy PED (IWK Health Center). Our approach enables a fine-grained indoor localization (14 locations) with good accuracy (ca. 90%), and supports coarser-grained localization (6 locations) at a higher accuracy (ca. 92%) that may still suit certain application needs. The proposed methodology is scalable to other health-care settings with location sensors and other classification methods (e.g. random forests).

Acknowledgements. This work was funded by an NSERC Discovery Grant.

References

1. Taylor, B.W., et al.: Implementation of a game-based information system and e-therapeutic platform in a pediatric emergency department waiting room: preliminary evidence of benefit. Procedia Comput. Sci. **63**, 332–339 (2015)
2. Roy, P.C., Abidi, S.R., Abidi, S.S.R.: Possibilistic activity recognition with uncertain observations to support medication adherence in an assisted ambient living setting. Knowl.-Based Syst. **133**, 156–173 (2017)
3. Calderoni, L., Ferrara, M., Franco, A., Maio, D.: Indoor localization in a hospital environment using Random Forest classifiers. Expert Syst. Appl. **42**, 125–134 (2015)
4. iPad mini - Technical Specifications. https://support.apple.com/kb/sp661?locale=en_US
5. iBeacon - Apple Developer. https://developer.apple.com/ibeacon/
6. Zafari, F., Gkelias, A., Leung, K.K.: A survey of indoor localization systems and technologies. CoRR. abs/1709.0 (2017)
7. Faragher, R., Harle, R.: Location fingerprinting with Bluetooth low energy beacons. IEEE J. Sel. Areas Commun. **33**, 2418–2428 (2015)
8. Biehl, J.T., Cooper, M., Filby, G., Kratz, S.: LoCo: a ready-to-deploy framework for efficient room localization using Wi-Fi. In: Proceedings of the 2014 ACM International Joint Conference on Pervasive and Ubiquitous Computing, pp. 183–187. ACM, New York (2014)
9. Zhang, L., Liu, X., Song, J., Gurrin, C., Zhu, Z.: A comprehensive study of Bluetooth fingerprinting-based algorithms for localization. In: 2013 27th International Conference on Advanced Information Networking and Applications Workshops, pp. 300–305. IEEE (2013)
10. Core Location|Apple Developer Documentation. https://developer.apple.com/documentation/corelocation
11. Spachos, P., Papapanagiotou, I., Plataniotis, K.N.: Microlocation for smart buildings in the era of the Internet of Things: a survey of technologies, techniques, and approaches. IEEE Signal Process. Mag. **35**, 140–152 (2018)
12. Gaddis, G.M., Gaddis, M.L.: Introduction to biostatistics: part 2, descriptive statistics. Ann. Emerg. Med. **19**, 309–315 (1990)
13. Murtagh, F., Legendre, P.: Ward's hierarchical agglomerative clustering method: which algorithms implement ward's criterion? J. Classif. **31**, 274–295 (2014)
14. Rousseeuw, P.J.: Silhouettes: a graphical aid to the interpretation and validation of cluster analysis. J. Comput. Appl. Math. **20**, 53–65 (1987)
15. Wu, X., et al.: Top 10 algorithms in data mining. Knowl. Inf. Syst. **14**, 1–37 (2007)

16. Frank, E., Witten, I.H.: Generating accurate rule sets without global optimization. In: Shavlik, J.W. (ed.) Proceedings of the Fifteenth International Conference on Machine Learning (ICML), pp. 144–151. Morgan Kaufmann Publishers Inc., San Francisco (1998)
17. Kifer, M., Yang, G., Wan, H., Zhao, C.: ERGOLite (a.k.a. Flora-2): User's Manual (2017)
18. Seitz, J., Vaupel, T., Jahn, J., Meyer, S., Boronat, J.G., Thielecke, J.: A Hidden Markov Model for urban navigation based on fingerprinting and pedestrian dead reckoning. In: 2010 13th International Conference on Information Fusion, pp. 1–8. IEEE (2010)

NONCADO: A System to Prevent Falls by Encouraging Healthy Habits in Elderly People

Elisa Salvi[1(✉)], Silvia Panzarasa[1], Riccardo Bagarotti[1], Michela Picardi[2], Rosangela Boninsegna[2], Irma Sterpi[2], Massimo Corbo[2], Giordano Lanzola[1], Silvana Quaglini[1], and Lucia Sacchi[1]

[1] University of Pavia, 27100 Pavia, Italy
elisa.salvi01@universitadipavia.it

[2] Department of Neurorehabilitation Sciences, Casa di Cura Privata del Policlinico (CCP), 20144 Milan, Italy

Abstract. Falls in the elderly are a known problem, leading to hospitalization, impaired life quality, and social costs. Falls are associated to multiple risk factors, related to the subject's health, lifestyle, and living environment. Living alone makes it difficult to detect a patient's decline, which increases the fall risk. In this paper, we present NONCADO, a project funded by the Lombardy Region (Italy), aimed at developing a system for preventing falls in the elderly living alone, by integrating data from a network of sensors (both wearable and environmental). The collected data are analyzed by a decision support system (DSS) that exploits advanced temporal data analysis techniques to detect behaviors known to increase the individual risk (e.g. moving within the house with inadequate lighting, or performing not enough physical activity). A daily report listing the detected risky behaviors is produced and delivered through a mobile app. Since we address long-term monitoring, it's important to detect as well the changes in a subject's habits that may increase fall risk. Such changes are summarized in a weekly report. A preliminary feasibility evaluation of the system was performed during a 2-weeks pilot study involving 16 patients treated at the Casa di Cura Privata del Policlinico hospital, in Milan, Italy. Patients were asked to perform 5 activities, and the system's ability to correctly detect them was assessed. The study results were encouraging, as the system reached an overall accuracy of 90%.

Keywords: Fall risk · Temporal data analysis · Home monitoring

1 Introduction

Falls in elderly people are a social problem, being a major cause of loss of independence, hospitalization (or increase of hospital stay), decreased quality of life, and increased social costs [1]. They are also associated with psychological and functional sequelae, independently from the injury severity. Falls are associated to a variety of risk factors, related to the subject's health status (e.g., neurological disorders, traumas, drug

D. Riaño et al. (Eds.): AIME 2019, LNAI 11526, pp. 227–232, 2019.
https://doi.org/10.1007/978-3-030-21642-9_28

therapies), lifestyle (e.g., lack or excess of physical activity), social and economic condition (possibly leading to malnutrition or impossibility of adapting the home to the subject's needs), and living environment (e.g., inadequate lighting, slippery floors).

Within NONCADO, a project funded by the Lombardy Region in Italy, we aim at preventing falls in elderly people living alone at home. Living alone implies a difficult or delayed detection of a possible decline that in turn may increase the subject's fall risk. Reasonably, a decline influences the patient's habits, concerning for example daily activity, sleep quality/quantity, time spent outside the house, and consumption of hot meals [1]. These considerations motivated the development of a monitoring system able to detect such changes, and to inform the patient's family of a possible decline. To detect changes and risky behaviors, NONCADO relies on a decision support system (DSS), which integrates data from (1) a network of environmental sensors, monitoring both the movement of the subject within the house and the quality of the environment, and (2) an activity tracker, monitoring the subject's activity and sleep. At the end of each day, the DSS produces a report, meant both for the monitored elderly subject and his/her family. This report warns against the potentially risky behaviors that have been detected during the day. An additional report is provided at the end of each week, to compare the considered week to the previous ones. The reports are delivered through a mobile app, and should allow the subject's family to remotely monitor the user, and to detect significant changes in his/her habits, to early identify a possible decline.

Early detection of changes in the patient's daily habits is the innovative feature of the NONCADO system, compared to the devices known in the literature [2, 3]. In fact, most of the existing projects are aimed at providing support when a fall occurs. Usually such systems are designed to detect the fall event, reach for the family member who is the closest to the patient's position, and possibly arrange an healthcare intervention [4]. Few systems focus on fall prevention. The majority of them assist the subject in performing specific exercises to maintain her/his walking ability [5]. Others [6] ask the subject to perform a specific action when receiving a specific audio/video signal. By analyzing the subject's reaction time, they detect changes in his physical/mental state, which could correspond to an increase in his/her risk of falling.

In this work we present the first prototype of the NONCADO system. First, we describe how it detects daily activities by integrating data from the network of environmental sensors. Furthermore, we present the results of its application in a preliminary evaluation study on a small group of neurological patients.

2 Methods

The NONCADO system includes two main components: a network of sensors to monitor the activities of daily living, and a DSS, implemented in java, to detect such activities and to identify possible unhealthy deviations occurring in time.

The sensors collect measurements of motion using passive infrared technology (PIR), temperature, humidity, and lighting within each room of the house. In addition, pressure mats can be placed under the mattress to detect a subject's presence in bed. For the prototype developed in the project, we used three pressure mats, each one positioned on a section of the bed (i.e., head, back, feet). Finally, one or more photocells positioned close to the door are used to detect when the subject passes through the door, and in which direction (enter/exit). The sensors are connected to Raspberry PI boards, which gather the collected measurements, and transmit them to a signal repository.

The DSS processes the raw signals to identify the patient's behavior and daily activities. To detect such activities, the DSS exploits Temporal abstraction (TA), a technique that allows converting a numeric time series (TS) to an interval-based representation, where each time interval has a label that summarizes the qualitative behavior of the TS in that interval [7]. To perform TA on our data, we use JTSA (Java Time Series Abstractor), a framework recently developed by the University of Pavia [8]. JTSA is modular: it provides several algorithms for TA, that can be personalized by tuning their parameters, and can be combined in workflows to detect user-defined patterns. Workflows must be formalized using a JTSA-specific XML file. In this work, we have identified the following daily activities to be detected by the system: *resting in bed, resting in bed with getting up, cooking, leaving the kitchen while cooking,* and *washing the dishes*. To detect such activities, we formalized four JTSA workflows, whose characteristics are shown in Table 1. For each workflow, we provide the input variables (Input TS) and a description of the pattern we aim to detect. The parameters of each JTSA algorithm in these workflows were tuned after functional tests aimed at observing the response of the involved sensors to stimuli designed to simulate the specific activity.

Table 1. Workflows formalized to summarize the TS provided by the environmental sensors.

Workflow ID	Input TS	Pattern to detect
Presence on a bed section	- Pressure (from one pressure mat)	Increasing pressure followed by stationary pressure
Motion in a room	- Motion (from multiple PIR sensors in the room)	Time intervals in which motion is detected by at least one motion sensor in the room
Presence in a room	- Pressure (from the pressure mats), if available - Motion (from all the PIR sensors in the room)	Time intervals that verify either *Presence on a bed section* or *Presence in a room* for the considered room
Cooking	- Temperature (from a sensor located near the stove)	Increasing temperature, followed by decreasing temperature
Washing the dishes	- Motion (from a PIR sensor next to the sink)	Time intervals in which motion is detected by the dedicated PIR sensor

For the detection of each activity, the DSS uses one or more workflows. In particular, to identify *resting in bed* and *resting in bed with getting up*, it first runs *Presence on a bed section* for each bed section. It then extracts all the time intervals in which at least one bed section is active. For each interval, the DSS distinguishes whether the subject is sitting (only one section is active) or lying (more than one section is active) on the bed. Multiple consequent episodes of *resting in bed,* separated by a time interval of absence from the bedroom (detected by running the *Presence in a room* workflow), verify *resting in bed with getting up. Cooking* is detected by running the *Cooking* workflow. If the photocell in the kitchen reports an exit occurring in an instant that verifies the *cooking* activity, also *exiting the kitchen while cooking* is verified. Finally, the *washing the dishes* activity is recognized by running the *Motion in a room* workflow to detect movements next to the sink faucet. In this case, the system selects as input for the workflow only the measurements provided by a dedicated motion sensor placed next to the sink. Given the complexity of the steps needed to define a specific activity, the DSS uses the Drools (https://www.drools.org/) framework to formalize into a set of rules the sequence of steps that must be completed to collect the necessary monitoring variables, to pre-process the signals to be compatible with the JTSA workflows, to run the workflows, and to save the obtained results (i.e., the time intervals in which the activity is detected) into a dedicated repository.

3 Results and Discussion

In September 2018, a prototype of the system was tested in a 2-weeks pilot study involving 16 patients (6 females, 10 males, aged 72.69 ± 8.53 years) with history of falls, treated at the Department of Neurorehabilitation Sciences, Casa di Cura del Policlinico (CCP), in Milan, Italy. The patients were asked to perform the 5 above-described activities during rehabilitation sessions held in the *"Living Lab"*, an environment specifically devoted to practice activities of daily living with a therapist. Every patient participated in two study sessions; in each session he/she completed each activity once. One patient interrupted one session early due to a concomitant visit, thus completing 3 activities out of 5. During each session, an observer was present in the room together with the patient and the therapist, and recorded the details (start time and end time) of the actions performed by the patient. The pilot study was approved by the Ethical Committee of Fondazione IRCCS Cà Granda Area 2, Milan, Italy (nr 570_2018bis).

To evaluate the performance of the system, the results of the DSS elaboration in terms of intervals of validity of each activity were compared to the details reported by the observer during each rehabilitation session. The accuracy of the system, computed as the ratio between the number of correctly identified actions and the number of actions actually performed by the patients was 31/32 (97%) for resting in bed, 30/31 (97%) for resting in bed with getting up, 30/31 (97%) for cooking, 25/32 (78%) for exiting the kitchen while cooking, and 26/32 (81%) for washing the dishes. Thus, considering all the activities, the system reached an overall accuracy of 142/158 (90%).

Identifying *exiting the kitchen while cooking* was the most challenging task, with 7 wrongly detected actions. In 3 out of 7 cases, the error was due to unreliable measures (i.e., −100 °C) provided by the sensor in the time intervals of interest. In 2 out of 7 cases, both occurring at the end of a test day, the photocell log was written with a delay, which caused the most recent data to be lost when the system was shut down. In 2 out of 7 cases, the photocell was able to detect the movement, but it was not able to determine its direction, probably due to the presence of the therapist who had to assist the patient on the way out of the living lab for safety reasons. This highlights one limitation of the system, which is designed to deal with a subject living alone. The other activity that showed low accuracy was *washing the dishes*. However, 5 out of the 6 errors happened during the same test day and were due to a cable throttling, which prevented the communication between the motion sensor and the Raspberry PI board. The sixth error was due to a temporary connection problem, causing the sensor to fail sending the collected data, as well. A connection problem also occurred for pressure mats, leading to the inability to detect one *resting in bed with getting up* activity. The system also failed in identifying one *resting in bed* activity. In this case, a spike in the pressure signal, probably due to an intervention of the therapist who needed to help the patient, prevented JTSA from detecting the increase in pressure triggered by the presence of the patient in the bed. Although spikes in the TS of pressure measurements occurred only in this case, it highlighted the need to remove them from the signal before analyzing it.

4 Conclusion

In this paper we presented NONCADO, a system aimed at preventing the risk of falling by identifying potentially unhealthy behaviors of its users through a network of sensors monitoring the daily activities within the house. A prototype of the system detecting 5 activities was tested during a pilot study. The system showed good performance in identifying the selected activities. In our laboratory we also tested the system ability to detect other patterns of interest, including movement within a room in poor lighting conditions. These activities were performed by healthy volunteers wearing a fitness tracker, to take into account the performed physical activity as well. In the future, we are planning a more extensive test, to be performed within real life settings.

References

1. Falls in older people: assessing risk and prevention. https://www.nice.org.uk/guidance/cg161
2. Majumder, S., et al.: Smart homes for elderly healthcare-recent advances and research challenges. Sensors **17**, E2496 (2017)
3. Nguyen, H., Mirza, F., Naeem, M.A., Baig, M.M.: Falls management framework for supporting an independent lifestyle for older adults: a systematic review. Aging Clin. Exp. Res. **30**, 1275–1286 (2018)
4. De Backere, F., et al.: Social-aware event handling within the FallRisk project. Methods Inf. Med. **56**, 63–73 (2017)
5. Choi, S.D., Guo, L., Kang, D., Xiong, S.: Exergame technology and interactive interventions for elderly fall prevention: a systematic literature review. Appl. Ergon. **65**, 570–581 (2017)

6. Ejupi, A., Gschwind, Y.J., Brodie, M., Zagler, W.L., Lord, S.R., Delbaere, K.: Kinect-based choice reaching and stepping reaction time tests for clinical and in-home assessment of fall risk in older people: a prospective study. Eur. Rev. Aging Phys. Act. **13**, 2 (2016)
7. Shahar, Y., Musen, M.A.: Knowledge-based temporal abstraction in clinical domains. Artif. Intell. Med. **8**, 267–298 (1996)
8. Sacchi, L., Capozzi, D., Bellazzi, R., Larizza, C.: JTSA: an open source framework for time series abstractions. Comput. Methods Programs Biomed. **121**, 175–188 (2015)

The Minimum Sampling Rate and Sampling Duration When Applying Geolocation Data Technology to Human Activity Monitoring

Yan Zeng[1], Paolo Fraccaro[1,2], and Niels Peek[1](✉)

[1] Centre for Health Informatics, Division of Informatics, Imaging and Data Science, The University of Manchester, Manchester, UK
niels.peek@manchester.ac.uk
[2] Hartree Centre, IBM Research UK, Daresbury, UK

Abstract. The availability of geolocation sensors embedded in smartphones introduces opportunities to monitor behaviours of individuals. However, sensing geolocation at high sampling rates can affect the battery life of smartphones. In this study, we sought to explore the minimum sampling rate of geolocation data required to accurately recognise out-of-home activities. We collected geolocation data from 19 volunteers sampled every 10 s for 8 non-consecutive days on average. These volunteers were also instructed to complete a paper-based activity diary to record all activities during each data collection day. For finding the minimum sampling rate, we derived datasets at lower sampling rates by down sampling the original data. A semantic analysis was applied using a previously published activity recognition algorithm. The impact of the sampling rates on accuracy of the algorithm was measured through the F_1 score. The best F_1 score was found at sampling intervals of 2 min and it did not drop substantially until the sampling intervals increased to 10 min. Our study proves the feasibility of monitoring activities at low sampling rates using smartphone-based geolocation sensing.

Keywords: Geolocation · Global positioning system · Smartphones · Sample frequency

1 Introduction

Personal digital devices with geolocation capabilities are ubiquitous nowadays, with three billion people estimated to use smartphones globally [1]. This introduces new possibilities of getting deeper understanding of human activities and behaviours based on geolocation data, at higher spatio-temporal resolution than traditional methods (e.g. questionnaires, activity diaries or interviews) and cheaper costs [2].

Currently, geolocation data are used for several applications: navigation systems to guide users in their journeys [3]; location-based recommender systems, suggesting nearby places based on the user's previous history and preferences [4, 5]; social networking services that based on users previous locations history connect individuals with similar interests [6, 7]; monitoring of criminal offenders based on their location [8]; monitoring of health-related activities and behaviours [9–11].

The original version of this chapter was revised: the family name has been corrected. The correction to this chapter is available at https://doi.org/10.1007/978-3-030-21642-9_53

D. Riaño et al. (Eds.): AIME 2019, LNAI 11526, pp. 233–238, 2019.
https://doi.org/10.1007/978-3-030-21642-9_29

Although the contexts might be different, similar challenges are encountered when processing geolocation data for human behaviour analysis. Geolocation data are subject to measurement error. This can be as low as 10 m, but on current smartphones the median is around 70 m [12]. Missing data is another issue faced when analysing geolocation data [13]. This can be caused by: signal loss in most buildings [14]; users not carrying their devices with them [13]; battery draining due to frequent geolocation data sampling [15, 16]. Even when the noise is removed, raw geolocation data is just the spatio-temporal positions of individuals, which is essentially meaningless. In applications where going beyond quantitative statements as "more activity" and "less activity" is needed, geolocation data trajectories are enriched with geographic and semantic information. This comes from databases such as Google Maps [17], Open-StreetMap (OSM) [18], and Foursquare [19], which often are user-editable and therefore might be inaccurate or incomplete [10].

This paper aims to explore the minimum sampling rate of geolocation data required to accurately recognise out-of-home activities. This is an essential aspect of geolocation data collection, since high sampling rate frequency leads to draining batteries and consequently missing values [15, 16]. This was investigated by applying a modified version of an algorithm originally developed by Difrancesco et al. [10] to detect out-of-home activities relevant to schizophrenia on data from 19 healthy volunteers that was progressively down-sampled to evaluate impact on the algorithm's performance.

2 Methods

We collected data from healthy volunteers (students and staff) at the University of Manchester in Spring 2017. The study received ethical approval from the Research Ethics Committee at the University of Manchester, with each participant who received £20 at the end of the study, as a compensation for their time and effort.

Participants were asked to collect data for ten randomly chosen days during a period of four weeks after entering the study. On data recording days, participants were instructed to collect geolocation data, while performing everyday activities as normal, using an application on their smartphone called GPSLogger for Android [20]. GPSLogger for Android collects raw geolocation data (i.e. geolocation timestamp, latitude, longitude) without profiling or analysing it. For this study, sample frequency was set up to one sample every 10 s, and to limit noise we set up GPSLogger for Android to record only data points with an accuracy of at least 40 m. As a gold standard for evaluating the accuracy of the algorithm we tested to identify daily activities from geolocation data, participants were asked to fill an activity diary containing places visited and activities undertaken for each data recording day.

To explore our research question, we improved an algorithm proposed by Difrancesco et al. [10], which was originally developed to identify out-of-home activities from geolocation data in the context of schizophrenia. The algorithm is composed of four steps. First, we found geolocations visited, such as where a participant stopped to perform an activity, by using two complementary approaches: a time-threshold method [21], which detects geolocations visited by the user by looking for signal loss when a participant enters into a building; a density-based method [22], which searches spatio-temporally dense areas to cluster geolocation data points, and extract their centroid as a geolocation that was visited. Second, we used a modified version of k-means [21] that is applied to all geolocations identified in the first step (using the two methods). The centroids of the different clusters identified are labelled as places visited by the user. The third step aims at associating each place visited to the most likely place of interest (POIs) in the real-world. This is done by applying a set of heuristic rules on the information retrieved from OSM [18]. Finally, to the found POIs we associated activities relevant to monitor schizophrenia (home, employment, shopping, sports, social activities, recreational activities, and other).

To find the minimum sampling rate to accurately detect out-of-home activities, we tested the performance of our algorithm on the original dataset (i.e. collected every 10 s) and down-sampled versions that we created by reducing the sampling rate to: 30 s, 1 min, 2 min, 3 min, 4 min, 5 min, 10 min and 20 min. The time threshold used by Difrancesco et al. [10] in the first step of their algorithm was kept constant at 10 min, except for the down-sampled dataset with a sampling rate of 20 min. This was done to avoid that each GPS data point would be labelled as a geolocation visited by the time-threshold method.

To evaluate algorithm performance, we compared the activities predicted by the algorithm to what was recorded by participants in their activity diaries. Note that no data were used to train the algorithm. We used two methods. First, we considered as correctly classified all unique activities that matched between the algorithm results and the activity diary on a given day. For example, if the algorithm found that someone went shopping that day, and the participant reported to have gone shopping at least once, this would be taken as a true positive. Second, since this approach is potentially too optimistic, we also accounted for the number of times an activity was performed. For example, if someone had reported to have gone shopping twice and our algorithm detected three shopping sessions that day, two would be considered as correctly classified (e.g. true positive) and one as incorrectly classified (e.g. false positive). We assessed performance with recall (i.e. proportion of correctly classified activities out of the ones recorded in the diary), precision (i.e. proportion of correctly classified activities out of the total number of activities identified) and F1 score (i.e. harmonic mean between precision and recall). Finally, we re-ran the algorithm with each set of resampled data, and evaluated diferenes in algorithm performance.

3 Results

Nineteen people volunteerd to take part in the study (5 males, 14 females), with a mean age of 25.4 years, (Standard Deviation [SD], 6.7 years). The vast majority (seveteen) were students. Five participants used their own smartphone for the study, all others used a smartphone that we handed out to them. The sampling duration varied from 4 days to 10 days.

Our algorithm showed moderate recall and good precision on activity recognition from individuals' geolocation data (see Table 1). As expected, the algorithm's performance decreased when accounting for the number of times an activity was recorded in the activity diary (i.e. method 2). However, results by the two methods for calculating performance are consistent in terms of trends. Particularly, recall decreased when the level of down-sampling increased. Conversely, precision increased when we used a higher degree of down-sampling. The F1 score peaked at a sampling frequency of 2-min, and progressively decreased with more down-sampled datasets.

Table 1. Results from testing on different down-sampled datasets. SD: Standard deviation.

Sampling frequency	Daily activities			Daily activity counts		
	Recall (SD)	Precision (SD)	F1 score (SD)	Recall (SD)	Precision (SD)	F1 score (SD)
10 s	0.65 (0.10)	0.76 (0.13)	0.70 (0.10)	0.52 (0.11)	0.65 (0.19)	0.57 (0.13)
30 s	0.64 (0.08)	0.80 (0.14)	0.71 (0.10)	0.52 (0.10)	0.73 (0.18)	0.60 (0.11)
1 min	0.62 (0.09)	0.81 (0.12)	0.70 (0.08)	0.49 (0.11)	0.75 (0.17)	0.58 (0.10)
2 min	0.64 (0.09)	0.84 (0.09)	0.72 (0.08)	0.51 (0.10)	0.79 (0.14)	0.61 (0.09)
3 min	0.59 (0.13)	0.84 (0.12)	0.68 (0.12)	0.46 (0.12)	0.78 (0.15)	0.57 (0.12)
4 min	0.57 (0.14)	0.80 (0.14)	0.66 (0.13)	0.45 (0.12)	0.74 (0.18)	0.55 (0.13)
5 min	0.56 (0.10)	0.81 (0.10)	0.66 (0.09)	0.43 (0.12)	0.75 (0.14)	0.54 (0.12)
10 min	0.56 (0.12)	0.84 (0.11)	0.66 (0.10)	0.42 (0.12)	0.81 (0.12)	0.54 (0.11)
20 min	0.50 (0.15)	0.84 (0.11)	0.61 (0.14)	0.35 (0.14)	0.81 (0.14)	0.47 (0.15)

4 Discussion and Conclusion

Few studies have considered the effect of a decreasing geolocation data sampling rate on performance, when inferring out-of-home behaviours. We found a similar experiment conducted by Zheng et al. [23]. By tracking more than 33,000 taxis, Zheng et al. compared the effect of sampling rate on four POI recognition algorithms. In their study, they also found that performance was dropping when the sampling rate was lower than 10 min.

This study has several limitations. First, OSM is a user-editable database. The large number of contributors enables the database to be complete, but it could also introduce inaccurate labels to places in the database. Second, the process of data collection relied on the compliance of participants and, as for any instrument of this kind, we expect the self-reported activity diary to be affected by recall bias. Finally, although sampling interval was predefined as 10 s, for some participants this sampling rate was not always maintained due to signal loss.

To conclude, the inclusion of geolocation sensors in smartphones introduces new opportunities for monitoring of individuals' activities in several domains. However, high-sampling-rate geolocation data collection affects the battery life of smartphones. We found the best performance (precision, recall and F1 score) for sampling intervals of 2 min. However, performance did not change substantially when down-sampling to intervals of 10 min. This indicates that it is probably feasible to accurately monitor activities at a low sampling rate using smartphone-based geolocation sensing. We recommend that future studies use a geolocation sampling rate of at least once per minute, and that more simulation studies are carried out on larger datasets to investigate whether further decrease in sampling frequency is warranted.

Acknowledgements. This project was funded by the Engineering and Physical Sciences Research Council (grant EP/P010148/1; The Wearable Clinic: Connecting Health, Self and Care).

References

1. Interim Ericsson Mobility Report (2018). https://www.ericsson.com/assets/local/mobility-report/documents/2018/emr-interim-feb-2018.pdf
2. Stanley, K., Yoo, E.-H., Paul, T., Bell, S.: How many days are enough?: capturing routine human mobility. Int. J. Geogr. Inf. Sci. **32**, 1485–1504 (2018). https://doi.org/10.1080/13658816.2018.1434888
3. Raine, J., Withill, A., Morecock, E.: Literature Review of the Costs and benefits of Traveller Information Projects (2014)
4. Guo, B., Fujimura, R., Zhang, D., Imai, M.: Design-in-play: improving the variability of indoor pervasive games. Multimed. Tools Appl. **59**, 259–277 (2012). https://doi.org/10.1007/s11042-010-0711-z
5. Royer, D., Deuker, A., Rannenberg, K.: Mobility and identity. In: Rannenberg, K., Royer, D., Deuker, A. (eds.) The Future of Identity in the Information Society, pp. 195–242. Springer, Heidelberg (2009). https://doi.org/10.1007/978-3-642-01820-6_5
6. Zheng, Y., Zhang, L., Ma, Z., Xie, X., Ma, W.-Y.: Recommending friends and locations based on individual location history. ACM Trans. Web. **5**, 5:1–5:44 (2011). https://doi.org/10.1145/1921591.1921596
7. Data Description: GeoLife User Guide 1.2. Microsoft Research Asia. **2**, 31–34 (2011)
8. Daubal, M., Fajinmi, O., Jangaard, L.: Safe step: a real-time GPS tracking and analysis system for criminal activities using ankle bracelets. In: Proceedings 21st ACM SIGSPATIAL International Conference Advances in Geographic Information System, pp. 502–505 (2013). https://doi.org/10.1145/2525314.2525316
9. Wahl, H.W., et al.: Interplay of cognitive and motivational resources for out-of-home behavior in a sample of cognitively heterogeneous older adults: findings of the SenTra project. J. Gerontol. Ser. B Psychol. Sci. Soc. Sci. **68**, 691–702 (2013). https://doi.org/10.1093/geronb/gbs106
10. Difrancesco, S., et al.: Out-of-home activity recognition from GPS data in schizophrenic patients. In: Proceedings of IEEE Symposium Computer Medical System, pp. 324–328, August 2016. https://doi.org/10.1109/cbms.2016.54
11. Torous, J., Kiang, M.V., Lorme, J., Onnela, J.-P.: New tools for new research in psychiatry: a scalable and customizable platform to empower data driven smartphone research. JMIR Ment. Heal. **3**, e16 (2016)
12. Bhattacharya, T., Kulik, L., Bailey, J.: Automatically recognizing places of interest from unreliable GPS data using spatio-temporal density estimation and line intersections. Pervasive Mob. Comput. **19**, 86–107 (2015). https://doi.org/10.1016/j.pmcj.2014.08.003
13. Krenn, P.J., Titze, S., Oja, P., Jones, A., Ogilvie, D.: Use of global positioning systems to study physical activity and the environment: a systematic review. Am. J. Prev. Med. **41**, 508–515 (2011) https://doi.org/10.1016/j.amepre.2011.06.046
14. Marmasse, N., Schmandt, C.: Location-aware information delivery with ComMotion. In: Thomas, P., Gellersen, H.-W. (eds.) HUC 2000. LNCS, vol. 1927, pp. 157–171. Springer, Heidelberg (2000). https://doi.org/10.1007/3-540-39959-3_12

15. Glasgow, M.L., et al.: Using smartphones to collect time–activity data for long-term personal-level air pollution exposure assessment. J. Expo. Sci. Environ. Epidemiol. **26**, 356 (2016)
16. Boonstra, W.T., Nicholas, J., Wong, J.J.Q., Shaw, F., Townsend, S., Christensen, H.: Using mobile phone sensor technology for mental health research: integrated analysis to identify hidden challenges and potential solutions. J. Med. Internet Res. **20**, e10131 (2018). https://doi.org/10.2196/10131
17. Google: Google Maps API (n.d.). http://www.webcitation.org/6ubEADyQl
18. OpenStreetMap API (n.d.). https://wiki.openstreetmap.org/wiki/API
19. Foursquare (n.d.). https://developer.foursquare.com/. Accessed 4 June 2018
20. GPSLogger for Android (n.d.). https://gpslogger.app/. Accessed 4 June, 2018
21. Ashbrook, D., Starner, T.: Using GPS to learn significant locations and predict movement across multiple users. Pers. Ubiquit. Comput. **7**, 275–286 (2003). https://doi.org/10.1007/s00779-003-0240-0
22. Fehér, M., Forstner, B.: Identifying and utilizing routines of human movement. In: Second Eastern European Regional Conference on the Engineering of Computer based System Identifying (2011). https://doi.org/10.1109/ecbs-eerc.2011.28
23. Zheng, K., Zheng, Y., Xie, X., Zhou, X.: Reducing uncertainty\ of low-sampling-rate trajectories. In: Proceedings of International Conference Data Engineering, pp. 1144–1155 (2012). https://doi.org/10.1109/icde.2012.42

Clustering, Natural Language Processing, and Decision Support

AI-Driven Pathology Laboratory Utilization Management via Data- and Knowledge-Based Analytics

Syed Sibte Raza Abidi[1](✉), Jaber Rad[1], Ashraf Abusharekh[1], Patrice C. Roy[1], William Van Woensel[1], Samina R. Abidi[1,2], Calvino Cheng[3], Bryan Crocker[3], and Manal Elnenaei[3]

[1] NICHE Research Group, Faculty of Computer Science, Dalhousie University, Halifax, Canada
ssrabidi@dal.ca
[2] Medical Informatics, Department of Community Health and Epidemiology, Dalhousie University, Halifax, Canada
[3] Department of Pathology and Laboratory Medicine, Nova Scotia Health Authority, Halifax, Canada

Abstract. Inappropriate pathology test orders are an economic burden on laboratories and compromise patient safety. We pursue a laboratory utilization management strategy that involves raising awareness amongst physicians regarding their test ordering behaviour. We are employing an AI-driven approach for laboratory utilization management, whereby we apply both machine learning and semantic reasoning methods to analyze pathology laboratory data. We are analyzing over 6-years of primary care physician's pathology test order 'big' data. Our analysis generates physician order profiles, based on their case-mix and orders-sets, to inform physicians about their laboratory utilization. We developed an AI-driven platform—i.e. Pathology Laboratory Utilization Scorecards (PLUS) that offers an interactive means for physicians to self-examine their test ordering pattern. PLUS aims to optimize the utilization of the Central Zone pathology laboratory of the Nova Scotia Health Authority.

Keywords: Machine learning · Data analytics · Semantic web · Pathology · Laboratory utilization · Big data

1 Introduction

Pathology laboratory testing is central to medical practice as most diagnostic and therapeutic decisions are guided by the patient's pathology test results. Pathology tests are routinely ordered by physicians and it has been observed that a significant number of tests ordered are *inappropriate*—i.e. the test is either redundant, clinically irrelevant or non-compliant with clinical guidelines. There are multiple reasons for the inappropriate ordering of pathology lab tests including inconsistencies in test nomenclature [1], poor implementation of evidence-based guidelines [2] and physician's discretionary behaviour when ordering tests [3]. A meta-analysis of 108 studies, examining 1.6 million results from 46 of the 50 most commonly ordered pathology tests, concluded that on average

D. Riaño et al. (Eds.): AIME 2019, LNAI 11526, pp. 241–251, 2019.
https://doi.org/10.1007/978-3-030-21642-9_30

30% of all tests ordered by physicians were likely to be inappropriate [4]. Inappropriate pathology test ordering [5] not only affects laboratory resource utilization, but it also compromises patient safety by producing falsely abnormal results which may require unnecessary interventions [6]. Given rising healthcare costs whilst the need to meet quality and efficiency targets, there is an awareness to minimize inappropriate pathology testing. In Canada, the 'Choosing Wisely' initiative aims to optimize healthcare services by reducing waste, and pathology test ordering is an area that needs innovative strategies to minimize inappropriate test ordering by physicians [7]. Utilization management is a strategy to evaluate the appropriateness and efficiency of healthcare services. As such, pathology laboratory utilization management aims to optimize pathology test ordering—i.e. the right test is ordered at the right time for the right patient—by reducing both over- and under-utilization of the pathology laboratory.

In our work, we pursue pathology laboratory utilization management by raising awareness amongst physicians about their inappropriate test ordering behaviour. Our approach is to provide physicians personalized insights into their laboratory utilization profile and peer comparisons via a self-auditing tool [8]. In this paper, we present an AI-based framework for laboratory utilization management that employs (1) machine learning methods to tackle overutilization of laboratory tests by (a) clustering physicians based on their patient case-mix for inter-physician peer comparisons; (b) using association rules to identify the unconventional order-sets of individual physicians with respect to their peers; and (2) knowledge-based reasoning to tackle underutilization of laboratory tests by implementing test appropriateness rules to recommend essential tests in response to the results of prior tests. We have implemented a *Pathology Laboratory Utilization Scorecard (PLUS)* platform that offers (i) *scorecards for physicians* to examine their test ordering pattern over time, and compare it with peers having the same case-mix; and (ii) *dashboards* for laboratory managers to assist with waste minimization. PLUS has been implemented to optimize the Central Zone pathology laboratory in Halifax that processes 8 million general pathology test orders yearly.

2 Laboratory Utilization Management Approaches: A Review

Current approaches for laboratory utilization management focus on reducing physician options in test ordering, physician education, decision support and peer comparisons.

Strategies to reduce physician options to tests include specialist vetting of lab orders [9] and at the CPOE level McDonald et al. removed the option to order daily tests beyond 2 days [10], Neilson et al. [11] utilized prompts to reduce the ease of repeating targeted tests, and Iturrate et al. [12] disallowed daily recurring tests entirely. In terms of physician education, Ryskina et al. [13] provided social comparison feedback linked to patients' EMR records, Bunting et al. [14] discussed with the physician their lab utilization and compared it with other physicians; Iams et al. [15] sent out weekly feedback e-mails to physicians comparing their lab ordering rates with the ordering rates of all others as well as the pre-set goal ordering rates; and. Srivastava et al. [16] illustrated the utility of reflex rules (and reflective testing) in lab test recommendation.

Pathology laboratory utilization management strategies have yielded encouraging results, for instance a saving of 19% of the total costs for genetic test orders by reviewing each ordered test [17], a 21% reduction in B-type natriuretic peptide test orders by employing a decision support system [18], unbundling of test panels and providing pocket cards with laboratory test costs to physicians resulted in a 21% reduction in costs [19], and an 8% test volume reduction was noted by leveraging social influence of opinion leaders, academic detailing, and giving test prices in newsletters to physicians [19]. The pathology department in Halifax provided written feedback to individual primary care physicians about their orders for specific tests and a 25% reduction in orders was noted [20]. In a Canadian study [14], physicians were provided feedback on their laboratory utilization rate along with peer-comparisons, and as a result a reduction in inappropriate ordering by physicians was noted.

We note from pathology laboratory utilization management strategies that addressing inappropriate test orders at the physician level can yield the highest impact given the significant variability in the test ordering pattern of physicians despite them treating patients with the same diagnosis. This alludes to physician's discretionary behaviour when ordering tests [3]—a behaviour that can be modified by providing physicians with education, self-audit of test ordering profile [21] and peer comparisons.

3 AI-Driven Laboratory Utilization Management Approach

We are targeting laboratory utilization optimization at the primary care physician level since they are the heaviest users of the pathology laboratory. To minimize inappropriate testing, our strategy is to engage physicians to (a) self-examine their test ordering pattern and its implications on laboratory utilization, (b) show how their laboratory utilization compares with their peers, and (c) recommend essential follow-up tests.

We argue that peer comparisons are meaningful when a physician is compared with similar physicians as opposed to all physicians. Typically, peer comparison of physician's pathology test ordering profile is based on the type, volume and frequency of tests they order [22]. However, such peer comparisons are inconclusive as it does not consider the physician's patient case-mix—i.e. if a physician is treating more elderly patients with chronic kidney and cardiac conditions then a higher volume of CBC and creatinine tests is not an inappropriate test ordering pattern; hence, they should not be flagged as a high laboratory user compared to a physician treating younger patients.

To determine the *laboratory overutilization* by a physician, our approach is to generate a multi-faceted physician test ordering profile that takes into account: (i) patient case-mix managed by the physician to justify the test orders based on medical necessity, and in turn to allow a fair comparison with peers having a similar case-mix; (ii) test co-occurrence pattern (i.e. order-sets) to determine the medical necessity of tests that are frequently ordered simultaneously; (iii) temporal variations in ordering patterns to account for seasonal needs, and also to detect changes in the physician's knowledge and guideline compliance over time, and (iv) geographical location of the physician to provide a fairer comparison with peers practicing in the same health zone. Machine learning methods, applied to a 'big' pathology laboratory order dataset, are used to generate a physician ordering profile to determine laboratory overutilization by the physician.

To handle *laboratory underutilization* by physicians our approach is to computerize diagnostic testing rules, derived from clinical guidelines and domain experts, that recommend follow-up laboratory tests that are essential for the diagnostic process in a timely and safe manner. We use semantic web based ontologies and decision rules to represent the test ordering protocols, and we apply logic-based reasoning to the decision rules to determine the follow-up tests based on the results of the ordered tests.

To engage physicians to perform a self-audit of their laboratory utilization, our approach is to provide them a web-based interactive *physician-specific scorecard* that they can securely access and privately view to examine their overall test ordering profile and its implications on the provincial laboratory's utilization.

4 Our Pathology Test Ordering Data

Our dataset comprises pathology test orders received by the Central Zone pathology lab (in Halifax) during the period 2011–2017. We analyzed 15 general tests—i.e. PT, CBC Auto Diff, Creatinine, Alkaline Phosphatase, Urea, Electrolyte Panel, AST, ALT, GGT, Glucose AC, Glucose Random, Cholesterol, HDL Cholesterol, Triglycerides, TSH. Note that a single test may comprise 1 or more procedures (i.e. the CBC test comprises 11 procedures, each generating an individual result), and a physician can order one or more tests for a patient in a single test order. The dataset covers around 2000 physicians and 250,000 patients. The annual breakdown of test orders is given in Table 1.

Table 1. Annual volume of general orders for 15 common tests.

2012	2013	2014	2015	2016	2017
1,447,798	1,481,788	1,567,964	1,569,422	1,583,671	1,679,531

5 Data Analytics for Laboratory Utilization Management to Minimize Laboratory Overutilization by Physicians

5.1 Data Clustering to Generate Physician Case-Mix Clusters

A tenet of our approach for minimizing overutilization of laboratory is to provide physicians with a comparison of their laboratory utilization with that of their peers. The key to peer comparison is that a physician is compared only with physicians that have a similar practice and case-mix of patients, and not with all physicians in the province.

We used machine learning based clustering methods to generate groups of physicians with similar case-mix of patients. Our dataset does not include the patient's diagnosis which is essential to determine a physician's case-mix. Given that specific tests are ordered to confirm the presence/absence of specific diseases, we can assume that physicians having patients with specific diseases will order more tests associated with those diseases and the test results will further confirm that the physician is treating a patient with a specific disease. For instance, high abnormal values for the potassium test and low abnormal values for the sodium and glucose AC tests are associated with

Addison's disease. Thus, physicians treating a high number of patients with Addison's disease will order the potassium, sodium, and glucose AC tests in higher proportions than a physician treating less patients with Addison's disease. Building on existing mappings between pathology tests and diseases, we developed a test-disease mapping between 26 diseases and 40 pathology tests which were validated by pathologists.

To generate the dataset for clustering physicians based on their case-mix, we retrieved the test orders of each physician for each of the 40 test types with their results, and then applied the test-disease mapping to assign a plausible disease diagnosis to each patient seen by a physician to determine his/her case-mix. We created ratios for each disease by using the total of ordered tests per physician, and used the 26 disease ratios along with physician attributes such as patient demographics, practice location, test results, test frequency per patient, etc. to generate the input vector for physician clustering. We standardized the data by centering (removing the mean) and scaling (dividing by standard deviation) the 26 diseases (ratios) to ensure that each disease contributes proportionately to determine similarity between two physicians. Next, we applied metric Multi-Dimensional Scaling (MDS), a non-linear dimension reduction approach, to reduce the inter-physician similarity from a 26-dimensional disease space to a 2-dimensional space. We used the Partitioning Around Medoids method with the Euclidian distance between physicians in the 2-dimensional MDS space to generate the physician clusters. To select the optimal number of physician clusters, we used the average silhouette width [23]. Based on the average silhouette widths, the best solution was $K = 4$ clusters (average silhouette width of 0.36). Figure 1 shows the silhouette widths of all physicians in the 4 clusters. Figure 2 shows the clusters of physicians on the 2-dimensional space. The physician clusters were annotated and validated by experts by comparing the inherent characteristics of physicians within a cluster. Since a physician's case-mix can vary over time, we also generated period-sensitive physician clusters at a 2-year interval, thus allowing peer comparisons across a given period, and across the overall study period (of 6 years). Our PLUS system applies the clustering results to group physicians, based on their case-mix, for peer comparisons.

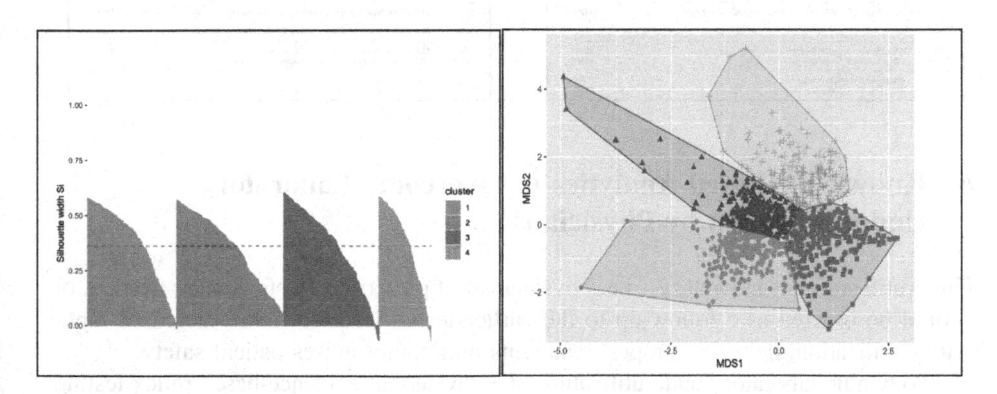

Fig. 1. Silhouette widths for the 4 clusters. **Fig. 2.** Clustering of physicians in 4 clusters.

5.2 Association Rule Mining to Generate Test Order-Sets

We examine a physician's order-sets to establish whether the high-volume of test order-sets justify clinical need or that the physician was overprescribing tests, maybe due to practice behaviour or lack of awareness of clinical guidelines. A frequent pattern refers to a set of items appearing as a pattern beyond a pre-specified frequency threshold. A frequent order-set illustrates a frequent pattern of tests ordered by a physician in a single test order. To analyze a physician's test ordering pattern—i.e. which tests are ordered simultaneously and which tests are ordered for specific patient groups. To generate frequent order-sets, we used the constrained association rule mining method to generate n-order association rules (where n = 2–15 tests) based on order frequency and test relevance at the physician cluster. To account for temporal changes, we generated order-sets for 3-year periods (Tables 2 and 3 show the 14, 13, 12, 10, 7, 5, 3 order-sets).

When comparing the order-sets over the two 3-year time periods, we noted that in general the most frequent order-sets remain the same over time—out of the 14 order-sets (i.e. comprising 1–14 items), 7 order-sets remained the most frequent. This suggests that the order-sets are well-defined with fluctuations in their ordering frequency over time. The identified order-sets were implemented within the PLUS system as a benchmark for a physician to compare his/her order-sets with peers, where the unit of comparison is test order-set as opposed to individual test orders.

Table 2. Frequent order-sets at the regional level for the period of Jul-2011 to Jun-2014

ORDER SET SIZE	CONCURRENTLY ORDERED TESTS (ORDER-SET)	ORDER SET VOLUME
F14 Tests	ALT, Alkaline Phosphatase, AST, CBC, Cholesterol, Creatinine, Electrolyte Panel, GGT, Glucose AC, HDL Cholesterol, PT, Triglycerides, TSH, Urea	1814
F13 Tests	ALT, Alkaline Phosphatase, AST, CBC, Cholesterol, Creatinine, Electrolyte Panel, GGT, Glucose AC, HDL Cholesterol, Triglycerides, TSH, Urea	24357
F12 Tests	ALT, AST, CBC, Cholesterol, Creatinine, Electrolyte Panel, GGT, Glucose AC, HDL Cholesterol, Triglycerides, TSH, Urea	13546
	ALT, Alkaline Phosphatase, AST, CBC, Cholesterol, Creatinine, Electrolyte Panel, Glucose AC, HDL Cholesterol, Triglycerides, TSH, Urea	10396
F10 Tests	ALT, CBC, Cholesterol, Creatinine, Electrolyte Panel, Glucose AC, HDL Cholesterol, Triglycerides, TSH, Urea	16310
F7 Tests	ALT, Alkaline Phosphatase, AST, CBC, Creatinine, Electrolyte Panel, Urea	4724
	CBC, Cholesterol, Creatinine, Glucose AC, HDL Cholesterol, Triglycerides, TSH	1992
F5 Tests	CBC, Creatinine, Electrolyte Panel, Glucose Random, Urea	6944
	CBC, Creatinine, Electrolyte Panel, PT, Urea	4099
F3 Tests	Creatinine, Electrolyte Panel, Urea	7896
	CBC, Creatinine, Electrolyte Panel	4873

Table 3. Frequent order-sets at the regional level for the period of Jul-2014 to May-2017

ORDER SET SIZE	CONCURRENTLY ORDERED TESTS (ORDER SET)	ORDER SET VOLUME
S14 Tests	ALT, Alkaline Phosphatase, AST, CBC, Cholesterol, Creatinine, Electrolyte Panel, GGT, Glucose AC, HDL Cholesterol, PT, Triglycerides, TSH, Urea	392
S13 Tests	ALT, Alkaline Phosphatase, AST, CBC, Cholesterol, Creatinine, Electrolyte Panel, GGT, Glucose AC, HDL Cholesterol, Triglycerides, TSH, Urea	4187
S12 Tests	ALT, Alkaline Phosphatase, AST, CBC, Cholesterol, Creatinine, GGT, Glucose AC, HDL Cholesterol, Triglycerides, TSH, Urea	7346
	ALT, AST, CBC, Cholesterol, Creatinine, Electrolyte Panel, GGT, Glucose AC, HDL Cholesterol, Triglycerides, TSH, Urea	2777
S10 Tests	ALT, CBC, Cholesterol, Creatinine, Electrolyte Panel, Glucose AC, HDL Cholesterol, Triglycerides, TSH	8734
S7 Tests	CBC, Cholesterol, Creatinine, Glucose AC, HDL Cholesterol, Triglycerides, TSH	3751
	ALT, Alkaline Phosphatase, AST, CBC, Creatinine, Glucose Random, Urea	2058
S5 Tests	ALT, Alkaline Phosphatase, AST, CBC, Creatinine	2457
	CBC, Creatinine, Electrolyte Panel, Glucose Random, Urea	1861
S3 Tests	CBC, Creatinine, Urea	6017
	Cholesterol, HDL Cholesterol, Triglycerides	2357

6 Knowledge-Based Analytics to Overcome Laboratory Underutilization by Physicians

Underutilization of laboratory—i.e. physicians not prescribing tests that are needed, or should be ordered as a follow-up to the earlier tests—leads to future increased laboratory utilization, delays in proper treatments and compromises patient safety.

To handle laboratory underutilization, we devised an evidence-based reflex testing strategy [24] that recommends (or "reflexes") follow-up tests in response to results of prior tests (shown in Fig. 3). We use knowledge-based analytics, employing semantic web methods, to represent and execute *reflex rules* that suggest follow-up pathology tests to confirm a diagnosis (e.g. early diagnosis of pituitary dysfunction). Responding to abnormal test results noted for certain elements (such as abnormal patterns of basal pituitary hormones), our reflex testing strategy [24] firstly identifies additional reflective pathology tests, and then directly conducts the follow-up tests if the patient's existing blood sample can be used, or recommends the tests to the patient's physician.

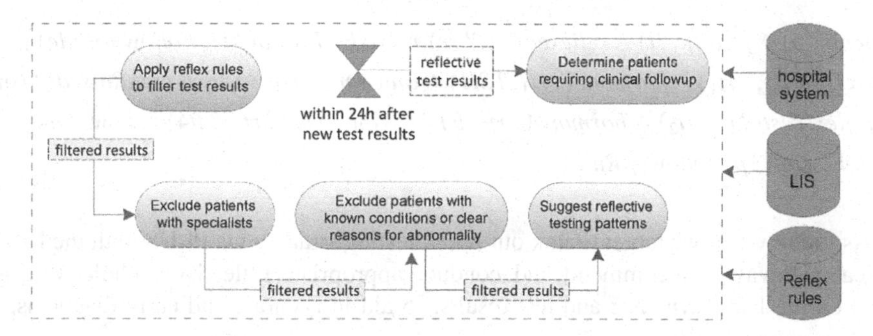

Fig. 3. Reflex strategy to identify abnormal lab result patterns and suggesting reflective testing.

To implement our reflective testing strategy, we developed an OWL ontology and a set of Description Logics (DL-safe) rules. We used the Protégé ontology engineering tool to construct the Reflex ontology and rules and utilized Hermit reasoner [25] for implementing the reasoning process. We explain reflex testing and its knowledge-based implementation using the example of diagnosing pituitary dysfunction. Based on pathology test results, as first step of our strategy we apply a set of context-sensitive *ReflexRules* to identify abnormal patterns in the pathology test results. For instance, the following rule "reflexes" when finding tests for women over 55 with a measurement of FSH (Follicle Stimulating Hormone) under 15:

$$\begin{gathered} Female(?p) \wedge age(?p, ?a) \wedge ?a \geq 55 \wedge test(?p, ?t) \wedge hormone(?t, FSH) \wedge \\ outcome(?t, ?fsh) \wedge ?fsh < 15 \rightarrow reflexed(?t, ReflexRule1) \end{gathered}$$

Note that, *ReflexRule1* is associated with a set of exclusion rules—based on clinical information, such rules exclude special cases from consideration. Following our example, this rule excludes cases where the patient is pregnant, or on HCT or HRT (Hormone Contraception/Replacement Therapy):

$$\begin{aligned} & Patient(?p) \wedge test(?p, ?t) \wedge reflexed(?t, ?rr) \wedge hasExclusion(?rr, ?excl) \wedge \\ & \quad (?excl = ExclRule31 \wedge isPregnant(?p, true)) \vee \\ & \quad (?excl = ExclRule32 \wedge followsTherapy(?p, HRT)) \vee \\ & \quad (?excl = ExclRule33 \wedge followsTherapy(?p, HC))) \rightarrow excluded(?p, ?excl) \end{aligned}$$

Once abnormal test patterns are identified and special cases are excluded, *ReflexRule1* suggests a series of reflective tests to confirm pituitary dysfunction; e.g., measuring Thyroid Stimulating Hormone (TSH) and Free Thyroxine (FT4). Once results from these new tests are available, a set of *FollowupRules*, related to the initial *ReflexRule1*, check whether follow-up with an endocrinologist is recommended. For instance, the below rule recommends follow-up if TSH is non-raised and FT4 is low:

$$Patient(?p) \wedge test(?p, ?t) \wedge reflexed(?t, ?rr) \wedge hasFollowup(?rr, FollowupRule1) \wedge reflexTest(?p, ?rt_1) \wedge hormone(?rt_1, TSH) \wedge outcome(?rt_1, ?tsh) \wedge NonRaised(?tsh) \wedge reflextTest(?p, ?rt_2) \wedge hormone(?rt_2, FT4) \wedge outcome(?rt_2, ?ft4) \wedge Low(?ft4) \rightarrow followup(?p, FollowupRule1)$$

As future work, we target to link our reflex testing strategy via PLUS with the LIS's clinical pathway to recommend and conduct appropriate reflective pathology tests, based on the latest evidence and test results, to aid in accurate and early diagnosis.

7 Visualization of Laboratory Utilization: The PLUS System

PLUS implements a web-based health data analytics system using machine learning methods [26, 27] for laboratory utilization management (shown in Fig. 4).

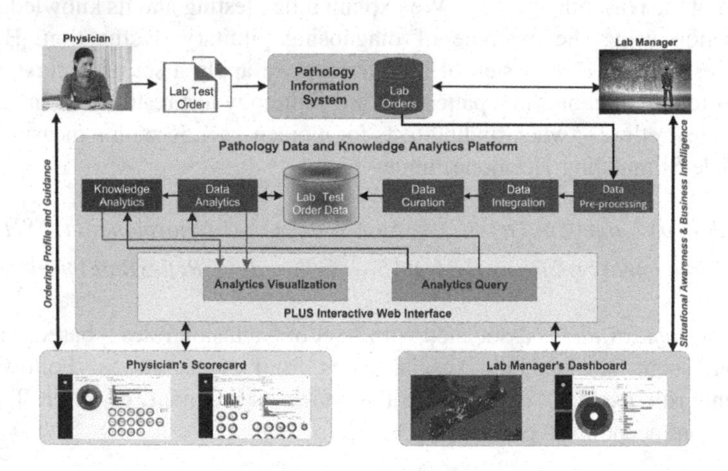

Fig. 4. PLUS functional architecture

Advance data visualization has been implemented for users to dynamically interact with the data analysis. PLUS provides a web-based (a) *scorecard* for physicians to understand their own test ordering profile over time and across different patient cohorts, and to compare their laboratory utilization (adjusted to case-mix) with similar peers. The scorecard presents physician's laboratory utilization in terms of abnormal results, test volumes and frequency over time, peer comparisons, cost incurred, case-mix and the rate and cost of inappropriate test orders. The physician's scorecard is private and cannot be viewed by practice auditing bodies; and (ii) *dashboard* for laboratory managers provides broad operational intelligence by aggregating the patient-level analytics to the regional level, displaying tests ordered, completion rates and flagged as inappropriate.

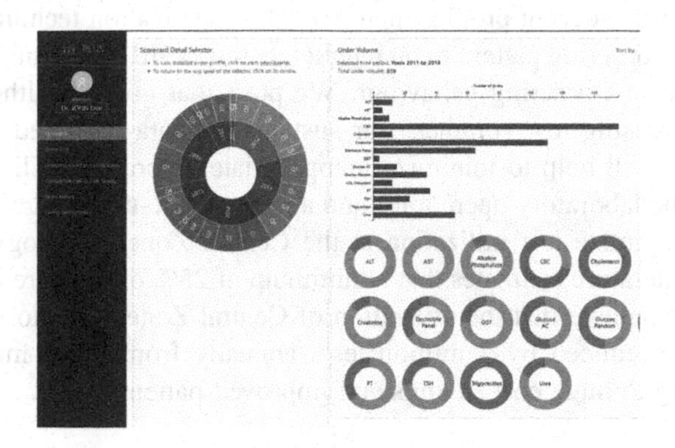

Fig. 5. Physician scorecard main page

We present a working example of PLUS use by a physician. Figure 5 illustrates the opening page of a physician's scorecard. The date selector (not shown) allows the physicians to select specific year(s) or quarter(s) within a year, following which the right-hand side visualizations are dynamically updated to show volume of tests ordered and the rate of abnormal and normal results. Physicians can hover over a visualization to get additional information. Figure 6 shows the physician's scorecard with peer comparisons (on a yearly basis) across all tests, with options for filtering the order-tests.

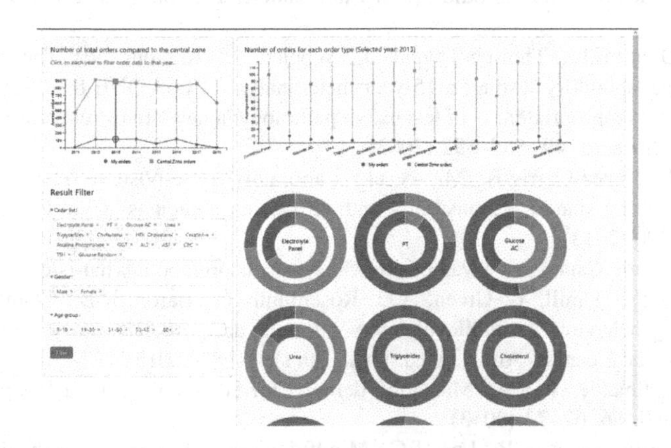

Fig. 6. Physician scorecard showing peer comparison across tests over a 6-year window

8 Concluding Remarks

As health care transitions from volume- to value-based care, there is an increasing need for efficient and effective laboratory utilization by physicians. In this paper, we presented an innovative and sustainable laboratory utilization management approach, targeting physicians, that leverages (a) data analytics methods to develop and understand each physician's test ordering profiles; and (b) data visualization techniques to display the physician's test ordering pattern as an interactive scorecard so that they can self-audit and -regulate their test ordering behaviour. We posit that given healthcare budgetary pressures and increasing test volumes, our sustainable, data-informed and physician-engaged approach will help to minimize inappropriate laboratory utilization, improve sustainability of the laboratory operations, and achieve value-based care. PLUS is being implemented to optimize test utilization at the Central Zone pathology laboratory in Nova Scotia. As literature estimates that a minimum of 25% of tests are inappropriately ordered [1–3], we believe that the utilization of Central Zone laboratories (in Halifax) can potentially be reduced by 2 million tests annually from the 8 million currently performed, leading to huge cost savings and improved patient safety.

Acknowledgements. We thank the NSHA Central Zone pathology lab for supporting the project, and Nova Scotia Health Research Foundation for giving the catalyst grant.

References

1. Freedman, D.B.: Towards better test utilization - strategies to improve physician ordering and their impact on patient outcomes. EJIFCC **26**, 15–30 (2015)
2. Misra, S., Barth, J.H.: Guidelines are written, but are they followed? Ann. Int. J. Biochem. Lab. Med. **50**, 400–402 (2013)
3. van Walraven, C., Goel, V., Austin, P.: Why are investigations not recommended by practice guidelines ordered at the periodic health examination? J. Eval. Clin. Pract. **6**(2), 215–224 (2000)
4. Zhi, M., Ding, E.L., Theisen-Toupal, J., Whelan, J., Arnaout, R.: The landscape of inappropriate laboratory testing: a 15-year meta-analysis. PLoS ONE **8**, e78962 (2013)
5. Smellie, W.S.: Appropriateness of test use in pathology: a new era or reinventing the wheel? Ann. Clin. Biochem. **40**, 585–592 (2003)
6. Salinas, M., Lopez-Garrigós, M., et al.: Laboratory false-positive results: a clinician responsibility or a shared responsibility with requesting clinicians? Clin. Chem. Lab. Med. **51**, e199–e200 (2013)
7. Choosing Wisely Canada. www.choosingwiselycanada.org/about/what-is-cwc/
8. Vidyarthi, A.R., Hamill, T., Green, A.L., Rosenbluth, G., Baron, R.B.: Changing resident test ordering behavior: a multilevel intervention to decrease laboratory utilization at an academic medical center. Am. J. Med. Qual. **30**(1), 81–87 (2015)
9. Fryer, A.A., Smellie, W.S.A.: Managing demand for laboratory tests: a laboratory toolkit. J. Clin. Pathol. **66**, 62–72 (2013)
10. McDonald, E.G., Saleh, R.R., Lee, T.C.: Mindfulness-based laboratory reduction: reducing utilization through trainee-led daily 'Time Outs'. Am. J. Med. **130**, e241–e244 (2017)

11. Neilson, E.G., Johnson, K.B., Rosenbloom, S.T., et al.: Resource utilization committee: the impact of peer management on test-ordering behavior. Ann. Intern. Med. **141**, 196–204 (2004)
12. Iturrate, E., Jubelt, L., Volpicelli, F., et al.: Optimize your electronic medical record to increase value: reducing laboratory overutilization. Am. J. Med. **129**, 215–220 (2016)
13. Ryskina, K., et al.: Effect of social comparison feedback on laboratory test ordering for hospitalized patients: a randomized controlled trial. J. Gen. Intern. Med. **33**, 1639–1645 (2018)
14. Bunting, P.S., van Walraven, C.: Effect of a controlled feedback intervention on laboratory test ordering by community physicians. Clin. Chem. **50**, 321–326 (2004)
15. Iams, W., Heck, J., Kapp, M., et al.: A multidisciplinary housestaff-led initiative to safely reduce daily laboratory testing. Acad. Med. **91**, 813–820 (2016)
16. Srivastava, R., Bartlett, W.A., Kennedy, I.M., et al.: Reflex and reflective testing: efficiency and effectiveness of adding on laboratory tests. Ann. Clin. Biochem. **47**, 223–227 (2010)
17. Dickerson, J.A., et al.: Improving the value of costly genetic reference laboratory testing with active utilization management. Arch. Pathol. Lab. Med. **138**(1), 110–113 (2014)
18. Levick, D.L., Stern, G., et al.: Reducing unnecessary testing in a CPOE system through implementation of a targeted CDS intervention. BMC Med. Inform. Decis. Mak. **13**, 43 (2013)
19. Vegting, I.L., van Beneden, M., Kramer, M.H., et al.: How to save costs by reducing unnecessary testing: lean thinking in clinical practice. Eur J Intern Med. **23**(1), 70–75 (2012)
20. Elnenaei, M.O., Campbell, S.G., Thoni, A.J., Lou, A., Crocker, B.D., Nassar, B.A.: An effective utilization management strategy by dual approach of influencing physician ordering and gate keeping. Clin. Biochem. **49**, 208–212 (2016)
21. Huang, Y., et al.: Improving serological test ordering patterns for the diagnosis of celiac disease through clinical laboratory audit of practice. Clin. Biochem. **45**(6), 455–459 (2012)
22. Yeh, D.D.: A clinician's perspective on laboratory utilization management. Clin. Chim. Acta **427**, 145–150 (2014)
23. Rousseeuw, P.: Silhouettes: a graphical aid to the interpretation and validation of cluster analysis. J. Comput. Appl. Math. **20**, 53–65 (1987)
24. Elnenaei, M., Minney, D., Clarke, D.B., Kumar-Misir, A., Imran, S.A.: Reflex and reflective testing strategies for early detection of pituitary dysfunction. Clin. Biochem. **54**, 78–84 (2018)
25. Glimm, B., Horrocks, I., et al.: An OWL 2 reasoner. J. Autom. Reason. **1**, 1–25 (2014)
26. Abusharekh, A., Stewart, S., Abidi, S.S.R.: H-DRIVE: a big data analytics platform for evidence-informed decision making. In: 4th IEEE International Congress on Big Data, New York (2015)
27. Cha, S., Abusharekh, A., Abidi, S.S.R.: Towards a 'big' health data analytics platform. In: 1st IEEE International Conference on Big Data Computing Service and Applications, San Francisco (2015)

Using Cluster Ensembles to Identify Psychiatric Patient Subgroups

Vincent Menger[1,2](✉), Marco Spruit[1], Wouter van der Klift[1], and Floor Scheepers[2]

[1] Department of Information and Computing Sciences, Utrecht University, Utrecht, The Netherlands
{v.j.menger,m.r.spruit,w.vanderklift}@uu.nl
[2] Department of Psychiatry, University Medical Center Utrecht, Utrecht, The Netherlands
f.e.scheepers-2@umcutrecht.nl

Abstract. Identification of patient subgroups is an important process for supporting clinical care in many medical specialties. In psychiatry, patient stratification is mainly done using a psychiatric diagnosis following the Diagnostic and Statistical Manual of Mental Disorders (DSM). Diagnostic categories in the DSM are however heterogeneous, and many symptoms cut across several diagnoses, leading to criticism of this approach. Data-driven approaches using clustering algorithms have recently been proposed, but have suffered from subjectivity in choosing a number of clusters and a clustering algorithm. We therefore propose to apply cluster ensemble techniques to the problem of identifying subgroups of psychiatric patients, which have previously been shown to overcome drawbacks of individual clustering algorithms. We first introduce a process guide for modelling and evaluating cluster ensembles in the form of a Meta Algorithmic Model. Then, we apply cluster ensembles to a novel cross-diagnostic dataset from the Psychiatry Department of the University Medical Center Utrecht in the Netherlands. We finally describe the clusters that are identified, and their relations to several clinically relevant variables.

Keywords: Cluster ensembles · Mental healthcare · Psychiatry · Patient subgroups · Patient stratification · Applied data science

1 Introduction

Identification of patient subgroups is an important process that is able to guide clinical treatment in many medical specialties. In psychiatry, the main construct for stratifying patients is a psychiatric diagnosis, typically performed using the Diagnostic and Statistical Manual of Mental Disorders (DSM). This manual describes various high level disorders such as depressive disorders, anxiety disorders, and developmental disorders, with sub-types for each category. It defines

D. Riaño et al. (Eds.): AIME 2019, LNAI 11526, pp. 252–262, 2019.
https://doi.org/10.1007/978-3-030-21642-9_31

clear diagnostic criteria based on symptoms—a major depressive disorder for instance can only be diagnosed after eight symptoms have been assessed, including depressed mood, weight loss, fatigue, and inability to concentrate, and at least five were observed in a two-week period. While the DSM is by far the most widely adopted standard for diagnosis, in recent years its rigid approach has been subject to criticism. Research for instance shows that the DSM has little biological validity (i.e. lack of connection to biomarkers), that diagnostic categories are not specific (i.e. large heterogeneity exists within groups), and that symptoms often cut across diagnostic categories [4].

This critique on the DSM has seeded data-driven approaches that seek interesting subgroups using relevant datasets rather than using expert elicited criteria. For this purpose, various clustering algorithms that are able to discover latent subgroups have been applied to patient data. One major downside of a clustering approach however is the need to select an appropriate number of clusters and an appropriate clustering algorithm, which both have been shown to provide challenges for researchers [12]. The majority of studies rely on a single metric for choosing the right number of clusters, and subsequently apply a single clustering algorithm [14], while both choices can have significant impact on the results that are obtained. Consequently, as of yet no consensus exists on either the number or nature of psychiatric patient subgroups that can be derived in this data-driven way.

In this work we therefore propose to apply cluster ensembles, i.e. combinations of multiple clustering algorithms, to this problem. This enables identification of distinct subgroups that can directly inform treatment, while overcoming the downsides of individual clustering algorithms. Previous work has already shown that cluster ensembles often improve robustness, stability and accuracy over individual clustering algorithms, both in general and in the medical domain, yet this approach is still rare in mental healthcare research [8,22].

The contribution of this work is twofold. First, we present a process guide for modelling and evaluating cluster ensembles in the form of a Meta Algorithmic Model, as introduced in [20]. This guide aims to support researchers in applying cluster ensembles in their particular (medical) domain. Second, we apply a cluster ensemble approach to a novel cross-diagnostic dataset of 1,098 Youth Self Report (YSR) questionnaires of adolescents that were treated at the University Medical Center Utrecht in the Netherlands. Since these questionnaires were routinely captured during treatment, using them to identify patient subgroups, if present, can have direct applicability in the psychiatric practice [16,17]. After applying the cluster ensemble approach, we examine key characteristics of the clusters we obtained, and assess their relation to several clinically relevant variables including DSM diagnosis.

2 Background and Related Work

Clustering algorithms have previously been used in mental healthcare research for stratifying patients with a common psychiatric diagnosis, such as schizophrenia, depression, or autism [14]. The number of clusters ranges from two to seven, typically selected based on a single measure such as Bayesian Information Criterion or Ward's method. Most researchers then apply one algorithm to their dataset, such as K-means Clustering, Hierarchical Clustering, or Latent Class Cluster Analysis. Clusters of various natures have been found, for instance based on differences in symptoms [5], treatment outcome and onset [3], and patient functioning [6]. A smaller number of studies focused on stratifying patients in a cross-diagnostic setting. A study by Olinio et al. for instance found six subtypes differing in presence of depression, anxiety, or a mixture of both [18], while Lewandowski et al. reported a neuropsychologically normal subtype, a globally impaired subtype, and two mixed cognitive profiles [13], and Kleinman et al. found a cluster with diminished sustained attention, inhibitory control and vigilance, and increased impulsiveness, and a second converse cluster [11]. So far, cluster ensembles have only been applied once in mental health research in a study by Shen et al. who used this technique to identify four subtypes of pervasive developmental disorders [19]. They reported differences in severity, in problems with language acquisition and impairment, and in aggressive behaviour.

To reduce variability in clustering outcomes, such as for example described above, cluster ensembles were proposed based on the principle that multiple weak partitions in combination can provide a more accurate and objective outcome than a single strongly optimized clustering [7]. This is analogous to ensemble learning techniques such as Boosting and Random Forests in the supervised domain. First, during the generation stage, a number of diverse partitions are obtained, ideally with strengths and weaknesses in different parts of the solution space [8]. This is for instance achieved by using multiple clustering algorithms and different algorithm parameters, by subsetting data, and by projecting data to subspaces [22]. The result of the generation stage on a dataset $X = \{x_i, ..., x_n\}$ with n observations is a partition set $P = \{p_1, ..., p_m\}$ of m partitions, where each $p_i = \{C_1^i, ..., C_k^i\}$ assigns every observation to a single cluster C_i out of k clusters. In the subsequent consensus stage, an optimal partitioning is obtained using partition set P. Various procedures have been proposed based on object co-occurrence in clusters, such as majority voting [23], or the graph-based Cluster-based Similarity Partitioning Algorithm (CSPA) [21]. Another type of approach finds the median partition in $p^* \in P$, for instance defined as the partition that maximizes similarity with all other partitions in P [24]. Cluster ensembles have recently successfully been applied in several biomedical domains [2].

3 Meta Algorithmic Model

To support researchers in applying cluster ensembles to their (medical) domain, we propose a Meta Algorithmic Model (MAM) of cluster ensemble modelling and evaluation (Fig. 1). Our MAM is an extension of the original work of Spruit and Jagesar [20], that was aimed at supervised learning tasks. In their words, MAMs are intended to provide "highly understandable and deterministic method fragments — i.e. activity recipes — to guide application domain experts without in-depth Machine Learning expertise". Method fragments are specified as a combination of a Unified Modelling Language (UML) activity diagram showing processes, and a UML class diagram showing concepts.

The cluster ensemble modelling process, shown on the left of Fig. 1, starts with loading a prepared dataset. Then, in the generation stage multiple methods for introducing diversity in the cluster portfolio are used, including observation and feature sampling, choosing clustering algorithms and selecting a number of repetitions. After a number of clusters and a distance measure are selected, the cluster portfolio is created. In the subsequent consensus stage a consensus function should be selected, and weak partitions can be trimmed from the cluster portfolio. During the evaluation stage, internal index criteria (e.g. Carlinski-Harabasz, Silhouette) can be evaluated, and clusters can be visualized after applying a dimension reduction algorithm to the dataset. Cluster characteristics can be identified based on the cluster assignments of the dataset, and an external evaluation (e.g. using expert evaluation, or comparison to a reference class)

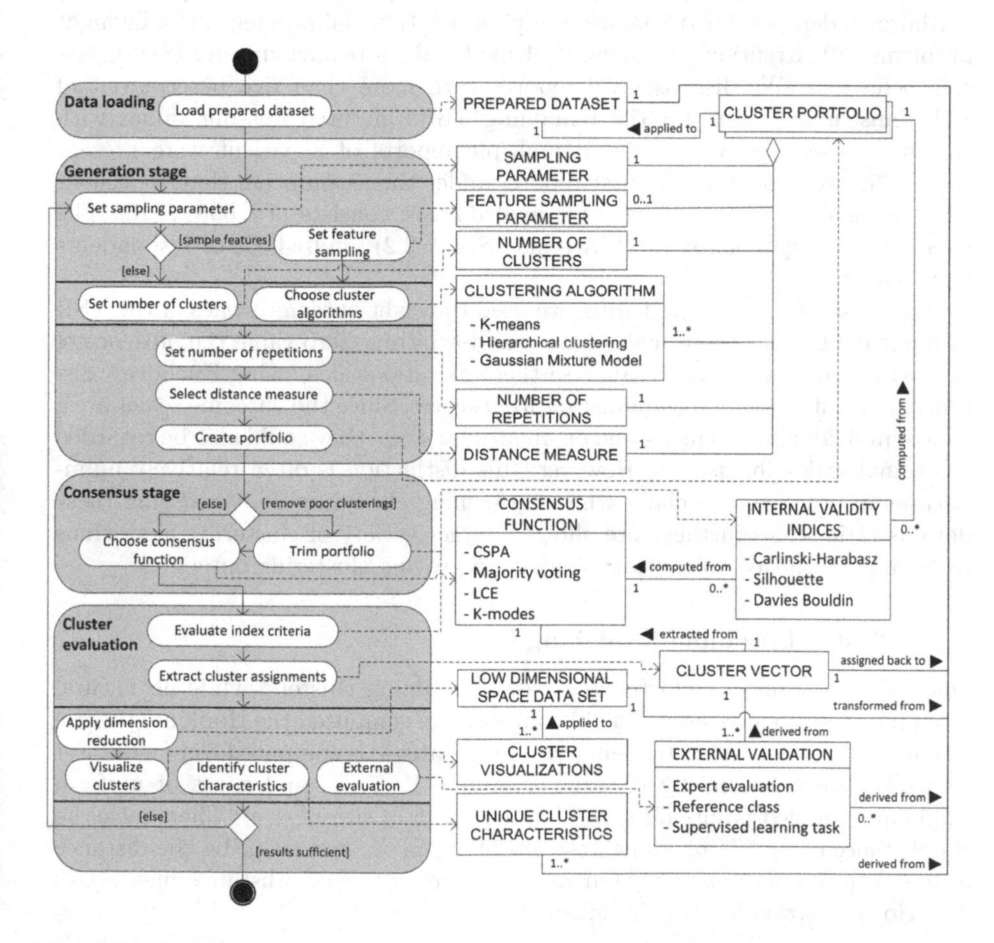

Fig. 1. Method fragment of the Meta Algorithmic Model for cluster ensemble modelling and evaluation.

can finally be performed. The class diagram on the right of Fig. 1 shows which concepts need to be instantiated in relation to each process step.

4 Applying Cluster Ensembles

4.1 Dataset

We applied the cluster ensemble modelling approach in Fig. 1 to a novel cross-diagnostic dataset of adolescent patients who were treated at the Psychiatry Department of the University Medical Center Utrecht in the Netherlands. The dataset consisted of Youth Self Report (YSR) questionnaires, a standardized checklist aimed at adolescents. It consists of 112 items in the form 'I am/have/feel *symptom/behaviour*', which a respondent can indicate as 'not true', 'somewhat or sometimes true', and 'very true or often true'. The YSR defines eight outcome scales by summing responses of specific item subsets: (1) Anxious depressed, (2) Withdrawn depressed, (3) Somatic complaints, (4) Social problems, (5) Thought problems, (6) Attention problems, (7) Rule breaking behaviour, and (8) Aggressive behaviour. We dismissed 50 reports with more than five percent out of 112 items missing, and for the remaining YSRs imputed missing values with the median score of that item. If multiple reports of a patient were present ($n = 175$), we used only the first report, under the assumption that treatment effect is smallest at this point. Our final dataset consists of 1,098 YSRs. The mean age of respondents was 14.7 years ($SD = 2.2$), and 44.5% of respondents were female.

For cluster ensemble modelling, we used the eight outcome scales of the YSR as input data. Since these scales have a non-arbitrary zero value (i.e. absence of any symptoms), we chose to analyse them as ratio scales, using Euclidean distance, implicitly assuming equidistant item scores. Since the outcome scales are a sum of individual items measured on an ordinal scale, they could also be regarded as ordinal scales themselves. However, this distinction is often relatively unimportant in practice, especially when performing clustering [9]. Analysing these data as ratio scales furthermore allows a larger variety of clustering algorithms to be applied to this dataset, most likely improving clustering outcomes.

4.2 Cluster Ensemble Modelling

One risk of performing cluster analysis is obtaining clusters, while no natural grouping exists in a dataset. For this purpose, we computed the Hopkins statistic as a measure of clustering tendency [1]. This statistic is computed from a dataset X with n observations by creating a sample $Y \subseteq X$, and a set of uniform randomly sampled points U, with U and Y both of size $m \ll n$. Then, let q_i be the distance of $u_i \in U$ to its nearest neighbour in X, and let p_i be the distance of $y_i \in Y$ to its nearest neighbour in X, according to some distance measure d. The Hopkins statistic is finally given by:

$$H = \frac{\sum_{i=1}^{m} q_i}{\sum_{i=1}^{m} p_i + \sum_{i=1}^{m} q_i} \tag{1}$$

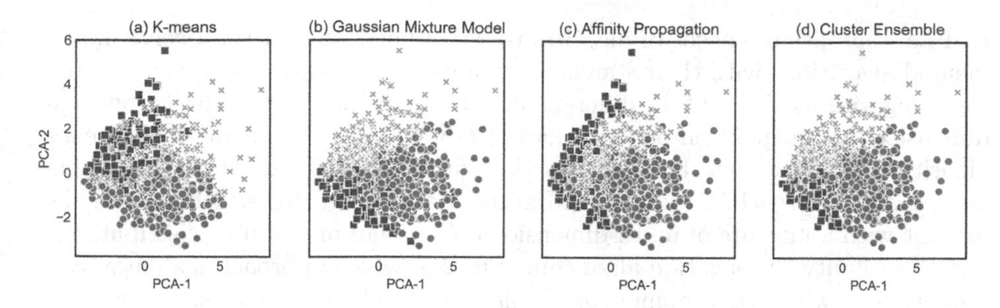

Fig. 2. Partitions of the dataset after applying Principal Component Analysis, both based on single algorithm (a–c), and combined in Cluster Ensemble (d).

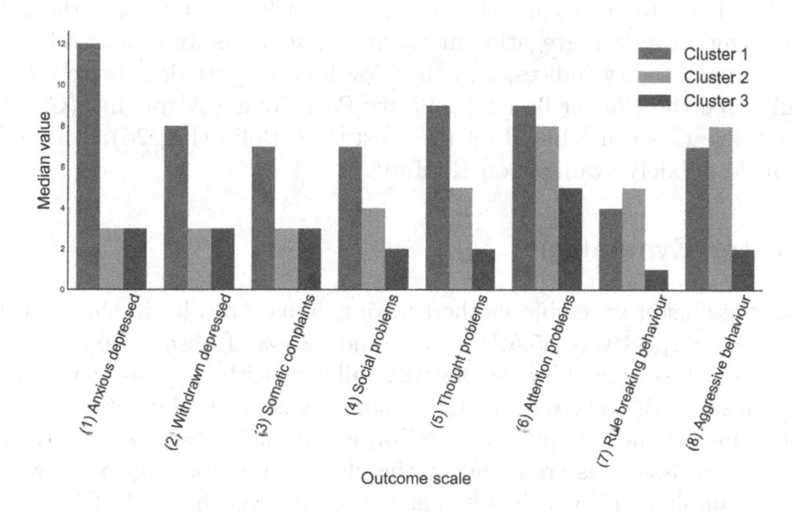

Fig. 3. Median YSR outcome scale value for each of the three identified clusters.

The Hopkins statistic ranges from 0 (uniformly distributed data), to 0.5 (randomly distributed data) to 1 (highly clusterable data). Computing this statistic for our dataset using Euclidean distance obtains $H = 0.71$. No definitive cut-off for cluster tendency has been established, but a value between 0.5 and 1 is regarded as indicative for high likelihood of significant clusters.

Next, an appropriate number of clusters k should be selected. Rather than rely on a single measure for determining this number, we used the R package `NbClust`, which computes 26 internal validity indices for several values of k, and proposes an optimal number of clusters based on a majority vote. We computed the validity indices in combination with both K-means clustering and hierarchical clustering, and set the number of clusters between two and seven. The majority vote shows that the optimal number of clusters $k = 3$ for our dataset, which we will use in all following steps.

For application of the cluster ensemble to our dataset, the R package `DiceR` was used, which implements various cluster ensemble techniques. In order to find an appropriate subset of algorithms, we applied each of the twelve implemented algorithms with their standard settings to the dataset. We then selected three algorithms that obtain different partitions of the dataset, based on two-dimensional Principal Component Analysis (PCA) plots. These are the K-means algorithm (Fig. 2a), which minimizes the within-cluster sum of squares using an iterative approach, a Gaussian Mixture Model (Fig. 2b), which models the dataset with a mixture of multi-dimensional Gaussian probability distributions, and the Affinity Propagation algorithm (Fig. 2c), which approaches a dataset as a network in which data points communicate with all other points.

To obtain a diverse cluster portfolio, we used five reruns for each of the three clustering algorithms with a random subset of 80% of all data. The number of clusters k is fixed to three, as determined previously. We trimmed the cluster portfolio using a Rank Aggregation method: all partitions were ranked based on several internal validity indices, and the 75% highest partitions were retained. We finally used the Cluster-based Similarity Partitioning Algorithm (CSPA) to obtain a single clustering based on the cluster portfolio (Fig. 2d). All analysis code is made publicly available on GitHub[1].

5 Cluster Evaluation

Applying the cluster ensemble method to our dataset results in three clusters, which contain respectively 55.5%, 32.1%, and 12.5% of observations (Fig. 2d). The ensemble clustering shows strongest similarity with the Gaussian Mixture model, with some differences in the two smallest clusters, and greater differences with the K-means and Affinity Propagation partitions. To assess statistical significance of the three clusters found by the cluster ensemble approach, we used the sigClust method [10] which tests against a null hypothesis of all data being from a single Gaussian distribution. This results in $p = 0.01$ when applied to our dataset, indicating presence of significant clusters at the $\alpha = 0.05$ level.

Figure 3 shows the median value of the eight YSR scales over the three clusters, where distinctions among the three clusters can be observed. Cluster 1, the largest cluster, has the highest overall scores, especially in the two depressed scales (1–2). Values of other scales are among the highest as well in Cluster 1, with Rule Breaking Behaviour being the lowest item. Clusters 2 and 3 on the other hand generally have lower scores, with equal median outcomes on the Anxious depressed, Withdrawn depressed, and Somatic problems scales (1–3). For the other five scales, Cluster 2 shows higher outcomes. For the Rule Breaking Behaviour and Aggressive Behaviour scales (7–8), Cluster 2 shows higher median values than Cluster 1 as well.

To identify clusters' distinguishing characteristics, we integrated clinical notes from the EHR, i.e. pieces of text written by caregivers about treatment, that were de-identified using the DEDUCE method [15], in the two weeks surrounding YSR response. We extracted the 1000 most frequent terms from these texts, and computed the Spearman correlation coefficient for each term and

[1] http://www.github.com/vmenger/cluster-ensembles.

each of the three clusters vs the other two clusters. A psychiatrist then selected three informative terms among those with the highest positive correlation coefficients. For Cluster 1, the selected terms are *depressive*, *dejected*, and *suicidal*, which is in line with high scores in the two depressed scales. For Cluster 2, the terms *behavioural problems*, *adhd* (attention deficit hyperactivity disorder), and *distracted* are identified, which is in line with high scores on the Attention Problems and Aggressive Behaviour scales. For Cluster 3, these terms are *speech*, *verbal*, and *individual*. Based on these terms and Fig. 3, we describe Cluster 1 as 'depressive symptoms', and Cluster 2 as 'behavioural problems'. A comprehensive description of Cluster 3 is less evident, we therefore describe it as 'low severity'.

Table 1 shows the three clusters versus the main DSM diagnosis, which had been made definitive within 12 weeks of YSR response for a subset of 665 patients. The most common diagnosis for the three clusters respectively are Anxiety Disorder, Attention Deficit Disorder, and Pervasive Developmental Disorder (PDD). Diagnoses are typically present in several clusters, although they are usually most prominent in one single cluster, with the exception of PDDs.

Table 1. Main DSM diagnoses per cluster, if finalized within 12 weeks.

Disorder	Cluster 1	Cluster 2	Cluster 3	Total
Anxiety disorder	76 (19.2%)	7 (3.5%)	13 (18.3%)	14.4%
Developmental disorder				
Attention deficit disorder	39 (9.8%)	82 (41.4%)	10 (14.1%)	19.7%
Pervasive developmental disorder	103 (26.0%)	51 (25.8%)	24 (33.8%)	26.8%
Other	15 (3.8%)	15 (7.6%)	2 (2.8%)	4.8%
Eating disorder	10 (2.5%)	2 (1.0%)	3 (4.2%)	2.3%
Mood disorder	65 (16.4%)	13 (6.6%)	7 (9.9%)	12.8%
Psychotic disorder	24 (6.1%)	7 (3.5%)	4 (5.6%)	5.3%
Personality disorder	27 (6.8%)	1 (0.5%)	0 (0.0%)	4.2%
Other	37 (9.3%)	20 (10.1%)	8 (11.3%)	9.8%
Total	**396 (100%)**	**198 (100%)**	**71 (100%)**	**100%**

We finally integrate several clinically relevant variables, including Global Assessment of Functioning (GAF) score at start and end of treatment, a seven-point burden of disease indicator, and length of treatment (Table 2). Although Cluster 1 has the highest overall YSR outcome scale scores, the GAF scores both at start and end of treatment are relatively low. The difference between these groups are assessed with a Kruskal-Wallis one-way analysis of variance test. Results show that significant differences in GAF score at start and end of treatment and in length of treatment exist at the $\alpha = 0.05$ level, but not in burden of disease. This indicates that clusters do not only differ in YSR outcome scales, but also in variables that are relevant in clinical practice.

Table 2. Average value of clinically relevant variables per cluster. P-value is assessed using a Kruskal-Wallis test, * indicates significance at the $\alpha = 0.05$ level. GAF = Global Assessment of Functioning.

Variable	Cluster 1	Cluster 2	Cluster 3	P-value
GAF at start of treatment	45.9	50.0	46.0	0.008*
GAF at end of treatment	53.6	56.7	51.5	0.012*
Burden of disease	4.5	4.5	4.8	0.550
Length of treatment (days)	132.6	175.3	160.7	0.003*

6 Discussion and Conclusion

In line with previous research, our results point out that different clustering algorithms indeed obtain different partitions. Cluster ensembles are a useful method to overcome such issues. By applying our proposed cluster ensemble approach to a dataset of YSR questionnaires, we obtained three distinct patient subgroups. Patients with the same DSM diagnosis are typically represented in multiple clusters, indicating that the three clusters are to some extent a novel stratification of adolescent patients. We furthermore identified significant differences in GAF both at start and end of treatment, and in length of treatment. Although absolute differences among clusters are modest, this shows that patient subgroups do not only differ in the YSR outcome scales.

The clustering outcomes of this study are limited by both the type of data and the specific patient population that reported it. The dataset includes eight outcome scales that are general, but may not capture all dimensions of patient well-being, and whether the clusters we obtained generalize to other populations should be the topic of further research. The main contribution of this research however lies in the cluster ensemble approach, and the process guide introduced with the Meta Algorithmic Model. Such cluster ensemble approaches are able to eliminate one source of variance in reported psychiatric patient subgroups, and can thereby in the future contribute to the identification of a more robust and objective stratification of psychiatric patients.

References

1. Banerjee, A., Dave, R.: Validating clusters using the Hopkins statistic. In: IEEE International Conference on Fuzzy Systems (2004)
2. Boongoen, T., Iam-On, N.: Cluster ensembles: a survey of approaches with recent extensions and applications. Comput. Sci. Rev. **28**, 1–25 (2018)
3. Cole, V.T., Apud, J.A., Weinberger, D.R., Dickinson, D.: Using latent class growth analysis to form trajectories of premorbid adjustment in schizophrenia. J. Abnorm. Psychol. **121**(2), 388–395 (2012)
4. Cuthbert, B.N., Insel, T.R.: Toward the future of psychiatric diagnosis: the seven pillars of RDoC. BMC Med. **11**(1), 126 (2013)

5. Dollfus, S., Everitt, B., Ribeyre, J.M., Assouly-Besse, F., Sharp, C., Petit, M.: Identifying subtypes of schizophrenia by cluster analyses. Schizophr. Bull. **22**(3), 545–555 (1996)
6. Fountain, C., Winter, A.S., Bearman, P.S.: Six developmental trajectories characterize children with autism. Pediatrics **129**(5), e1112–e1120 (2012)
7. Fred, A.: Finding consistent clusters in data partitions. In: Kittler, J., Roli, F. (eds.) MCS 2001. LNCS, vol. 2096, pp. 309–318. Springer, Heidelberg (2001). https://doi.org/10.1007/3-540-48219-9_31
8. Ghaemi, R., Sulaiman, N., Ibrahim, M., Mustapha, N.: A survey: clustering ensembles techniques. Int. J. Comput. Inf. Eng. **3**(2), 365–374 (2009)
9. Harpe, S.E.: How to analyze likert and other rating scale data. Curr. Pharm. Teach. Learn. **7**(6), 836–850 (2015)
10. Huang, H., Liu, Y., Yuan, M., Marron, J.S.: Statistical significance of clustering using soft thresholding. J. Comput. Graph. Statist. **24**(4), 975–993 (2015)
11. Kleinman, A., et al.: Attention-based classification pattern, a research domain criteria framework. Aust. N. Z. J. Psychiatry **49**(3), 255–265 (2014)
12. Kuncheva, L., Hadjitodorov, S.: Using diversity in cluster ensembles. In: IEEE International Conference on Systems, Man and Cybernetics (2004)
13. Lewandowski, K.E., Sperry, S.H., Cohen, B.M., Ongur, D.: Cognitive variability in psychotic disorders: a cross-diagnostic cluster analysis. Psychol. Med. **44**(15), 3239–3248 (2014)
14. Marquand, A.F., Wolfers, T., Mennes, M., Buitelaar, J., Beckmann, C.F.: Beyond lumping and splitting: a review of computational approaches for stratifying psychiatric disorders. Biol. Psychiatry Cogn. Neurosci. Neuroimaging **1**(5), 433–447 (2016)
15. Menger, V., Scheepers, F., van Wijk, L.M., Spruit, M.: DEDUCE: a pattern matching method for automatic de-identification of Dutch medical text. Telematics Inform. **35**(4), 727–736 (2018)
16. Menger, V., Spruit, M., de Bruin, J., Kelder, T., Scheepers, F.: Supporting reuse of EHR data in healthcare organizations: the CARED research infrastructure framework. In: Proceedings of the 12th International Joint Conference on Biomedical Engineering Systems and Technologies (2019)
17. Menger, V., Spruit, M., Hagoort, K., Scheepers, F.: Transitioning to a data driven mental health practice: collaborative expert sessions for knowledge and hypothesis finding. Comput. Math. Methods Med. **2016**, 1–11 (2016)
18. Olino, T.M., Klein, D.N., Lewinsohn, P.M., Rohde, P., Seeley, J.R.: Latent trajectory classes of depressive and anxiety disorders from adolescence to adulthood: descriptions of classes and associations with risk factors. Compr. Psychiatry **51**(3), 224–235 (2010)
19. Shen, J., Lee, P.H., Holden, J., Shatkay, H.: Using cluster ensemble and validation to identify subtypes of pervasive developmental disorders. In: AMIA Annual Symposium Proceedings, vol. 2007, pp. 666–670 (2007)
20. Spruit, M., Jagesar, R.: Power to the people! - Meta-algorithmic modelling in applied data science. In: Proceedings of the 8th International Joint Conference on Knowledge Discovery, Knowledge Engineering and Knowledge Management (2016)
21. Strehl, A., Ghosh, J.: Cluster ensembles – a knowledge reuse framework for combining multiple partitions. J. Mach. Learn. Res. **3**, 583–617 (2003)
22. Topchy, A., Jain, A.K., Punch, W.: Combining multiple weak clusterings. In: Proceedings of the Third IEEE International Conference on Data Mining, p. 331 (2003)

23. Topchy, A.P., Law, M.H.C., Jain, A.K., Fred, A.L.: Analysis of consensus partition in cluster ensemble. In: Proceedings of the Fourth IEEE International Conference on Data Mining, ICDM 2004, pp. 225–232 (2004)
24. Vega-Pons, S., Ruiz-Shulcloper, J.: A survey of clustering ensemble algorithms. Int. J. Pattern Recognit Artif Intell. **25**(03), 337–372 (2011)

Connection Between the Parkinson's Disease Subtypes and Patients' Symptoms Progression

Anita Valmarska[1(✉)], Dragana Miljkovic[1], Marko Robnik–Šikonja[2], and Nada Lavrač[1,3]

[1] Jožef Stefan Institute, Jamova 39, Ljubljana, Slovenia
{anita.valmarska,dragana.miljkovic,nada.lavrac}@ijs.si
[2] Faculty of Computer and Information Science, University of Ljubljana, Ljubljana, Slovenia
marko.robnik@fri.uni-lj.si
[3] University of Nova Gorica, Vipavska 13, Nova Gorica, Slovenia

Abstract. Parkinson's disease (PD) is a neurodegenerative disease characterized by a variety of symptoms. Clinicians studying movement disorders have tried to connect the variability of symptoms to underlying subtypes of PD, the two most common being tremor dominant (TD) and postural instability and gait difficulty dominant (PIGD). This paper investigates the connection between the Parkinson's disease PIGD and TD subtypes, and patients' symptoms progression. We present a set of symptoms closely related to each subtype as well as the patients' statuses that indicate a switch in subtype classification. Detection of patients' symptoms that possibly lead towards subtype change can contribute to the more personalized treatment of PD patients. The results of experiments on Parkinson's Progression Markers Initiative (PPMI) data suggest the connection of the PIGD subtype to non-motor symptoms associated with decreased quality of life.

1 Introduction

Patients suffering from Parkinson's disease (PD) experience a variety of symptoms whose severity can significantly affect the quality of life of both the patients and their families. As a step to better understanding the illness, the clinicians have tried to connect the variability of symptoms to some underlying subtype(s) of PD. The commonly used subtype classification, proposed by Jankovic et al. [2], is the division of PD patients into the tremor dominant (TD), postural instability and gait difficulty (PIGD), and indeterminate (Indeterminate) subtypes.

The classification of PD patients in the TD/PIGD subtypes can be derived from the assessment of patients' symptoms severity using the well-established MDS-UPDRS scale [5]. Classification into TD and PIGD is actually based on the calculation of the ratio of the TD and the PIGD scores. Usually, the classification into PIGD/TD subtypes is done only at the beginning of the patient's

D. Riaño et al. (Eds.): AIME 2019, LNAI 11526, pp. 263–268, 2019.
https://doi.org/10.1007/978-3-030-21642-9_32

diagnostic/treatment process. However, this classification can change as the disease progresses and the patient's symptoms are affected by both the natural progression of the disease and their symptomatic treatment.

The definition of consistent subtypes of PD is still an open issue and is key to better understand the underlying disease mechanisms, predict disease course, and design clinical trials [1]. To better understand the properties of the current subtype categorization, this paper investigates how the PIGD/TD subtypes are related to the several aspects of experiences of daily living of PD patients. We investigate which non-motor symptoms are closely related to each subtype and which symptoms mostly influence the change of subtype classification between two consecutive patient's visits to their clinician.

The paper is structured as follows. Section 2 presents the data used in the analysis. Section 3 presents the proposed methodology and Sect. 4 outlines the experimental results, followed by the conclusions and plans for further work outlined in Sect. 5.

2 Data

In this paper, we use clinical data from the Parkinson's Progression Markers Initiative (PPMI) data collection [3]. The PPMI data collection records data for over 400 PD patients, involved in the study for up to 5 years. During their involvement, the patients made regular visits on every 3–6 months to their assigned clinicians where their symptoms are assessed. The clinical data used in our work was gathered using several standardized questionnaires:

- MDS-UPDRS (Movement Disorder Society-sponsored revision of Unified Parkinson's Disease Rating Scale) is the most widely used, four-part questionnaire (65 questions) addressing both motor and non-motor aspects of the patients' life.
- MoCA (Montreal Cognitive Assessment) is a rapid screening instrument for mild cognitive dysfunction.
- SCOPA-AUT (Scales for Outcomes in Parkinson's disease - Autonomic) is a specific scale for assessing autonomic dysfunction in PD patients.
- PASE (Physical Activity Scale for the Elderly) is a questionnaire which is a practical and widely used approach for physical activity assessment in epidemiological investigations.

Answers to the questions from each questionnaire form the vectors of attribute values. All of the considered questions have ordered values, and—with the exception of questions from MoCA and PASE—increased values suggest higher symptom severity and decreased quality of life.

The classification of patients into the PIGD/TD subtypes is done in accordance with the guidelines for PIGD/TD classification using the patients' MDS-UPDRS scores, where the PIGD/TD subtype classification of patients is dependent on the ratio between the *TD score* and the *PIGD score*. The *TD score* reflects the mean severity of symptoms describing different types of tremors.

The *PIGD score* reflects the mean severity of symptoms concerning patients' gait and postural instability. Detailed information about the relevant symptoms and the ratio boundaries used for subtype classification is available in [5].

3 Methodology

In the PPMI data collection the patients' symptoms were regularly updated during their visits to their clinicians. In our study, we transformed the original PPMI data set into a set of 1,345 instances, each identifiable as a pair (p_i, v_{ij}), where p_i is the identification of a patient, and v_{ij} identifies the j-th visit the i-th patient has made to the clinician. Each instance (p_i, v_{ij}) is a record of all the symptoms severity assessments for patient p_i on their j-th visit.

As the patients' symptoms change over time as a reaction to the natural progression of the disease and their symptomatic treatment, we expect their assigned subtype to change over time. With the aim to investigate how the PIGD/TD classification of PD patients is connected to their overall quality of life, we investigate which symptoms are important for the classification of an instance as PIGD or TD. In this work we also investigate which symptoms influence the change of patients' subtype classification between two consecutive visits. The methodology is composed of consecutively applying two methods.

- Since the classification into subtypes is based only on motor symptoms, while the patients' overall status actually depends on the severity of both the motor and non-motor symptoms, we are interested in finding out which non-motor symptoms affect the classification of an instance (p_i, v_{ij}) into subtype PIGD or TD. To study the influence of non-motor symptoms on the classification of instances as PIGD/TD we applied the EXPLAIN methodology [4], which decomposes the classification model into the individual contribution of attributes using weighted evidence (WE). We used the random forest as the classifier.
- Since we hypothesize that there is a connection between the patients' overall status and their subtype classification, the improvement or degradation of the patient's overall status or their subtype classification will share a subset of influential symptoms. To investigate the symptoms' influence on PD subtype change, we used our previously proposed methodology [6], where influential symptoms are detected as those whose severity changes most frequently as the PD subtype changes. The calculation of influential symptoms is done for all symptoms identified in Sect. 2.

4 Results

Figure 1 outlines the importance of non-motor symptoms for classifying an instance as PIGD. The blue lines suggest attributes' positive influence towards classifying an instance as PIGD. The red lines indicate the opposite. The results show that patients, classified as PIGD, experience also non-motor symptoms

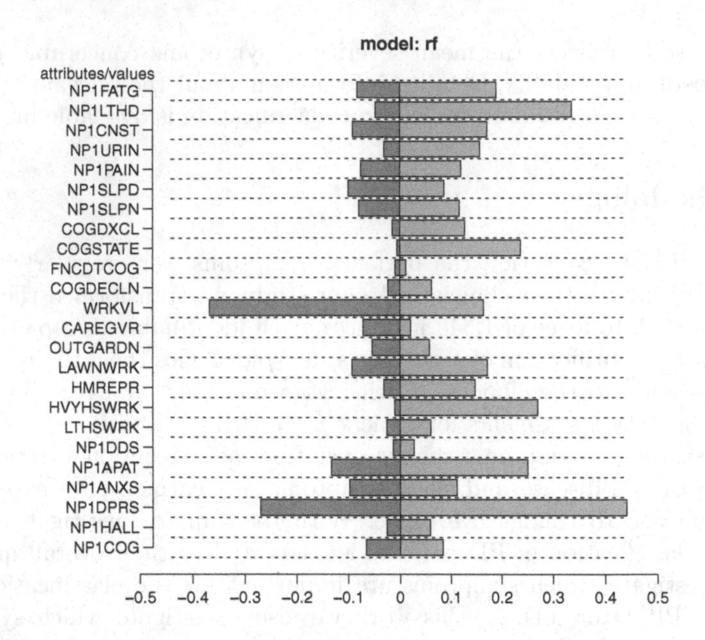

Fig. 1. Non-motor symptoms importance for the classification of instances as PIGD. Classification model: random forest, method: EXPLAIN, type: WE. (Color figure online)

indicating a declined quality of life (NP1FATG - fatigue, NP1LTHD - lightheadedness, NP1CONST - constipation, NP1APAT - apathy, etc.).

Table 1 presents the top 5 attributes whose severity improved or worsened most frequently when a subtype change between two consecutive patient's visits was detected. We present the lists of most influential symptoms for subtype change from PIGD to TD, and from TD to PIGD, respectively.

The results are not surprising. When the patients' subtype changes from PIGD to TD, the most obvious improvement is noted in symptoms that are representative of PIGD: gait, stability, apathy. At the same time, there is worsening of many types of tremor, i.e. the symptoms that are significant for the classification of TD. Constancy of rest, hand pronation/supination, urinary problems, and fatigue are among the most influential symptoms for subtype change and are also among the ones most affected by the changes of the patients' overall status, as defined in [6]. This is an indication of the possible association between the subtype classification and the overall quality of life of Parkinson's disease patients. On the other hand, when the subtype classification between two visits has changed from TD to PIGD, the symptoms that have most frequently improved are various types of tremor, which are highly associated with subtype TD. When the patient's subtype has changed to PIGD, patients most frequently experience increased severity of their gait, fatigue, speech, stability symptoms, and also their inability to commit time to their hobbies.

Table 1. Top 5 symptoms with most frequently improved/worsened severity when a subtype change is detected between two consecutive visits, uncovered using the method presented in [6]. Symptoms are ordered according to their influence, in descending order of influence. For readability, symptoms descriptions are used without PPMI codes.

Subtype change	Symptoms improvement	Symptoms worsening
From PIGD to TD	Gait	Constancy of rest
	Saliva	Resting tremor
	Stability	Tremor
	Apathy	Hand pronation/supination
	Lightheadedness	Urinary problems
From TD to PIGD	Resting tremor	Gait
	Constancy of rest	Fatigue
	Postural tremor	Speech
	Kinetic tremor	Stability
	Tremor	Hobby

5 Conclusion

This paper investigates the connection between the subtypes of PD patients and the progression of patients' symptoms. It combines two methodological threads to (a) explore the influence of non-motor symptoms on the classification of patients as PIGD and (b) to investigate the symptoms influence on the change of patients' subtype classification between two consecutive visits. The experimental results reveal a subset of non-motor symptoms whose increased severity is associated with both the PIGD classification and the declined quality of life of patients. The comparison of our results to previous research has also identified a subset of symptoms with high influence on the change of patients' subtype and the change of the patients' overall status.

In further work, we will take advantage of patients' MR and medication treatment data to improve subtype classification.

Acknowledgments. We acknowledge the support of the Slovenian Research Agency research core funding program P2-0103. The data used were obtained from the Parkinson's Progression Markers Initiative (PPMI) (www.ppmi-info.org/data).

References

1. Fereshtehnejad, S.M., Postuma, R.B., Dagher, A., Zeighami, Y.: Clinical criteria for subtyping Parkinson's disease: biomarkers and longitudinal progression. Brain **140**(7), 1959–1976 (2017)
2. Jankovic, J., et al.: Variable expression of Parkinson's disease - a base-line analysis of the DATATOP cohort. Neurology **40**(10), 1529–1529 (1990)

3. Marek, K., et al.: The Parkinson's Progression Markers Initiative (PPMI). Prog. Neurobiol. **95**(4), 629–635 (2011)
4. Robnik-Šikonja, M., Kononenko, I.: Explaining classifications for individual instances. IEEE Trans. Knowl. Data Eng. **20**(5), 589–600 (2008)
5. Stebbins, G.T., Goetz, C.G., Burn, D.J., Jankovic, J., Khoo, T.K., Tilley, B.C.: How to identify tremor dominant and postural instability/gait difficulty groups with the movement disorder society unified Parkinson's disease rating scale: comparison with the unified Parkinson's disease rating scale. Mov. Disord. **28**(5), 668–670 (2013)
6. Valmarska, A., Miljkovic, D., Konitsiotis, S., Gatsios, D., Lavrač, N., Robnik-Šikonja, M.: Symptoms and medications change patterns for Parkinson's disease patients stratification. Artif. Intell. Med. **91**, 82–95 (2018)

Interpretable Patient Subgrouping Using Trace-Based Clustering

Antonio Lopez Martinez-Carrasco[1], Jose M. Juarez[1(✉)], Manuel Campos[1], Antonio Morales[1], Francisco Palacios[1], and Lucia Lopez-Rodriguez[2]

[1] AIKE Research Group, Computer Science Faculty, Universidad de Murcia, Murcia, Spain
jmjuarez@um.es

[2] Intensive Care Unit, Hospital University of Getafe, Getafe, Spain

Abstract. Antibiotic resistance in hospitals is a general problem whose solution includes the adaptation of antimicrobial therapies to the local epidemiology. The identification of groups of the population with a common phenotype (by means of their clinical histories) is essential if a hospital is to establish a policy regarding antibiotics. Descriptive pattern mining is effective when carrying out exploratory analyses, such as the design of clustering and subgroup algorithms in order to generate groups of interest. However, the researchers have paid little attention to how these types of algorithms are combined and supervised in medical research. We believe that the implication of clinicians in the process, the interpretability of algorithms and patient traceability of the results obtained are also key requirements. In this work, we propose: (1) to adapt well-known clustering algorithms in order to identify subgroups of patients and (2) a man-in-the-loop methodology so as to carry out this task, thus fulfilling the abovementioned requirements. This proposal is evaluated in the context of a hospital's antimicrobial stewardship problem.

1 Introduction

The characterisation and identification of groups of patients of special interest (phenotype) is a core issue in medical research [7]. For example, the loss of efficacy of antibiotics in antimicrobials is a growing problem, which requires not only global actions but also local measurements in hospitals [1]. In this context, subgroups of patients with a common unexpected response to antibiotics could be identified in order to review specific administration protocols.

In Machine Learning, the exploratory analysis of datasets is confronted using descriptive pattern-mining approaches. In unsupervised learning, traditional clustering methods divide the dataset into groups (clusters), thus establishing a partition. In supervised learning, Subgroup Discovery (SD) is a relatively new approach whose purpose is to find subgroup patterns when given an attribute target and (often) a rule format to represent those patterns [6]. Tailored algorithms has recently been used to solve medical subgrouping [5].

D. Riaño et al. (Eds.): AIME 2019, LNAI 11526, pp. 269–274, 2019.
https://doi.org/10.1007/978-3-030-21642-9_33

However, little attention has been paid to how pattern-mining methods are adapted in practical medical research. In particular, we have identified two fundamental issues that should be dealt with. First, the difficulty involved in stating, a-priori, the attributes of interest and the scarce availability of inter-department hospital data limits the quality of patterns (rules) found. In this respect, clinicians carry out a post-mining task to trace patients' records in order to evaluate their clinical relevance. Second, the trust placed in the mining method is still an open problem in the clinical community. The use of interpretable outcomes, the methods themselves and the implication of the clinicians in the process (man-in-the-loop) are necessary in order to resolve this problem [3].

2 Trace-Based Clustering

Our objective is to discover and evaluate subgroups of patients, while combining machine learning (ML) techniques and the involvement of the medical experts in the process, and we, therefore, propose the use of trace-based clustering. In particular, we deal with the subgroup discovery problem by considering unsupervised learning.

The proposal has two parts: we propose (a) an algorithm that extends well-known clustering techniques and (b) a methodology based on the man-in-the-loop strategy in order to select the best candidates from the clinical point of view.

Formal Aspects. We propose to tackle the patient subgroup discovery problem for unlabelled data by combining clustering algorithms. Our proposal is based on the simple idea of considering interesting subgroups as elements that often/always remain together in different clusters when executing clustering algorithms iteratively. This proposal is, therefore, based on the evaluation of clusters between different partitions of the same dataset. The fundamental concepts of the proposal are the following:

Def. Partition (C_x): given a dataset C, C_x is a *partition* of C if $C_x \subseteq \mathcal{P}(C)$ with $|C_x| = x$ where $C_x = \{C_{x1}, \ldots, C_{xx}\}$ and $C_{x1} \cup \ldots \cup C_{xx} = C$.

Def. Cluster (C_{xi}): the elements of partition C_x are called *clusters*, meeting that $\forall C_{xi}, C_{xj} \in C_x, C_{xi} \cap C_{xj} = \emptyset$.

Using this terminology, when $x \neq y$, C_{xi} and C_{yi} are two clusters of different partitions, C_x and C_y.

We denote $\mathcal{P}(C)_k$ as the set of all possible partitions of C with k clusters.

Def. Clustering Function: given a dataset C and a positive integer value of k, the *clustering function* obtains a partition of C with k clusters, expressed as follows: $Clustering : C \times \mathbb{Z}^+ \rightarrow \mathcal{P}(C)$, where $Clustering(C, k) \in \mathcal{P}(C)_k$.

For the sake of clarity but without a loss of generality, in this research we use classic clustering algorithms (understanding that the objective of these algorithms is the partition of a dataset in k clusters and the value of k parameter is set *a-priori*).

An essential aspect in the study of clustering algorithms is the evaluation of their partitions by using *Cluster Validity Indexes* (CVIs), such as Rand or Silhouette indexes and various criteria are involved in the direct evaluation of clusters. In this paper, we focus on measures with which to evaluate clusters of different partitions. We, therefore, generalise this kind of metrics by introducing the *matching function.*

Def. Matching Function (M): given two partitions C_x and C_y, the *matching function* between the clusters C_{xi} and C_{yj} measures the similarity between such clusters in terms of the elements they contain. Formally: $M : \mathcal{P}(C)_a \times \mathcal{P}(C)_b \rightarrow [0, 1]$.

The intuitive idea of *trace* eases the task of tracking the elements of a cluster that also remain grouped in the clusters of other partitions.

Def. Trace: Let us suppose a dataset C and a set of partitions $\{C_2, \ldots, C_K\}$ (as a result of computing iteratively $Clustering(C, i), i \in \{2, \ldots, K\}$). Given a cluster C_{Ki} of the partition C_K, the *trace* of this cluster is the set of clusters of partitions from C_2 to C_{K-1} that maximize the Matching function in relation to the cluster C_{Ki}.

Def. Trace Function: The *trace function* calculates the trace of a cluster, given a set of partitions, as follows: $Trace : C_K \times \{\mathcal{P}(C)_2, \ldots, \mathcal{P}(C)_{K-1}\} \rightarrow C_{2*} \times \ldots \times C_{K-1*}$.

In Algorithm 1, we propose a method by which to implement the Trace function.

Algorithm 1. Trace

```
Input C_xi: cluster ; {C_2, ..., C_x}: set of partitions ; M: matching function
Output T % vector of selected clusters
  T ← ∅
  for k = x − 1 ... 2 do
     candidate ← C_k1
     for y = 1 ... k do
        if M(C_xi, C_ky) > M(C_xi, candidate) then
           candidate ← C_ky
        end if
     end for
     T_k ← candidate
  end for
  return T
```

In Algorithm 1, M is a matching function and T stores the set of selected clusters that trace C_{xi} (input cluster). Note that, if there are x partitions $(C_1, \ldots, C_x)$, then $k \in [2, x - 1]$. This is because: (1) C_1 is never taken into account because it is a partition with only one cluster (signifying that $C_{xi} \subseteq C_{11}$ and $C_1 = C$) and (2) C_{xi} is a cluster of C_x and, according to the definition, $C_{xi} \cap C_{xj} = \emptyset$.

Def. M-Traces Function: given a dataset C and a positive integer value of K, the $M - Traces$ function obtains a matrix of traces for the partitions $C_2, \ldots, C_K$, computing the corresponding vectors through the $Trace$ function for each cluster C_{Ki} of the C_K partition. Formally, $M - Traces : C \times Z^+ \rightarrow C_2 \times \ldots \times C_{K-1}$

In Algorithm 2, we propose a specific implementation of the $M - Traces$ function.

Algorithm 2. M-Traces: Matrix of traces

Input C: dataset ; $K \in Z^+$, M: matching function
Output $\mathcal{T}$ *% matrix of selected clusters*
  $\mathcal{C} = \emptyset$
  $\mathcal{T} \leftarrow \emptyset$
  for $i = 2 \ldots K$ **do**
    $C_i \leftarrow Clustering(C, i)$
    $\mathcal{C} = \mathcal{C} \cup \{C_i\}$
  end for
  for $i = 1 \ldots K$ **do**
    $\mathcal{T}_i \leftarrow Trace(C_{Ki}, \mathcal{C}$,M)
  end for
  return $\mathcal{T}$

Methodology. We follow a methodology comprising the following steps: (1) Extraction of data and selection of attributes: In our case, this process is accomplished by means of WASPSS [1,4], a tool that integrates data from hospital departments for antibiotic surveillance. (2) Selection of clustering algorithms and parameters: We select the clustering algorithm (*clustering function*) and we estimate the maximum number of expected clusters (K parameter). (3) Automatic generation of candidates (subgroups): One key aspect in this step is the adoption of a specific *matching function* M. In this work, we adapted the Jaccard coefficient. (4) Visual support of candidate selection: Once matrix $\mathcal{J}$ has been computed, we create a visual representation in order to ease the selection of the subgroups using *heat-maps*. (5) Evaluation by clinical experts: The patients related to the clusters eventually selected will be supervised by the clinicians by revising their personal records.

3 Experiments

Our experiments have been carried out in the context of the rational use of antibiotics in a hospital with the specific objective of identifying groups of patients with similar antibiotic resistance behaviour. More precisely, we focus on the study of patients treated with *Vancomycin*, considering their antibiogram and regarding the following target Gram positive bacteria: *Staphylococcus Aureus*, *Enterococcus Faecalis*, *Staphylococcus Epidermidis*, *MRSA* (*Methicillin-Resistant Staphylococcus Aureus*), *Negative Staphylococcus Coagulasa* and *Enterococcus Faecium.*

In our experiments, the dataset was collected from 4 different sources: clinical records, microbiology department, pharmacy department and laboratory. The initial dataset has 169 attributes and 1778 records, collected between 2015 and 2016 using WASPSS platform [1]. After carrying a data transformation process, we obtain a final mining view containing 83 attributes and 1768 records.

The experiment described was carried out using $k-medoids$ clustering algorithms: PAM and CLARA [2]. We chose these classic algorithms because of the evidence provided in clinical literature and in order to facilitate the traceability of the process by the clinical expert. Taking into account the total number of patients and the local epidemiology, the K parameter will be in range $[2,\ldots,20]$ (number of possible subgroups). The $M-Traces$ function has been implemented in R (version 3.3.2) using clustering algorithms from the *cluster* package (version 2.0.5).

Note that the number of candidate clusters is high, and that they are, in most cases, irrelevant. After generating the $\mathcal{J}$ matrix (visualization step) for the previous results, we, therefore, select the $C_{20,x}$ clusters with statistics: $Mean(\sum_{n=2}^{19}\mathcal{J}_{x,n}) > 0.7$ and $Median(\sum_{n=2}^{19}\mathcal{J}_{x,n}) > 0.7$. This additional prune reduces 92% of the number of clusters and avoids the expert having to carry out a manual study of an excessive number of clusters that are of no interest.

Once the processes of candidate generation and pruning has been completed, the information from matrix $\mathcal{J}$ for each experiment is displayed visually.

The 11 relevant groups of patients obtained were reviewed by 2 medical doctors.

All experts agreed the outstanding and unexpected nature of the subgroup found. According to their opinion, conventional profiles of patients usually include the type of sample and the length of stay. However, this unexpected outcome mostly focuses on elderly patients with cardiopathy or ischemic peripheral vasculopathy. That is, the potential existence of Vancomycin resistance in patients with atherosclerosis and a 'clean' medical surgery from an infectious point of view. More specifically, this non-trivial outcome may suggest that patients sharing procedures such as venous access and prostheses have a high risk of infection by Gram-positive bacteria.

4 Conclusions

The aim of this paper is to support clinicians in the identification of relevant subgroups of patients by paying attention to the interpretability and traceability of the process. We propose a methodology, denominated as trace-based clustering.

We have evaluated the suitability of our proposal by studying Gram-positive infections treated with Vancomycin (treatment of choice) and the changes in the minimum inhibitory concentration in groups of patients at a hospital. The results obtained helped the clinicians to characterise a specific group of patients, currently under supervision at the hospital.

Acknowledgement. This work was supported by the WASPSS project (Ref: TIN2013-45491-R), funded by the national research programme from the Spanish Ministry of Economy and Competitiveness and the European Regional Development Fund (ERDF, FEDER).

References

1. Cánovas-Segura, B., Campos, M., Morales, A., Juarez, J.M., Palacios, F.: Development of a clinical decision support system for antibiotic management in a hospital environment. Prog. Artif. Intell. **5**(3), 181–197 (2016)
2. Jin, X., Han, J.: K-medoids clustering. In: Sammut, C., Webb, G.I. (eds.) Encyclopedia of Machine Learning and Data Mining, pp. 697–700. Springer, Boston (2017). https://doi.org/10.1007/978-1-4899-7687-1
3. Mühlbacher, T., Piringer, H., Gratzl, S., Sedlmair, M., Streit, M.: Opening the black box: strategies for increased user involvement in existing algorithm implementations. IEEE Trans. Visual. Comput. Graph. **20**(12), 1643–1652 (2014)
4. Palacios, F., et al.: A clinical decision support system for an antimicrobial stewardship program. In: HEALTHINF 2016–9th International Conference on Health Informatics, Proceedings; Part of 9th International Joint Conference on Biomedical Engineering Systems and Technologies, BIOSTEC 2016, pp. 496–501. SciTePress (2016)
5. Valmarska, A., Miljkovic, D., Konitsiotis, S., Gatsios, D., Lavrač, N., Robnik-Šikonja, M.: Symptoms and medications change patterns for Parkinson's disease patients stratification. Artif. Intell. Med. **91**, 82–95 (2018)
6. Ventura, S., Luna, J.M.: Supervised Descriptive Pattern Mining. Springer, Cham (2018). https://doi.org/10.1007/978-3-319-98140-6
7. Wojczynski, M.K., Tiwari, H.K.: Definition of phenotype. In: Genetic Dissection of Complex Traits, Volume 60 of Advances in Genetics, pp. 75–105. Academic Press, Cambridge (2008)

Extracting Food-Drug Interactions from Scientific Literature: Tackling Unspecified Relation

Tsanta Randriatsitohaina[1(✉)] and Thierry Hamon[1,2]

[1] LIMSI, CNRS, Univ. Paris-Sud, Universit Paris-Saclay, 91405 Orsay, France
{tsanta,hamon}@limsi.fr

[2] Université Paris 13, Sorbonne Paris Cité, 93430 Villetaneuse, France

Abstract. This paper tackles the problem of mining scientific literature to extract Food-Drug Interaction (FDI). This problem is viewed as a relation extraction task which can be solved with classification method. Since FDI need to be described in a very fine way with many relation types, we face the data sparseness and the lack of examples per type of relation. To address this issue, we propose an effective approach for grouping relations sharing similar representation into clusters and reducing the lack of examples. Since unspecified relations represent more than half the data, we propose to contrast supervised and unsupervised methods to identify the specific relation involved in these examples. The performance of our classification-based labeling approach is twice better than on initial dataset and the data imbalance is significantly reduced. Besides, how learning models combine relations can be interpreted to more effectively group relations.

Keywords: Clustering · Classification · Medical text · Food-Drug Interaction

1 Introduction

Food-drug interactions (FDI) correspond to various types of adverse drug effects and lead to harmful consequences on the patient's health and well-being. They are also less known and studied and consequently very sparse in the scientific literature. Similarly to interactions between drugs, FDI is the appearance of an unexpected effect, e.g. grapefruit is known to have an inhibitory effect on the metabolism of several drugs [5]. Other foods may affect the absorption of a drug or its distribution in the organism [3].

In this article, we address the automatic identification of interaction statements between drug and food in abstracts of scientific articles issued from the

This work was supported by the MIAM project and Agence Nationale de la Recherche through the grant ANR-16-CE23-0012 France.

D. Riaño et al. (Eds.): AIME 2019, LNAI 11526, pp. 275–280, 2019.
https://doi.org/10.1007/978-3-030-21642-9_34

Medline database. To extract this information from the abstracts, we face several difficulties: (1) drugs and foods are very variable in the abstracts. (2) the interactions are described in a rather precise way in the texts, which leads to a limited number of examples; (3) the available set of annotations does not include the different types of interaction homogeneously and the learning set is often unbalanced.

Our contributions focus on FDI extraction and improvement of previous classification results by proposing a relation representation which addresses the lack of data, applying clustering method on type of relations, and using cluster labels in a classification step for identification of FDI type.

2 Related Work

Several approaches have been proposed to extract relations from biomedical texts. [6] combine patterns and CRF methods for symptom recognition. [10] learn lexical patterns based on a multiple sequential alignment to identify similar contexts. In [11], sentence's verb are compared to a list of verb known as indicating relation to determine the relation between entities. To extract Drug-drug Interaction (DDI) [8] focus on pharmacokinetic evidence identification in relevant sentences and abstracts. [1] propose a SVM-based approach combining features describing context and composite kernels using dependency trees. This method first detects potential DDI and then classify previously identified relations. [7] built a binary classifier to extract interacting drug pairs. A DDI type classifier is then applied to assign pairs to predefined relation categories. [2] consider the extraction of protein localization relation as a binary classification. In contrast, we use multi-class classification for relation type recognition. [9] propose a CNN-based method for DDI extraction with normalization of drug mentions. Other works use recurrent neural network model with multiple attention layers for DDI classification [12].

Our FDI extraction method is similar to the DDI extraction approach proposed by [1] even if we need to identify much more types of relations (see Sect. 3). Our method joins their two steps approach for DDI detection and classification in which we added a relevant sentences selection step as proposed in [8].

3 Dataset

FDI have been already considered in the POMELO corpus [4]. This corpus consists of 639 abstracts of medical articles (269,824 words, 5,752 sentences) collected from the PubMed portal[1] by the query: `("FOOD DRUG INTERACTIONS"[MH] OR "FOOD DRUG INTERACTIONS*") AND ("adverse effects*")`. The abstracts have been annotated according to 9 types of entities and 21 relation types by a pharmacy resident. From the POMELO corpus, we collect the sentences containing couples of *drug* and *food* or *food-supplement*. The resulting dataset is composed

[1] https://www.ncbi.nlm.nih.gov/pubmed/.

of 831 sentences labeled with 13 types of relations: unspecified relation (476, 57.3%), no effect on drug (109, 13.1%), decrease absorption (49, 5.9%), improve drug effect (6, 0.7%), positive effect on drug (19, 2.3%), without food (9, 1.1%), negative effect on drug (85, 10.2%), speed up absorption (1, 0.1%), increase absorption (38, 4.6%), worsen drug effect (5, 0.6%), slow elimination (15, 1.8%), new side effect (4, 0.5%), slow absorption (15, 1.8%).

4 Grouping Types of Relation

To solve the lack of examples per relation type, we propose to contrast an intuitive grouping method based on the definition of the relation types and an unsupervised clustering based on the instances of each relation type.

Intuitive Grouping. First, we propose a very intuitive grouping of Food-Drug relations, defined as the following: **Unspecified relation**. Instances labeled with '*unspecified relation*' do not give more precision about the relation involved. **No effect**. '*No effect on drug*' instances represent food-drug relations in sentences where it is explicitly expressed that the considered food have no effect on the drug. **Reduction**. Instances labeled with '*decrease absorption*', '*slow absorption*', '*slow elimination*' express diminution of action of drug under the influence of a food. **Augmentation**. Instances labeled with '*increase absorption*', '*speed up absorption*'. **Negative**. *negative effect on drug* express explicitly a negative effect of food on drug, '*worsen drug effect*' express a negative effect of the drug, *side effect* is generally an drug adverse effect drug that coins a negative connotation, the same to '*without food*' that prevents from taking food with the considered drug. **Positive**. By analogy, '*positive effect on drug*', '*improve drug effect*' are grouped to form the *positive* relation. Henceforth, the intuitive grouping method is named ARNP (Augmentation, Reduction, Negative and Positive).

Relation Clustering. We propose to use unsupervised clustering method to group Food-Drug relations. To achieve this purpose, each type of relation should be represented by a set of features. The most natural way to get these features is to group every sentences labeled by the relation in the initial dataset. In order to capture more accurately the expression of the relation in a sentence, we propose to use as features, lemmas of words before the first argument of the relation (*Before*), lemmas of words between the two arguments (*Between*), and lemmas of words after the second argument (*After*). The data is given to a unsupervised clustering to group the relations given the associated features. Relation labeled with the same cluster are assumed to belong to a same group.

5 Unspecified Relation Labeling

We propose to identify which relation type among those annotated can be invoked in such example. We use the entire abstract to represent the example instead of the sentence, then we apply classifiers to label the example with a specific relation type.

Representation for Clustering Method. Since the unsupervised clustering of examples involving the same relation can not be constrained to be in the same cluster, we propose to group together the sentences involving the same relation in a single example by applying the relation representation method described in the Sect. 4.

Representation for Classification Method. We train a classifier on examples labeled with specified relation and apply the resulting model to the unspecified relation examples to predict which relation is involved. The representation is used for learning classifiers such as SVM or Logistic Regression.

6 Experiments, Results and Discussion

We first apply KMeans algorithm in order to obtain an automated grouping scheme. We assume that unspecified relation examples should belong to the 12 manually annotated specific relations, so following the method in Sect. 5 the data is composed of 488 examples that are 12 examples as representations of the 12 specific relations and the 476 unspecified examples to be clustered by considering the whole abstract instead of the sentence as context of the relations.

We note that for any number of clusters, *decrease absorption*, *increase absorption* are always in the same cluster, showing that the two relations are differentiated from the others according to the pharmacokinetics point of view. *no effect on drug* is represented individually in 6 cases when using 7 clusters or more, which confirms the ARNP hypothesis. *slow elimination* is represented individually in 5 cases (3 and 6 to 9 cluster grouping), and grouped with *positive effect on drug* in 4 cases (5, 10, 11, 12 cluster grouping). This clustering result suggests possible interaction between the 2 types of relation and can be a support for medical specialists analysis. 9-clusters grouping is most in agreement with other assignments, that we keep as a new grouping scheme: (1) *decrease absorption*, *increase absorption*, (2) *improve drug effect*, *new side effect*, *worsen drug effect*, which refer to effect of drug, *speed up absorption*, *slow absorption*, *without food*, *positive effect on drug*, (3) *negative effect on drug*, (4) *no effect on drug*, (5) *slow elimination*. However, the specific relations are grouped in 5 clusters, which means that 4 other relations have emerged among the unspecified relation examples. And if we look at 5-clusters assignments, only 4 clusters appear, so one other have emerged. These emerging relations can be analyzed better by a specialist to agree if the new relations are relevant or just mislabeled. We also note that the average V-measure of ARNP grouping is quite low, which means that the automated grouping does not follow the same scheme as ARNP.

Once specific relations clustered, we keep the best grouping scheme obtained with 9 clusters for classification-based labeling of the unspecified examples: we use the cluster label of each relation type as the new label of examples labeled with the considered relation. Then we train the 5 classification models on these examples and apply these models on the 476 unspecified relation examples to obtain a predicted label, i.e. the group to which the example belongs. Finally, the results are contrasted with the assignments obtained with KMeans clustering in the previous step on FDI extraction task using all the 831 examples with their new labels. 9-clusters-classification-based labeling outperforms KMeans-based labeling and ARNP grouping when labeling the unspecified examples, and the resulting F_1score is twice better than on initial dataset. Among the 6 classifiers, Decision Tree classifier leads to the best performance. The difference between micro-score and macro-score decreases from 0.236 with initial dataset to 0.005 with 9-clusters-based-classification, suggesting an important reduction of the data imbalance. We analyze the agreement score of the methods for labeling the 476 unspecified examples: the agreement between clustering-based and classification-based labeling methods is quite low. This phenomenon probably occurs because 4 new labels appear in cluster-based labeling. It represents a difference compared to the classification methods that only use the 5 pre-defined labels in the grouping scheme.

7 Conclusion and Future Work

Our paper contributes to the task of extraction of FDI from scientific literature, that we address as a relation extraction task. When applying supervised machine learning to this purpose, we face an issue of lack of examples because of the high number of types of relation. To address this issue, we propose to represent each relation by words before, between and after the arguments of the relation in order to group these relations into clusters. The resulting cluster labels are used to label unspecified relation examples. The KMeans-based automated clustering shows relation grouping scheme which differs from an intuitive grouping based on the definition of the relation types. It also allows emergence of new group of relations. This computed scheme is used as new labels for specific examples and combined with whole abstract to identify the specific relations involved in unspecified relation examples by labeling these examples based on clustering and classification methods. Our results show that labeling unspecified examples improve significantly the performance on FDI extraction. Indeed, we obtain a F1-score twice better with classification-based labeled unspecified examples while grouping data in 5 clusters than on initial dataset. Besides, the decrease in the difference between micro- and macro-score suggests an important reduction of the data imbalance. For future work, we will use features such as semantic category of terms, go through emerging relations, and consider a more domain-based labeling following the ADME classes [3] of Drug-Drug Interaction by transfer learning.

References

1. Abacha, A.B., Chowdhury, M.F.M., Karanasiou, A., Mrabet, Y., Lavelli, A., Zweigenbaum, P.: Text mining for pharmacovigilance: using machine learning for drug name recognition and drug-drug interaction extraction and classification. J. Biomed. Inform. **58**, 122–132 (2015)
2. Cejuela, J.M., et al.: LocText: relation extraction of protein localizations to assist database curation. BMC Bioinform. **19**(1), 15 (2018)
3. Doogue, M., Polasek, T.: The ABCD of clinical pharmacokinetics. Ther. Adv. Drug Saf. **4**, 5–7 (2013)
4. Hamon, T., Tabanou, V., Mougin, F., Grabar, N., Thiessard, F.: POMELO: medline corpus with manually annotated food-drug interactions. In: Proceedings of Biomedical NLP Workshop Associated with RANLP 2017, Varna, Bulgaria, pp. 73–80, September 2017
5. Hanley, M., Cancalon, P., Widmer, W., Greenblatt, D.: The effect of grapefruit juice on drug disposition. Expert Opin. Drug Metabol. Toxicol. **7**(3), 267–286 (2011)
6. Holat, P., Tomeh, N., Charnois, T., Battistelli, D., Jaulent, M.-C., Métivier, J.-P.: Weakly-supervised symptom recognition for rare diseases in biomedical text. In: Boström, H., Knobbe, A., Soares, C., Papapetrou, P. (eds.) IDA 2016. LNCS, vol. 9897, pp. 192–203. Springer, Cham (2016). https://doi.org/10.1007/978-3-319-46349-0_17
7. Kim, S., Liu, H., Yeganova, L., Wilbur, W.J.: Extracting drug-drug interactions from literature using a rich feature-based linear kernel approach. J. Biomed. Inform. **55**, 23–30 (2015)
8. Kolchinsky, A., Lourenço, A., Wu, H.Y., Li, L., Rocha, L.M.: Extraction of pharmacokinetic evidence of drug-drug interactions from the literature. PloS One **10**(5), e0122199 (2015)
9. Liu, S., Tang, B., Chen, Q., Wang, X.: Drug-drug interaction extraction via convolutional neural networks. Comput. Math. Methods Med. **2016**, 8 (2016)
10. Meng, F., Morioka, C.: Automating the generation of lexical patterns for processing free text in clinical documents. J. Am. Med. Inform. Assoc. **22**(5), 980–986 (2015)
11. Song, M., Chul Kim, W., Lee, D., Eun Heo, G., Young Kang, K.: PKDE4J: entity and relation extraction for public knowledge discovery. J. Biomed. Inform. **57**, 320–332 (2015)
12. Yi, Z., et al.: Drug-drug interaction extraction via recurrent neural network with multiple attention layers. In: Cong, G., Peng, W.-C., Zhang, W.E., Li, C., Sun, A. (eds.) ADMA 2017. LNCS (LNAI), vol. 10604, pp. 554–566. Springer, Cham (2017). https://doi.org/10.1007/978-3-319-69179-4_39

Bayesian Clustering for HIV1 Protease Inhibitor Contact Maps

Sandhya Prabhakaran[1(✉)] and Julia E. Vogt[2]

[1] Computational and Systems Biology Program, Memorial Sloan Kettering Cancer Center, New York, USA
sandhyaprabhakaran@gmail.com

[2] Department of Computer Science, ETH Zurich, Zurich, Switzerland

Abstract. We present a probabilistic model for clustering which enables the modeling of overlapping clusters where objects are only available as *pairwise distances*. Examples of such distance data are genomic string alignments, or protein contact maps. In our clustering model, an object has the freedom to belong to one or more clusters at the same time. By using an IBP process prior, there is no need to explicitly fix the number of clusters, as well as the number of overlapping clusters, in advance. In this paper, we demonstrate the utility of our model using distance data obtained from HIV1 protease inhibitor contact maps.

Keywords: Bayesian nonparametrics · Clustering · Medical informatics

1 Introduction

Traditional clustering methods partition objects into mutually exclusive clusters. In many applications, however, it is more realistic that objects may belong to multiple, *overlapping clusters* [5,9]. If, for instance, a gene has many different functions, it might belong to more than one cluster and therefore be potentially identified with multiple functional pathways rather than just one. However, if the data is available in the form of pairwise distances, there exists no probabilistic clustering model which allows overlapping clustering of objects. The general strategy in this situation is to first embed the distance data into a (lower-dimensional) Euclidean space [8] and then cluster the embedded data. This introduces unnecessary noise and bias, as was shown in [11]. Therefore, it would be advantageous to have a model which can cater to distance data directly. In this work, we present a method which enables the modeling of overlapping clusters. The model is able to deal with objects that are available as *pairwise distance data*, and one object is not restricted to just lie within one cluster but is allowed to belong to multiple clusters. In this short paper, we concentrate on the applications of our model to clustering HIV1 protease inhibitor contact maps.

D. Riaño et al. (Eds.): AIME 2019, LNAI 11526, pp. 281–285, 2019.
https://doi.org/10.1007/978-3-030-21642-9_35

2 A Brief Introduction into the Model

We developed a model that allows inferring multiple clusters per object where objects are represented by their pairwise distances. In order to be able to model overlapping clusters, we consider an Indian Buffet Process (IBP) in our model [4,5]: the IBP is a generative process with an analogy that n customers are lined up in a sequence in an Indian buffet restaurant that offers infinitely many dishes. The first customer chooses some of the $Poisson(\xi)$ dishes before sitting down. Every incoming customer then helps themselves to some of the buffet dishes in proportion to their popularity, such that customer i serves herself dish k with probability m_k, where m_k is the proportion of previous customers having chosen that dish. After passing by every dish previously sampled, customer i tries $Poisson(\xi/i)$ new dishes. The binary cluster assignment matrix Z has n rows denoting customers and C columns for C buffet dishes. An entry z_{ik} indicates whether customer i chose dish k, and objects (customers) can be assigned to more than one cluster (dish). The model is not restricted *a priori* to having a fixed number of clusters, and allows the data to determine how many clusters are required. Further, since we consider an exchangeable distribution, neither the number of identical columns nor the column sums are affected by the ordering on objects. The assumption is that the (suitably pre-processed) input matrix D contains *squared Euclidean distances* with components $D_{ij} = K_{ii} + K_{jj} - 2K_{ij}$ and K represents the pairwise similarities between objects. The likelihood $P(D|Z)$ is chosen to be the translation and rotation invariant generalized Wishart distribution W_d. This implies that in the inference, we do not need to consider any geometry in vector spaces, as the likelihood solely depends on the pairwise distances D. Inference is performed using Gibbs Sampling.

3 Results

3.1 Performance of the Model on Simulated Data

Distance data D is simulated as per the generative model we introduced above. We set the following parameters: $n = 30$ objects, $d = 400$ dimensions, $C_{max} = 100$ as the maximum number of clusters allowed. $Z_{(n \times C_{max})}$ is generated via the beta-binomial prior. The similarity matrix $K_{(n \times n)} \sim \mathrm{W}_d(\Sigma_Z)$ where $\Sigma_Z = \alpha I + \beta B_Z$ with $B_Z = ZZ^T$. K is centered to K_c using *kernel* centering. The distance $D_{(n \times n)}$ is computed as $D_{(ij)} = K_{c(ii)} + K_{c(jj)} - 2K_{c(ij)}$. We use D as input to our model to infer both Z and B_Z. The columns in the inferred Z matrix can be permuted and therefore we cannot directly compare Z_{est} with Z_{true}. Instead, we compare B_{true} with B_{est}. These matrices are invariant to column permutations, and contain the number of shared clusters between each pair of objects in the data set. Comparing B_{est} and B_{true} is a column invariant way to determine how well the true underlying cluster assignment structure is discovered. Since we obtain many Gibbs sampled B_{est} matrices, we compute an average matrix $B_{average}$ for the comparison with B_{true}. We deploy the following error

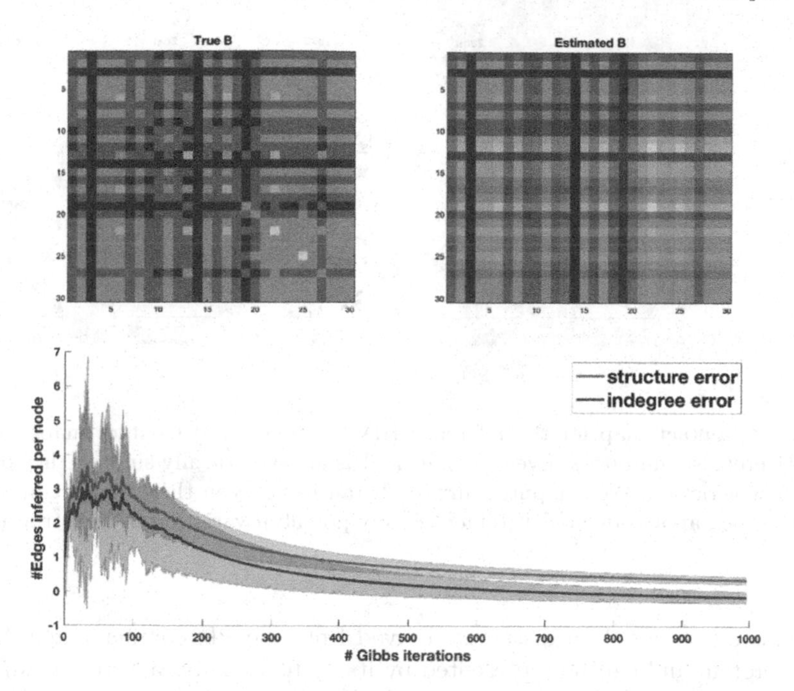

Fig. 1. Top: B_{true} (left) and inferred B_{est} (right). **Bottom**: Error statistics from simulated D: the in degree error (blue) and structure error (red) between B_{true} and $B_{average}$. We display the error for the first 1000 independent iterations. (Color figure online)

statistics to quantify the errors obtained in the diagonal and off-diagonal elements between B_{true} and $B_{average}$: *in degree error* $= \sum \left|\text{diag}(B_{true} - B_{average})\right|$ and *structure error* $= \sum \left(\text{lowtri}(B_{true}) - \text{lowtri}(B_{average})\right)$. In Fig. 1, we see that the inferred B_{est} (top right panel) has captured the underlying partition structure of that of B_{true} (top left panel).

3.2 Clustering Protein Contact Maps from HIV Protease Inhibitors

Amongst the 26 US Food and Drug Administration (FDA) approved anti-HIV drugs, 10 are HIV protease inhibitors (PIs). These are saquinavir (3TK9), indinavir (2B72), ritonavir (1HXW), nelfinavir (2PYM), amprenavir (3NU3), fosamprenavir (3S85), lopinavir (2O4S), atazanavir (2AQU), tipranavir (2O4L) and darunavir (3TTP) with the unique identifiers as provided in the RCSB Protein Data Bank (PDB) [2], in brackets. The HIV PIs exhibit similar behaviour which can be attributed to their similarity in chemical structures as well as binding sites but these PIs are not readily available and have high toxicity when administered as anti-HIV drug concoctions [7]. It is therefore of utmost importance to identify alternative PIs for therapy which calls for identifying any potential structural dissimilarities that stand out amongst the already known HIV PIs. We use our

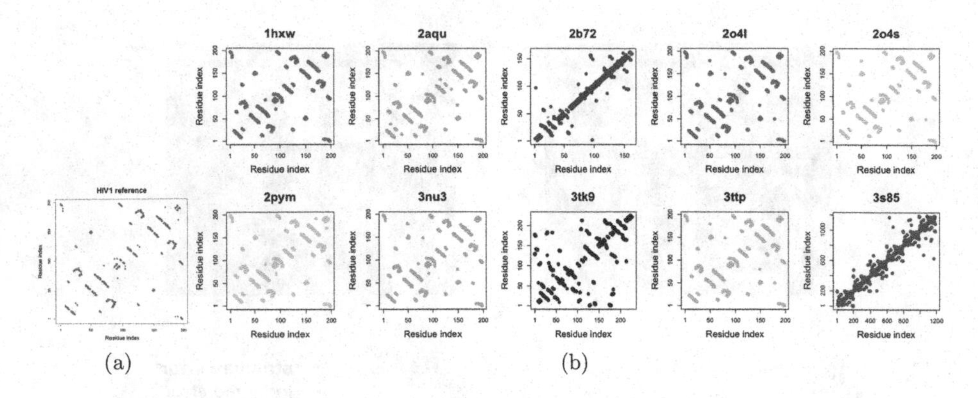

Fig. 2. (a) Contact map for the reference HIV1 protease. (b) Contact maps for the 10 HIV1 protease inhibitors. Even though the PIs are structurally similar, the contact maps show variation. We compute pairwise distances between these contact maps via NCD and then apply our model to identify any potential variation in the structure of PIs.

novel clustering model on distances derived from the PIs contact maps. Every PI is a protein and can be represented by its vectorized contact map computed from their PDB files. A contact map contains the distances between all possible amino acid residue pairs for a given protein. For the HIV1 reference protein with T amino acid residues, the contact map (see Fig. 2(a)) would be a $T \times T$ binary matrix CM where $CM_{ij} = 1$ if residues i and j are similar or 0 otherwise. The contact maps for the 10 PIs are shown in Fig. 2(b). Next, we flatten the contact map representation into row-wise vectors to construct a string. To obtain the pairwise distances between strings in these contact maps, we compute the Normalized Compression Distance (NCD) [6] which is an approximation to the Normalized Information Distance (NID) [10]. Given strings x and y, NID is proportional to the length of the shortest program that computes $x|y$ and $y|x$. $NID(x,y) = \frac{Q(xy)-min[Q(x),Q(y)]}{max[Q(x),Q(y)]}$ where Q(x) is the Kolmogorov complexity of the string x. The real-world approximated version of NID is given by NCD where $Q(xy)$ now represents the size of the file obtained by compressing the concatenation of x and y.

We compute NCD for all the 10 PIs as well as the reference HIV1 protease (1HIV) and apply our model to the resulting $D_{11\times 11}$. We observe that the inferred B_Z primarily has 3 blocks: *block 1* groups 1HXW (ritonavir) and 2AQU (atazanavir) together and the efficacy of this drug combination has been highlighted in [1,3], *block 2* has a singleton 2AQU (atazanavir) showing its difference to the rest of the PIs. In fact, atazanavir is distinguished from other PIs in that it can be given once-daily (rather than multiple daily-doses) and has lesser effects on the patient's cholesterol and fat amounts present in the blood [3] and *block 3* finds the rest of the PIs clustering together with the reference HIV1. Even though this test considered only 11 PIs, our model was able to identify established differences amongst themselves. We believe that by extending this test to involve all the 26 FDA approved anti-HIV drugs, the subsequent identification of structural differences would aid in suggesting potential drug candidates.

4 Conclusion

We present a probabilistic model, for inferring overlapping clusters where objects are only available as *pairwise distance data*. Here, an object may belong to one or more clusters concurrently. Examples of such pairwise distance data are genomic string alignments, protein contact maps or pairwise patient similarities. So far, to the best of our knowledge, there is no probabilistic model that can handle identifying overlapping clusters in distance data. The main contributions of our work are the following: **(i)** we propose a probabilistic model that is able to partition data into overlapping groups and **(ii)** we demonstrate the usefulness of our model by clustering 11 HIV1 Protease inhibitor protein contact maps where the model is able to tease apart potential structural variations that were inherent amongst the 11 PIs and that these could be used as cues to identify drug candidates.

References

1. Achenbach, C.J., Darin, K.M., Murphy, R.L., Christine, K.: Atazanavir/ritonavir-based combination antiretroviral therapy for treatment of HIV-1 infection in adults. Future Virol. **6**(2), 157–177 (2011)
2. Berman, H.M., et al.: The protein data bank. Nucleic Acids Res. **28**, 235–242 (2000)
3. Bernardino, J.I., Arribas, J.R.: Antiviral therapy. Infect. Dis. **4**, 918–926 (2011)
4. Griffiths, T.L., Ghahramani, Z.: Infinite latent feature models and the Indian buffet process, May 2005
5. Heller, K.A., Ghaharamani, Z.: A nonparametric Bayesian approach to modeling overlapping clusters. In: AISTATS (2007)
6. Li, M., Vitányi, P.: An Introduction to Kolmogorov Complexity and Its Applications. Texts in Computer Science. Springer, New York (2008). https://doi.org/10.1007/978-0-387-49820-1
7. Zhengtong, L., Chu, Y., Wang, Y.: HIV protease inhibitors: a review of molecular selectivity and toxicity. HIV AIDS (Auckl.) **7**, 95 (2015)
8. Schölkopf, B., Smola, A.J., et al.: Learning with kernels: support vector machines, regularization, optimization, and beyond (2002)
9. Streich, A.P., Frank, M., Buhmann, J.M.: Multi-assignment clustering for Boolean data. In: ICML (2009)
10. Vitányi, P.M.B., Balbach, F.J., Cilibrasi, R.L., Li, M.: Normalized information distance. In: Emmert-Streib, F., Dehmer, M. (eds.) Information Theory and Statistical Learning, pp. 45–82. Springer, Boston (2009). https://doi.org/10.1007/978-0-387-84816-7_3
11. Vogt, J.E., Prabhakaran, S., Fuchs, T.J., Roth, V.: The translation-invariant Wishart-Dirichlet process for clustering distance data. In: ICML, pp. 1111–1118 (2010)

Time-to-Birth Prediction Models and the Influence of Expert Opinions

Gilles Vandewiele[1](✉), Isabelle Dehaene[2], Olivier Janssens[1], Femke Ongenae[1], Femke De Backere[1], Filip De Turck[1], Kristien Roelens[2], Sofie Van Hoecke[1], and Thomas Demeester[1]

[1] IDLab, Ghent University – imec, Technologiepark-Zwijnaarde 126, Ghent, Belgium
{gilles.vandewiele,olivier.janssens,femke.ongenae,femke.debackere, filip.deturck,sofie.vanhoecke,thomas.demeester}@ugent.be

[2] Department of Gynecology and Obstetrics, Ghent University Hospital, Corneel Heymanslaan 10, Ghent, Belgium
{isabelle.dehaene,kristien.roelens}@uzgent.be

Abstract. Preterm birth is the leading cause of death among children under five years old. The pathophysiology and etiology of preterm labor are not yet fully understood. This causes a large number of unnecessary hospitalizations due to high–sensitivity clinical policies, which has a significant psychological and economic impact. In this study, we present a predictive model, based on a new dataset containing information of 1,243 admissions, that predicts whether a patient will give birth within a given time after admission. Such a model could provide support in the clinical decision-making process. Predictions for birth within 48 h or 7 days after admission yield an Area Under the Curve of the Receiver Operating Characteristic (AUC) of 0.72 for both tasks. Furthermore, we show that by incorporating predictions made by experts at admission, which introduces a potential bias, the prediction effectiveness increases to an AUC score of 0.83 and 0.81 for these respective tasks.

Keywords: Preterm birth · Clinical decision support · eHealth

1 Introduction

Preterm birth, defined as birth before 37 weeks of gestational age, is the leading cause of death among children younger than five years, according to the World Health Organization (WHO) [8]. In Flanders, the average prevalence rate amounts to 7%. Furthermore, for tertiary care centers in Flanders this can be significantly higher, as for example 18% of the deliveries in Ghent University Hospital are preterm. Today, a patient at risk of preterm birth, is often hospitalized in order to take measures that ameliorate the short- and long-term outcome for the neonate. These measures include the administering of antenatal corticosteroids (ACS) for fetal lung maturation, often under tocolysis for labor arrest. Unfortunately, these measures can have short- and long-term maternal and offspring side effects and should therefore only be taken when imminent birth is

D. Riaño et al. (Eds.): AIME 2019, LNAI 11526, pp. 286–291, 2019.
https://doi.org/10.1007/978-3-030-21642-9_36

Table 1. Predictive modeling studies for tasks related to birth prediction.

Study	Dataset	Target	Results
[17]	170 patients w/ preterm labor; 24 w ≤ gest. ≤ 34 w; singleton preg.	birth ≤ 37 w	AUC: 0.81
[5, 18]	58,807 singleton pregnancies; 20 w ≤ gest. ≤ 25 w	4 categories	AUC: 0.65 (mild) – 0.92 (extreme)
[16]	3 million singleton, 105,000 twin pregnancies and 4,000 triplet pregnancies	birth ≤ 32 w	AUC: 0.65 (> 1 fetus) – AUC 0.73 (singleton)
[2]	906 patients (2 datasets) w/ preterm labor; 22 w ≤ gest. ≤ 32 w	birth ≤ 32 w & birth in 48 h	AUC: 0.73 (within 48 h), AUC 0.72 (≤ 32 w)
[19]	142 singleton pregn. w/ preterm labor & intact membranes; 22 w ≤ gest. < 34 w	birth within 7 d	AUC: 0.88
[15]	1.5 million singleton pregnancies; 22 w ≤ gest	birth ≤ 37 w	AUC: 0.63
[4]	33,370 singleton pregnancies w/ preterm labor ≤ 34 w; data collected 11 w ≤ gest. ≤ 13 w	birth ≤ 34 w	AUC: 0.67
[14]	31,834 singleton pregnancy; data collected 11 ≤ gest. ≤ 14 w; 24 w ≤ gest. ≤ 37 w	birth ≤ 37 w	18.4% sensitivity, 97.1% specificity
[1]	2,699 patients; data from both first and second trimester; birth after 20 w	birth ≤ 37 w	AUC: 0.70
[12]	191 singleton pregnancies; only biomarker data; 24 w ≤ gest	birth ≤ 36 w	AUC: 0.66–0.89
[22]	3,073 singleton pregnancies; multiple admissions	birth ≤ 37 w	45.4%–59.4% sensitivity, 71.9%–84% specificity at third admission
[9]	617 patients with preterm labor; 22 w ≤ gest. ≤ 32 w	birth in 48 h & birth ≤ 32 w	AUC: 0.8 (birth within 48 h), 0.85 (birth ≤ 32 w)
[20]	600 singleton pregnancies from 10 centers; 24 w ≤ gest. ≤ 34 w	birth within 7 d	AUC: 0.95
[7]	166 women with preterm labor; 24 w ≤ gest. ≤ 31 w	birth within 7 d and 14 d	AUC: 0.63
[6]	3,012 symptomatic women; 24 w ≤ gest. ≤ 28 w or readmission before 35 w	birth within 7 d	AUC: 0.724
[23]	prospective; 355 women with preterm labor; 24 w ≤ gest. ≤ 34 w	birth within 7 d	100% sensitivity, 92.3% specificity
[10]	2,540 women; 24 w ≤ gest.; data collected gest. ≤ 16 w	birth ≤ 37 w & birth ≤ 34 w	AUC: 0.54–0.67 (37 w); AUC: 0.56–0.70 (34 w)
This study	**1,243 high-risk admissions; 24 w ≤ gest. ≤ 37 w**	**birth within 48 h and 7 d**	**AUC: 0.83 (48 h); AUC: 0.81 (7 d)**

expected [13]. Moreover, the societal and personal psychological and economical burden related to the hospitalization of these patients, should not be underestimated. For example, the costs associated with preterm birth in the USA in 2005 were estimated around $51,600 per infant [3]. Currently, the pathophysiology and etiology of preterm labor are not yet fully understood, making it hard for experts to accurately determine whether a patient will give birth at term or not. As is often the case in medical domains, the sensitivity of a policy is typically considered more important than its specificity, resulting in a high number of false positives, or unnecessary hospitalizations.

Predictive machine learning models have been applied to numerous medical use cases [21]. The prior research on predictive models for preterm birth risk is shown in Table 1. The most important difference w.r.t. the presented study is the incorporation of expert opinions within the model. We assess the added value of a model that predicts the time-to-birth, based on a simple, interpretable logistic regression model. The input to the model consists of structured clinical variables which are available shortly after the patient's admission to the hospital, e.g. the gestational age at intake, the patient's BMI and how long membranes have been ruptured, and indications given by domain experts at admission.

2 Methodology

2.1 Data Collection and Filtering

The dataset used within this study consists of data collected from patients admitted to the Department of Gynecology and Obstetrics at Ghent University Hospital, between 2012 and 2017. In total, 3,611 women were admitted in that period, corresponding to 4,332 pregnancies and 5,030 admissions. From these, 1,243 pregnancy-related admissions, corresponding to 1,145 high-risk pregnancies of 1,056 women, occurring between 24 and 37 weeks of gestation, were used in the proposed pipeline. The reason for excluding other admissions is because the clinical use of our model is limited for these type of admissions. Patients at a gestational age less than 24 weeks are not included, since neonatal intensive care is not started before this term in Ghent University Hospital. Patients arriving at the hospital after 37 weeks of gestation are no longer at risk for preterm birth and thus do not require potential preventive measures.

2.2 Predictive Variables

From the data we extract: number of fetuses, age (mother), gravidity, parity, length (mother), weight (mother), BMI, gestational age at admission, duration ruptured membranes, method of conception, smoking history, alcohol usage, drug usage, history of cesarean section, race (mother), and admission indications. This list of variables has been constructed in consultation with domain experts. The admission indications are keywords that can either be objective observations including 'blood loss' and 'stomach ache', or more subjective keywords of experts such as 'imminent partus prematurus'. The latter type of keywords can include indications of what the expert expects to happen, i.e. an expert opinion. Hence, such keywords introduce a potential bias which could cause the model to simply repeat the predictions of an expert. This however does not need to be the case, especially if it turns out that the prediction implied by such keywords does not always hold true. In fact, these subjective expert predictions are recorded directly after the patient's intake, and we therefore propose to investigate the predictions of a model in which these expert opinions are actually used as highly informative features.

2.3 Data Processing and Modeling

Before feeding the data to a machine learning model, all variables were first transformed to a numerical form. To achieve this, categorical variables were one-hot-encoded and a bag-of-terms was constructed for each patient based on her listed keywords. This bag-of-terms is a k-dimensional binary vector, with k being the number of available keywords in the training set, and each value indicating the presence of a certain keyword. In our study, k is equal to 30. Afterwards, the processed data was fed to logistic regression classifiers to solve two tasks, corresponding to threshold values chosen in consultation with experts, as they are the bounds between which the effect of corticosteroids is thought to be optimal [11]. On the one hand, we will predict whether birth will occur within 48 h after admission (***Task 1***), while on the other hand we make the prediction for birth within 7 days (***Task 2***).

3 Results

To assess the predictive performance of the proposed model, we measured different metrics using five-fold cross-validation, based on the patient identifiers rather than individual admissions. First, we report the classifier's accuracy. As accuracy does not provide a good indication of the model's predictive performance in the scenario of imbalanced data, we also report the specificity, sensitivity and Diagnostic Odds Ratio (DOR) obtained from the confusion matrices of our classifiers, and the Area Under the Curve (AUC) of the Receiver Operating Characteristic curve (ROC) score. Table 2 summarizes the predictive performances for the models with and without inclusion of the expert predictions. The table lists the mean $\pm$ std over the five folds. From these results, we can conclude that ***Task 1*** seems to be a slightly easier prediction task. We hypothesize that this is due to the fact that patients that would give birth within 48 h often have more distinctive symptoms. Further, we see that incorporating the biased expert predictions in our model results in a considerable increase in predictive performance. It should be noted that the sensitivity values are rather low, since we did not optimize specifically for this and kept the default decision boundary, i.e., at a predicted probability level of 0.5.

Table 2. The predictive performances of a model that predicts whether a patient will give birth (i) within 48 h, or (ii) within 7 days. Model without and model with inclusion of admission indication keywords as input features.

Pred.	Task	Accuracy	Specificity	Sensitivity	DOR	AUC
Without	≤48 h	0.74 ± 0.02	0.94 ± 0.02	0.32 ± 0.05	9.3 ± 5.3	0.72 ± 0.02
	≤7 d	0.67 ± 0.02	0.83 ± 0.05	0.48 ± 0.08	5.1 ± 1.5	0.72 ± 0.04
With	≤48 h	0.80 ± 0.02	0.90 ± 0.01	0.60 ± 0.05	14.5 ± 4.3	0.83 ± 0.03
	≤7 d	0.76 ± 0.02	0.83 ± 0.03	0.67 ± 0.02	10.7 ± 2.7	0.81 ± 0.02

4 Conclusion

A simple and interpretable logistic regression model was presented to assess the time-to-birth of a patient upon admission to the hospital. Preliminary results show the positive impact of incorporating expert opinions within the model. Future work includes applying survival analysis to directly predict the time-to-birth, as opposed to a dichotomized target.

Acknowledgements. This research is funded by imec, the PRETURN clinical trial (B670201836255, EC/2018/0609) and a Ph.D. SB fellow scholarship of FWO (1S31417N).

References

1. Alleman, B.W., et al.: A proposed method to predict preterm birth using clinical data, standard maternal serum screening, and cholesterol. AJOG **208**(6), 472-e1 (2013)
2. Allouche, M., et al.: Development & validation of nomograms for predicting ptb. AJOG **204**(3), 242 (2011)
3. Behrman, R.E., Butler, A.S., et al.: Societal costs of preterm birth (2007)
4. Beta, J., et al.: Prediction of spontaneous preterm delivery from maternal factors, obstetric history and placental perfusion and function at 11–13 weeks. Prenat. Diagn. **31**(1), 75–83 (2011)
5. Celik, E., et al.: Cervical length and obstetric history predict sptb: development & validation of a model to provide individualized risk assessment. UOG **31**(5), 549–554 (2008)
6. De Silva, D.A., et al.: Timing of delivery in a high-risk obstetric population: a clinical prediction model. BMC Pregnancy Childbirth **17**(1), 202 (2017)
7. García-Blanco, A., et al.: Can stress biomarkers predict preterm birth in women with threatened preterm labor? Psychoneuroendocrinology **83**, 19–24 (2017)
8. Liu, L., et al.: Global, regional, and national causes of under-5 mortality in 2000–15: an updated systematic analysis with implications for the sustainable development goals. Lancet **388**(10063), 3027–3035 (2016)
9. Mailath-Pokorny, M., et al.: Individualized assessment of preterm birth risk using two modified prediction models. Eur. J. Obstet. Gynecol. Reprod. Biol. **186**, 42–48 (2015)

10. Meertens, L.J., et al.: Prediction models for the risk of spontaneous preterm birth based on maternal characteristics: a systematic review and independent external validation. Acta Obstet. Gynecol. Scand. **97**(8), 907–920 (2018)
11. Melamed, N., et al.: Association between antenatal corticosteroid administration-to-birth interval and outcomes of preterm neonates. Obstet. Gynecol. **125**(6), 1377–1384 (2015)
12. Menon, R., et al.: Multivariate adaptive regression splines analysis to predict biomarkers of spontaneous preterm birth. Acta Obstet. Gynecol. Scand. **93**(4), 382–391 (2014)
13. Roelens, K., et al.: Prevention of preterm birth in women at risk: Selected topics. Technical report, Belgian Health Care Knowledge Centre (KCE) (2014)
14. Sananes, N., et al.: Prediction of spontaneous preterm delivery in the first trimester of pregnancy. Eur. J. Obstet. Gynecol. Reprod. Biol. **171**(1), 18–22 (2013)
15. Schaaf, J.M., et al.: Development of a prognostic model for predicting spontaneous singleton preterm birth. Eur. J. Obstet. Gynecol. Reprod. Biol. **164**(2), 150–155 (2012)
16. Tan, H., et al.: Early prediction of preterm birth for singleton, twin, and triplet pregnancies. Eur. J. Obstet. Gynecol. Reprod. Biol. **131**(2), 132–137 (2007)
17. Tekesin, I., et al.: Evaluation and validation of a new risk score (cleopatra score) to predict the probability of premature delivery for patients with threatened preterm labor. UOG **26**(7), 699–706 (2005)
18. To, M., et al.: Prediction of patient-specific risk of early preterm delivery using maternal history and sonographic measurement of cervical length. UOG **27**(4), 362–367 (2006)
19. Tsiartas, P., et al.: Prediction of spontaneous preterm delivery in women with threatened preterm labour: a prospective cohort study of multiple proteins in maternal serum. BJOG **119**(7), 866–873 (2012)
20. Van Baaren, G.J., et al.: Risk factors for preterm delivery: do they add to fetal fibronectin testing and cervical length measurement in the prediction of preterm delivery in symptomatic women? Eur. J. Obstet. Gynecol. Reprod. Biol. **192**, 79–85 (2015)
21. Vandewiele, G., et al.: A decision support system to follow up and diagnose primary headache patients using semantically enriched data. BMC Med. Inform. Decis. Mak. **18**(1), 98 (2018)
22. Vovsha, I., et al.: Predicting preterm birth is not elusive: machine learning paves the way to individual wellness. In: 2014 AAAI Symposia (2014)
23. Watson, H., et al.: Quipp app: a safe alternative to a treat-all strategy for threatened preterm labor. UOG **50**(3), 342–346 (2017)

Modelling the Impact of AI for Clinical Decision Support

Mariana R. Neves(✉) and D. William R. Marsh

School of Electronic Engineering and Computer Science, Queen Mary University of London, London, UK
{m.r.neves,d.w.r.marsh}@qmul.ac.uk

Abstract. Many AI (or ML) systems have been proposed for clinical decision support. Clinical usefulness is assessed in an 'Impact Study', a form of trial of a completed system. In development, in contrast, the focus is on AI accuracy measures, such as the AUC. To improve impact and to justify the cost of a study, the impact of a proposed AI system should be modelled during its development. We show that an Influence Diagram can be used for this and provide a small set of generic models for diagnostic AI systems. We show that how the AI interacts with clinical decision makers is at least as important as its predictive accuracy.

Keywords: Impact analysis · Clinical decision support · Influence diagram · AI

1 Introduction

1.1 Is AI for Decision Support Clinically Beneficial?

AI (or ML) can be used in a clinical decision-support system (CDSS) to improve medical care. It is important to distinguish the 'prediction' (which uses AI) from the CDSS, into which the prediction goes and it is well understood that prediction accuracy does not ensure clinically utility. Wyatt and Altman [1] point out the need for evidence of clinical effectiveness, making the key distinction between accuracy (or validation – see e.g. [2]) and effectiveness or, more commonly, impact. A CDSS has impact when its implementation benefits care, by improving decision-making or reducing costs. In frameworks for the implementation of prediction systems, for example [3], impact is evaluated in an experimental study. We argue that impact should be considered before the experimental phase of an impact study and an estimate of impact made, covering:

- The way the use the prediction interacts with clinical decision making, and
- The hoped-for benefits, whether in costs or workload or better decision-making.

The frameworks consider these issues: for example, in [3] the 'preparation' phase including 'determining how the *[prediction system]* will be integrated into the clinical workflow'. Similarly, [2] suggests using simulated decisions, pointing out that: "clinicians will not always follow the rule's recommendations". Building on these ideas, we propose a method to estimate impact during the development of a CDSS and

D. Riaño et al. (Eds.): AIME 2019, LNAI 11526, pp. 292–297, 2019.
https://doi.org/10.1007/978-3-030-21642-9_37

draw attention to the relationship between prediction accuracy, proposed used and impact.

2 Approach to Modelling Different Uses of AI System

2.1 How an AI System Can be Used Clinically

A prediction (the 'AI' element) can be integrated into the clinical workflow in many different ways to form a CDSS. Table 1 shows three simple uses that we will analyze, representing a much wider range of real possibilities.

Table 1. Possible uses of AI in a diagnostic CDSS.

Name	Description of use
Replace	The AI diagnoses all patients, replacing the clinician
Filter	The AI diagnoses some patients; the clinician sees fewer
Assist	The AI and the clinician work together

The first two uses 'Replace' and 'Filter' are broadly directive while the other is assistive. Many papers on predictors for clinical still say little about intended use; see for example [4, 5] in a recent collection of papers about clinical AI systems.

2.2 Utility Models

The impact of an AI system may include both benefit and harm. We can model this using a utility model (see Sect. 3) given estimates of these benefits and harms which become parameters in the utility model. We focus on the forms of information needed and how it varies with use, including:

- Healthcare cost: to illustrate, we assume that the cost of a consultation (100 units) exceeds the cost of operating the AI (0 units).
- Patient outcome: this is typically measured in quality-adjusted life-year (QALY) and depends the disease state and the diagnosis. Table 2 shows some illustrative values.

Table 2. Example patient outcomes.

Disease	False		True	
Diagnosis	False	True	False	True
Patient Outcome	0	−50	−500	5000

These outcomes measures depend on the consequences of both false positive and false negative error. We have assumed less harm from a false negative compared to the benefit of a correct diagnosis; other assumptions can be modelled in the same way. In a utility models, a (small) deterioration in outcome may be compensated for by a cost reduction but this might not be permitted by a regulator.

The benefits also depend on the way that the AI is used. Table 3 summarizes the potential areas of benefit for difference AI system uses.

Table 3. Summaries of expected benefit by use.

Name	Potential expected benefits
Replace	Primarily cost saving, with equivalent patient outcome
Filter	Cost reduction, by reducing the number of patients to be seen. Outcome may also change, however
Assist	Primarily improved patient outcome; since all patients are still seen and the healthcare costs do not reduce (given our assumptions)

2.3 Performance Assumptions

In this section, we introduce some assumptions about the performance of both the AI system and clinician, as it will become clear that we cannot estimate the impact of an AI system without knowledge of the performance of an existing clinical care system.

Disease Prevalence: We assume disease is present in 30% of cases.

Receiver Operating Curve (ROC) and Confusion Matrix. The performance of the AI system can be represented by its ROC, showing the tradeoff between sensitivity (or TPR) and specificity (or 1 - FPR). Figure 1 shows a simulated ROC with 5 possible operating points P1-5 for the AI system, all lying on the curve. Two points are shown for the clinical decision maker, one (B) above the AI's ROC and the other (A) below.

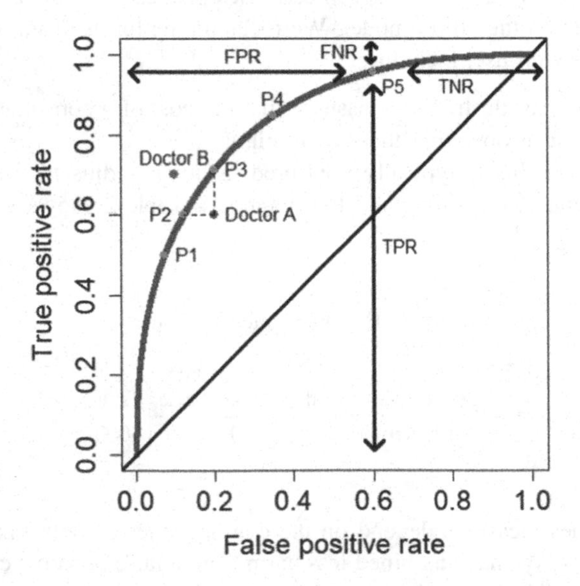

Fig. 1. Simulated ROC curve for the AI in which Area Under the Curve (AUC) is 84%.

Given an operating point (and the prevalence), decision accuracy can be summarized in a confusion matrix; we assume this is known for the clinician as shown below.

		Actual	
		False	**True**
Predicted	False	TN	FN
	True	FP	TP

		Actual	
		False	**True**
Predicted	False	63%	9%
	True	7%	21%

3 Utility of an AI System in Different Uses

3.1 Modelling an AI System Replacing a Clinician

Figure 2 shows a utility model (as an influence diagram, implemented using the AgenaRisk toolset [6]) for an AI system replacing a clinician. If the performance of the clinical decision-maker matches point A in Fig. 1 then the use of AI is beneficial (has greater utility), anywhere between point P2 and P3, as both the FNR and FPR are reduced. The clinician's utility (at point A) is 833.0; the utility of the AI system at point P2 is 835.71 and at P3 is 1017.6.

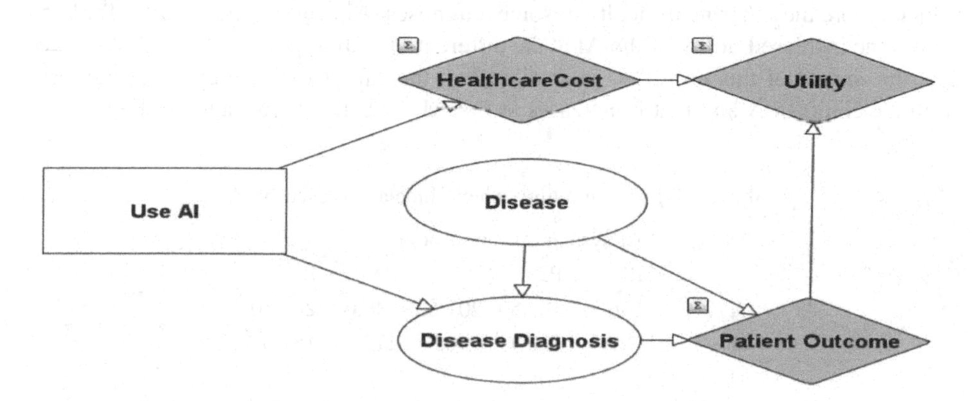

Fig. 2. Influence diagram for the case in which the clinician is replaced by the AI.

3.2 Modelling an AI System Filtering Patients Seen by a Clinician

In a diagnostic problem, an AI system can be used to filter out some cases so that a clinician sees fewer cases, saving clinical time (and money). Using a similar utility model, we consider (a) no disease filter, where the AI only filters out the cases it is confident have no disease and (b) disease present filter, the opposite to examine whether AI used in this way can benefit even if its performance does not dominate the clinician's. Comparing the AI to clinician B in Fig. 1 we obtain Table 4.

Table 4. Change in utility when AI filters patients seen by the clinician.

Type of filter	AI system operation point				
	P1	P2	P3	P4	P5
(a) No disease	−494.3	−385.5	−265.3	−124.9	−17.6
(b) Disease	225.1	266.6	310.7	358	384.3

We expected that the 'no disease' filter would be beneficial, particular operated at P5 (to minimize FNR). The model does not support our intuition: the problem is that both AI and clinician have an FNR, so together they miss more case than either operating alone. On the other hand, the 'disease present' filter is beneficial.

3.3 Modelling an AI Used to Assist a Clinician

An AI system could assist a clinician, who has access to the AI's predictions before a final diagnosis. If there is a conflict, the clinician can re-evaluate her diagnosis. A model of this requires some assumption about how the clinician and AI system interact. As an example, we optimistically assume that the diagnosis is revised in 50% of the cases when the clinician was wrong (but never when she was right) but this revision has the same cost as the original consultation (100 monetary units). With these assumptions, which ignore the differing difficulty of some diagnoses, AI cannot reduce costs. Table 5 shows the increased utility of the AI at the different operating points P1 to P5. We note that the impact of this use of AI is smaller than the earlier cases, greater for the less accurate clinician A and that it increases as the AI is operated with a lower FNR.

Table 5. Increase in utility when clinician assisted by AI.

Clinician	AI system operation point				
	P1	P2	P3	P4	P5
A	136.21	167.50	201.60	240.00	266.10
B	99.3	122.4	146.9	173.2	187.8

4 Conclusions and Further Work

We have shown that the same AI system used in different ways can have different impact. In some uses, the accuracy of the AI system must exceed that of the clinical decision maker; in other cases, it can still be beneficial despite lower accuracy. Making an estimate of impact requires the AI developer to consider both the potential benefits and the proposed use; we believe more systems would have more impact if this was done. In all cases we need some information about the performance of the clinical decision maker, though our use of a simple confusion matrix for this is simplistic as, for example, performance may vary between individuals. We know little about the interaction between a clinical decision maker and AI intended to assist with decision-making. We could consider the AI to behave like a 'second opinion' and the effects of this have been studied, for example in [7]. Some decisions are harder than others and it is likely that those that the AI system gets wrong may also be incorrectly diagnosed by the clinician. We will investigate this correlation between the two decision makers in future.

Acknowledgements. Support is acknowledged from EPSRC project EP/P009964/1 PAM-BAYESIAN for MR and WM and for WM from the Institutional Links grant 352394702, funded by the UK Department of Business, Energy and Industrial Strategy.

References

1. Wyatt, J.C., Altman, D.G.: Commentary: prognostic models: clinically useful or quickly forgotten? BMJ **311**(7019), 1539–1541 (1995)
2. Altman, D.G., Vergouwe, Y., Royston, P., Moons, K.G.: Prognosis and prognostic research: validating a prognostic model. BMJ **338**, b605 (2009)
3. Wallace, E., et al.: Framework for the impact analysis and implementation of Clinical Prediction Rules (CPRs). BMC Med. Inform. Decis. Mak. **11**(1), 62 (2011)
4. Huang, C., et al.: Enhancing the prediction of acute kidney injury risk after percutaneous coronary intervention using machine learning techniques: a retrospective cohort study. PLOS Med. **15**(11) (2018)
5. Hae, H., et al.: Machine learning assessment of myocardial ischemia using angiography: development and retrospective validation. PLOS Med. **15**(11), e1002693 (2018)
6. Yet, B., Neil, M., Fenton, N., Constantinou, A., Dementiev, E.: An improved method for solving hybrid influence diagrams. Int. J. Approx. Reason. **95**, 93–112 (2018)
7. Elmore, J.G., et al.: Evaluation of 12 strategies for obtaining second opinions to improve interpretation of breast histopathology: simulation study. BMJ **353**, i3069 (2016)

Generating Positive Psychosis Symptom Keywords from Electronic Health Records

Natalia Viani[1](✉), Rashmi Patel[1,2], Robert Stewart[1,2], and Sumithra Velupillai[1]

[1] IoPPN, King's College London, London, UK
{natalia.viani,rashmi.patel,robert.stewart,sumithra.velupillai}@kcl.ac.uk
[2] South London and Maudsley NHS Foundation Trust, London, UK

Abstract. The development of Natural Language Processing (NLP) solutions for information extraction from electronic health records (EHRs) has grown in recent years, as most clinically relevant information in EHRs is documented only in free text. One of the core tasks for any NLP system is to extract clinically relevant concepts such as symptoms. This information can then be used for more complex problems such as determining symptom onset, which requires temporal information. In the mental health domain, comprehensive vocabularies for specific disorders are scarce, and rarely contain keywords that reflect real-world terminology use. We explore the use of embedding techniques to automatically generate lexical variants of psychosis symptoms into vocabularies, that can be used in complex downstream NLP tasks. We study the impact of the underlying text material on generating useful lexical entries, experimenting with different corpora and with unigram/bigram models. We also propose a method to automatically compute thresholds for choosing the most relevant terms. Our main contribution is a systematic study of unsupervised vocabulary generation using different corpora for an under-studied clinical use-case. Resulting lexicons are publicly available.

Keywords: Natural language processing · Electronic health records · Embedding models · Schizophrenia

1 Introduction and Background

Secondary healthcare sources such as electronic health records (EHRs) contain a large proportion of text with clinically relevant information. To analyze this

RS, RP and SV are part-funded by the NIHR Biomedical Research Centre at South London and Maudsley NHS Foundation Trust and King's College London. RP has received support from a Medical Research Council (MRC) Health Data Research UK Fellowship (MR/S003118/1) and a Starter Grant for Clinical Lecturers (SGL015/1020) supported by the Academy of Medical Sciences, The Wellcome Trust, MRC, British Heart Foundation, Arthritis Research UK, the Royal College of Physicians and Diabetes UK. NV and SV have received support by the Swedish Research Council (2015-00359), Marie Skodowska Curie Actions, Cofund, Project INCA 600398.

D. Riaño et al. (Eds.): AIME 2019, LNAI 11526, pp. 298–303, 2019.
https://doi.org/10.1007/978-3-030-21642-9_38

information for clinical research, natural language processing (NLP) techniques are needed. In recent years, NLP systems have been developed to process clinical texts and extract relevant information [1]. An essential step for such systems is the identification of relevant entities, such as medications, symptoms, and time expressions, which can be linked to extract more complex constructs (e.g., treatment and symptom onset). Symptom onset extraction is important in the field of mental health, as a longer duration of untreated symptoms can be associated to worse intervention outcomes [2]. In EHRs related to schizophrenia patients, this information is documented in textual notes in a variety of ways. The first step towards extracting symptom onset is the identification of symptom mentions, which can be achieved using a domain-specific vocabulary. In the mental health domain, however, few standardized vocabularies are available for specific diseases and they rarely contain entries that reflect real-world terminology use.

To develop more comprehensive vocabularies, word embedding techniques can be exploited [3] which rely on neural models to automatically learn word representations (in the form of numeric vectors) from large collections of texts. Given their ability to capture semantic similarity, embedding models have been increasingly used to enhance NLP development, especially for general-domain applications and datasets. In the clinical domain, Ye and Fabbri created embedding models trained on multiple types of EHR clinical notes (e.g., Prescription, Problem List), proposing a method to combine them to enhance term discovery [4]. In the mental health domain, Velupillai et al. compared three approaches for vocabulary generation (dictionary search, linguistic rules, and embedding models) from intensive care unit EHRs [5]. Jackson et al. trained embeddings on mental health EHRs from the Clinical Record Interactive Search (CRIS) database [6], to identify concepts related to mental illness symptomatology.

In this paper, we explore unsupervised embedding models to automatically generate variants of psychosis symptoms indicative of disease onset. Our aim is to generate comprehensive use-case specific lexicons that could be used to solve complex information extraction tasks. Our long-term goal is to identify symptom onset in clinical notes for patients with a diagnosis of schizophrenia. In particular, we study how the choice of the underlying text material impacts the generation of useful terms, comparing four different input corpora and experimenting with bigram models (where frequent word pairs are mapped to a single vector). Moreover, we propose a method to automatically compute appropriate thresholds for choosing the most relevant terms from each model.

2 Materials and Methods

We use data from the CRIS database[1], which gathers anonymized patient information from the EHR system used at the South London and Maudsley NHS Foundation Trust (SLaM) [7].

In our embedding experiments, we trained different (unigram and bigram) models on: (1) Use-case specific EHR texts (*CRIS_specific*, 23.3m words) from

[1] Ethical approval for secondary analysis: Oxford REC C, reference 18/SC/0372.

early psychosis intervention services; (2) Institution-specific discharge summaries (*CRIS_general*, 23.6m words) for all mental health disorders (not restricted to psychosis); (3) External clinical texts from MIMIC II [8] (*MIMIC*, 187.4m words), i.e., an intensive care unit setting. To train embeddings, we used the gensim implementation of Word2Vec, with the CBOW model[2]. We also experimented with (4) Pre-trained embeddings from MEDLINE/PubMed (*PubMed*, 3.6bn words): we used the available models off-the-shelf (only unigram), without re-training [9].

We considered an initial list of keywords from a comprehensive mental health vocabulary [6]. Two psychiatrists reviewed the most frequent vocabulary terms found in *CRIS_specific*, and selected only those that were relevant to identify symptom onset, e.g., *positive* psychosis symptoms. This led to a list of 26 terms - 7 unigrams (e.g., hallucinations), 14 bigrams (e.g., persecutory ideas), and 5 trigrams (e.g., loosening of associations) - which were used for vocabulary generation. For each model, we considered the most similar terms with respect to these keywords (highest vector cosine similarity), and the terms with low Levenshtein distance, i.e., the edit difference (to capture misspellings).

To automatically compute a similarity threshold for each model, we relied on the "elbow" method proposed by Ye and Fabbri [4], which searches for a keyword-specific cutoff point. Given the top **K** similar terms (with decreasing similarity), the method selects the point with maximum distance from the curve connecting the two end-point similarities. In our case, the method was applied to the overall list resulting from all keywords, thus obtaining a model-specific threshold. Since the elbow threshold can change depending on **K**, we automatically computed an optimal value for each model: we tested all **K** values from 50 to 200 and looked at the greatest drop in the resulting elbow threshold.

To evaluate the generated vocabularies, two psychiatrists manually classified terms as: (1) Relevant psychosis symptom term (RT); (2) Potentially relevant term (PT); (3) Not relevant term (NT). As real-world clinical text is likely to contain errors, we also manually assessed the amount of misspelled terms (MSP) per vocabulary. To measure agreement between the raters, we computed the proportion of terms classified with the same label (Ac). Given the nature of this evaluation, other agreement measures (e.g., Cohen's κ) were deemed inappropriate.

3 Results

Figure 1 shows Venn diagrams for the vocabularies generated by each model. Table 1 reports model-specific results: the number of found original keywords (Keywords), the selected **K**, the vocabulary size, and the number of misspellings (MSP). We report the number of terms classified as relevant by both (RT) or at least one (RT*) rater (the most useful terms for our use-case), and the number of

[2] From: https://pypi.org/project/gensim/. Implementation details (preprocessing, parameters) available at: https://github.com/medesto/psychosis-symptom-keywords.

terms classified as PT/NT by both. We also report accuracy on all terms (Ac-all) vs on RT terms only (Ac-RT). Examples of RT were variants (*hallucinatory*), misspellings (*hallicinations*), and specific bigrams (*auditory hallucinations*).

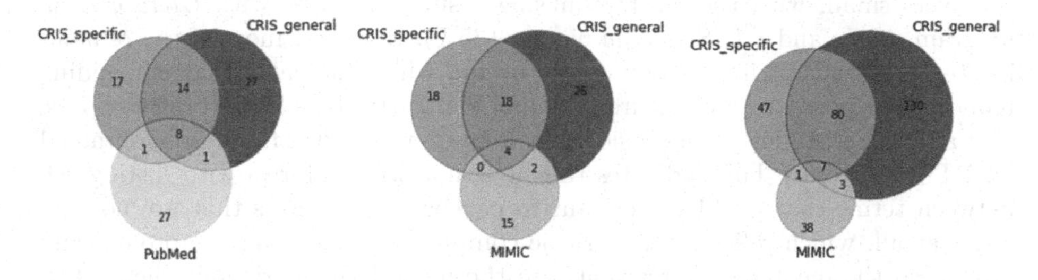

Fig. 1. Venn diagrams for unigram models (center and left) and bigram models (right).

Table 1. Manual evaluation results

Corpus	Model	Keywords	K	All terms	MSP	RT*	RT	PT	NT	Ac-all	Ac-RT
CRIS_specific	unigram	7/7	90	40	0	4	3	12	6	53%	75%
CRIS_general	unigram	7/7	60	50	10	15	13	15	4	64%	87%
MIMIC	unigram	5/7	60	21	1	5	4	2	0	29%	80%
PubMed	unigram	7/7	90	37	4	8	6	11	3	54%	75%
CRIS_specific	bigram	21/21	100	135	0	47	40	44	9	69%	85%
CRIS_general	bigram	19/21	160	220	8	70	58	57	20	61%	83%
MIMIC	bigram	6/21	120	49	1	7	7	6	5	37%	100%

4 Discussion and Conclusion

Generating vocabularies that reflect real-world terminology use is needed to facilitate complex NLP tasks. Moreover, sharing comprehensive lexical resources is an important step to support research in the NLP community. Our main contribution is a systematic study of unsupervised vocabulary generation using different corpora for an understudied clinical use-case. In addition, we proposed a method to automatically compute thresholds to select useful terms from embedding models. All developed resources (vocabularies and evaluations) are made publicly available on our github repository.

A first observation on the impact of corpus selection regards the size of generated vocabularies (Fig. 1). Despite *PubMed* being the largest corpus, the resulting list of terms was comparable to the other models. Also in the case of *MIMIC*, generated vocabularies were relatively small - partially due to some missing original keywords. This observation confirms that larger corpus sizes do not necessarily lead to more useful embedding models [10]. When comparing the two CRIS datasets, the *CRIS_general* model led to a larger vocabulary, especially in the bigram setting (220 vs 135). Interestingly, the proportion of bigram terms was actually higher in the *CRIS_specific* vocabulary (60.7% vs 50.9%). As for misspellings, while the *CRIS_specific* model did not find any entry of interest, the other considered datasets were useful (in particular *CRIS_general*).

As regards the manual evaluation process, the proportion of RT* terms was relatively small, with the most promising results obtained with *CRIS_general* unigram (30%) and *CRIS_specific* bigram (35%). However, most of the remaining terms were classified as potentially useful, which indicates that embedding models hold potential to capture semantic similarity. It is important to notice that agreement values when considering all labels were lower than those obtained on RT values only. This indicates that it is not straigthforward to distinguish between terms that could be relevant to psychosis and terms that are not relevant at all, which reflects the intrinsic complexity of defining symptoms (and in general the meaning of "relevant") in the mental health domain, hence terminologies. As a starting point, the new RT terms could be successfully reused in support of symptom onset extraction. To improve the proposed methodology/classification, further analysis will be performed on the terms that caused disagreements, with the final aim of developing a psychosis terminology to be linked to SNOMED CT.

As a main limitation of this work, we did not consider different embedding configurations nor n-gram models beyond bigrams. This could have impacted on the small size of RT lists, as single words or word pairs might not be sufficient to identify psychosis symptoms in a definite way (e.g., beliefs, anxiety attacks). More generally, given the intrinsic complexity of this domain, embedding models alone might not be the ideal choice to generate new concepts for specific use-cases. In future work, we will extend our study to take into account more complex models, and we will investigate other ways of modelling the extraction problem.

References

1. Wang, Y., Wang, L., Rastegar-Mojarad, M., et al.: Clinical information extraction applications: a literature review. J. Biomed. Inf. **77**, 34–49 (2018)
2. Kisely, S., Scott, A., Denney, J., Simon, G.: Duration of untreated symptoms in common mental disorders: association with outcomes. Br. J. Psychiatry **189**(1), 79–80 (2006)
3. Mikolov, T., Chen, K., Corrado, G., Dean, J.: Efficient estimation of word representations in vector space. arXiv preprint arXiv:1301.3781 (2013)
4. Ye, C., Fabbri, D.: Extracting similar terms from multiple EMR-based semantic embeddings to support chart reviews. J. Biomed. Inf. **83**, 63–72 (2018)
5. Velupillai, S., Mowery, D.L., Conway, M., et al.: Vocabulary development to support information extraction of substance abuse from psychiatry notes. In: Proceedings of BioNLP 2016, pp. 92–101 (2016)
6. Jackson, R., Patel, R., Velupillai, S., et al.: Knowledge discovery for deep phenotyping serious mental illness from electronic mental health records. F1000Res. **7** (2018). https://doi.org/10.12688/f1000research.13830.2

7. Perera, G., Broadbent, M., Callard, F., et al.: Cohort profile of the South London and Maudsley NHS Foundation Trust Biomedical Research Centre (SLaM BRC) Case Register: current status and recent enhancement of an Electronic Mental Health Record-derived data resource. BMJ Open **6**(3) (2016). https://doi.org/10.1136/bmjopen-2015-008721
8. Saeed, M., Villarroel, M., Reisner, A.T., et al.: Multiparameter intelligent monitoring in intensive care II (MIMIC-II): a public-access intensive care unit database. Crit. Care Med. **39**(5), 952–960 (2011)
9. McDonald, R., Brokos, G.I., Androutsopoulos, I.: Deep relevance ranking using enhanced document-query interactions. In: Proceedings EMNLP 2018 (2018)
10. Chiu, B., Crichton, G., Korhonen, A., Pyysalo, S.: How to train good word embeddings for biomedical NLP. In: Proceedings BioNLP 2016, pp. 166–174 (2016)

Text-to-Movie Authoring of Anatomy Lessons

Vaishnavi Ameya Murukutla[1(✉)], Elie Cattan[2], Olivier Palombi[1], and Remi Ronfard[1]

[1] Univ. Grenoble Alpes, LJK, Inria, CNRS, 38000 Grenoble, France
vaishnavi-ameya.murukutla@inria.fr
[2] Anatoscope, 38330 Montbonnot-Saint-Martin, France

Abstract. With popular use of multimedia and 3D content in anatomy teaching there is a need for a simple yet comprehensive tool to create and edit pedagogical anatomy video lessons. In this paper we present an automated video authoring tool created for teachers. It takes text written in a novel domain specific language (DSL) called the Anatomy Storyboard Language (ASL) as input and translates it to real time 3D animation. Preliminary results demonstrates the ease of use and effectiveness of the tool for quickly drafting video lessons in realistic medical anatomy teaching scenarios.

1 Introduction

Anatomy is the cornerstone of medical education. Fundamental knowledge of the human body is essential for understanding other subjects in medical and para medical fields. Traditionally anatomy is taught using visual aids such as chalk board drawings and slide presentations. Previous studies have shown that 3D graphics and animation make anatomical learning more engaging [1] and effective [3,6] but they suffer from a content creation bottleneck. If teachers choose to incorporate 3D animation in their lessons they either have to use the content already available to them or invest resources to create new content with the help of a graphic designer. In the first case content may not match the learning objectives of the class and the second case offers very little control to the teachers over the finished video. The solution to this would be to enable anatomy experts to generate their own 3D animations using innovative authoring tools.

Text-to-movie (or text-to-scene) authoring is a general class of methods that have been proposed for automatically generating 3D graphics and animation from text written by a domain expert. Recently, Hassani and Lee have proposed a review of text-to-movie research focusing on natural language [2]. While they provide a useful conceptual framework for our work, we chose to use a special-purpose authoring language, rather than natural language, in order to better address the needs of medical education.

In our authoring system (Fig. 1(a)) scripts for the lessons are written in a new formal language called the Anatomy Storyboard Language (ASL). The scripts are

D. Riaño et al. (Eds.): AIME 2019, LNAI 11526, pp. 304–308, 2019.
https://doi.org/10.1007/978-3-030-21642-9_39

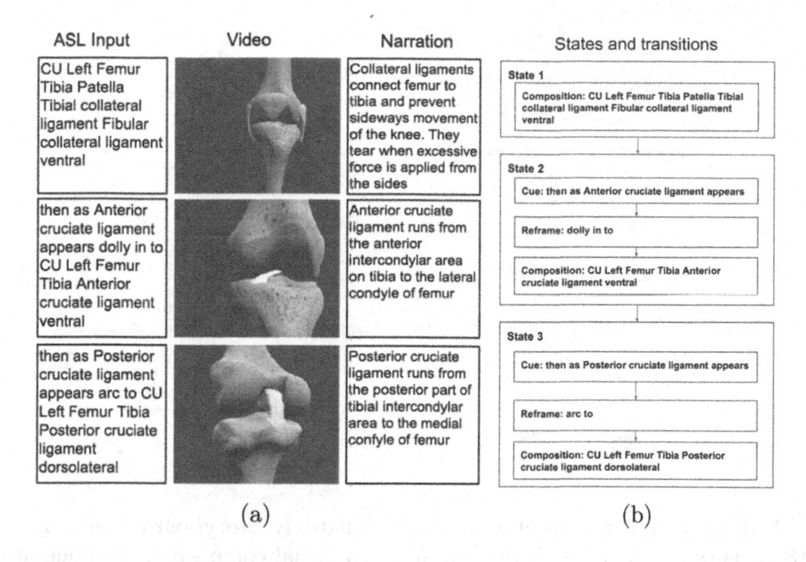

Fig. 1. Text-to-movie generation example with HFSM representation.

then parsed and translated part by part into hierarchical finite state machines (HFSM). Finally, state machines are executed in Unity 3D game engine to produce the desired animation at runtime.

2 Anatomy Storyboard Language

It is a domain specific language that is both machine and human readable. The video to be produced is written as a set of unique sentences. Each sentence describes all the visual elements, camera actions and animations seen from the start of the recording till the camera stops. ASL is an anatomical extension of the Prose Storyboard Language [5] that was designed for annotating and directing movies. As each sentence is capable of generating a complete shot it must have all the information necessary to transition into the shot, build composition, direct camera movements and record changes in composition as subjects in the video perform actions. The complete And/Or graph of the ASL grammar is presented in Fig. 2.

ASL is a context free language with terminals (anatomical entities and cinematographic terms) and non terminals (initial *composition* and subsequent *development*). The terminals are either generic terms used for camera movement or animation, or specific terms referring to the *subject* described in the shot such as anatomical parts and regions. The nomenclature of these specific terminals is derived from My Corporis Fabrica [4] (MyCF), an extensive ontology that describes structural and functional relations of different parts of human body. *Composition* is a description of all the elements that are seen in a particular frame. It needs to be comprehensive in detailing the size (Fig. 3(b)), angle, plane of view, anatomical location or specification, profile and relative screen position

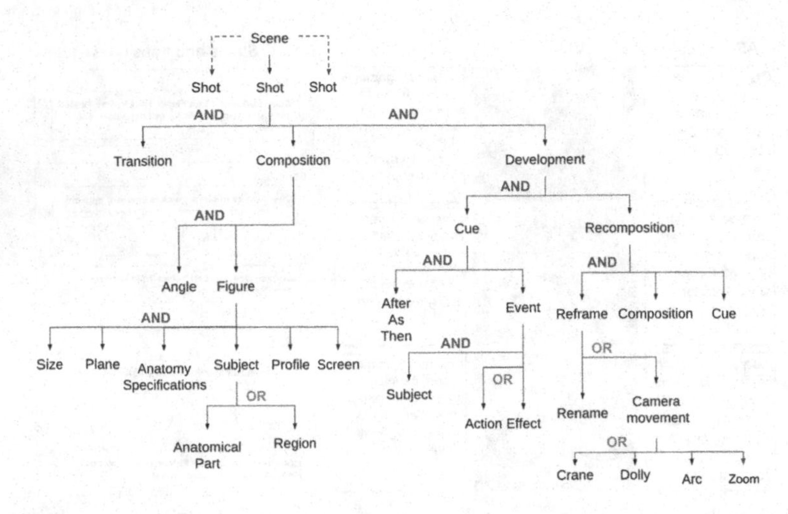

Fig. 2. And/Or Graph representation of the Anatomy Storyboard Language grammar. ASL scenes are made of shots containing an initial composition and one or more optional developments.

of the subjects viewed. The subjects in our case are anatomical parts and regions as seen in the complete 3D male Zygote model for human body[1].

The most important descriptive elements that are essential in building the *composition* are *plane*, *anatomical specification*, *profile*. *Plane* refers to the hypothetical planes that divides the human body. In ASL they define the view in which we see the anatomical parts and direct camera position accordingly. If a plane is not mentioned in a composition then the system will automatically assign a plane in the vertical axis *(sagittal or frontal)* based on the *profile* information but if the desired composition is in horizontal axis *(transverse)* then it must be mentioned in composition.

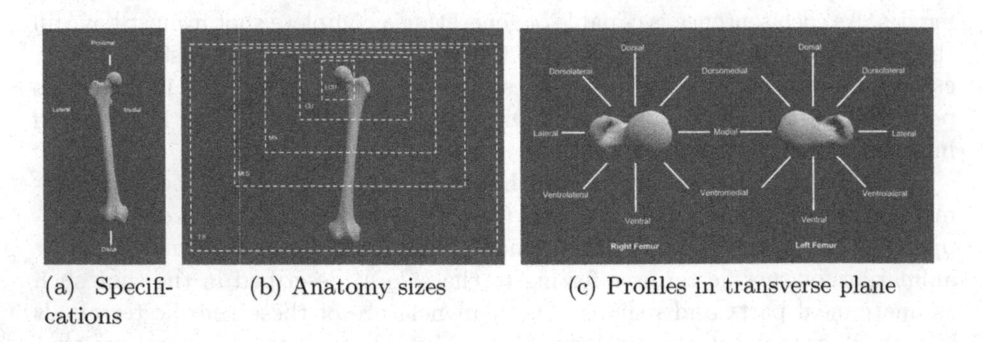

(a) Specifications (b) Anatomy sizes (c) Profiles in transverse plane

Fig. 3. ASL specifications, sizes and profiles.

[1] https://www.zygote.com.

Anatomical specification (Fig. 3(a)) further defines the part of anatomy most in focus in the current composition. *Profile* describes the orientation of the part in relation to the camera. It is the side of the subject that is viewed by the camera. The anatomical profile of left and right femurs are shown in Fig. 3(c). As the shot develops there will be changes in the composition. These changes can be due to *Actions* or *Effects* in *Cues*, or *Camera movements*, or both.

Sentences written in ASL are parsed via the Parsimonious[2] parser in the Python language. The parsed sentences are then translated to a hierarchical finite state machine (HFSM), with one state per composition or development. The HFSM is described in a XML format with specific tags to define each state of the machine. Finally, the HFSM is interpreted and executed to generate the desired animation. We now describe each of those steps separately.

3 From ASL to Animation

3.1 HFSM Generation

The different elements written in ASL are organised into states and transitions of a HFSM. Particular tags are used to describe the state machine. Our Python based HFSM generator creates a *scenario* tag for each complete ASL sentence. A *scenario* tag contains a list of *states* and *transitions*. Each *state* is given a unique *name* and an *anatomy* list of parts present in the current composition. A *state* also describes a *camera* with several tags that characterise its positioning such as *orientation*, *angle_up*, *angle_side*, *up*. In particular, a *lookat* tag lists the objects the camera should look at. A *transition* is a change from a *start* state to an *end* state. Currently the transitions are executed automatically between two consecutive states with a preset time delay that is specified in the *delay* tag.

Actions in ASL are translated into animations in the HFSM. An additional *animation* tag is added in *state* to trigger animations of anatomical elements (e.g. a knee flexion). In the current state of the application, animations are pre-made and cyclical. This is done to avoid editing glitches that could arise if the body position at the end of animation in one state does not match the body position in the next state.

Some descriptive terms of the ASL need to be converted to numerical values in HFSMs. For example, for the lessons written in this paper we specified that the ASL term *high angle* will be translated to a 45° bird's eye view. This numerical value of 45 is defined in an animation style sheet along with other global values that change the camera position and total run-time of the video. This style sheet can be edited by the teachers depending on their preferences thereby giving them more nuanced control over the video making process.

3.2 Animation Generation

We developed an application using the Unity 3D game engine to generate the desired animation at runtime from the HFSM obtained from the ASL script. The

[2] https://github.com/erikrose/parsimonious.

application is thus an interpreter, from a specific XML format to 3D videos of anatomy. The description of the *camera* in the HFSM is given by a view plane (frontal, sagittal or transverse), an object or a group of object to look at and a up vector to orient the rotation of the camera. With these pieces of information, the application computes the bounding boxes of the objects to look at and deduces the position and orientation to reach. The camera then moves from its previous position to the new one according to the other parameters translated from the ASL (e.g. type and speed of camera movement). If an *animation* tag is present, it executes an animation from the database that is registered under the same name.

4 Results and Future Work

We used our text to movie authoring system to create short videos based on scripts written by two anatomy professors. The teachers were given a brief introduction to ASL using the And/Or chart. Examples of compositions in ASL and their corresponding frames in Unity player were shown to get them familiarised with the system and ASL grammar. They initially started writing very short scripts with initial composition and one development. Progress was made one composition at a time during which they tested different animations and decided on the best viewing positions. The most liked features of the system are that it allows the user to build the video state by state and has immediate visualisation of the video made so far which facilitates easier editing. After some practice the teachers were able to write three lessons on the knee joint and one lesson on the forearm.

In future work, we would like to extend our approach to non-linear content generation by taking into account user-triggered transitions between states. This would make our approach applicable to mixed reality. Another promising direction for future research would be to generate ASL scripts directly from audio narrations, using a combination of speech processing and natural language understanding.

References

1. Estai, M., Bunt, S.: Best teaching practices in anatomy education: a critical review. Ann. Anat. **208**, 151–157 (2016)
2. Hassani, K., Lee, W.S.: Visualizing natural language descriptions: a survey. ACM Comput. Surv. **49**(1), 17 (2016)
3. Hoyek, N., Collet, C., Rienzo, F.D., Almeida, M.D., Guillot, A.: Effectiveness of three-dimensional digital animation in teaching human anatomy in an authentic classroom context. Anat. Sci. Educ. **7**(6), 430–437 (2014)
4. Palombi, O., Ulliana, F., Favier, V., Léon, J.C., Rousset, M.C.: My Corporis Fabrica: an ontology-based tool for reasoning and querying on complex anatomical models. J. Biomed. Semant. **5**(1), 20 (2014)
5. Ronfard, R., Gandhi, V., Boiron, L.: The prose storyboard language: a tool for annotating and directing movies. CoRR abs/1508.07593 (2015)
6. Vernon, T., Peckham, D.: The benefits of 3D modelling and animation in medical teaching. J. Audiov. Media Med. **25**(4), 142–148 (2002)

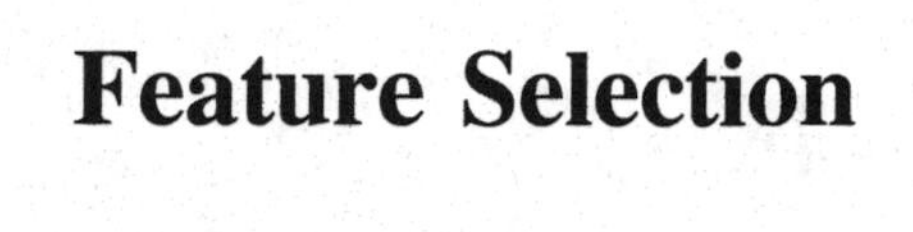
Feature Selection

Machine Learning and Feature Selection for the Classification of Mental Disorders from Methylation Data

Christopher L. Bartlett[1(✉)], Stephen J. Glatt[2], and Isabelle Bichindaritz[1]

[1] Intelligent Bio Systems Laboratory, Biomedical and Health Informatics, State University of New York at Oswego, 7060 NY-104, Oswego, NY 13126, USA
cbartle3@oswego.edu

[2] Psychiatric Genetic Epidemiology & Neurobiology Laboratory (PsychGENe Lab), Department of Psychiatry and Behavioral Sciences, SUNY Upstate Medical University, Syracuse, NY, USA

Abstract. Psychiatric disorder diagnoses are heavily reliant on observable symptoms and clinical traits, the skill level of the physician, and the patient's ability to verbalize experienced events. Therefore, researchers have sought to identify biological markers that accurately differentiate mental disorder subtypes from psychiatrically normal comparison subjects. One such putative biomarker, DNA methylation, has recently become more prevalent in genetic research studies in oncology. This paper proposes to apply this paradigm in a study of the diagnostic accuracy of DNA methylation signatures for classifying schizophrenia, bipolar disorder, and major depressive disorder. Very high classification performance measures were obtained from differentially methylated positions and regions, as well as from selected gene signatures. This work contributes to the path toward the identification of biological signatures for mental disorders.

Keywords: Machine learning · Feature selection · Bioinformatics · Psychiatry

1 Introduction

Psychiatric disorder diagnoses are heavily reliant on the behavioral criteria enumerated in the Diagnostic and Statistical Manual of Mental Disorders' listing of observable symptoms and clinical traits, as well as the skill level of the physician, and the patient's ability to verbalize experienced events. As stated in Demkow and Wolańczyk [1], the patient's ability to consistently verbalize their experiences coupled with varying degrees of perceptive awareness in the health professional exaggerate complications in proper diagnosis. This sentiment is echoed in the mission of the National Institute of Mental Health's (NIMH) Research Domain Criteria (RDoC) initiative. In commentary for the initiative, Insel [4] suggests that "While we can improve psychiatric diagnostics by more precise clustering of symptoms,

D. Riaño et al. (Eds.): AIME 2019, LNAI 11526, pp. 311–321, 2019.
https://doi.org/10.1007/978-3-030-21642-9_40

diagnosis based only on symptoms may never yield the kind of specificity that we have begun to expect in the rest of medicine." Therefore, researchers have sought to identify biological markers that accurately differentiate mental disorder subtypes from psychiatrically normal individuals. Among the'omics range of data, one particular biomarker, DNA methylation, has recently raised attention in genetic research studies in oncology. Following on these tracks, this paper proposes to classify a subset of mental disorders based on their methylome with the goal of determining whether DNA methylation data alone can be successful at diagnosing a mental disorder. Machine learning has been chosen for this task as a method of choice to complement statistical data processing.

2 Epigenetics and Methylation

The term epigenetics was first introduced into modern biology by Conrad Waddington as a means of defining interactions between genes and their products that result in phenotypic variations. Waddington's landscape presents a cell becoming more differentiated as time goes on. What causes them to differentiate, however? What sort of forces are at play? One of many such events is methylation. Methylation is a covalent attachment of a methyl group to cytosine. Cytosine (C) is one of the four bases that construct DNA and one of only two bases that can be methylated. While adenine can be methylated as well, cytosine is typically the only base that's methylated in mammals. Once this methyl group is added, it forms 5-methylcytosine where the 5 references the position on the 6-atom ring where the methyl group is added. Under the majority of circumstances, a methyl group is added to a cytosine followed by a guanine (G) which is known as CpG. While the methyl group is added onto the DNA, it doesn't alter the underlying sequence but it still has profound effects on the expression of genes and the functionality of cellular and bodily functions. Methylation at these CpG sites has been known to be a fairly stable epigenetic biomarker that usually results in silencing the gene. Further, the amount of methylation can be increased (known as hypermethylation) or decreased (known as hypomethylation) and improper maintenance of epigenetic information can lead to a variety of human diseases.

3 Methylation in Mental Disorders

Methylation's influence in cancer has been introduced with great success which encourages its application in psychiatry. There have been a number of studies disclosing the impact of methylation levels on overall psychiatric health that are both broad and narrow in scope. One potential precursor to the onset of a mental disorder is the presence of chronic stress. It has been linked to the development of schizophrenia and bipolar disorder [14], major depressive disorder [12], and addiction [11]. Klengel, Pape, Binder and Mehta [5] performed a review of literature and noted how stress induces long-term changes in DNA methylation. Specifically, they report differential methylation among genes and promoters for post-traumatic stress disorder (PTSD), major depressive disorder (MDD), depressive symptoms and suicide. Radtke et al., [9] also found that maternal exposure to intimate partner violence had a sustained increase in methylation of the GR promoter and altered the hypothalamic-pituitary-adrenal axis (HPA-axis), which has been linked to several mental disorders. Methylation of the GR promoter has not only been implicated in internalizing behavioral problems in preschoolers [7] but methylation of this promoter has been found in those suffering from PTSD and from depressive, anxiety and substance-abuse disorders [13]. Further, DNA methylation increases throughout the lifespan but this was found to be 8-fold greater in those who have committed suicide [2].

4 Materials

4.1 Datasets

Three publicly available Gene Expression Omnibus (GEO) datasets were selected from Array Express; GSE80417, GSE41169 and GSE44132. Selection criteria were based on the availability, affordability, measurement apparatus and sample location. These three datasets were Illumina HumanMethylation450 (HM450) BeadChip data derived from whole blood samples. Data from 322 psychiatrically normal control subjects was extracted from GSE80417 and added to 33 control subjects from GSE41169. The 322 control subjects from GSE80417 were screened for an absence of mental health problems, and interviewed. The interview ensured that these subjects did not have a personal history of a mental disorder, or a family history of schizophrenia, bipolar disorder or alcohol dependence.

Whole blood HM450 data from 62 subjects with schizophrenia (SCZ) were also used from GSE41169 and added to HM450 data from 34 subjects with major depressive disorder (MDD) and 21 subjects with bipolar disorder (BP) from the GSE44132 dataset (Table 1). These three datasets were merged, resulting in 402,723 observations of CpG sites and 472 subjects. Each dataset was β-quartile normalized. β values are an estimation of the methylation levels between 0 and 1 with 0 being completely non-methylated and 1 being completely methylated.

Table 1. Number of subjects per disorder for each dataset

Dataset	Normal	SCZ	MDD	BP
GSE44132	0	0	34	21
GSE41169	33	62	0	0
GSE80417	322	0	0	0
TOTAL	355	62	34	21

4.2 Data Preprocessing

The GEO datasets from Array Express were stored in an ExpressionSet object within RStudio. These objects contain the phenotype data, assay data, metadata, protocol and feature data. Before beginning the surrogate variable analyses, the phenotype and assay data were extracted and stored in their own dataframe objects in RStudio. The majority of the phenotype data pertained to the laboratory and these variables were removed while subject-specific variables were maintained.

4.3 Surrogate Variable Analyses

Genetic and epigenetic factors are highly-dimensional with thousands of potential influencers. Surrogate variable analyses (SVA) attempts to estimate some of these unobserved influencers for their effect on a variable of interest. The effects of the subject's age and gender on methylation levels were tested at this stage, as well as variations in equipment and extraction methodologies. Commonly referred to as a "batch effect," different laboratories having different equipment and extraction procedures can alter obtained methylation levels was therefore accounted for.

Surrogate variables and batch effects were handled through the R package SVA. SVA operates by using the iteratively least weighted squares approach to estimate surrogate variables. Upon estimation of the surrogate variables, SVA calculates the probability that each probe is associated with the variable of interest (disorder classification, in this case). Parametric F-test p-values are calculated for differential methylation in regards to disorder status, adjusted for multiple testing with the Benjamin-Hochberg (BH) method, and further adjusted for the influence of the surrogate variables. Through this stage, probes that were associated with age, gender, and batch were removed from each of the three datasets prior to re-assembly. The resulting SVA set was comprised of 472 subjects and 10,890 remaining CpG sites and was utilized in each of the following analyses. Additionally, smoking has been shown to be a confounding factor. As the selected datasets did not screen for smokers, an additional check was performed to ensure that probes significantly associated with smoking were not in the dataset after SVA.

5 Methods

The first step was to take an overall glance at the differences in disorders and the control group. This entailed viewing the average total methylation per disorder, percent hypermethylation and percent hypomethylation. On average, the sites in the control subjects were 35.5% methylated, SCZ subjects were 35.3%, MDD subjects were 39.6% and BP subjects were 36.9%. This was performed simply by computing the average β value per subject and calculating an average for all subjects for each disorder.

Computing the percent of hypermethylation was performed by filtering for all sites within each subject that had a β greater than 0.8 and dividing by 10,890 (total number of sites in the dataset). After the percent of hypermethylated sites were determined for each subject, an average for each disorder was computed. 15.7% of sites were hypermethylated on average for the control subjects, 23.8% for SCZ, 30.3% for MDD, and 22.2% for BP.

Calculating the percent of hypomethylation was done by filtering for all sites within each subject that had a β lower than 0.2 and dividing by 10,890. 56.7% were hypomethylated for the control subjects, 58.7% for SCZ, 57.6% for MDD and 57.3% for BP (Fig. 1).

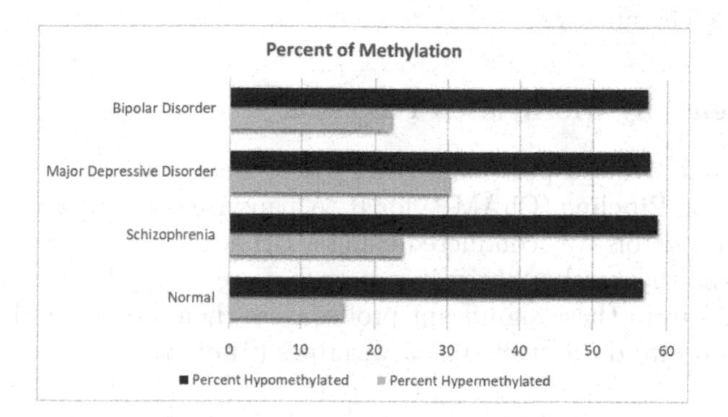

Fig. 1. Percent of hyper and hypomethylation per disorder.

A scatterplot was produced in Fig. 2 that plotted the percent of hypermethylation (X-axis) versus the percent of hypomethylation (Y-axis) for each subject. The plot is arranged based on disorder class with the psychiatrically normal control group (N) being unfilled squares, SCZ being the letter X, MDD being unfilled circles and BP being filled circles.

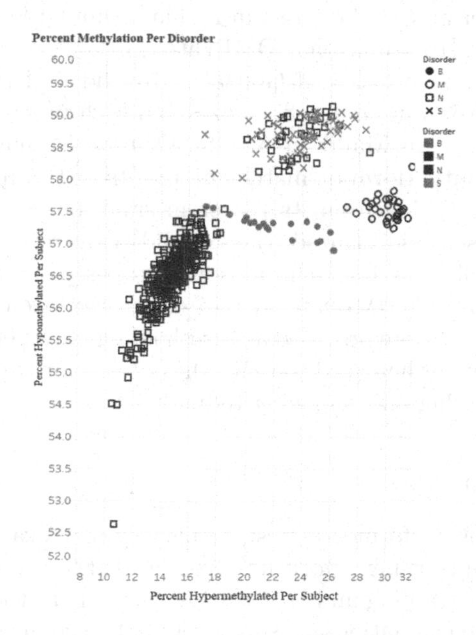

Fig. 2. A plot of the percent of hyper and hypomethylation for 455 subjects in 4 disorder classes.

An important issue to address is that 33 of the normal subjects were clustered in with the schizophrenia subjects. These 33 subjects were determined to be those in the GSE41169 dataset, so it is possible that equipment calibration procedures may have contributed to the higher hypomethylated positions. However, since these subjects were screened to be psychiatrically normal they were left in the dataset during classification.

5.1 Differentially Methylated Positions

Differentially methylated positions (DMP) were identified using the Chip Analysis Methylation Pipeline (ChAMP) for R. A pairwise comparison of all possible disorders and controls was conducted using a BH adjusted p-value of 0.05. Each comparison and the number of significant probes are available in Table 2. Methylation β-values from these significant probes were then merged and compiled to form one composite dataset of 10,585 variables (Table 3).

5.2 Differentially Methylated Regions

The DMRcate method within ChAMP was used to extract the differentially methylated regions (DMR). Regions are clusters of probes that serve a similar function in gene transcriptional regulation. ChAMP allows for three different methods of locating DMRs: Bumphunter, Probe Lasso and DMRcate. Probe Lasso can only compare two phenotypical categories, and thus wouldn't be functional without methodical pairwise comparisons. Upon comparison, Peters et al., [8] found DMRcate to have superior predictive performance compared to Bumphunter and Probe Lasso in real and simulated data though Ruiz-Arenas and Gonzalez [10] found that DMRcate had low power for smaller effect sizes. In contrast, Ruiz-Arenas and Gonzalez [10] did find that DMRcate had high power and precision in larger effect sizes and high precision for small effect sizes which was echoed in Mallik et al., [6]. Of note is that Mallik et al., [6] found DMRcate to outperform or match the results of Bumphunter and Probe Lasso in all instances aside from its power in small effect sizes. These led to DMRcate being chosen and utilized, though DMRcate required the removal of cross-hybridizing probes and sex-chromosome probes prior to operation. This could be handled through DMRcate. A false-discovery rate of 0.05 and a minimum probe number of 3 were provided as thresholding parameters. Probes within the located regions were then used to build the dataset for this stage. 494 probes remained after compiling and removing redundancies.

5.3 Classification

The classification was performed almost exclusively in Weka [3], an open source collection of machine learning algorithms. We decided to use leave-one-out cross validation and a 90% training and 10% testing split due to the small sample size in each disorder. The evaluation measures were balanced accuracy and the area

Table 2. Number of significant probes for each pairwise comparison

Comparison	Probes
Control and SCZ	10,265
Control and MDD	9,520
Control and BP	7,401
SCZ and MDD	10,178
SCZ and BP	9,024
MDD and BP	10,452

Table 3. Number of significant regions for each pairwise comparison

Comparison	Regions
Control and SCZ	493
Control and MDD	480
Control and BP	397
SCZ and MDD	493
SCZ and BP	490
MDD and BP	493

under the ROC curve due to the substantially uneven classes. Balanced accuracy was computed by applying Formula 1 to locate the accuracy of each class with i representing one of the four classes. Formula 2 was then applied to determine the overall balanced accuracy, where N is the number of classes.

$$acc_i = 1/2(TruePositiveRate_i + (1 - FalsePositiveRate_i)) * 100 \quad (1)$$

$$Acc = \frac{\sum_{i=1}^{N} acc_i}{N} \quad (2)$$

The first stage was to classify the disorders using the dataset after SVA had been conducted but prior to any other stages. This stage was meant to establish a baseline through which to compare the results of any subsequent stages. Following were the utilization of the DMP and DMR datasets. The next sections disclose the classification results. Among machine learning algorithms, Nearest Neighbor (NN), Support Vector Machines (SVM), Naïve Bayes (NB), and Random Forest (RF) were compared on all 10,890 probes (after SVA), 10,585 differentially methylated positions, and 494 differentially methylated regions.

5.4 Feature Selection

Feature selection was carried out on the dataset after surrogate variable analyses were performed, as well as on the data after differentially methylated position analyses. Five algorithms consisting of the Information Gain Attribute Evaluation, Correlation Attribute Evaluation, SMO Classifier Attribute Evaluation and Naïve Bayes Classifier Attribute Evaluation were performed. An ensemble was then created using all of the results by tallying the rankings for each probe in the results of each algorithm. In each list, the best probe would be ranked first and the second best would be ranked second and so forth. The 15 with the lowest overall tallied score were mapped to their nearest gene and used to classify sequentially using Naïve Bayes. 15 was selected as subsequent tests to 50 features revealed no significant changes in accuracy. Balanced accuracy and the ROC area were used as performance measures.

6 Results

6.1 All Probes

Classification results on all probes are very high (see Table 4).

Table 4. Classification results for the overall 10,890 probes at the probe level

Leave-one-out cross validation			90% Training, 10% Testing		
Algorithm	Balanced accuracy	ROC	Algorithm	Balanced accuracy	ROC
NN	89.2%	0.952	NN	91.2%	0.976
SVM	95.45%	0.947	SVM	94.08%	0.935
NB	97.52%	0.955	NB	94.08%	0.914
RF	97.58%	0.970	RF	98.77%	0.990

6.2 Differentially Methylated Positions

Classification results on differentially methylated positions are also high (see Table 5).

Table 5. Classification results using 10,585 differentially methylated positions

Leave-one-out cross validation			90% Training, 10% Testing		
Algorithm	Balanced accuracy	ROC	Algorithm	Balanced accuracy	ROC
NN	90.77	0.958	NN	94.9	0.970
SVM	95.83	0.952	SVM	93.46	0.928
NB	97.52	0.955	NB	96.33	0.931
RF	96.51	0.971	RF	94.9	0.953

6.3 Differentially Methylated Regions

Moving from probes to regions reduces significantly the number of features while classification performance shows a trend of improvement (see Table 6).

Table 6. Classification results using 494 differentially methylated regions

Leave-one-out cross validation			90% Training, 10% Testing		
Algorithm	Balanced accuracy	ROC	Algorithm	Balanced accuracy	ROC
NN	95.67	0.972	NN	98.75	0.996
SVM	93.06	0.920	SVM	98.75	0.984
NB	96.17	0.939	NB	98.75	0.967
RF	95.88	0.972	RF	98.75	0.995

6.4 Feature Selection

Balanced accuracy and the ROC area were computed for each iteration and the results are available in Table 7 from 1 to 15 top ranked genes.

When comparing performances with Naïve Bayes (see Table 8), one notices that both accuracy and area under the ROC curve reach similar performance on 15 selected genes, which is an impressive result. Having a small number of genes to differentiate between the four classes makes the results of the classification much more explainable and interpretable.

Table 7. Accuracy and ROC for the top ranked genes

Analysis stage	Split	Top 1	Top 5	Top 10	Top 15
SVA	LOOC	80.05% (0.957)	85.45% (0.965)	86.48% (0.967)	97.3% (0.974)
SVA	90/10	80.69% (0.983)	90.08% (0.972)	90.08% (0.98)	97.26% (0.964)
DMP	LOOC	79.81% (0.916)	95.95% (0.912)	96.83% (0.911)	96.44% (0.912)
DMP	90/10	75.79% (0.916)	94.16% (0.912)	94.16% (0.911)	94.16% (0.912)

Table 8. Comparison of overall performance with and without feature selection

Analysis stage	Number of features	LOOC	90/10
SVA	10,890	97.52% (0.955)	94.08% (0.914)
DMP	10,585	97.52% (0.955)	96.33% (0.931)
DMR	494	96.17% (0.939)	98.75% (0.967)
Ensemble feature selection (SVA)	15	97.3% (0.974)	97.26% (0.964)
Ensemble feature selection (DMP)	15	96.44% (0.912)	94.16% (0.912)

7 Conclusion

This paper outlines the application of four classification algorithms for determining the accuracy of identifying three disorders and a control class using DNA methylation signatures. Differentially methylated positions and regions were detected and utilized, and potential associated genes were notated for possible further investigation. Classification accuracy and area under the ROC curve show very high figures that demonstrate an excellent performance particularly by Random Forest and Naïve Bayes. It is notable that the reduction in number of features from 10,890 to 494, then to 15 when moving from probe to region level then to the gene level did not lower the classification performance while greatly improving the execution speed and interpretability. These results are all the more encouraging and robust given that patients were evaluated by different teams of psychiatrists, which could have conflated the diagnostic categories; however, we were limited by the availability of data. Future work will involve focusing on annotating the genetic signature at the pathway level, performing more thorough analyses on larger datasets as they become available, and in particular on independent test sets to test the replicability of results.

Acknowledgements. We thank the State University of New York EIPF grant #172 for their support of this work and Dr. Renaud Seigneuric for his advice.

References

1. Demkow, U., Wolańczyk, T.: Genetic tests in major psychiatric disorders-integrating molecular medicine with clinical psychiatry–why is it so difficult? Transl. Psychiatry **7**(6), e1151 (2017). https://doi.org/10.1038/tp.2017.106. http://www.nature.com/doifinder/10.1038/tp.2017.106
2. Haghighi, F., et al.: Increased DNA methylation in the suicide brain. Dialogues Clin. Neurosci. **16**(3), 430–438 (2014). https://doi.org/10.3760/cma.j.issn.0366-6999.2010.15.032
3. Hall, M., Frank, E., Holmes, G., Pfahringer, B., Reutemann, P., Witten, I.H.: The WEKA data mining software: an update. SIGKDD Explor. **11**(1), 10–18 (2009)
4. Insel, T.: The NIMH Research Domain Criteria (RDoC) Project: implications for genetics research. Mamm. Genome **25**(1–2), 23–31 (2014). https://doi.org/10.1007/s00335-013-9476-9
5. Klengel, T., Pape, J., Binder, E.B., Mehta, D.: The role of DNA methylation in stress-related psychiatric disorders. Neuropharmacology **80**, 115–132 (2014). https://doi.org/10.1016/j.neuropharm.2014.01.013
6. Mallik, S., Odom, G.J., Gao, Z., Gomez, L., Chen, X., Wang, L.: An evaluation of supervised methods for identifying differentially methylated regions in Illumina methylation arrays. Brief. Bioinform. (2018). https://doi.org/10.1093/bib/bby085
7. Parade, S.H., et al.: Change in FK506 binding protein 5 (FKBP5) methylation over time among preschoolers with adversity. Dev. Psychopathol. **29**(5), 1627–1634 (2017). https://doi.org/10.1017/s0954579417001286
8. Peters, T.J., et al.: De novo identification of differentially methylated regions in the human genome. Epigenetics Chromatin **8**(1), 6 (2015). https://doi.org/10.1186/1756-8935-8-6
9. Radtke, K.M., et al.: Transgenerational impact of intimate partner violence on methylation in the promoter of the glucocorticoid receptor. Transl. Psychiatry **1**(7), e21–6 (2011). https://doi.org/10.1038/tp.2011.21
10. Ruiz-Arenas, C., González, J.R.: Redundancy analysis allows improved detection of methylation changes in large genomic regions. BMC Bioinform. **18**(1) (2017). https://doi.org/10.1186/s12859-017-1986-0
11. Sinha, R., Jastreboff, A.M.: Stress as a common risk factor for obesity and addiction. Biol. Psychiatry **73**(9), 827–835 (2013). https://doi.org/10.1016/j.biopsych.2013.01.032
12. Slavich, G.M., Irwin, M.R.: From stress to inflammation and major depressive disorder: a social signal transduction theory of depression. Psychol. Bull. **140**(3), 774–815 (2014). https://doi.org/10.1037/a0035302

13. Tyrka, A.R., et al.: Methylation of the leukocyte glucocorticoid receptor gene promoter in adults: associations with early adversity and depressive, anxiety and substance-use disorders. Transl. Psychiatry **6**(7), e848–9 (2016). https://doi.org/10.1038/tp.2016.112
14. Van Winkel, R., Van Nierop, M., Myin-Germeys, I., Van Os, J.: Childhood trauma as a cause of psychosis: linking genes, psychology, and biology. Can. J. Psychiatry **58**(1), 44–51 (2013). https://doi.org/10.1177/070674371305800109

A Machine Learning Approach to Predict Diabetes Using Short Recorded Photoplethysmography and Physiological Characteristics

Chirath Hettiarachchi(✉) and Charith Chitraranjan

Department of Computer Science and Engineering, University of Moratuwa, Moratuwa, Sri Lanka
{chirath, charithc}@cse.mrt.ac.lk

Abstract. Diabetes is a global epidemic, which leads to severe complications such as heart disease, limb amputations and blindness, mainly occurring due to the inability of early detection. Photoplethysmography (PPG) signals have been used as a non-invasive approach to predict diabetes. However, current methods use long, continuous signals collected in a clinical setting. This study focuses on predicting Type 2 Diabetes from short ($\sim$2.1s) PPG signals extracted from smart devices, and readily available physiological data such as age, gender, weight and height. Since this type of PPG signals can be easily extracted using mobile phones or smart wearable technology, the user can get an initial prediction without entering a medical facility. Through the analysis of morphological features related to the PPG waveform and its derivatives, we identify features related to Type 2 Diabetes and establish the feasibility of predicting Type 2 Diabetes from short PPG signals. We cross validated several classification models based on the selected set of features to predict Type 2 Diabetes, where Linear Discriminant Analysis (LDA) achieved the highest area under the ROC curve of 79%. The successful practical implementation of the proposed system would enable people to screen themselves conveniently using their smart devices to identify the potential risk of Type 2 Diabetes and thus avoid austere complications of late detection.

Keywords: Machine Learning · Diabetes Type II · Photoplethysmography · Feature selection

1 Introduction

Diabetes is a severe global phenomenon that has been the root cause of millions of deaths worldwide. This study focuses on detecting Type 2 Diabetes, which is caused by the body's inefficient use of insulin, resulting in abnormally high levels of sugar in the blood. The symptoms of the disease are less marked and is often detected several years after the onset through complications, which can result in premature heart disease, blindness, limb amputations and kidney failure. Hence, the utmost importance of early detection and continuous monitoring. Individuals at risk are advised to undergo regular

D. Riaño et al. (Eds.): AIME 2019, LNAI 11526, pp. 322–327, 2019.
https://doi.org/10.1007/978-3-030-21642-9_41

medical tests such as Hb1Ac to detect diabetes, which is intrusive and often neglected due to the busy lifestyles and the associated costs.

Our research focuses on developing a system capable of predicting Type 2 Diabetes using readily available data of general users. We use physiological characteristics and short($\sim$2.1s) recorded Photoplethysmography (PPG) signal measurements of the users towards Diabetes prediction. Photoplethysmography can be identified as a non-invasive, inexpensive, optic technique that measures the blood volume changes in blood vessels through which, oxygen saturation, blood pressure, cardiac output could be measured [1, 2]. Recent research studies have identified that PPG is a promising technique towards early screening of diseases as the PPG waveform possess significant information embedded within. The recent advancement in hardware has enabled the integration of high-quality PPG sensors within mobile phones & smart watches, which has enabled easy access to regular measurement of required health parameters motivating towards early disease prediction and continuous health monitoring, which is recommended for diseases such as Diabetes.

The objective of this study is to predict Diabetes using a single instance of a short-recorded PPG signal combined with physiological characteristics, which is uncommon. However, there have been studies that have focused on analyzing the PPG signals and other features of the users towards the prediction of Type 2 Diabetes. Reddy et al. [3] have focused on analyzing features related to PPG derived Heart Rate Variability (HRV) and shape information of the PPG waveform towards diabetes prediction. Ballinger et al. [4] focused on predicting a range of diseases including diabetes using features such as the user medical history, step count and continuous optical heart rate measured using PPG and was able to achieve an accuracy of 84.51%. Moreno et al. [5] achieved an accuracy of 69.4% using HRV and Cepstral Analysis. In contrast the core focus of this research relies on developing a system capable of using basic user known physiological characteristics and a single instance of a PPG signal recorded for at most 2–3 s to be used towards diabetes prediction. The successful research in this domain would ensure great value addition in the fields of smart wearables and healthcare.

2 Methodology

2.1 Data Description

The research was conducted based on a de-identified open clinical trial dataset for non-invasive detection of cardiovascular diseases by Liang et al. [6], which contains physiological characteristics, short recorded PPG signals and information related to the presence of Diabetes and Hypertension in patients. The final extracted data for evaluation purposes eliminating the erroneous signals, contained a total of 150 subjects, comprising of 51 healthy, 39 prehypertension, 28 hypertension, 9 diabetes, 16 diabetes with prehypertension and 7 diabetes with hypertension subjects. Subjects with cerebral infarction and cerebrovascular disease were excluded from the study. It should be noted that the number of subjects in the target group Diabetes was comparatively less and there were diabetes subjects who were also suffering from both Hypertension and Prehypertension. Hypertension and Diabetes are strongly interconnected diseases, both affecting the cardiovascular system of the human body and thus the PPG signals. Hence, considering the practical disease distribution in the environment, it is important to clearly identify unique features to develop a system that is able to clearly predict diabetes when healthy, and subjects with a form of hypertension are present.

2.2 Data Preprocessing and Feature Extraction

The PPG signals related to the above identified subjects were analyzed in order to extract features related to the cardiovascular system. For each patient, three segments of short recorded PPG waveforms had been captured in the dataset, and the best signal was selected based on the Skewness Signal Quality Index (SSQI) [7] for the study. The selected signal was passed through a 4th Order Chebyshev II filter in order to eliminate noise [8]. The final processed signal and its second derivative referred to as the Accelerated Photoplethysmography (APG) signal, was used for the extraction of identified features related to the cardiovascular system using the MATLAB software. The list of extracted features is presented in Table 1.

Table 1. Features selected for the analysis

APG signal features	PPG signal features	Physiological features
b/a Ratio	Systolic amplitude	Gender
c/a Ratio	Pulse area	Age
d/a Ratio	Inflexion Point Area (IPA) ratio	Height
e/a Ratio	Pulse Interval (PI)	Weight
(b-c-d-e)/a Ratio	PI/Systolic amplitude ratio	BMI
(b-e)/a Ratio	Augmentation Index (AI)	Body fat (%)
(b-c-d)/a Ratio	Adjusted AI	
(c + d – b)/a Ratio	Large artery stiffness index	
a-a Interval	Rise time	
(-d/a) Ratio		

2.3 Feature Selection

The above extracted signal features have been clearly identified and described in Elgendi et al. [1] and Allen et al. [2]. The identified features were analyzed in previous researchers mainly towards the understanding and interpretation of the cardiovascular system. The cardiovascular system is affected by a range of diseases such as Hypertension, Diabetes and Renal failure. Hence, it is important to uniquely identify the features related to diabetes that affect the vascular system. An ANOVA test with a 95% confidence interval was carried out focusing on the healthy subjects (51) and the subjects with only diabetes (9). Through the test it was identified that the Age, Augmentation Index (AI), Adjusted AI, e/a Ratio and the ratio between the Pulse Interval to the Systolic Amplitude are suitable features for the prediction of type 2 diabetes. It is important to note that other features analyzed related to vascular ageing and arterial stiffness were not identified as predictors towards diabetes, even though they represent the vascular system. A second ANOVA test with a 95% confidence interval was conducted focusing on the healthy (51), prehypertension (39) and hypertension (28) subjects. The results of the earlier test were justified as the e/a ratio, AI, Adjusted AI and the ratio of pulse interval to its systolic amplitude were not identified as prominent features towards the prediction of hypertension. Hence it can be clearly identified that the e/a ratio, AI, Adjusted AI and the ratio of pulse interval to its systolic amplitude are unique features towards the prediction of type 2 diabetes.

2.4 Machine Learning Models

Supervised Machine Learning techniques were used to carry out binary classification where the evaluation metric was set to the area under the ROC curve. The study evaluated the Naive Bayes classifier, Linear Discriminant Analysis (LDA), Decision Trees, Random Forest, AdaBoost Classifier, Logistic Regression and Support Vector Machine (SVM) towards the classification of Diabetes. The selected models were initially tuned to identify the optimum hyper parameters for each model using random search and stratified 10-fold cross validation. All the classification models were tuned ensuring the selection of suitable hyper parameters to avoid overfitting due to the relatively small number of data samples. Upon successful tuning of the algorithms a final 10-fold cross validation was run on the entire dataset in order to select the best performing model. A binary classification ensuring equal samples for each class was carried out towards the prediction of diabetes. The evaluation focused upon three experiments. Healthy versus diabetes only subjects, healthy versus diabetes subjects with prehypertension, healthy versus diabetes subjects with prehypertension and hypertension. A final experiment was carried out using only the selected PPG signal features (excluding physiological features) to evaluate the robustness of using PPG for diabetes prediction. The subjects below 30 years were excluded from the test ensuring an even distribution of healthy and diabetes subjects across age groups. The classification results for the aforementioned four experiments are presented in Table 2.

Table 2. Diabetes prediction results.

Experiment	Classification algorithm	ROC	F1
Healthy (9) vs Diabetes Only (9)	Decision Trees LDA	0.83 ± 0.14 0.82 ± 0.14	0.79 ± 0.21 0.76 ± 0.13
Healthy (25) vs Diabetes with Prehypertension (25)	SVM (Linear Kernel) LDA	0.71 ± 0.20 0.69 ± 0.21	0.55 ± 0.30 0.68 ± 0.16
Healthy (32) vs Diabetes with Prehypertension & Hypertension (32)	LDA SVM (Linear Kernel)	0.79 ± 0.15 0.74 ± 0.17	0.71 ± 0.15 0.69 ± 0.10
Using only PPG Signal Features. Healthy (31) vs Diabetes with Prehypertension & Hypertension (31)	Decision Trees LDA	0.70 ± 0.20 0.64 ± 0.15	0.56 ± 0.28 0.56 ± 0.15

3 Evaluation and Results

Decision Trees and LDA achieved accuracies of 83% and 82% respectively in classifying diabetes only subjects from healthy subjects. However, in order to enhance the confidence of this finding, the test should be carried out with additional samples from diabetes only subjects. In contrast diabetes prediction for subjects with prehypertension and hypertension provided more confident results due to the relatively larger number of subjects. SVM performed well predicting diabetes with the presence of prehypertension providing an accuracy of 71%, whereas LDA was able to obtain an accuracy of 79% for predicting diabetes with the presence of hypertension and prehypertension. It should be noted that this test case portrays the real-world scenario of a robust diabetes prediction system encompassing all possible diabetes and hypertension combinations. The evaluation of the robustness of using PPG for the diabetes prediction is of utmost importance to ensure that the predictions are not biased from physiological features such as the age, which is used in this study. The conducted control test achieved an accuracy of 70% for the Decision Trees classifier which establishes the suitability of using the PPG signal for diabetes prediction.

4 Conclusion and Future Work

In this paper we have focused on integrating both physiological and short recorded PPG signal characteristics, which can be easily extracted through the help of smart devices, in order to predict Type 2 Diabetes. Through the two ANOVA tests, we were able to identify the best features for the classification of diabetes. It was identified that the e/a ratio, AI, Adjusted AI and the ratio of pulse interval to its systolic amplitude are unique features towards the prediction of type 2 diabetes. The accuracy of predicting diabetes without the presence of hypertension or prehypertension was 83%. The confidence of the obtained results can be further improved through data collection and validation from additional subjects, which is the current focus of our research. According to our knowledge, this is the first research focusing on analyzing the easily obtainable short ($\sim$2.1s) photoplethysmography signals in order to predict diabetes in a practical setting with the presence of Hypertension & Prehypertension. The approach demonstrates good potential through the achievement of an accuracy of 79%. Hence, it is evident that there exists a potential in using PPG signals to develop intelligent systems for diabetes prediction through the utilization of smart devices.

Acknowledgement. This research was partially funded by the Senate Research Council grant number SRC/LT/2019/33 of University of Moratuwa, Sri Lanka.

References

1. Elgendi, M.: On the analysis of fingertip photoplethysmogram signals. Curr. Cardiol. Rev. **8** (1), 14–25 (2012)
2. Allen, J.: Photoplethysmography and its application in clinical physiological measurement. Physiol. Meas. **28**(3), R1 (2007)
3. Reddy, V.R., et al.: DMSense: a non-invasive diabetes mellitus classification system using photoplethysmogram signal. In: IEEE International Conference on Pervasive Computing and Communications Workshops (PerCom Workshops). IEEE (2017)
4. Ballinger, B., et al.: DeepHeart: semi-supervised sequence learning for cardiovascular risk prediction. In: Thirty-Second AAAI Conference on Artificial Intelligence (2018)
5. Moreno, E.M., et al.: Type 2 diabetes screening test by means of a pulse oximeter. IEEE Trans. Biomed. Eng. **64**(2), 341–351 (2017)
6. Liang, Y., et al.: A new, short-recorded photoplethysmogram dataset for blood pressure monitoring in China. Sci. Data **5**, 180020 (2018)
7. Elgendi, M.: Optimal signal quality index for photoplethysmogram signals. Bioengineering **3** (4), 21 (2016)
8. Liang, Y., et al.: An optimal filter for short photoplethysmogram signals. Sci. Data **5**, 180076 (2018)

Identification of Patient Prescribing Predicting Cancer Diagnosis Using Boosted Decision Trees

Josephine French[1,2(✉)], Cong Chen[1,2], Katherine Henson[2], Brian Shand[1,2], Patrick Ferris[1,2], Josh Pencheon[2], Sally Vernon[2], Meena Rafiq[4], David Howe[1,2], Georgios Lyratzopoulos[2,3], and Jem Rashbass[1,2]

[1] Health Data Insight CIC, Cambridge, UK
josephine.french@phe.gov.uk

[2] National Cancer Registration and Analysis Service, Public Health England, London, UK

[3] ECHO (Epidemiology of Cancer Healthcare and Outcomes) Group, Department of Behavioural Science and Health, University College London, London, UK

[4] University College London Institute of Health Informatics, London, UK

Abstract. Machine learning has potential to identify patterns in pre-diagnostic prescribing that act as an early signal of cancer diagnosis. Danish studies using classical regression models have shown that prescribing of particular drugs increases in the months prior to lung and colorectal cancer diagnosis. The aim of this case-control study is to assess the potential for machine learning to extend these findings to identify combinations of prescriptions that might act as pre-cancer signals. We use a boosted trees approach to analyse prescriptions data from NHS Business Services Authority linked to English cancer registry data to classify individuals into two classes: cancer patients and controls. We then identify the drugs that contributed the most to the classification decisions in the models. To the best of our knowledge, this is the first study utilising machine learning to find pre-diagnostic primary-care-prescription-related indicators of cancer diagnosis in England. We assess two feature selection approaches using text categorisation methods alone and in combination with clinical domain knowledge. Matched samples of controls (ten controls for each patient) to control for age are used throughout. We train models for matched cohorts of 6,770 lung cancer patients and 5,869 colorectal cancer patients starting the cancer pathway for the first time between January and March 2016. The models outperform classical methods by AUC, AUC-PR, and $F_{0.5}$ score, showing strong potential for

Supported by a Cancer Research UK Pioneer Award. Data for this study is based on patient-level information collected by the NHS, as part of the care and support of cancer patients. The data is collated, maintained and quality assured by the National Cancer Registration and Analysis Service, which is part of Public Health England (PHE). Dr. Meena Rafiq is funded by a National Institute for Health Research (NIHR) in-practice clinical fellowship (IPF-2017-11-011). This article presents independent research funded by the NIHR. The views expressed are those of the author(s) and not necessarily those of the NHS, the NIHR or the Department of Health.

D. Riaño et al. (Eds.): AIME 2019, LNAI 11526, pp. 328–333, 2019.
https://doi.org/10.1007/978-3-030-21642-9_42

using machine learning to extract signals from this dataset to aid earlier diagnosis. Our findings confirm the Danish studies.

Keywords: Cancer · Boosted trees · Feature selection · Clinical input

1 Introduction

Cancer is the second most common cause of death worldwide, with lung and colorectal cancer the leading causes of cancer deaths. Further, lung and colorectal cancers are the two most common forms of cancer that do not primarily affect one sex. Early diagnosis of cancer can increase survival, improve quality of life, and decrease healthcare costs [1]. Identification of cancer is a difficult problem, and the NHS in England considers it worth investigating 100 patients with a suspicion of cancer for 3 diagnoses of cancer. This means that algorithms with moderate machine learning performance can add high clinical value.

Patterns in pre-diagnostic prescribing could provide an early signal of cancer diagnosis. Danish studies [2–4] have observed increases in prescribing of particular drugs prior to cancer diagnosis, including COPD drugs, antibiotics, and opioids prior to lung cancer diagnosis, and haemorrhoid drugs, laxatives, oral iron, and opioids prior to colorectal or colon cancer diagnosis. These studies use classical approaches including generalised linear models. While these methods have been effective in showing prescribing increases of some drugs prior to diagnosis, we are not aware of studies using modern machine learning methods to try to find signals in prescriptions prior to cancer diagnosis. Machine learning has the potential to identify complex, non-linear relationships between high numbers of variables. Our study aims to assess whether there is potential utility for a machine learning approach to build on the published traditional approaches to identify combinations of prescriptions indicative of subsequent cancer diagnosis.

In this work, we apply a boosted trees classifier to prescription data prior to cancer pathway start date to classify individuals as cancer patients or matched controls and extract a list of drugs that contributed most to the decisions in the model. We compare our models to classical methods. As drugs prescribed for comorbidities add noise to the data, and to increase interpretability, we assess two feature selection approaches: a data-driven approach using text categorisation methods, and these drugs grouped by a domain expert.

2 The Approach

The Data: We use population-level anonymised data of all prescriptions dispensed in community pharmacies in England provided by NHS Business Services Authority linked to data held by Public Health England's National Cancer Registration and Analysis Service (NCRAS) where a patient has received a cancer diagnosis [5]. In England, drugs are categorised by levels of the British National Formulary (BNF) [6], a multi-level hierarchy. Drugs in the prescription data are identified at the lowest level, presentation code, which includes how the chemical is administered. In this analysis, drugs within the same BNF subparagraph are combined to form one drug. For example, a presentation code may be 'Amoxicillin' with the dosage information of a 250 mg capsule. This would be classed under the wider class of 'broad-spectrum penicillins'.

We evaluate our method on two cohorts consisting of all patients diagnosed with a first cancer of lung (N = 6,770) or colorectal cancer (N = 5,869) who started the patient pathway between January and March 2016 with at least one prescription in the previous nine months. This cohort was chosen as the earliest cohort with nine prior months of data available. We define cancer pathway start as the first referral or diagnosis recorded by the cancer registry for the specified tumour. As early exploratory analysis found age to be a strong confounder, we case-match each patient to ten controls to ensure we identify drugs linked to cancer rather than comorbidities in older patients, and to control for seasonal variation in prescribing. We randomly sample controls from the population who have no history of cancer matched on age and having received a prescription in the same nine-month period as the patient. Prescriptions for both patients and their matched controls are taken from the nine months prior to the patient pathway start date. Sex was not available in the data.

Data Preprocessing and Feature Selection: After linking tables, the data has the form *(patient, age, drug code, month of prescription, other variables)*, which we aggregate for each patient to take the form $(age, d_1, \ldots, d_n)$, where d_i is the number of prescriptions of drug i received by the patient in the nine-month period. There were 563 (lung) and 534 (colorectal) drugs before feature selection. The first feature selection method we use is an effective method used in text categorisation, the *Entropy based Category Coverage Difference (ECCD)* criterion [7]. In this, we view patients as "documents", and the drugs they receive as the "words" within them. This method was chosen because it ranks drugs using both the frequency of prescriptions and the proportions of individuals receiving the drugs for patients compared with controls. We select the highest and lowest 25 drugs for the 50-drug feature selection (respectively, highest and lowest 50 drugs for the 100-drug method), ranked by ECCD score for the cancer patient cohort. Further, as several drugs (and types of drugs) are used for the same clinical indication, the 100-drug ECCD-identified drugs are then grouped into 54 drug groups, each representing distinct clinical presentations. This classification was developed by two medically qualified co-authors (MR-general practitioner, GL-public health physician).

The Models: We train the models using XGBoost [8] in R, with evaluation metric AUC. Data is split into training (80%) and testing sets (20%). We tune parameters using 10-fold cross-validation applied to the training set with metric AUC, with the learning rate in $\{0.01, 0.1, 0.3\}$ and maximum tree depth in $\{3, 6, 10\}$, with early stopping after 10 rounds with no increase in test AUC (capped at 500 rounds) and all other parameters set to the default values. Models are evaluated using AUC and AUC-PR. The threshold probability for classifying cases as cancer patients was chosen to maximise the $F_{0.5}$ score, to balance identifying sufficiently many cases for the test to be useful with prioritising minimising misclassified healthy controls as a high PPV makes a test of clinical value. For comparison, we fit a LASSO logistic regression model with the same variables as the data for our machine learning model with no feature selection (the choice of LASSO regularisation is to handle the high dimensionality).

Table 1. Results of the models.

Classifier	Lung cancer					
	AUC	AUC-PR	Accuracy	$F_{0.5}$	PPV	Sensitivity
XGBoost: No feature selection	0.735	0.257	0.903	0.319	0.412	0.167
XGBoost: ECCD (50 drugs)	0.729	0.251	0.871	0.314	0.309	0.335
XGBoost: ECCD (100 drugs)	0.731	0.248	0.886	0.312	0.331	0.253
XGBoost: ECCD (clinical groups)	0.716	0.227	0.889	0.297	0.330	0.212
Logistic model	0.715	0.223	0.880	0.284	0.298	0.236
	Colorectal cancer					
XGBoost: No feature selection	0.637	0.177	0.897	0.261	0.334	0.138
XGBoost: ECCD (50 drugs)	0.639	0.177	0.896	0.267	0.335	0.145
XGBoost: ECCD (100 drugs)	0.643	0.182	0.899	0.272	0.354	0.139
XGBoost: ECCD (clinical groups)	0.631	0.181	0.894	0.255	0.314	0.144
Logistic model	0.595	0.131	0.884	0.181	0.218	0.106

3 Implementation

The results of the models on the testing sets are shown in Table 1. We compare the models by the AUC, AUC-PR, and maximum $F_{0.5}$ scores, and show the accuracy, PPV, and sensitivity for the prediction threshold maximising $F_{0.5}$. For lung cancer, the machine learning models outperform classical logistic regression by all three metrics, and the data-driven approaches outperform the clinical approach. The model with no feature selection outperforms any feature selection method for these metrics. In the case of colorectal cancer, again the machine learning models outperform the logistic regression by all three metrics. Feature selection improves the model, and the ECCD algorithm (100 drugs) performs the best by the three metrics. Clinical groupings still reduce performance compared to the other machine learning models by AUC and max $F_{0.5}$. For both sites, all models outperform chance (AUC $= 0.5$, AUC-PR $= 0.091$, max $F_{0.5} = 0.111$).

For each site, the most informative five drugs coincide for all XGBoost models, bar the order of analgesics for lung cancer. Those for lung cancer, with clinical interpretations by MR, are given in Table 2. Those for colorectal cancer are: oral iron, antispasmodics and other drugs altering gut motility, haemorrhoidal preparations with corticosteroids, osmotic laxatives, and non-opioid analgesics. The drugs obtained are consistent with [2–4] and clinical understanding.

Table 2. The drugs of highest importance to the lung cancer model.

Drug	Clinical interpretation
Antimuscarinic Bronchodilators	Treatment of COPD and asthma. Overlapping symptoms with lung cancer
Selective Beta(2)-Agonists	Treatment of COPD and asthma. Overlapping symptoms with lung cancer
Broad-Spectrum Penicillins	Treatment of respiratory tract infections. Symptoms overlap with lung cancer
Tetracyclines	Treatment of respiratory tract infections. Symptoms overlap with lung cancer
Opioid/Non-Opioid Analgesics	Treatment of pain. Could be from cancer/metastases

The efficacy of feature selection for colorectal cancer data, but not lung cancer, could be due to fewer data points and higher sparsity. The ECCD algorithm was run on the full dataset for the clinical grouping, which will introduce positive bias due to the feature selection seeing the test data. In further work, optimising the models using AUC-PR may better account for the imbalanced classes.

4 Conclusion

Earlier diagnosis of lung and colorectal cancers could be aided by signals arising from prescription history. We test the potential for machine learning to identify such signals. We assess the impact of two feature selection approaches based on an effective text categorisation method and clinical domain knowledge. Classical methods are able to classify patients more effectively than chance alone, showing that prescribing patterns may provide pre-cancer signals. Machine learning methods outperform classical approaches and identify clinically meaningful drugs, demonstrating potential for machine learning to find pre-diagnostic prescribing patterns. The data-driven approaches outperform feature selection with domain knowledge. Further machine learning work could identify more complex prescribing patterns, explore feature choices in more depth, and potentially improve model performance.

References

1. World Health Organisation Cancer Fact Sheet. https://www.who.int/news-room/fact-sheets/detail/cancer. Accessed 3 Jan 2019
2. Pottegård, A., Hallas, J.: New use of prescription drugs prior to a cancer diagnosis. Pharmacoepidemiol. Drug Saf. **26**, 223–227 (2017)
3. Guldbrandt, L.M., Møller, H., Jakobsen, E., Vedsted, P.: General practice consultations, diagnostic investigations, and prescriptions in the year preceding a lung cancer diagnosis. Cancer Med. **6**, 79–88 (2017)

4. Hansen, P.L., Hjertholm, P., Vedsted, P.: Increased diagnostic activity in general practice during the year preceding colorectal cancer diagnosis. Int. J. Cancer **137**, 615–624 (2015)
5. Henson, K., et al.: Cohort profile: prescriptions dispensed in the community linked to the national cancer registry in England. BMJ Open **8**, e020980 (2018)
6. Joint Formulary Committee: British National Formulary Vol 68 and 72. BMJ Group and Pharmaceutical Press, London (2014, 2016)
7. Largeron, C., Moulin, C., Géry, M.: Entropy based feature selection for text categorization. In: ACM Symposium on Applied Computing TaiChung, pp. 924–928 (2011)
8. Chen, T., Guestrin, C.: XGBoost: a scalable tree boosting system. In: Proceedings of the 22nd ACM SIGKDD Conference on Knowledge Discovery and Data Mining, San Francisco, pp. 785–794 (2016)

Image Processing

Automatic Image-Derived Estimation of the Arterial Whole-Blood Input Function from Dynamic Cerebral PET with ^{18}F-Choline

Carlos González[1], Pedro Bibiloni[1,2](✉), Manuel González-Hidalgo[1,2], Arnau Mir[1,2], and Sebastià Rubí[2,3]

[1] SCOPIA Research Group, Universitat de les Illes Balears, 07122 Palma, Spain
p.bibiloni@uib.es

[2] Balearic Islands Health Research Institute (IdISBa), 07010 Palma, Spain

[3] Department of Nuclear Medicine, University Hospital Son Espases, 07010 Palma, Spain

Abstract. The arterial concentration of the radiopharmaceutical ^{18}F-choline is needed to estimate its absorption by tumors and other tissues. The blood concentration of ^{18}F-choline changes as it interacts with tissues, and so it is represented as a function with respect to time, the so-called Input Function (IF). In this paper, we present the estimation of an arterial whole-blood Image-Derived Input Function (IDIF) from the PET image, needed to model its absorption. The sagittal and transverse brain venous sinuses are automatically segmented based on the top-hat morphological transform. Such segmentation provides an estimation of the venous whole-blood IDIF. It is then corrected to obtain the arterial whole-blood IDIF by relating the amount of radioactivity material entering the brain region with the amount leaving it and the amount remaining. We compare the automatic venous whole-blood IDIF with a whole-blood venous IDIF from a region manually segmented. Also, we compare the automatic arterial whole-blood IDIF with the arterial IF obtained with serial blood samples on the radial artery. Quantitative measures indicate the overall accuracy of the estimation.

Keywords: PET · ^{18}F-choline · Image processing · Vessel segmentation · Input function · Image-derived input function · Top-hat morphological transform

1 Introduction

Almost 50% of cerebral primary tumors diagnosed are gliomas, from which the majority of them are of high grade (grades III and IV according to the WHO

This work was supported by the project 2015 PI15-01653 of the Carlos III Health Institute and by the project TIN 2016-75404-P AEI/FEDER.

D. Riaño et al. (Eds.): AIME 2019, LNAI 11526, pp. 337–346, 2019.
https://doi.org/10.1007/978-3-030-21642-9_43

classification) [4]. High-grade gliomas are malignant tumors, which frequently grow rapidly. Their diagnosis is based on Magnetic Resonance Imaging (MRI) with gadolinium-based contrast medium. The clinical treatment of high-grade gliomas is their surgical excision if possible, usually completed with radiotherapy and adjuvant chemotherapy. Different prognostic indicators that may help to categorize tumors and guide therapeutic decisions have been previously described [7].

The work presented in this paper is framed in a research project aimed at identifying noninvasive biomarkers based on neuroimaging for high-grade gliomas. More specifically, we study the pharmacokinetics of ^{18}F-choline with positron-emission tomography (PET) imaging to characterize its transport and metabolization in the tumoral tissue [5]. The radiotracer ^{18}F-choline contains a short-lived isotope of fluor, ^{18}F, incorporated into choline, which is an essential nutrient precursor for the synthesis of phospholipids, used to build cell membrane. The PET imaging technology is able to capture a 3D map of the distribution of ^{18}F-choline, which indicates how much of it has been absorbed by the target tissue.

The concentration of radiotracer in blood during the duration of the study is also an essential measure. In order to quantify the absorption of ^{18}F-choline in tissues, we must measure the quantity absorbed by the tissue, shown in the PET image, and the quantity available to be absorbed by the target tissue, which is the concentration of ^{18}F-choline in the blood irrigating the tissue. In addition, both of these concentrations depend on time. Instead of a single measure, we must obtain them in the form of Time-Activity Curves (TAC), which is a function with respect to time. A vascular TAC is also called Input Function (IF) due to its role in kinetic modeling. Currently, we measure the blood concentration with serial blood sampling on the radial artery. In particular, with 27 blood samples non-uniformly distributed during the duration of the study [5]. However, this is an invasive procedure for patients that will also undergo surgery.

This paper aims at avoiding the invasive serial blood sampling by facing two objectives: (i) to introduce a method to estimate the arterial whole-blood IF (AIF) from the image, and (ii) to validate the method. To attain (i), we first estimate the venous whole-blood IF (VIF) with image-processing techniques, obtaining the so-called Venous Image-Derived IF (VIDIF). Afterwards, the Arterial Image-Derived IF (AIDIF) is computed based on the conservation of radiotracer in the brain region. Objective (ii) is met by comparing the data obtained from the semi-automatic methods with measured data. First, by comparing the automatic VIDIF with a reference VIDIF computed from a manually segmented region. And second, by comparing the semi-automatic AIDIF with the arterial blood sampling measures.

Automatic vessel segmentation has been approached with a multitude of image processing techniques [1]. The manual segmentation of vessels and blood structures has also been used, *e.g.* the venous sinuses by Wahl *et al.* [9], the left ventricle or other large arterial blood pools by Verwer *et al.* [8], etc. Schiepers *et al.* [6] used factor analysis, a model that splits each voxel's contribution as vessel, tumor and residual. More works were reviewed by Zanotti *et al.* [10], who focus on image-derived input functions for brain PET studies with different radiotracers. In contrast to our proposed methodology, almost all the works they review are based on manually segmenting the carotid artery, a population-based input function or simultaneous estimation of input functions and the kinetic model parameters.

2 Materials and Methods

As previously discussed, we express each Time-Activity Curves (TAC) as a mathematical function, the so-called Input Function, $\text{IF}(t)$. It provides the concentration at the time specified by its only parameter, the time t. The IF can be estimated from the PET image, which in this case is known as Image-Derived Input Function (IDIF). We also consider the TAC of the cranial cavity, also represented by a mathematical function, $\text{INTRACRANIAL}(t)$, which is measured in kBq. The kinetic modeling of tumors requires differentiating between the whole-blood concentration of radiotracer, the plasma concentration and the metabolite fraction. However, on the following we focus on whole-blood concentrations only: they are the only concentrations we can measure in the PET image.

The IF used for the pharmacokinetic analysis is a mathematical model adjusted to the experimental or imaging data. With the radiotracer used, ^{18}F-choline, the tri-exponential model has proved to accurately represent the arterial IF (AIF) [5]. Such model represents the first seconds—corresponding to the radiotracer injection—with linear interpolation starting at $\text{IF}(0) = 0$, and the tail of the IF as the sum of three different exponential functions, which has 6 parameters:

$$\text{IF}(t) = \sum_{i=1}^{3} A_i \cdot e^{-t \cdot \frac{\log(2)}{T_i}}, \text{ for } t > t_0,$$

where $A_i \in \mathbb{R}$ is the amplitude, $T_i \in \mathbb{R}$ the semi-life of the i-th exponential, and t_0 the end of the linear interpolation interval.

All the methods presented in the following have been executed with PMOD and Matlab. The manual segmentations were also carried out with the PMOD software.

2.1 Data Acquisition

The data used to compute the arterial whole-blood IDIF (AIDIF) is the following. The study comprises 8 patients. Each PET image is acquired with a GE-Discovery-PET/CT-600, in dynamic mode, to capture 36 temporal frames with different duration (we emphasize that the majority of absorption takes place at the beginning of the study, as can be observed in Fig. 2). Each frame is a $256 \times 256 \times 47$-voxel image, accounting for 4.49 mm^3 per voxel. Also, a coregistered low-dose PET-TAC and a high-resolution MR image are captured.

To validate the method, we also have blood samples and the manual segmentation of part of the venous sine. 27 blood samples are obtained from the radial artery, also at different times [5], which provides a good estimator of the AIF. The superior sagittal and transverse venous sines are manually segmented as follows. We averaged the first 14 temporal frames, corresponding to the first 90 s of the study. Manually, we selected a voxel on the frontier of the superior sagittal venous sine. This was used as a seed to a region-growing algorithm that iteratively adds connected voxels if their activity is *higher* to the activity of the original seed voxel. Also manually, we checked that the mask delineating this Volume of Interest (VOI) was overall correct and did not intersect with any tissue. This venous sine VOI is used to compute a venous whole-blood IDIF (VIDIF)—which we emphasize that is just an estimator, mainly due to lack of resolution in both time and space.

2.2 Venous Image-Derived Input Function

In this section, we introduce an automatic algorithm to automatically extract a region corresponding to veins to then compute a VIDIF. This algorithm and the one to estimate the AIDIF are illustrated in Fig. 1. We aim at capturing the veins that appear to be more visible in the PET image, again the superior sagittal and transverse sinuses. To do so, we select voxels whose concentration with respect to time resembles that of a generic VIF, iteratively favouring the groups of voxels disposed in thin structures and adjusting the reference VIF to the PET image.

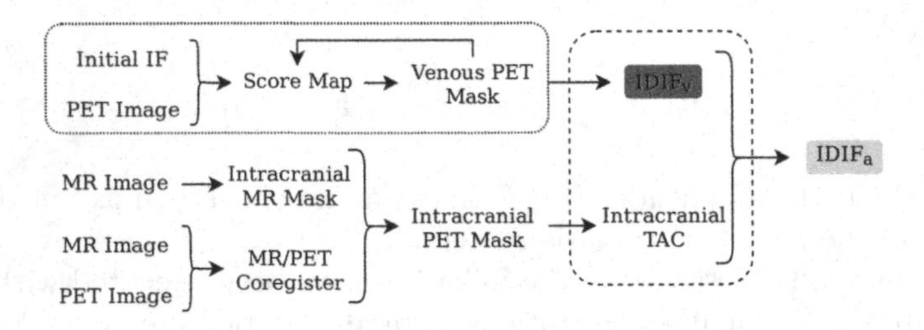

Fig. 1. Flow diagram of our algorithms, highlighting the IDIF_v estimation from Sect. 2.2 (dotted frame) and the IDIF_a estimation from Sect. 2.3 (dashed frame).

First, we initialize our reference VIF as a pre-defined tri-exponential model, based on the VIF from the manual segmentation of the superior sagittal and the transverse sinuses, as previously described. In particular, we use a model whose exponentials have an amplitude of 4, 1 and 0.3 kBq/ml, and a respective semi-life of 5, 5 and 20 s. We emphasize that this is valid since the amount of radiotracer was similar for all the patients, and was always injected at a constant speed, for 60 s beginning at the study start. Let $\text{IF}_1(t)$ be this reference model, as a function with respect to time.

Second, we score each voxel based on its similarity to the reference VIF obtained. To do so, we consider the correlation between the TAC of each voxel and the current reference, IF_1, both as one-dimensional signals.

$$\text{SCORE}_1(\boldsymbol{x}) = \sum_{\tau=1}^{T} \text{PET}(\boldsymbol{x}, \tau) \cdot \text{IF}_1(\tau),$$

where $\text{PET}(\boldsymbol{x}, t)$ is a function that gives the concentration in kBq/cc (with $\boldsymbol{x}$ representing the 3D coordinates of a voxel, and t the temporal slice).

Third, the score, interpreted as a parametric 3D image, is filtered with the grayscale top-hat morphological transform to enhance curvilinear structures [1]. As structuring element, we use a $3D$ sphere with radius equal to 5 voxels.

Fourth, we select the best voxels from which we will estimate the new reference VIF, to iteratively improve it. We leave out the anatomically lower part of the filtered score 3D image, which does not contain a structure of interest but may contribute with voxels belonging to the carotid artery. We emphasize that we are interested in the VIF. Then, we select the voxels whose score is at least 50% of the maximum score minus the minimum one. Although the method is robust to this threshold, we observe that a 50% already selects the candidates adapting very well to the VIF shape. These voxels, interpreted as a 3D image, are a mask of the structure of interest, M_1. The new reference VIF will be the average of the voxels in the mask, $\text{IF}_2(t)$, from which we will estimate a new tri-exponential model and iteratively improve from the beginning.

The iterative method stops, at the iteration ξ, when the number of voxels that were added or excluded to the mask is less than 1% of the total amount of voxels in the mask itself. The final mask, $M = M_\xi$, is consistently a region, not necessarily connected, within the transverse and superior sagittal sinuses. An example of such mask is found in Fig. 2, which shows how the biggest blood pools tend to be included in the mask. The computed venous IDIF is then $\text{IDIF}_\text{v} = \text{IF}_{\xi+1}$.

2.3 Arterial Image-Derived Input Function

This section introduces a method to compute the AIDIF. It is based on the following observation: the amount of radiotracer that enters the intracranial region minus the amount of radiotracer that leaves it must be equal to the amount captured. We emphasize that, given the size of the carotid artery and the limited resolution, an automatic segmentation-based strategy as the one in Sect. 2.2 is not well suited. Since we can measure both the amount leaving, the VIF, and the amount absorbed within the intracranial region, reflected in the PET image, we are able to infer the amount entering it, the arterial IDIF, which we will denote as $\text{IDIF}_\text{a}(t)$. Let us note the following relation holding for all $t \in \mathbb{R}$:

$$\int_{-\infty}^{t} \text{FLUX}_a(\tau) \cdot \text{IF}_\text{a}(\tau)\, \mathrm{d}\tau = \text{INTRACRANIAL}(t) + \int_{-\infty}^{t} \text{FLUX}_v(\tau) \cdot \text{IF}_\text{v}(\tau)\, \mathrm{d}\tau,$$

where $\text{INTRACRANIAL}(t)$ is the total activity in the intracranial region in kBq; and $\text{FLUX}_a(t)$, $\text{FLUX}_v(t)$ are, respectively, the arterial and venous blood flow in ml/s.

Before describing the method, let's study the assumptions between the simplified relation of $\text{IF}_\text{a}(t)$, $\text{IF}_\text{v}(t)$, and $\text{INTRACRANIAL}(t)$ that will allow us to compute the former. Although the arterial blood flow is pulsatile, given the time scale of our study we will assume that both arterial and venous blood flow are constant. Let us also assume that, in the intracranial region, the total arterial blood flow and the total venous blood flow are equal: $\text{FLUX}_a(t) = \text{FLUX}_v(t) = F$ for all t. Also, let us assume that, at each instant t, $\text{IF}_\text{a}(t)$ is the radiotracer concentration averaged across all the arteries irrigating the intracranial region; and that $\text{IF}_\text{v}(t)$ is the radiotracer concentration averaged among all blood drained through veins. In this case, we obtain the following simplified relation:

$$\text{IF}_\text{a}(t) \approx \text{IF}_\text{v}(t) + \frac{1}{F} \cdot \frac{\text{d}}{\text{d}t}\text{INTRACRANIAL}(t), \text{ for all } t \in \mathbb{R},$$

which gives an estimate of the AIF based on the image, $\text{IDIF}_\text{a}(t)$:

$$\text{IDIF}_\text{a}(t) = \text{IDIF}_\text{v}(t) + \frac{1}{F} \cdot \frac{\text{d}}{\text{d}t}\text{INTRACRANIAL}(t), \text{ for all } t \in \mathbb{R}.$$

First, we obtain $\text{INTRACRANIAL}(t)$ by segmenting such region based on a high-quality MR image. The Hammers N30R83 brain atlas [2], with 1 mm sampling, is adjusted to each MR image. All the atlas' VOIs are merged into a single brain VOI. Afterwards, we compute the transform that coregisters the PET-TAC image to the MR image of the same patient. The transform is obtained with a multi-modality iterative rigid matching algorithm that optimizes the normalized mutual information, without previously smoothing either of the images. Since the PET-TAC and the PET are already coregistered, the inverse transform of the abovementioned transforms the brain VOI into the PET image space. This VOI is finally postprocessed to fill the holes within different brain regions, employing a 3D morphological closing with a spherical structuring element of radii equal to 6 voxels. This provides an intracranial VOI, whose averaged TAC is our estimate of the TAC of the intracranial region, $\text{INTRACRANIAL}(t)$, measured in kBq. This first step was carried out with the PMOD software.

Let us assume that we have somehow estimated F, so we are able to estimate $\text{IDIF}_\text{a}(t)$ at some instances t, then followed by adjusting it to a tri-exponential model. Let us note that we only compute $\text{IDIF}_\text{a}(t)$ for each t being the mid-time of a PET temporal frame, since these are the instants at which we can compute both $\text{IDIF}_\text{v}(t)$ and $\frac{\text{d}}{\text{d}t}\text{INTRACRANIAL}(t)$. With them, we can obtain the tri-exponential model that best fits these data points. On the following, we will consider that $\text{IDIF}_\text{a}(t)$ is such tri-exponential model.

Finally, we actually compute $\text{IDIF}_\text{a}(t)$ using a mean blood flow F, that is, estimated from all the patients in the study. Let us recall that we have a good estimator of $\text{IF}_\text{a}(t)$, provided by the blood serial sampling. Thus, we use the same relation to infer the value of F that best fits all the patients' TACs simultaneously. In particular, we choose the value of F that minimizes the sum of mean squared distances between the arterial image-derived input function, $\text{IDIF}_\text{a}(t)$, and the sampled arterial input function $\text{IF}_\text{a}(t - \Delta)$, where the delay Δ accounts for the time difference when measuring on the brain arteries and the radial artery

$$F = \arg\min \left\{ \sum_{\text{patient } p} \text{MSE}(p) \right\}, \text{ where}$$

$$\text{MSE}(p) = \min_{\Delta} \left\{ \frac{1}{T} \left(\int_0^T (\text{IDIF}_\text{a}(\tau) - \text{IF}_\text{a}(t - \Delta))^2 \, d\tau \right)^{\frac{1}{2}} \right\}.$$

We emphasize that we estimate a single flow F for all patients, but a different delay Δ for each of them. We also emphasize that the estimated F is *not* a good estimator of the blood flow, but a value that has been experimentally proved to be consistent among patients (see discussion in Sect. 4).

3 Validation of the Method

In this section, we introduce quantitative metrics and visual plots to validate the methods introduced in Sects. 2.2 and 2.3.

To compare the similarity of two estimations of the same TAC we quantitatively compare the differences between the two curves. We are interested in these TAC to later infer the parameters of a compartmental kinetic model that describes the absorption of radiotracer by the tissue. Thus, if the parameters obtained when using both TACs were similar, they would be *close enough*. However, as a stronger requirement, we require that both TACs represent a very similar function. To do so, we focus on an interpretable indicator, the mean absolute error between a reference curve $f(t)$, and an estimated curve, $g(t)$,

$$\text{MAE}(f, g) = \frac{1}{T} \int_0^T |f(\tau) - g(\tau)| \, d\tau.$$

We provide the absolute mean error of IF, which is measured in kBq/ml, and in relative terms (that is, the ratio between $\text{MAE}(f, g)$ and $\text{MAE}(f, 0)$, where 0 is a function that always returns zero).

The first comparison is between the automatic VIDIF, presented in Sect. 2.2; and a second VIDIF, obtained from a manually segmented VOI (see Sect. 2.1 for details). To do so, we compare the tri-exponential models adjusted to the data points extracted from the image in each case. The results are shown in Table 1, were we emphasize that the mean absolute error was 0.73 kBq/ml.

Second, we compare the automatic AIDIF (see Sect. 2.3) with the sampled AIF (see Sect. 2.1). Since the automatic AIDIF method makes use of a poblational model to infer the value of F, we perform this comparison in a leave-one-out fashion: to obtain the automatic AIDIF of a patient, the value of F is inferred from all the other patients, excluding the current one. This procedure is designed to avoid the reference data—the sampled AIF in our case—to be used

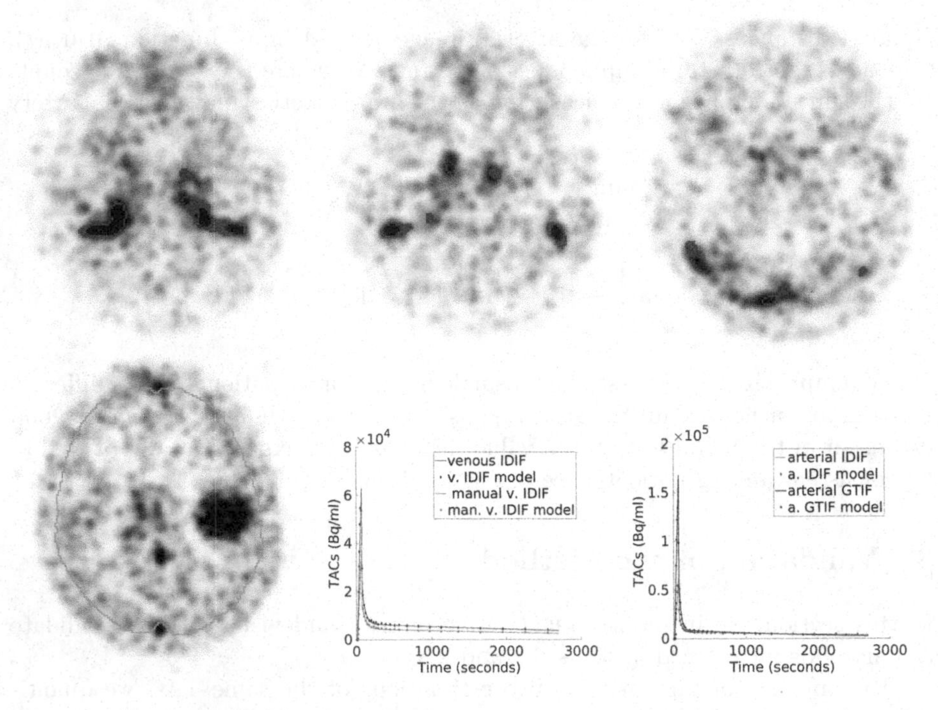

Fig. 2. Different axial slices showing the automatic venous mask overimposed on the PET frame at $t = 60\,\mathrm{s}$ (top), an anatomically higher slice showing the tumor and the intracranial mask (bottom left) and visual results of the automatic VIDIF and the automatic AIDIF. All visualizations correspond to patient №4, which yields the worst quantitative results.

to infer the output of the method, ensuring that the comparison is not positively biased.

This second comparison is also carried out using the tri-exponential models adjusted to the data points. However, these TACs correspond to slightly different measures, one based on the cerebral arterial flow and the other on samples taken from the radial artery. This produces a temporal delay between them. We avoid this effect by finding the delay that makes the two curves more similar in terms of their mean squared error, computed as in Sect. 2.3, since this is a trade-off between the mean absolute error and the maximum error. In other words, Table 1 shows the comparison between $\text{IDIF}_\text{v}(t)$ and $\text{IF}_\text{v}(t-\Delta)$, with Δ as the one that minimizes the the mean squared error, MSE, between $\text{IDIF}_\text{v}(t)$ and $\text{IF}_\text{v}(t-\Delta)$.

4 Analysis and Conclusion

In this section, we analyze the results and we introduce the strengths and limitations of the automatic arterial IDIF method and its validation.

The automatically estimated TACs tend to be similar to the original ones, as seen in Fig. 2, as well as in Table 1. The automatic VIDIF tends to be very similar

Table 1. Quantitative results of the automatic VIDIF and the automatic AIDIF methods. The latter, validated in a leave-one-out fashion, also shows the estimated cerebral flow F.

	Mean absolute error Reference: manual VIDIF Input: automatic VIDIF		Reference: sampled AIF Input: automatic AIDIF		
Patient	Absolute [kBq/ml]	Relative	Absolute [kBq/ml]	Relative	Estimated F [ml/min]
№1	1.23	25.4%	1.82	37.5%	402.3
№2	0.11	1.9%	1.89	31.4%	448.9
№3	0.33	5.6%	1.29	21.2%	401.5
№4	1.98	33.4%	2.16	36.0%	401.6
№5	0.58	17.1%	1.08	31.0%	401.5
№6	0.54	10.5%	1.69	32.1%	407.2
№7	0.14	2.8%	0.66	13.1%	448.9
№8	0.90	19.6%	0.84	17.9%	401.5
Mean	0.73	14.5%	1.43	27.5%	414.2
Std	0.63	11.3%	0.54	9.0%	21.5

to the manually-segmented VOI-based VIDIF. Its mean error ranges, in relative terms, between 1.9% and 33.4%, having an average of 14.5% of dissimilarity. The automatic AIDIF has a larger error, from 13.1% to 36%, with an average of 27.5%. We observe that all these errors tend to be higher than in the VIDIF case. This could be caused by a bias of our method as well as by the harder nature of the task.

Although the quantitative results seem acceptable, they have to be correctly interpreted. First, the automatic and manual VIDIF methods are very similar because our gold reference also suffers some of the same biases. All the noise that the image contains affects both methods, such as the spill-out effect. Also, the error is averaged during the whole study duration, but the measure is more accurate after the first 10–15 min. Therefore, the longer the study is, the lowest error measures we would obtain.

The methods presented are robust. We emphasize that they depend on different pre-defined parameters, such as the population-based loose estimator of the cerebral flow or the size of the structuring element to fill holes within the intracranial region. However, the results are not very sensitive to them, which encourages its use with other patients and slightly different settings. Provided the IDIF introduces an error that can be assumed, most blood samples could be avoided, avoiding the invasive arterial blood samples.

Several limitations of the method must be mentioned. First, some of its underlying assumptions are not strictly verified. In particular, we assumed that the average activity in the sagittal and transverse sinuses is equal to the real average of blood leaving the intracranial region. However, we are not able to measure the jugular veins and, furthermore, some intracranial regions may not drain through them. Also, we have assumed that the cerebral flow, F, relates the total activity in the intracranial region and the concentration entering and leaving it. While this assumption seems correct, the estimated cerebral flow tends to be ~400 ml/min, while young adults present a cerebral flow of ~750 ml/min [3]. The cause of this difference may be the lack of accuracy and the biases in the data used to estimate F, as well as the demographic differences between our patients and the average young adult. Rather than aiming at accurately measuring a physiological process, we experimentally found that F tends to be a parameter that is consistently estimated to be within the same range (see Table 1), and this fact can be leveraged to infer the AIDIF.

The validation of the method is also somewhat limited. First, the comparison is not done with other methods to infer the IDIF. We recall that IDIF methods seem to be very tracer specific [10]. Also, most of them are manual extractions of the IDIF, and others such as factor analysis do not always distinguish between arterial and venous IDIF. Also, the comparison of the TACs is done based on the similarity with a reference TAC, rather than studying its effect on the kinetic models used to study the radiotracer absorption. Finally, we recall that this study is limited to only eight patients.

References

1. Bibiloni, P., González-Hidalgo, M., Massanet, S.: A survey on curvilinear object segmentation in multiple applications. Pattern Recognit. **60**, 949–970 (2016)
2. Hammers, A., et al.: Three-dimensional maximum probability atlas of the human brain, with particular reference to the temporal lobe. Hum. Brain Mapp. **19**(4), 224–247 (2003)
3. Lassen, N.A.: Normal average value of cerebral blood flow in younger adults is 50 ml/100 g/min (1985)
4. Louis, D.N., et al.: The 2007 who classification of tumours of the central nervous system. Acta Neuropathol. **114**(2), 97–109 (2007)
5. Rubí, S., Bibiloni, P., Galmés, M., et al.: PET kinetic modeling with arterial sampling of 18F-choline uptake in patients with a suspected initial diagnosis of high grade glioma. In: 30th Annual Congress of the European Association of Nuclear Medicine (EANM 2017) (2017)
6. Schiepers, C., Chen, W., Dahlbom, M., Cloughesy, T., Hoh, C.K., Huang, S.C.: 18F-fluorothymidine kinetics of malignant brain tumors. Eur. J. Nucl. Med. Imaging **34**(7), 1003–1011 (2007)
7. Siegal, T.: Clinical impact of molecular biomarkers in gliomas. J. Clini. Neurosci. **22**(3), 437–444 (2015)
8. Verwer, E.E., et al.: Quantification of 18F-fluorocholine kinetics in patients with prostate cancer. J. Nucl. Med. **56**(3), 365–371 (2015)
9. Wahl, L.M., Asselin, M.C., Nahmias, C.: Regions of interest in the venous sinuses as input functions for quantitative PET. J. Nucl. Med. **40**(10), 1666–1675 (1999)
10. Zanotti-Fregonara, P., Chen, K., Liow, J.S., Fujita, M., Innis, R.B.: Image-derived input function for brain pet studies: many challenges and few opportunities. J. Cereb. Blood Flow Metab. **31**(10), 1986–1998 (2011)

A Semi-supervised Approach to Segment Retinal Blood Vessels in Color Fundus Photographs

Md. Abu Sayed[1], Sajib Saha[2](✉), G. M. Atiqur Rahaman[1], Tanmai K. Ghosh[1], and Yogesan Kanagasingam[2]

[1] Computational Color and Spectral Image Analysis Lab, Computer Science and Engineering Discipline, Khulna University, Khulna, Bangladesh
sayed4931@gmail.com, gmatiqur@gmail.com, ghosh.tanmoi@gmail.com

[2] Australian e-Health Research Centre, CSIRO, Floreat, Australia
{Sajib.Saha,Yogi.Kanagasingam}@csiro.au

Abstract. Segmentation of retinal blood vessels is an important diagnostic procedure in ophthalmology. In this paper we propose an automated blood vessels segmentation method that combines both supervised and un-supervised approaches. A novel descriptor named Local Haar Pattern (LHP) is proposed to describe retinal pixel of interest. The performance of the proposed method has been evaluated on three publicly available DRIVE, STARE and CHASE_DB1 datasets. The proposed method achieves an overall segmentation accuracy of 96%, 96% and 95% respectively on DRIVE, STARE, and CHASE DB1 datasets, which are better than the state-of-the-art methods.

Keywords: Color fundus photographs · Vessel segmentation · Haar feature · Multiscale line detector · Random forest

1 Introduction

Segmentation of retinal blood vessel is an important step in several retinal image analysis tasks including automated pathology detection and registration of retinal images [1]. Manual segmentation of retinal blood vessels is a long and tedious task. That is why, over the last two decades a large number of methods have been proposed to automatically segment retinal blood vessels. However, still there are challenges to address.

Ricci et al. [2] proposed a simple yet efficient segmentation method based on basic line operators and support vector machine. Despite addressing many of the important challenges in vessel segmentation [3], the method fails in the presence of central vessel reflex, at bifurcation and crossover regions. To overcome these problems, Nguyen et al. [3] proposed a method based on multi-scale line detector. While the method is one of the best in its category, the method fails to segment blood vessels accurately in the presence of pathology. On that perspective, in this work we aim to augment the method proposed by Nguyen et al. [3], so that blood vessels can be segmented accurately even with the presence of pathology.

D. Riaño et al. (Eds.): AIME 2019, LNAI 11526, pp. 347–351, 2019.
https://doi.org/10.1007/978-3-030-21642-9_44

2 Proposed Method

A preliminary segmentation of the blood vessels is performed relying on multi-scale line detector approach [3]. Each of the pixels that are determined as vessels (that also contain misclassified pathology pixels) are then defined using Local Haar Pattern (LHP) descriptor (proposed here; described below). A random forest (RF) classifier [4] then determines a pixel as true vessel or not depending on its LHP description. While training the RF classifier, manual labeling (done by expert grader) of the actual vessel and pathology pixels were made available. A diagram of the proposed system is shown in Fig. 1.

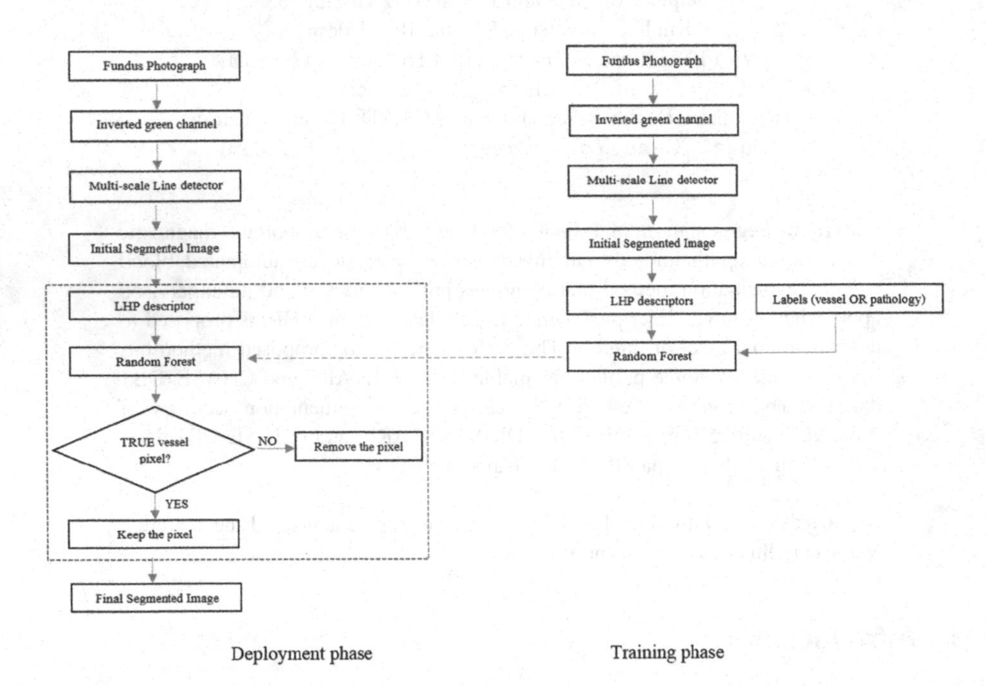

Fig. 1. Diagram of the proposed system. Operations shown within the dotted box are performed pixel-wise.

2.1 Local Haar Pattern (LHP) to Describe Retinal Pixel of Interest

A novel descriptor named Local Haar Pattern (LHP) is proposed here. LHP is inspired by the earlier works of Saha et al. in [5]. Rather than comparing the intensity of two groups of pixels to generate one bit of the descriptor as in [5], in this work, we compute and store the actual intensity difference, which is to some extent similar to Speeded Up Robust Feature (SURF) [6]. In order to perform pixels grouping, we define a set of 16-pixel patterns depicted in Fig. 2, which are reminiscent of Haar basis function [7].

In order to compute the LHP descriptor a patch p of size 32 × 32 is consider around the pixel of interest, and vector of size 128 bytes is calculated that represents the patch.

Each byte of the vector is computed based on the intensity comparisons of two-pixel groups as defined below:

$$T(p, X, Y) = \overline{I_X} - \overline{I_Y}. \quad (1)$$

Here, $\overline{I_X}$ and $\overline{I_Y}$ represent the mean intensities of two different pixel groups X and Y belonging to the patch p. 128 bytes vector is generated in three steps. At the first step, all the 16 patterns are considered to perform intensity comparisons (Eq. 1) on the whole patch, that results 16 bytes vector. In the second step, the patch is divided into 4 sub-patches of size 16 × 16. All the 16 patterns are considered and intensity comparisons are performed on each of these sub-patches, which results 64 (=4 × 16) bytes vector. In third stage, each of the sub-patches is further divided into 4 sub-patches of size 8 × 8 and the first three of 16 patterns are considered to perform intensity comparisons, which results 48 (=16 × 3)-bytes vector.

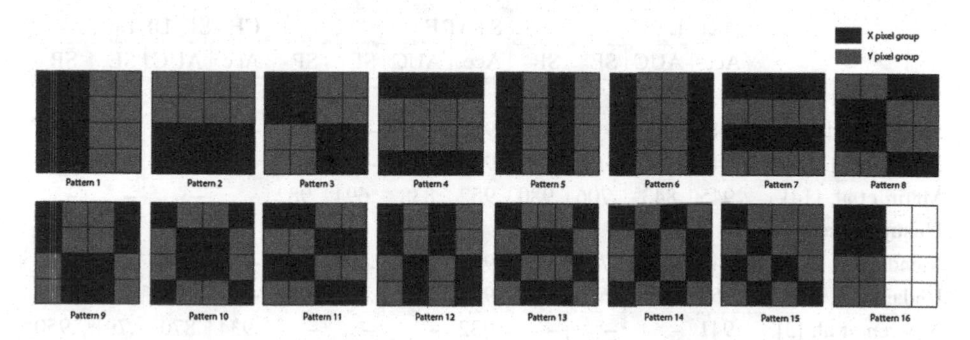

Fig. 2. All of the 16-pixel patterns used to compute LHP descriptor.

All these vectors are concatenated at the end and a feature vector of size 128-bytes is formed. Finally, the feature vector is normalized and LHP descriptor is formed.

3 Experiments and Results

Experiments are conducted on three publicly available datasets: DRIVE [8], STARE [9], and CHASE_DB1 [10]. 90% of these images are used for training and the rest 10% are used for testing (10-fold cross validation approach). Some sample outputs produced by the proposed method and Nguyen et al.'s method is shown in Fig. 3.

Sensitivity, specificity, accuracy and area under the curve (AUC) as computed in [11] are used to quantitatively the measure the performance of the proposed and the state-of-the-art methods. Table 1 compares the performance of the proposed method with the state-of-the-art methods on DRIVE, STARE, and CHASE_DB1 datasets.

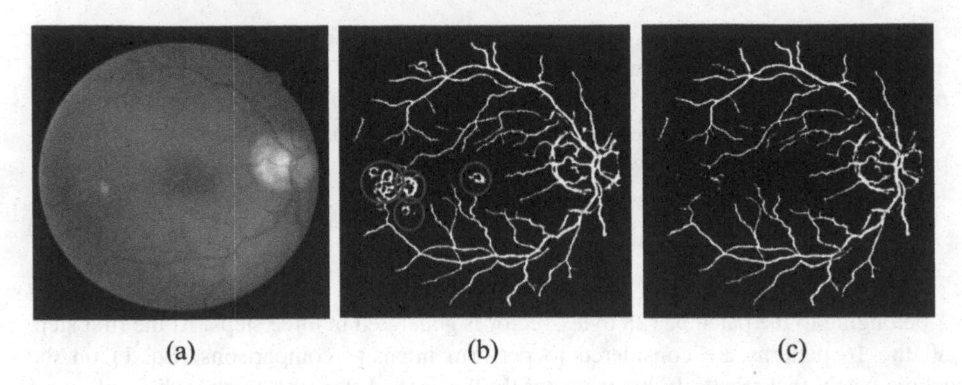

(a) (b) (c)

Fig. 3. Sample outputs. (a) Original image, (b) segmentation by Nguyen et al.'s method, (c) segmentation by proposed method. Misclassified pathology pixels are circled in blue. (Color figure online)

Table 1. Comparison of performance on DRIVE, STARE, and CHASE_DB1 datasets.

Methods	Datasets											
	DRIVE				STARE				CHASE_DB1			
	Acc	AUC	SE	SP	Acc	AUC	SE	SP	Acc	AUC	SE	SP
Supervised												
Staal et al. [12]	.944	–	–	–	.952	–	–	–	–	–	–	–
Soares et al. [13]	.946	–	–	–	.948	–	–	–	–	–	–	–
Marin et al. [14]	.945	.843	.706	.980	.952	.838	.694	.982	–	–	–	–
Unsupervised												
Mendoca et al. [15]	.945	.855	.734	.976	.944	.836	.699	.973	–	–	–	–
Budai et al. [16]	.957	.816	.644	.987	.938	.781	.580	.982	–	–	–	–
Nguyen et al. [3]	.941	–	–	–	.932	–	–	–	.934	.870	.791	.950
Proposed	**.961**	**.847**	**.711**	**.983**	**.960**	**.878**	**.790**	**.973**	**.951**	**.854**	**.742**	**.967**

(Acc = Accuracy, SE = sensitivity, SP = specificity)

4 Conclusion

In this paper, a semi-supervised method for retinal blood vessels segmentation is proposed. The method augments the multi-scale line detector approach [3] of Nguyen et al. by incorporating a supervised step with it. A novel descriptor named LHP is proposed to describe retinal pixels of interest. The descriptor encodes rich texture information around the pixel of interest. LHP descriptor together with random forest classifier is applied to separate vessel pixels from pathology pixels. Experimental results have shown that the proposed method produces higher accuracy (0.961 for DRIVE, 0.960 for STARE, and 0.951 for CHASE DB1), than the state-of-the-art methods, with comparable or higher sensitivity, specificity, and AUC. Future work will be focused on identifying more optimized pixel patterns to compute the descriptor and designing of more effective segmentation model. Ensemble learning may be a way for boosting the performance of the classifiers.

References

1. Saha, S.K., Xiao, D., Bhuiyan, A., Wong, T.Y., Kanagasingam, Y.: Color fundus image registration techniques and applications for automated analysis of diabetic retinopathy progression: a review. Biomed. Signal Process. Control **47**, 288–302 (2019)
2. Ricci, E., Perfetti, R.: Retinal blood vessel segmentation using line operators and support vector classification. IEEE Trans. Med. Imaging **26**(10), 1357–1365 (2007)
3. Nguyen, U.T., Bhuiyan, A., Park, L.A., Ramamohanarao, K.: An effective retinal blood vessel segmentation method using multi-scale line detection. Pattern Recogn. **46**(3), 703–715 (2013)
4. Breiman, L.: Random forests. Mach. Learn. **45**(1), 5–32 (2001)
5. Saha, S., Dmoulin, V.: ALOHA: an efficient binary descriptor based on Haar features. In: 19th IEEE International Conference 2012, ICIP, Lake Buena Vista, Orlando, FL, USA, pp. 2345–2348. IEEE (2012)
6. Bay, H., Ess, A., Tuytelaars, T., Van Gool, L.: Speeded-up robust features (SURF). Comput. Vis. Image Underst. **110**(3), 346–359 (2008)
7. Viola, P., Jones, M.: Rapid object detection using a boosted cascade of simple features. In: Computer Vision and Pattern Recognition, CVPR 2001, Proceedings of the 2001 IEEE Computer Society Conference, vol. 1, pp. I-511–I-518. IEEE (2001)
8. DRIVE Homepage. https://www.isi.uu.nl/Research/Databases/DRIVE/. Accessed 08 July 2018
9. STARE Homepage. http://cecas.clemson.edu/~ahoover/stare/. Accessed 29 Nov 2018
10. CHASE DB1 Homepage. https://blogs.kingston.ac.uk/retinal/chasedb1/. Accessed 15 Nov 2018
11. Fan, Z., Lu, J., Wei, C., Huang, H., Cai, X., Chen, X.: A hierarchical image matting model for blood vessel segmentation in fundus images. IEEE Trans. Image Process. (2018)
12. Staal, J., Abrmoff, M.D., Niemeijer, M., Viergever, M.A., Van Ginneken, B.: Ridge-based vessel segmentation in color images of the retina. IEEE Trans. Med. Imaging **23**(4), 501–509 (2004)
13. Soares, J.V.B., Leandro, J.J.G., Cesar, R.M., Jelinek, H.F., Cree, M.J.: Retinal vessel segmentation using the 2-D Gabor wavelet and supervised classification. IEEE Trans. Med. Imaging **25**(9), 1214–1222 (2006)
14. Marin, D., Aquino, A., Gegndez-Arias, M.E., Bravo, J.M.: A new supervised method for blood vessel segmentation in retinal images by using gray-level and moment invariants-based features. IEEE Trans. Med. Imaging **30**(1), 146 (2011)
15. Mendonca, A.M., Campilho, A.: Segmentation of retinal blood vessels by combining the detection of centerlines and morphological reconstruction. IEEE Trans. Med. Imaging **25**(9), 1200–1213 (2006)
16. Budai, A., Bock, R., Maier, A., Hornegger, J., Michelson, G.: Robust vessel segmentation in fundus images. Int. J. Biomed. Imaging **2013** (2013)

General Machine Learning

A Critical Look at Studies Applying Over-Sampling on the TPEHGDB Dataset

Gilles Vandewiele[1(✉)], Isabelle Dehaene[2], Olivier Janssens[1], Femke Ongenae[1], Femke De Backere[1], Filip De Turck[1], Kristien Roelens[2], Sofie Van Hoecke[1], and Thomas Demeester[1]

[1] IDLab, Ghent University – imec, Technologiepark-Zwijnaarde 126, Ghent, Belgium
{gilles.vandewiele,olivier.janssens,femke.ongenae,femke.debackere, filip.deturck,sofie.vanhoecke,thomas.demeester}@ugent.be

[2] Department of Gynaecology and Obstetrics, Ghent University Hospital, Corneel Heymanslaan 10, Ghent, Belgium
{isabelle.dehaene,kristien.roelens}@uzgent.be

Abstract. Preterm birth is the leading cause of death among young children and has a large prevalence globally. Machine learning models, based on features extracted from clinical sources such as electronic patient files, yield promising results. In this study, we review similar studies that constructed predictive models based on a publicly available dataset, called the Term-Preterm EHG Database (TPEHGDB), which contains electrohysterogram signals on top of clinical data. These studies often report near-perfect prediction results, by applying over-sampling as a means of data augmentation. We reconstruct these results to show that they can only be achieved when data augmentation is applied on the entire dataset prior to partitioning into training and testing set. This results in (i) samples that are highly correlated to data points from the test set are introduced and added to the training set, and (ii) artificial samples that are highly correlated to points from the training set being added to the test set. Many previously reported results therefore carry little meaning in terms of the actual effectiveness of the model in making predictions on unseen data in a real-world setting. After focusing on the danger of applying over-sampling strategies before data partitioning, we present a realistic baseline for the TPEHGDB dataset and show how the predictive performance and clinical use can be improved by incorporating features from electrohysterogram sensors and by applying over-sampling on the training set.

Keywords: Preterm birth · Electrohysterogram (EHG) · Imbalanced data · Over-sampling

1 Introduction

Giving birth before 37 weeks of pregnancy, which is referred to as preterm birth, has a significant negative impact on the expected outcome of the neonate.

D. Riaño et al. (Eds.): AIME 2019, LNAI 11526, pp. 355–364, 2019.
https://doi.org/10.1007/978-3-030-21642-9_45

According to the World Health Organization (WHO), preterm birth is one of the leading causes of death among young children, and its' prevalence ranges from 5% to 18% globally [23]. As preterm labor is currently not yet fully understood, gynecologists are experiencing difficulties in assessing whether a patient recently admitted to the hospital will deliver at term or not. In order to support experts in their assessment, several studies have already investigated the added value of a predictive model [6,13,24,35]. These models are based on a large number of variables extracted from clinical sources such as the electronic health record. These variables include the gestational age, results of a biomarker, cervical length, clinical history, and more. In this study, we provide a thorough and extensive overview of related work on a public dataset and discuss many of the overly optimistic results. These results are often obtained by introducing a large bias through over-sampling the dataset, before partitioning the data, in order to combat the class imbalance, i.e., the fact that it contains many more pregnancies with term deliveries than preterm. Afterwards, we set a realistic baseline and assess the impact of correct over-sampling and of incorporating features extracted from the electrohysterogram data.

2 The Impact of Over-Sampling Prior to Data Partitioning

In this section, we highlight the impact of applying over-sampling prior to the data partitioning on an artificially generated dataset. We generated a binary classification problem with 100 samples. Twenty samples were marked positive (red circles), and the others negative (blue squares). The generated dataset is depicted on the left of Fig. 1 (step 0). We now compare the effect of over-sampling data after partitioning with the effect of over-sampling prior to partitioning. In the former approach, we first partition our data into two mutually exclusive sets (step 1). Then, we create artificial samples (red, unfilled circles) that are highly correlated to the training samples of the minority class (step 2) in order to have a similar number of samples for both classes in our training set. On the other hand, if we over-sample the data prior to partitioning, we generate train samples that are highly correlated with original data points that will end up in the test set (step 1). Moreover, some of the generated artificial samples will be distributed to the test set as well (step 2). These two consequences result in highly optimistic results that merely reflect the model's capability to memorize samples seen during training, rather than its predictive performance if it were applied in a real-world setting on unseen data.

3 A Critical Look on Studies Reporting Near-Perfect Results on the TPEHGDB Dataset

In 2008, a public dataset, called TPEHGDB (Term/Preterm ElectroHysteroGram DataBase), containing 300 records, which correspond to 300 pregnancies,

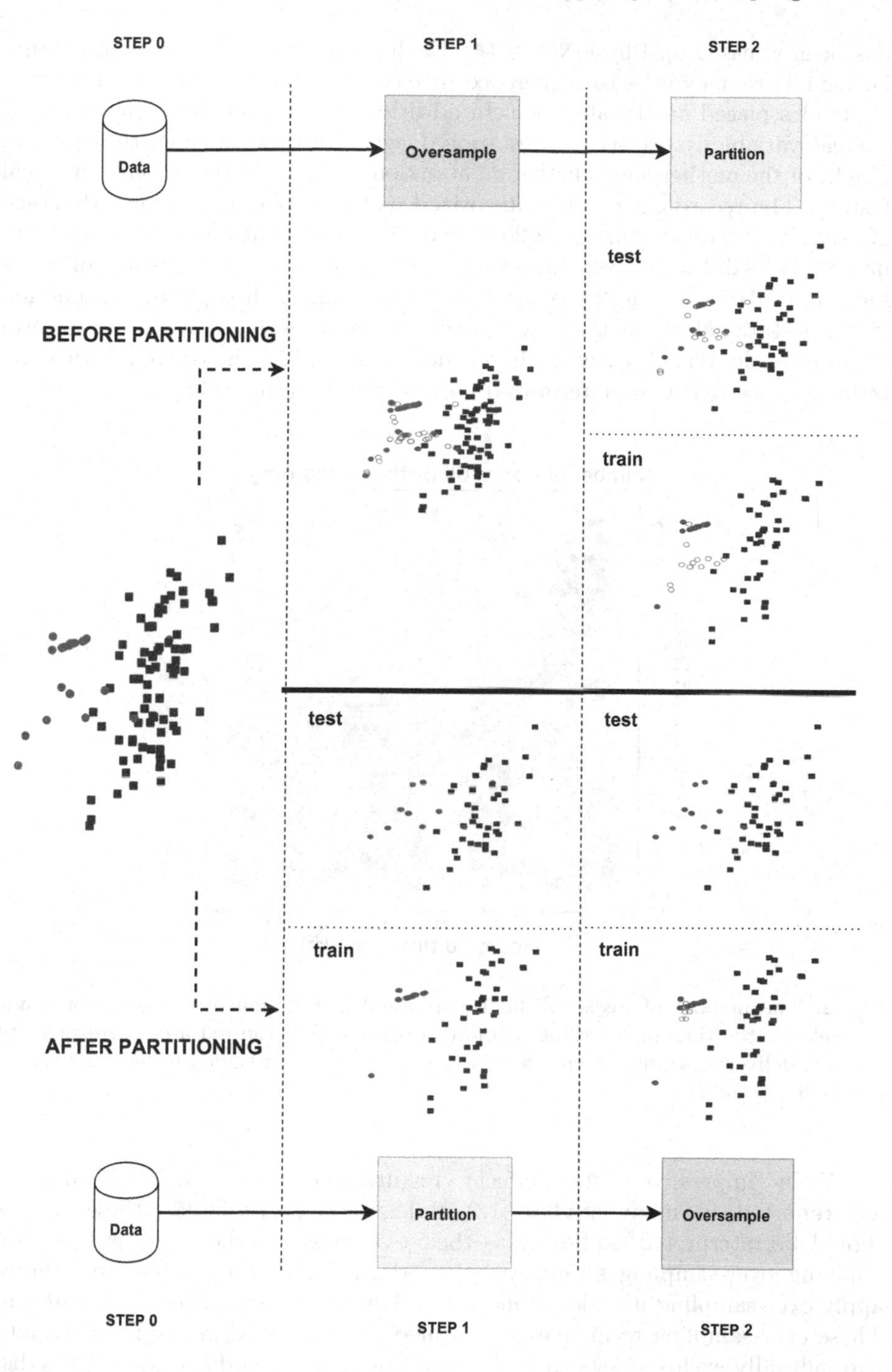

Fig. 1. Comparing the impact of applying over-sampling prior to data partitioning to applying over-sampling after data partitioning on a two-dimensional classification problem.

has been released on PhysioNet [9,14]. Each record consists of three raw bipolar signals that express the difference in electric potentials, measured by four electrodes placed on the abdomen. In addition, each record is accompanied by clinical variables, such as the gestational age at recording time, the age and weight of the mother, and whether an abortion occurred in the patients' medical history. The recordings can be categorized as being captured at an early stage of pregnancy (gestational age of 23.11 ± 0.77 weeks) or at a later stage of pregnancy (31.09 ± 1.05 weeks). Recordings were captured at a frequency of 20 Hz for about 30 min. In Fig. 2 the number of weeks till birth is plotted in function of the gestational age at the time of recording and displayed according to term or preterm delivery. Clearly, an imbalance is present in the dataset with more term (green area) than preterm (red area) deliveries (262 vs 38).

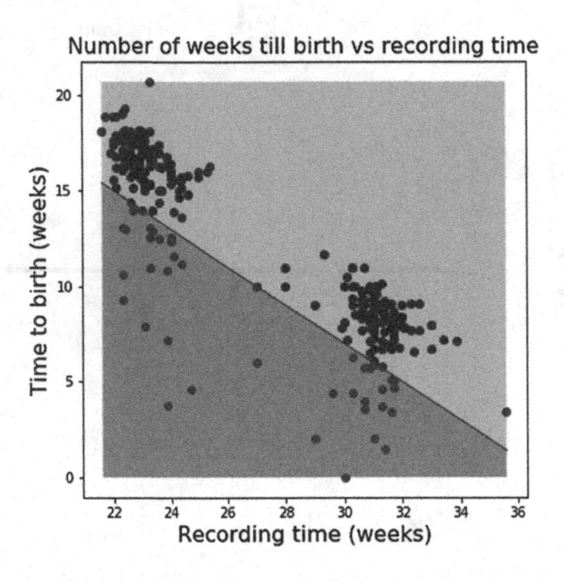

Fig. 2. The number of weeks till birth expressed in function of the gestational age in weeks at the time of recording. All data points within the red area correspond to preterm deliveries, while the ones within the green area correspond to term deliveries. (Color figure online)

While impressive (near-perfect) results on the TPEHGDB dataset are reported in many studies [1,2,10–12,15–21,27,29,33,34], these results should be interpreted cautiously as their evaluation methodology is based on applying over-sampling strategies before data partitioning. All these studies apply over-sampling in order to make the distribution of classes more uniform. These over-sampling techniques are applied prior to partitioning the data into two mutually exclusive sets (referred to as the training and testing set). As discussed earlier, this causes the predictive performance metrics to be overly optimistic.

Nevertheless, a significant number of studies on the TPEHGDB dataset do not apply any over-sampling technique. However, in these studies, certain decisions concerning the evaluation were often made which raises serious questions concerning the credibility of the provided results [3,4,8,25,26,32]. In many of these studies, results were either not obtained through cross-validation, or cross-validation was applied on a subset of data subsampled from the original dataset. Performing this kind of pre-processing, in a machine learning context, without any kind of argumentation, raises doubts since it drastically increases the variance of the obtained results and avoids the problem of imbalanced data, which does not reflect reality in terms of potential applications. In other studies, segments are extracted from the original signals, which are highly correlated with each other, and then partitioned into training and testing set [7,31], which again results in highly optimistic results.

At the time of writing (December 2018), within all the 153 citations to the original paper, which introduced the TPEHGDB dataset, we have found three machine learning studies that were accessible and, to the best of our knowledge, had a sound evaluation methodology [22,28,30]. In the study of Sadi-Ahmed et al. [30], all records taken before 26 weeks of gestation were filtered away from the dataset, resulting in a dataset of 138 recordings taken after the 26th week of gestation. All of these signals were processed in order to detect contractions through Auto-Regressive Moving Averages (ARMA). From the detected contractions, features were extracted such as the total number of contractions, average duration and average time between contractions. Unfortunately, only an accuracy score of 0.89 to distinguish between term and preterm pregnancies was achieved within this study, making it hard to assess the clinical use of such a model. It is important to note that, on this filtered dataset, an accuracy score of 0.86 can be achieved by always predicting term birth, precisely because of the aforementioned class imbalance, with a fraction of 119 term deliveries on 138 records from 26 weeks onwards. Janjarasjitt et al. proposed a new type of feature, based on a wavelet decomposition of the signals [22]. The feature was evaluated by tuning a threshold on a single feature in a leave-one-out cross-validation scheme. A sensitivity and specificity of 0.6842 and 0.7133 are achieved. While these scores are very promising, it should be noted, that they are rather optimistic due to the fact that the evaluation happened in a leave-one-out scheme. As such, the performance of the sample entropy feature, provided along with the original data, closely matches, and sometimes even outperforms, that of the proposed feature. Nevertheless, the wavelet-based feature may be an interesting addition to the feature set. In the work of Ryu et al. [28] a similar study is performed in which a feature based on Multivariate Empirical Mode Decomposition (MEMD) is proposed. They evaluate the added value of their feature, by subsampling a balanced dataset of 38 term and 38 preterm records, 100 times, from the original dataset. They found that the AUC improved from 0.5698 to 0.6049 by adding their feature to the dataset. While this subsampling strategy again avoids the problem of imbalanced data, which is reflected in the original dataset, it does show an improvement in AUC and thus indicates that adding the MEMD-based feature to the dataset could be beneficial for the predictive performance. Moreover, due to the many repetitions of the experiment, the sample mean better reflects the real mean.

4 Setting a Realistic Baseline for the TPEHGDB Dataset

In this section, we will assess the effects of incorporating information from raw EHG signals, and of over-sampling the data after partitioning, on the predictive performance of the resulting model. Moreover, we will show that predictive performances similar to the aforementioned studies can be only be achieved through over-sampling before data partitioning.

Seven machine learning algorithms were trained on the original dataset consisting of clinical features and four features extracted from the raw EHG signals, i.e.: the root mean square value & entropy of the raw signal and the median and peak frequency from the spectral information of each signal. The seven different classification techniques, and their corresponding abbreviations, are: (1) Logistic Regression (LR), (2) Decision Trees (DT), (3) Linear Discriminant Analysis (LDA), (4) Quadratic Discriminant Analysis (QDA), (5) K-Nearest Neighbors (KNN), (6) Random Forests (RF), and (7) Support Vector Machines (SVM). All reported results are generated using five-fold stratified cross-validation. Hyperparameters were tuned using grid search. Moreover, to solve the issue of imbalanced data, and to improve the clinical use of the different classifiers, we apply over-sampling, using SMOTE [5], on the train set. We compare these results to when SMOTE is applied on the entire dataset, to show that near-perfect predictive performance can only be achieved by introducing label leakage.

In total, we evaluate four different approaches: (i) clinical features and no over-sampling, (ii) clinical and EHG features and no over-sampling, (iii) clinical and EHG features and over-sampling in a correct fashion, and finally (iv) clinical and EHG features and over-sampling in an incorrect fashion. The first two approaches are compared in Table 1. As can be seen, the AUC scores drastically improve when features, extracted from the EHG signals, are incorporated. Nevertheless, the clinical use of both approaches is very limited, as all the models almost always predict that someone will deliver at term (which is reflected in the low sensitivity scores), which is a typical problem that arises when dealing with imbalanced data. The performance for both over-sampling approaches is listed in Table 2. We can conclude that the near-perfect performances from the studies mentioned in Sect. 3 can only be closely matched by applying over-sampling prior to data partitioning. If we apply over-sampling on the training set, we see that the clinical use of a predictive model for preterm birth prediction, based on the TPEGHDB dataset, is still limited, with a maximum AUC score of 63.20%.

Table 1. The results obtained with seven different classifiers, on (i) a dataset constructed using solely clinical variables and (ii) a dataset with clinical variables concatenated to four features extracted from the EHG data. No over-sampling is applied for both approaches.

Algorithm	Sensitivity (%)		Specificity (%)		AUC (%)	
	Clinical	All	Clinical	All	Clinical	All
LR	0 ± 0	0 ± 0	100 ± 0	100 ± 0	48 ± 6	58 ± 7
DT	3 ± 5	0 ± 0	96 ± 4	96 ± 3	47 ± 6	62 ± 9
LDA	0 ± 0	0 ± 0	97 ± 3	96 ± 4	54 ± 9	59 ± 5
QDA	28 ± 34	11 ± 11	67 ± 36	90 ± 7	48 ± 5	62 ± 4
KNN	0 ± 0	0 ± 0	100 ± 1	98 ± 2	50 ± 8	57 ± 7
RF	0 ± 0	0 ± 0	99 ± 2	95 ± 4	52 ± 8	58 ± 5
SVM	0 ± 0	0 ± 0	100 ± 0	100 ± 0	52 ± 8	56 ± 9

Table 2. The results obtained with seven different classifiers, on the entire TPEHGDB dataset, constructed using clinical features and features extracted from the 3 filtered EHG signals. Oversampling with SMOTE is applied before data partitioning (column *correct*) versus after data partitioning (column *incorrect*).

Algorithm	Sensitivity (%)		Specificity (%)		AUC (%)	
	Correct	Incorrect	Correct	Incorrect	Correct	Incorrect
LR	39 ± 26	74 ± 3	68 ± 19	66 ± 6	59 ± 6	78 ± 3
DT	40 ± 16	81 ± 3	71 ± 10	84 ± 5	59 ± 3	86 ± 4
LDA	53 ± 14	73 ± 1	59 ± 10	69 ± 7	59 ± 5	78 ± 3
QDA	51 ± 36	100 ± 0	58 ± 24	41 ± 7	61 ± 6	79 ± 2
KNN	48 ± 16	99 ± 1	64 ± 3	73 ± 5	58 ± 6	92 ± 2
RF	42 ± 13	91 ± 4	68 ± 6	95 ± 2	63 ± 5	98 ± 1
SVM	43 ± 31	99 ± 2	64 ± 23	86 ± 4	58 ± 6	98 ± 1

5 Conclusion and Future Work

This study tackles the problem of preterm birth risk prediction, based on the publicly available dataset TPEHGDB. Our contributions are two-fold. First, in the light of a significant body of recent literature, we show that applying over-sampling for data-augmentation purposes, prior to partitioning the data into separate parts for training and evaluation, leads to overly optimistic results. To evaluate a model's predictive performance, the data partitioning needs to be performed before applying over-sampling. Second, a realistic baseline was set in which it was shown how an increase in AUC score can be obtained by using features extracted from electrohysterogram recordings, besides clinical observations. This confirms the potential added value of such recordings. In future work we will investigate whether deep learning techniques can improve the predictive performance by directly training on the raw recordings, as opposed to manually extracting features. Unfortunately, for this, a larger dataset may be required.

Acknowledgements. Gilles Vandewiele is funded by a scholarship of FWO (1S31417N). This study has been performed in the context of the 'Predictive health care using text analysis on unstructured data project', funded by imec, and the PRETURN (PREdiction Tool for prematUre laboR and Neonatal outcome) clinical trial (EC/2018/0609) of Ghent University Hospital.

Reproducibility and Dataset Availability. In order to allow reproduction of the reported results on this public dataset, we host all code, required to reproduce the results reported in this paper, on a public GitHub repository[1]. The dataset is available from that repository, or from the original hosting location[2].

References

1. Acharya, U.R., et al.: Automated detection of premature delivery using empirical mode and wavelet packet decomposition techniques with uterine electromyogram signals. Comput. Biol. Med. **85**, 33–42 (2017)
2. Ahmed, M.U., Chanwimalueang, T., Thayyil, S., Mandic, D.P.: A multivariate multiscale fuzzy entropy algorithm with application to uterine EMG complexity analysis. Entropy **19**(1), 2 (2016)
3. Baghamoradi, S.M.S., Naji, M., Aryadoost, H.: Evaluation of cepstral analysis of EHG signals to prediction of preterm labor. In: 2011 18th Iranian Conference of Biomedical Engineering (ICBME), pp. 81–83. IEEE (2011)
4. Beiranvand, M., Shahbakhti, M., Eslamizadeh, M., Bavi, M., Mohammadifar, S.: Investigating wavelet energy vector for pre-term labor detection using EHG signals. In: Signal Processing: Algorithms, Architectures, Arrangements, and Applications (SPA), 2017, pp. 269–274. IEEE (2017)
5. Chawla, N.V., Bowyer, K.W., Hall, L.O., Kegelmeyer, W.P.: SMOTE: synthetic minority over-sampling technique. J. Artif. Intell. Res. **16**, 321–357 (2002)
6. De Silva, D.A., Lisonkova, S., von Dadelszen, P., Synnes, A.R., Magee, L.A.: Timing of delivery in a high-risk obstetric population: a clinical prediction model. BMC Pregnancy Childbirth **17**(1), 202 (2017)
7. Despotović, D., Zec, A., Mladenović, K., Radin, N., Turukalo, T.L.: A machine learning approach for an early prediction of preterm delivery. In: 2018 IEEE 16th International Symposium on Intelligent Systems and Informatics (SISY), pp. 000265–000270. IEEE (2018)
8. Far, D.T., Beiranvand, M., Shahbakhti, M.: Prediction of preterm labor from EHG signals using statistical and non-linear features. In: 2015 8th Biomedical Engineering International Conference (BMEiCON), pp. 1–5. IEEE (2015)
9. Fele-Žorž, G., Kavšek, G., Novak-Antolič, Ž., Jager, F.: A comparison of various linear and non-linear signal processing techniques to separate uterine emg records of term and pre-term delivery groups. Med. Biol. Eng. Comput. **46**(9), 911–922 (2008)

[1] https://github.com/IBCNServices/TPEHGDB-Experiments/.
[2] https://physionet.org/physiobank/database/tpehgdb/.

10. Fergus, P., Cheung, P., Hussain, A., Al-Jumeily, D., Dobbins, C., Iram, S.: Prediction of preterm deliveries from EHG signals using machine learning. PloS ONE **8**(10), e77154 (2013)
11. Fergus, P., Hussain, A., Al-Jumeily, D., Hamdan, H.: A machine learning system for automatic detection of preterm activity using artificial neural networks and uterine electromyography data. Int. J. Adapt. Innov. Syst. **2**(2), 161–179 (2015)
12. Fergus, P., Idowu, I., Hussain, A., Dobbins, C.: Advanced artificial neural network classification for detecting preterm births using EHG records. Neurocomputing **188**, 42–49 (2016)
13. García-Blanco, A., Diago, V., De La Cruz, V.S., Hervás, D., Cháfer-Pericás, C., Vento, M.: Can stress biomarkers predict preterm birth in women with threatened preterm labor? Psychoneuroendocrinology **83**, 19–24 (2017)
14. Goldberger, A.L., et al.: PhysioBank, PhysioToolkit, and PhysioNet: components of a new research resource for complex physiologic signals. Circulation **101**(23), e215–e220 (2000). https://doi.org/10.1161/01.CIR.101.23.e215
15. Hoseinzadeh, S., Amirani, M.C.: Use of electro hysterogram (EHG) signal to diagnose preterm birth. In: Iranian Conference on Electrical Engineering (ICEE), pp. 1477–1481. IEEE (2018)
16. Hussain, A.J., Fergus, P., Al-Askar, H., Al-Jumeily, D., Jager, F.: Dynamic neural network architecture inspired by the immune algorithm to predict preterm deliveries in pregnant women. Neurocomputing **151**, 963–974 (2015)
17. Idowu, I.O.: Classification Techniques Using EHG Signals for Detecting Preterm Births. Ph.D. thesis, Liverpool John Moores University (2017)
18. Idowu, I.O., Fergus, P., Hussain, A., Dobbins, C., Al Askar, H.: Advance artificial neural network classification techniques using EHG for detecting preterm births. In: 2014 Eighth International Conference on Complex, Intelligent and Software Intensive Systems (CISIS), pp. 95–100. IEEE (2014)
19. Idowu, I.O., et al.: Artificial intelligence for detecting preterm uterine activity in gynecology and obstetric care. In: 2015 IEEE International Conference on Computer and Information Technology; Ubiquitous Computing and Communications; Dependable, Autonomic and Secure Computing; Pervasive Intelligence and Computing (CIT/IUCC/DASC/PICOM), pp. 215–220. IEEE (2015)
20. Jager, F., Libensek, S., Gersak, K.: Characterization and automatic classification of preterm and term uterine records. bioRxiv, p. 349266 (2018)
21. Janjarasjitt, S.: Evaluation of performance on preterm birth classification using single wavelet-based features of EHG signals. In: 2017 10th Biomedical Engineering International Conference (BMEiCON), pp. 1–4. IEEE (2017)
22. Janjarasjitt, S.: Examination of single wavelet-based features of EHG signals for preterm birth classification. IAENG Int. J. Comput. Sci. **44**(2), 212–218 (2017). https://www.researchgate.net/publication/317749466_Examination_of_single_wavelet-based_features_of_EHG_signals_for_preterm_birth_classification
23. Liu, L., et al.: Global, regional, and national causes of under-5 mortality in 2000–15: an updated systematic analysis with implications for the sustainable development goals. Lancet **388**(10063), 3027–3035 (2016)
24. Meertens, L.J., et al.: Prediction models for the risk of spontaneous preterm birth based on maternal characteristics: a systematic review and independent external validation. Acta Obstet. Gynecol. Scand. **97**(8), 907–920 (2018). https://www.ncbi.nlm.nih.gov/pmc/articles/PMC6099449/

25. Naeem, S., Ali, A., Eldosoky, M.: Kl. comparison between using linear and non-linear features to classify uterine electromyography signals of term and preterm deliveries. In: 2013 30th National Radio Science Conference (NRSC), pp. 492–502. IEEE (2013)
26. Naeem, S.M., Seddik, A.F., Eldosoky, M.A.: New technique based on uterine electromyography nonlinearity for preterm delivery detection. J. Eng. Technol. Res. **6**(7), 107–114 (2014)
27. Ren, P., Yao, S., Li, J., Valdes-Sosa, P.A., Kendrick, K.M.: Improved prediction of preterm delivery using empirical mode decomposition analysis of uterine electromyography signals. PloS ONE **10**(7), e0132116 (2015)
28. Ryu, J., Park, C.: Time-frequency analysis of electrohysterogram for classification of term and preterm birth. IEIE Trans. Smart Process. Comput. **4**(2), 103–109 (2015)
29. Sadi-Ahmed, N., Kacha, B., Taleb, H., Kedir-Talha, M.: Relevant features selection for automatic prediction of preterm deliveries from pregnancy electrohysterograhic (EHG) records. J. Med. Syst. **41**(12), 204 (2017)
30. Sadi-Ahmed, N., Kedir-Talha, M.: Contraction extraction from term and preterm electrohyterographic signals. In: 2015 4th International Conference on Electrical Engineering (ICEE), pp. 1–4. IEEE (2015)
31. Shahrdad, M., Amirani, M.C.: Detection of preterm labor by partitioning and clustering the EHG signal. Biomed. Signal Process. Control. **45**, 109–116 (2018)
32. Sim, S., Ryou, H., Kim, H., Han, J., Park, K.: Evaluation of electrohysterogram feature extraction to classify the preterm and term delivery groups. In: Goh, J. (ed.) The 15th International Conference on Biomedical Engineering. IP, vol. 43, pp. 675–678. Springer, Cham (2014). https://doi.org/10.1007/978-3-319-02913-9_172
33. Smrdel, A., Jager, F.: Separating sets of term and pre-term uterine EMG records. Physiol. Meas. **36**(2), 341 (2015)
34. Subramaniam, K., Iqbal, N.V., et al.: Classification of fractal features of uterine EMG signal for the prediction of preterm birth. Biomed. Pharmacol. J. **11**(1), 369–374 (2018)
35. Watson, H., Carter, J., Seed, P., Tribe, R., Shennan, A.: QuiPP app: a safe alternative to a treat-all strategy for threatened preterm labor. Ultrasound Obstet. Gynecol. **50**(3), 342–346 (2017)

Perspectives on Assurance Case Development for Retinal Disease Diagnosis Using Deep Learning

Chiara Picardi(✉) and Ibrahim Habli

University of York, York YO10 5DD, UK
{chiara.picardi,ibrahim.habli}@york.ac.uk

Abstract. We report our experience with developing an assurance case for a deep learning system used for retinal disease diagnosis and referral. We investigate how an assurance case could clarify the scope and structure of the primary argument and identify sources of uncertainty. We also explore the need for an assurance argument pattern that could provide developers with a reusable template for communicating and structuring the different claims and evidence and clarifying the clinical context rather than merely focusing on meeting or exceeding performance measures.

Keywords: Assurance case · Machine learning · Retinal disease · Safety

1 Introduction

Justifying the use of machine learning in critical healthcare applications is currently a significant technological and societal challenge [5]. The developers and clinical users of the technology have to assure, prior to deployment, different critical properties such as safety, performance, usability and cost-effectiveness [6]. This challenge can be refined further into 2 parts. Firstly, there is no consensus on the assurance criteria or specific properties that machine learning systems have to exhibit for them to be accepted by the public or by the clinical and regulatory authorities, i.e. what is good enough? Secondly, there is very little guidance, e.g. standards, on accepted means for achieving such properties [6].

In this paper, we investigate the extent to which an explicit assurance case could inform a decision concerning the use of machine learning in clinical diagnosis. An assurance case is *"a reasoned and compelling argument, supported by a body of evidence, that a system, service or organisation will operate as intended for a defined application in a defined environment"* [1]. An assurance case is considered as a generalisation of a safety case, i.e. where safety claims are the focus of the assurance.

We build on the results of De Fauw et al. [2] on the use of a deep learning system for diagnosis and referral in retinal disease. This system comprises 2 different neural networks. The first network, called Segmentation Network,

D. Riaño et al. (Eds.): AIME 2019, LNAI 11526, pp. 365–370, 2019.
https://doi.org/10.1007/978-3-030-21642-9_46

takes as input three-dimensional Optical Coherence Tomography (OCT) scans and creates a detailed device-independent tissue-segmentation map. The second network examines the segmentation map and outputs one of the four referral suggestions in addition to the presence or absence of multiple concomitant retinal pathologies.

Through an assurance case, our objectives are to (1) clarify structure of the primary argument and the clinical context and (2) identify sources of uncertainty. The contribution of the paper is that it provides a self-contained assurance case for a deep learning system, thereby highlighting assurance issues that have to be considered explicitly beyond merely exceeding a specific performance measure.

2 Assurance Case

The assurance case is represented using the Goal Structuring Notation (GSN) [1]. GSN is a generic argument structuring language that is widely used in the safety-critical domain. The reader is advised to consult the publicly available GSN standard [1] for a more detailed description of the notation. Due to the space limitation, we focus the discussion on 2 assurance argument fragments:

1. Segmentation network assurance argument (Fig. 1, Sect. 2.1)
2. Classification network assurance argument (Fig. 2, Sect. 2.2)

These arguments capture the essence of the justification based on performance against clinical experts. The clinical context in the assurance case is the ophthalmology referral pathway at Moorfields Eye Hospital, from which the training, validation and test data is provided. At this stage, the scope of the claims is limited to this clinical setting with no evidence for generalisation (despite the wide and diverse population served). It is important to note that the assurance case focuses exclusively on the chain of reasoning and evidence based on the data in the original study [2]. The extent to which this assurance case could be improved, or its scope extended, is discussed in Sect. 3.

2.1 Segmentation Network Assurance Argument

Figure 1 shows the assurance argument fragment concerning the performance and transparency of the segmentation network. The argument makes a distinction between the scans that include ambiguous and unambiguous regions. The context is important here, referencing the data used for training, testing and validation. It also clarifies the profile of the clinical experts involved in the segmentation experiment. Evidence of sufficient performance is provided based on two different scanning devices (99.21% and 99.93%). The argument clarifies further that for unambiguous regions, the network produces tissue-segmentation maps that are comparable to manual segmentation. For scans with ambiguous regions, the network provides different (but plausible) interpretations of the low quality regions, i.e. similar to how the different human experts might produce different interpretations. The evidence is represented by supplementary videos that show the multiple hypotheses of the segmentation maps produced by the network. An important aspect of creating a separate network for segmentation is greater transparency. By being able to inspect the tissue-segmentation map (and not just referral decisions), clinicians have clearer means for understanding the basis for the final clinical decision. What is less clear, however, is the effectiveness of this visualisation, i.e. degree of acceptance by clinical experts. As such, this is labelled as *'to be developed'* (small diamond below the claim).

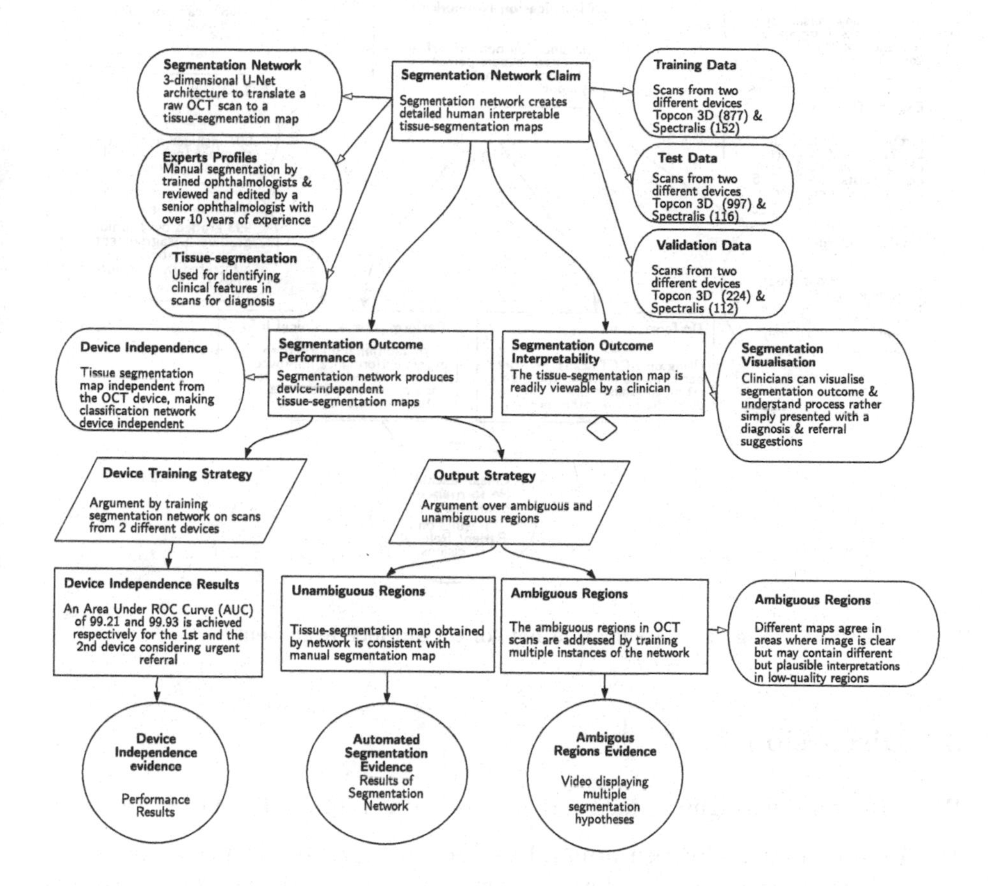

Fig. 1. Segmentation network assurance argument

2.2 Classification Network Assurance Argument

The argument in Fig. 2 states the primary claim that the system achieves or in some cases exceeds human expert performance in retinal disease diagnosis and referral. Experts comprise 4 retina specialists with respective 21, 21, 13 and 12 years of experience and 4 optometrists with respective 15, 9, 6 and 3 years of experience. Two sessions were organised. In the first session experts were required to give the referral suggestions using the OCT scans only. In the second session they were also able to use fundus images and clinical notes. Similar to the segmentation network assurance argument, this argument communicates clearly the training, test and validation data as well as the benchmark against which performance is assessed (i.e. gold standard and expert profiles).

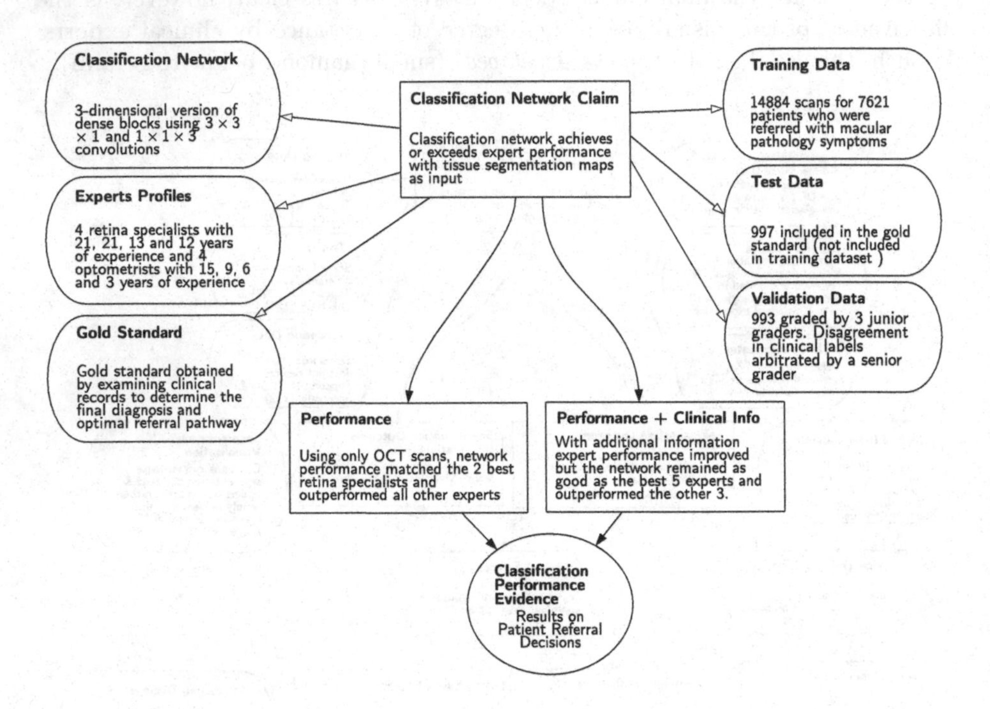

Fig. 2. Classification network assurance argument

3 Discussion

We reflect on the insights gained and lessons learned from different perspectives.

Performance-Based Arguments. Evidence in machine learning studies tends to focus on meeting or exceeding certain performance criteria. The assurance argument above is consistent with this approach. Importantly, it ensures that the different training, test and validation datasets are explicitly referenced in addition to the performance results. It clarifies, particularly to non-technical reviewers and decision makers, the importance of appraising the quality of these datasets and the extent to which the data used is relevant to the context in which the performance claims are made. The argument also prompts the reviewers to question the performance criteria used.

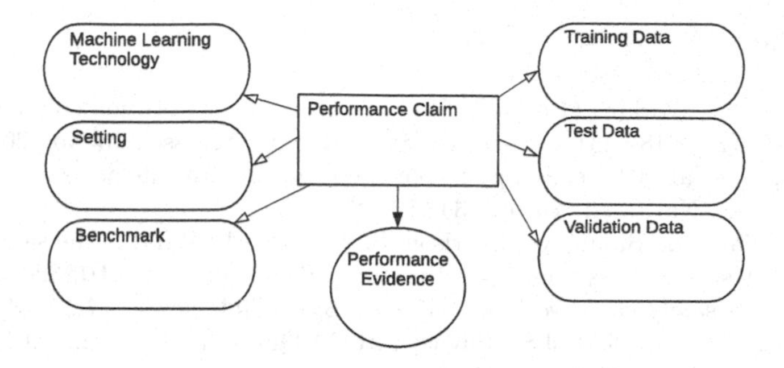

Fig. 3. Preliminary machine learning assurance argument pattern

Assurance Case Pattern. By looking at the argument fragments for the Segmentation and Classification Networks, a *pattern* of reasoning seems to emerge (Fig. 3). Such a pattern could prompt the developers and assessors of machine learning to more explicitly consider the relevance and appropriateness of the contextual and evidential data, i.e. ensuring sufficient confidence in the quality and relevance of the data and models, by scrutinising the *links* in the argument in Fig. 3, rather than merely exceeding a specific performance measure.

Assumptions and Transparency. An assurance case can help ensure that the assumptions made are explicitly listed. For example, the reviewers of the case can question the profiles and representativeness of the clinical experts involved in the experiments and the extent to which further clarification might be necessary. Transparency in how the machine makes clinical decisions is also important. Here, the assurance case clarifies that transparency is limited to the output of the segmentation and not the classification network, i.e. prompting the reviewer to question the need for transparency in the final diagnosis and referral decision.

Safety and Regulations. Although our assurance case does not directly address patient safety [3], there remain fundamental questions as what is deemed as good enough for assuring the safety of machine learning. For example, are arguments based on exceeding human equivalence or appealing to risk-benefit evidence acceptable? How do we address non-quantifiable factors such as those related to human or organisational factors? Another issue is the readiness of the regulators to review, challenge and approve machine learning evidence. Kelly in [4] talks about the *Imbalance of Skills* between the developers and the independent assessors of novel technologies as a major hurdle for effective assurance case practices. The readiness of regulators to appraise machine learning algorithms, evaluation evidence and deployment constraints is an ongoing concern.

References

1. Assurance Case Working Group [ACWG]. Goal structing notation community standard version 2 (2018). https://scsc.uk/r141B:1?t=1. Accessed 13 Nov 2018
2. De Fauw, J., et al.: Clinically applicable deep learning for diagnosis and referral in retinal disease. Nat. Med. **24**(9), 1342 (2018)
3. Habli, I., White, S., Sujan, M., Harrison, S., Ugarte, M.: What is the safety case for health it? A study of assurance practices in england. Saf. Sci. **110**, 324–335 (2018)
4. Kelly, T.: Are safety cases working. Saf. Crit. Syst. Club Newsl. **17**(2), 31–33 (2008)
5. Maddox, T.M., Rumsfeld, J.S., Payne, P.R.O.: Questions for artificial intelligence in health care. JAMA **321**(1), 31–32 (2019)
6. Shortliffe, E.H., Sepúlveda, M.J.: Clinical decision support in the era of artificial intelligence. JAMA **320**(21), 2199–2200 (2018)

Rapid Detection of Heart Rate Fragmentation and Cardiac Arrhythmias: Cycle-by-Cycle rr Analysis, Supervised Machine Learning Model and Novel Insights

Ananya Rajagopalan[1(✉)] and Marcus Vollmer[2]

[1] Redmond High School, Redmond, WA, USA
ananya.rajagopalan@gmail.com
[2] Institute of Bioinformatics, University Medicine Greifswald, Greifswald, Germany
marcus.vollmer@uni-greifswald.de

Abstract. Heart rate dynamics are a macroscopic indicator of cardiac health. Sino-atrial degradation manifested as heart rate fragmentation (HRF) are analyzed using rr values (relative-RR intervals) derived from the inter-beat-intervals of ECGs. The rr-value is useful for the analysis of cycle-by-cycle variations such as HRF and arrhythmias. Three novel metrics developed in this work: CM20, Z3e20 and sPIP, along with two conventional metrics: SDNN and LFHF ratio are used for the detection of HRF and arrhythmias. The supervised machine learning technique of random forests is applied to develop the classification model. For this, we used a balanced dataset of 300 cases comprising of arrhythmic, non-arrhythmic coronary artery disease, and individuals without any medically significant cardiac conditions. The model was tested on 104 independent cases. The F1 score of the classifier is 91.1% without any adjustments for age, gender, prior medical conditions, etc. Insight into threshold values of heart rate dynamics for arrhythmic, heart rate fragmentation and normal cases are obtained from a single decision tree model.

Keywords: Heart rate fragmentation · Arrhythmia detection · Random forest · Machine learning

1 Introduction

Heart disease is responsible for 31% of deaths worldwide [1]. Thus, it is necessary to develop methods that accurately detect medically significant incipient cardiac conditions such as arrhythmias and heart rate fragmentation (HRF). HRF has been shown to be a dynamical biomarker of the neuroautonomic-electrophysiologic system breakdown: patients with greater HRF have been shown to be at an increased risk of coronary artery disease (CAD) [2]. HRF has been recently determined as an anomalous increase in heart rate due to short-term acceleration/deceleration [2, 10]; it has the potential to provide, falsely, values of heart rate variability (such as SDNN) that are deemed normal. Prior work [2] utilizes univariate predictors of HRF with limited discrimination; one

D. Riaño et al. (Eds.): AIME 2019, LNAI 11526, pp. 371–375, 2019.
https://doi.org/10.1007/978-3-030-21642-9_47

objective of this work was thus to improve the discrimination ability of HRF. In this work, heart rate fragmentation is synonymous with non-arrhythmic individuals who have been diagnosed as having CAD in the THEW database [4].

Work on the detection of arrhythmias outside of the clinical setting has shown tremendous progress in recent years, such as the 2017 Physionet/CinC Challenge [3] on single-lead atrial fibrillation detection. The work presented here seeks to improve upon existing research in the following ways: (1) accurate identification of heart rate fragmentation with high discrimination capability (2) separate classification of heart rate fragmentation, arrhythmias, and "normal" patterns (3) emphasis on the application of Random Forest (RF) for a computationally non-intensive methodology with straightforward implementation into battery-operated wearable devices.

2 Datasets and Methodology

2.1 Datasets

A total of 300 cases, representing a set of balanced class examples, was used for statistical analysis and ML model development: 90 cases of non-arrhythmic CAD from the University of Rochester THEW-project database [4], 115 examples of normal heart rate rhythm data from THEW-project database [4] and MIT-BIH nsr2db (normal sinus rhythm) database [5], and 95 examples of arrhythmic cases from MIT-BIH mitdb and afdb [5]. To test the model, 104 (30 non-arrhythmic CAD, 30 normal, 44 arrhythmic) new cases from the same databases were used [4, 5].

2.2 Data Visualization

In this work, rr (moving average window normalized RR intervals, [6] called relative RR intervals) is used, as it overcomes limitations of RR [7, 8]. The novelty of the usage of these rr-values for this work comes from the application to HRF and non-conduction type arrhythmia detection.

Data visualization of successive rr-values in a scatter plot [6] has yielded interesting insights: most cases of arrhythmias exhibited rr excursions >20%. Thus, the "Zones of Cardiac Activity" concept was established: It was observed that individuals having an $|rr| < 5\%$ (Zone 1) most of the time is a warning of arrhythmia especially if the SDNN value is low (<75 ms); most healthy individuals usually have $5\% < |rr| < 20\%$ (Zone 2); rr-variations greater than 20% existing for >1% of the time (Zone 3) indicative of arrhythmias. Non-arrhythmic individuals with HRF from the THEW-Coronary Artery Disease datasets do not exhibit any discernible excursion signature on the rr scatter plot.

Table 1 shows the group statistics of the three patient classes and corresponding metrics summary represented by the median and interquartile range. To investigate the group specific distribution of the parameters, Random Forest was applied as the suitable classification technique (Fig. 1).

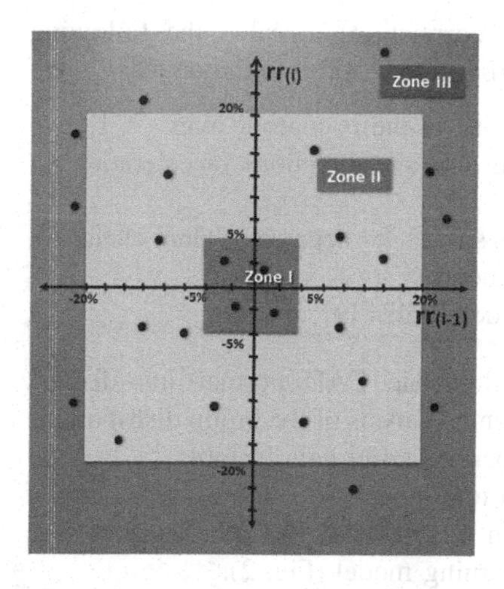

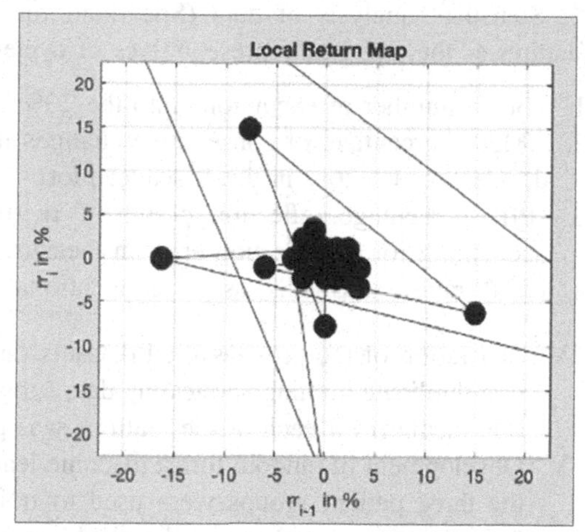

Fig. 1. Concept of zones of heart rate dynamics, rr-values in contour map (left); 60 s contour map of arrhythmic heart rate dynamics

Table 1. Group statistics (median and IQR) of training data

Class	Count	CM20	Z3e20	SDNN	sPIP	LFHF
Arrhythmic	95	1.91 [0.46,9.26]	5.90 [1.96,18.7]	133.17 [97,169]	60.16 [55,67]	0.50 [0.36,0.68]
CAD	90	0.06 [5e-3,0.16]	0.31 [0.09,0.58]	111.19 [86,129]	56.84 [52,62]	0.82 [0.61,1.03]
Normal	115	2.1e–3 [8e–4,0.11]	0.14 [0.04,0.38]	142.24 [115,168]	47.78 [45,50]	1.22 [0.93,1.33]

2.3 Methodology and Features Used for Machine Learning Model

I. Analysis of datasets to determine critical rr-derived metrics that characterize arrhythmias; this step implies arrhythmia detection as detectable RR interval fluctuations (such as sinus arrhythmia, sinus tachycardia, sinus bradycardia, etc.).

II. Extraction of features from the datasets (arrhythmia, THEW-cardiac and THEW healthy). In addition to conventional metrics such as SDNN and pNN50, the magnitude of cycle-to-cycle rr variation and consecutiveness of this magnitude was also analyzed.

III. Elimination of statistically insignificant/redundant features using Spearman rank correlation through a pairwise analysis of the features.

Statistical analysis of data (Spearman rank correlation) revealed the following features as the five best representatives of cycle-by-cycle heart rate dynamics:

1. Z3e20: number of excursions into the 20% zone in the rr-contour map
2. CM20: Percentage of consecutive changes in opposite directions (acceleration ↔ deceleration), >20% in the rr scatter plot
3. sPIP: percentage inflection points of rr (positive ↔ negative value changes), accounting for quantization error in measurements
4. LFHF ratio and SDNN: as per conventional definitions [9]

IV Separation of the pre-classified datasets (arrhythmic, CAD, normal) into distinct non-duplicate training and testing data subsets. Analysis of the group distributions (i.e. median) for each of the features was performed to gain insight.

V. Development of random forest machine learning model: 300 datasets representing the three patient groups were used to train the models. 104 new datasets (see Sect. 2.1) were used to test the machine learning model (Fig. 2).

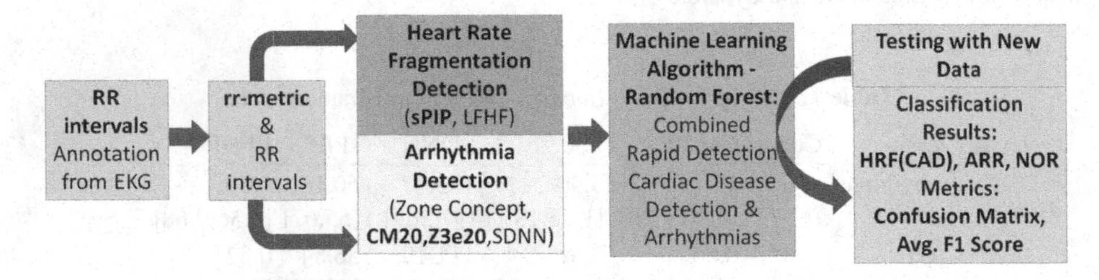

Fig. 2. Workflow methodology for HRF and arrhythmia detection

The combined incidence of HRF and arrhythmia in American adults is ~10% [12, 13]; training the detection model with ~90% "normal" patients for the sake of simulating real world scenarios is, thus, not desirable over the balanced datasets used in this work and hence the above balance of classes was chosen.

3 Results and Insights

A 30-tree ensemble of a random forest implementation [11] showed out-of-bag error leveling and was chosen. From Table 2, the F1_ARR score is 0.967, the F1_CAD score is 0.871 and F1_NOR score = 0.900; the overall F1 score is 0.911.

Table 2. Classification results of random forest classifier

Class	Actual ARR	Actual CAD	Actual NOR
Pred ARR	44	3 (FP)	0
Pred CAD	0	27	5 (FP)
Pred NOR	0	0	25 (TN)

Using a single decision tree algorithm, insights into thresholds for the various classes were identified: ARR cases were characterized by Z3e20 > 1.1%; NOR and CAD cases were separated by a sPIP threshold of 54% and LFHF of ~1.00.

4 Limitations and Future Work

The limitations of this work are analyzed by reviewing the mis-classifications among the results (Table 2). The classifier has a bias towards false positive results: 5 NOR cases are wrongly classified as CAD (sPIP > 54% and LFHF tending towards 1.00 in these cases); 3 CAD cases are wrongly classified as ARR – these cases have a Z3e20 > 1.1%, sPIP > 54% and LFHF < 1.00. The wrongly classified cases while justified by thresholds generated by the decision tree, are medically classified differently. Additional morphological features from the EKG might be required to decrease the false positive bias of the classifier.

References

1. Leading Causes of Deaths, Centers for Disease Control and Prevention (CDC). https://www.cdc.gov/nchs/fastats/leading-causes-of-death.htm
2. Costa, M.D., Redline, S., Davis, R.B., Heckbert, S.R., Soliman, E.Z., Goldberger, A.L.: Heart rate fragmentation as a novel biomarker of adverse cardiovascular events: the multi-ethnic study of atherosclerosis. Front. Physiol. (2018). https://doi.org/10.3389/fphys.2018.01117
3. PhysioNet/CinC Challenge (2017). https://physionet.org/challenge/2017/
4. Thew-project, University of Rochester, Coronary Artery Patients database – http://thew-project.org/Database/E-HOL-03-0271-002.html, Healthy Individuals database – http://thew-project.org/Database/E-HOL-03-0202-003.html
5. Goldberger, A.L., et al.: PhysioBank, PhysioToolkit, and PhysioNet: components of a new research resource for complex physiologic signals. Circulation **101**(23), e215–e220 (2000). https://doi.org/10.1161/01.CIR.101.23.e215
6. Vollmer, M.: Arrhythmia classification in long-term data using relative RR intervals, Computing in Cardiology (CinC), September 2017. http://www.cinc.org/archives/2017/pdf/213-185.pdf
7. Vollmer, M.: Ph.D. Dissertation, p. 63, Sect. 2.4.2. https://d-nb.info/1124413723/34
8. Lin, C.C., Yang, C.-M.: Heartbeat classification using normalized rr intervals and morphological features. Math. Probl. Eng. **2014**. http://dx.doi.org/10.1155/2014/712474
9. Schaffer, F., Ginsberg, J.P.: An overview of heart rate variability metrics and norms. Front. Public Health **5**, 258 (2017). https://www.frontiersin.org/articles/10.3389/fpubh.2017.00258/full
10. Costa, M.D., Davis, R.B., Goldberger, A.L.: Heart rate fragmentation: a new approach to the analysis of cardiac interbeat interval dynamics. Front. Physiol. **8**, 255 (2017). https://doi.org/10.3389/fphys.2017.00255
11. Mathworks: Matlab ver. 2018a, Random Forest Tree Bagger algorithm
12. Scripps Health. https://www.scripps.org/sparkle-assets/documents/heart_rhythm_facts.pdf
13. Sanchis-Gomar, F., Perez-Quillis, C., Leischik, R., Lucia, A.: Epidemiology of coronary heart disease and acute coronary syndrome. Ann. Transl. Med. **4**(13), 246 (2016). https://doi.org/10.21037/atm.2016.06.33

Evaluation of Random Forest and Ensemble Methods at Predicting Complications Following Cardiac Surgery

Linda Lapp[1(✉)], Matt-Mouley Bouamrane[1], Kimberley Kavanagh[1], Marc Roper[1], David Young[1], and Stefan Schraag[2]

[1] University of Strathclyde, Glasgow G1 1XH, UK
{linda.lapp,mattmouley.bouamrane,kimberley.kavanagh,marc.roper,david.young}@strath.ac.uk

[2] Golden Jubilee National Hospital, Clydebank G81 4DY, UK
stefan.schraag@gjnh.scot.nhs.uk

Abstract. Cardiac patients undergoing surgery face increased risk of postoperative complications, due to a combination of factors, including higher risk surgery, their age at time of surgery and the presence of co-morbid conditions. They will therefore require high levels of care and clinical resources throughout their perioperative journey (i.e. before, during and after surgery). Although surgical mortality rates in the UK have remained low, postoperative complications on the other hand are common and can have a significant impact on patients' quality of life, increase hospital length of stay and healthcare costs. In this study we used and compared several machine learning methods – random forest, AdaBoost, gradient boosting model and stacking – to predict severe postoperative complications after cardiac surgery based on preoperative variables obtained from a surgical database of a large acute care hospital in Scotland. Our results show that AdaBoost has the best overall performance (AUC = 0.731), and also outperforms EuroSCORE and EuroSCORE II in other studies predicting postoperative complications. Random forest (Sensitivity = 0.852, negative predictive value = 0.923), however, and gradient boosting model (Sensitivity = 0.875 and negative predictive value = 0.920) have the best performance at predicting severe postoperative complications based on sensitivity and negative predictive value.

Keywords: Postoperative complications · Machine learning · Cardiac surgery

1 Introduction

The 2011 National Confidential Enquiry into Patient Outcome and Death (NCEPOD) estimated that there are between 20,000–25,000 deaths among people undergoing a surgical procedure every year in the UK [1]. Approximately 80% of these deaths occur amongst a minority of 'high risk' patients, who make up approximately 10% of the overall surgical population. In addition to facing higher mortality rates, these patients also have increased risk of postoperative complications, and therefore require high levels of care and clinical resources before, during and after surgery [1].

D. Riaño et al. (Eds.): AIME 2019, LNAI 11526, pp. 376–385, 2019.
https://doi.org/10.1007/978-3-030-21642-9_48

Over the last two decades, an increasing number of hospitals have developed preoperative clinics and services [2] designed to triage patients well in advance of their surgery into 'low risk patients', suitable for day-care surgery, and 'high-risk patients', requiring additional management and admission as inpatients [3]. Data-driven risk scoring systems are now an integral component of these surgical pre-assessment clinics, and most of these generally focus specifically on predicting patients' risks of mortality [4].

According to the Society of Cardiothoracic Surgery in Great Britain and Ireland, the in-hospital mortality rate after cardiac surgery has remained low: i.e. under 3% over the past five years [5]. Although surgical mortality rates are low, complications after surgery are common, and can have an important impact on patients' quality of life [6]. Surgical complications can also increase hospital length of stay [7] and healthcare costs [8]. Hence, a robust and reliable predictive model for postoperative complications would prove extremely useful for managing patient flows and clinical resources in surgical care.

Although there are numerous preoperative risk models predicting mortality, such as EuroSCORE [9] and EuroSCORE II [10], there are currently no validated surgical preoperative risk scoring systems available which can predict generic surgical complications and their severity [4, 11]. In order to explore the feasibility of developing such a scoring system, we have previously explored various machine learning methods, such as logistic regression, random forest, naïve Bayes and bootstrap aggregated classification and regression trees at predicting severe postoperative complications in our patient population. As the percentage of patients with severe postoperative complications is relatively small compared to no or other complications, we are facing an imbalanced classification problem, which is one of the biggest challenges in prediction modeling due to its presence in many real-world classification tasks [12]. There are various methods available to approach this, including modifying existing algorithms to take into account the significance of positive examples and using methods to balance datasets, such as Synthetic Minority Over-sampling Technique (SMOTE) [12].

In this paper we are presenting our results from another approach: the use of ensembles of classifiers, which has been shown to have a better performance when approaching class imbalance problems [13]. Ensembles are designed to increase the accuracy of a single classifier by training several different classifiers and combining their decisions to output a single class label [13]. The range of methods which were evaluated and compared include: random forest and ensemble methods.

This paper is structured as follows: we describe our methods in Sect. 2, provide our results in Sect. 3 and discuss the relevance of our findings in Sect. 4.

2 Methods

2.1 Study Setting, Cardiac Surgery Data and Categorization of Complications

Setting. This project was conducted with the Golden Jubilee National Hospital (GJNH),[1] Clydebank, Scotland. GJNH is a state-of-the art tertiary referral center, carrying out a range of major surgical procedures (general, cardiac, orthopedic and thoracic surgery) with a commitment to reducing patient national waiting times across the National Health Service (NHS) in Scotland, while striving to deliver the highest quality of care. The hospital has 15 operating theatres. In 2016/17 GJNH carried out a total of 40,929 inpatients, day cases and diagnostic examinations.

Study Ethics and Data. This study was approved by our Institution's Research and Development Review Board and classified as an anonymized data study covered by Caldicott status. Data about cardiac procedures were obtained from a clinical audit database called the Cardiac, Cardiology and Thoracic Health Information system (CaTHI). The database consists of cardiac, cardiology and thoracic patients' diagnostics, surgical procedures and discharge information. All admissions in cardiac surgery between 1st April 2012 and 31st March 2016 were recorded in the CATHI database, adding up to a total of n = 3838 admissions. All patients reported in the CaTHI database received a treatment. In the analysis, only patients undergoing coronary artery bypass graft (CABG), valve and combined CABG and valve surgery were included in the study, the final study sample being n = 3700 clinical records.

Being a clinical audit database, most variables in the CaTHI database were consistently recorded. In cases where categorical variables had missing data, the blank fields were coded as "Unknown". The variables with "Unknown" entries included renal impairment (43.38% unknown), rhythm (7.97%), smoking status (36.24%), and left main stem disease (48.76%). If a numerical variable was not recorded consistently, the variable was excluded from the analysis. The only variable excluded for that reason was preoperative hemoglobin level.

Therefore, the final dataset used for our analysis consists of 25 preoperative variables,[2] including patient characteristics, preoperative variables about patients' cardiac status and comorbidities, as well as other surgical variables.

Categorization of Complications. With the assistance of a panel of consultant cardiac anesthetists and surgeons in GJNH, we categorized complications reported in the CaTHI database into four discrete categories (no/mild/moderate/severe) based on their impact on hospital length of stay, patients' quality of life and cost of care. For example,

[1] https://www.nhsgoldenjubilee.co.uk/.

[2] Attributes are: Age, sex, diabetes, body mass index, smoking status, surgical priority, critical preoperative state, procedure, left main stem, extracardiac arteriopathy, pulmonary disease, creatinine level, renal impairment, New York Heart Association grade, angina status, rhythm, left ventricular function, neurological dysfunction, congestive cardiac failure, previous myocardial infarction, active endocarditis, hypertension, previous cardiac surgery, previous percutaneous coronary intervention.

urinary retention was categorized as "mild", elevated creatinine as "moderate", and renal failure as "severe". The categorization was subsequently cross-referenced with findings from a literature review we have conducted in relation to risk scoring of perioperative complications. Based on the categorization task, our patient population was recorded to have 17 types of mild complications, 42 moderate complications and 19 severe complications.[3]

2.2 Model Development

In this study, we have focused on developing a predictive model for solving a binary classification task: i.e. whether a patient is likely to have a severe postoperative complication ('yes' or 'no or other'). The reason why we chose to focus on predicting severe complications in the first instance is due to the fact that these have the most detrimental impact to patients and on the use of clinical resources (e.g. such as requiring additional procedures to manage the complication or increasing hospital length of stay).

As this is an imbalanced classification problem involving both categorical and numerical variables, we used machine learning methods shown to be effective [14] for this kind of data analysis: *random forest (RF), AdaBoost (AB), gradient boosting model (GBM)* and *two stacked models*.

All analysis was conducted with statistical package R version 3.5.0.

Random Forest, AdaBoost and Gradient Boosting. The RF, AB and GBM were developed using *k*-fold cross-validation, where the training data (n = 2479 records) was randomly partitioned into *k* sub-sets of approximately equal sizes. At each *k* iteration one of the folds is chosen as the test set and the remaining *k-1* are used for the training.

This method often results in a less biased and less optimistic estimate of the model than other methods. In this study we use 5-fold cross-validation, as is generally recommended in the literature [15].

For RF, the package 'randomForest' version 4.6-14 was used with the number of trees set at n = 200. The AB model was developed using the package 'fastAdaboost' version 1.0.0, which implements Freund and Schapire's Adaboost.M1 algorithm [16], and for which we conducted n = 40 iterations. For the GBM, the package 'gbm' version 2.1.5 was used, which uses the Friedman's gradient boosting algorithm [17]. The number of trees was chosen to be n = 1000 and the shrinkage parameter as 0.01. For these three models, we evaluated the performance using a separate set of testing data (n = 1221 records).

Stacked Models. The base learners included in our stacked models were *generalized linear model* [18], *random forest* [19], *naïve Bayes* [20] and *bootstrap aggregated classification and regression trees (Bagging CART)* [21]. These learners are appropriate

[3] Severe complications in this study include: Acute renal failure, deep sternal wound infection, septicemia, transient stroke, tracheostomy, cardiac arrest, permanent stroke, severe heart failure, adult respiratory distress syndrome, multi-organ failure, mesenteric infarction, required laparotomy, severe pulmonary edema, left ventricular wall dissection, hepatic failure, reopening requiring coronary artery bypass graft, paraparesis, and amputation.

for our data due to allowing both numerical and categorical features in the dataset. We firstly generated k-fold cross-validated predicted values from the base learners to generate the training data for the metalearner algorithm. The training set (n = 1850 records) was used to develop our base learners. Then a validation set (n = 925 records) was used to create the level one dataset. The base learners and the ensemble were then evaluated using the testing dataset (n = 925 records). In this study we compared two different metalearner algorithms: random forest (n = 3000 trees) and generalized linear model. All analysis for the stacked models was done using the package 'caret' version 6.0–81.

2.3 Model Evaluation and Performance Measures

The models were evaluated based on the area under the receiver operating characteristic (ROC) curve (AUC), sensitivity (a.k.a. recall), specificity (a.k.a. true negative rate), and positive (PPV) and negative predictive value (NPV). We also look at the Gini importance measure and relative influence for the variables in RF and GBM model, respectively. As this is an imbalanced classification problem, where the prevalence of severe postoperative complications is small compared to *'no or other'* complications, using these performance measures help us avoid the accuracy paradox [22].

As the aim of this study is to predict severe complications, we are aiming for the highest sensitivity and negative predictive value as possible. This is to ensure that the model recognizes as many patients with severe complications as possible *(i.e. sensitivity)* and in case of negative testing: to ensure that the probability that the patient actually does not have a severe complication is high *(i.e. negative predictive value).* However, in order to compare our results with existing literature, we also discuss AUC as a performance measure.

3 Results

3.1 Population Characteristics

In our study sample of n = 3700 clinical records and using the classification of complications described earlier in Sect. 2.1, 48.65% of the patients had a recorded postoperative complication. Of these: 7.05% had mild complications, 36.65% moderate complications, and 4.95% severe complications after cardiac surgery. The prevalence for severe postoperative complications indicates that this is a highly imbalanced classification task.

Of all patients, 59.65% had a CABG, 26.49% had a valve surgery, and 13.86% had a combined CABG and valve surgery. The mean age was 66.7, with a median of 68 years. Most of the patients were men (73.22%). The majority of the patients (77.54%) had an elective surgery, 14.27% had an urgent surgery, 7.49% had a prioritized surgery and a small number of patients (0.70%) had an emergency surgery.

3.2 Performance of the Models

Table 1 shows that in terms of AUC, AB outperforms RF, GBM and the stacked models with an AUC of 0.731. However, as our end goal is to develop a clinical decision support system predicting severe postoperative complications, our aim is to have the highest possible sensitivity and negative predictive value. Based on that, the GBM has the highest sensitivity of 0.875, meaning that the model recognizes patients with severe complications 87.5% of the time. The GBM also has a very high negative predictive value of 0.920, which means that if the test is negative, the probability that the patient actually does not have a severe complication is 92.0%.

Table 1. Area under the curve (AUC), sensitivity, specificity, positive (PPV) and negative predictive value (NPV) for the models.

Algorithm	AUC (95% CI)	Sensitivity (95% CI)	Specificity (95% CI)	PPV (95% CI)	NPV (95% CI)
RF	0.724 (0.650–0.798)	0.852 (0.790–0.912)	0.462 (0.390–0.534)	0.017 (0.010–0.023)	0.923 (0.876–0.969)
AB	0.731 (0.658-0.804)	0.738 (0.665–0.811)	0.629 (0.552–0.706)	0.021 (0.013–0.029)	0.905 (0.854–0.956)
GBM	0.718 (0.644–0.792)	0.875 (0.818–0.932)	0.465 (0.392–0.538)	0.014 (0.008–0.020)	0.920 (0.873–0.967)
Stacked with RF	0.659 (0.583–0.735)	0.804 (0.737–0.871)	0.472 (0.399–0.545)	0.026 (0.017–0.035)	0.911 (0.861–0.961)
Stacked with GLM	0.655 (0.579–0.731)	0.643 (0.566–0.720)	0.639 (0.562–0.716)	0.035 (0.024–0.046)	0.897 (0.844–0.950)

Surprisingly, both stacked models had a considerably worse performance in terms of AUC compared to the other models. Even though the stacked model with RF has a high sensitivity, in order to reach such high level of sensitivity, the model was built using 3000 trees. A high number of trees, however, results in the model being computationally expensive, making it difficult to apply to clinical practice. Compared to other models, the GLM stacked model had a considerably worse performance in terms of sensitivity compared to the other models.

As the RF and GBM have the highest sensitivities and negative predictive values, we further investigated these two models. To assess which variables are the most important, for RF we calculated at the Gini importance measure and for GBM we calculated the relative influence (Table 2).

Both of these models show—with some differences in ordering—that preoperative creatinine, BMI, age, angina status and smoking are the most important variables when predicting severe complications. These results are also supported by findings from the literature: elderly patients are at a greater risk of postoperative complications, especially for bleeding, infections, neurologic and renal problems [23]. Patients with a higher BMI have increased risk of wound infection, blood loss and acute kidney injury [24]. Angina status is shown to be a significant predictor of long-term mortality [25]. Persistent smokers have a higher incidence of pulmonary complications [26] and also slower wound healing following CABG surgery [27].

Table 2. The importance measures for the top five variables of Gradient Boosting Model (GBM) and Random Forest (RF).

Variable	GBM (relative influence)	RF (Gini importance)
Pre.Op. Creatinine	17.93	31.89
BMI	16.41	35.24
Age	10.25	28.90
Angina status	6.90	13.27
Smoking	6.57	12.04

4 Discussion

Our study found that postoperative complications are common (48.65% in our study population) and the most severe of these—although less frequent at 4.95%—can have a significant impact on episodes of care and use of clinical resources as well as being potentially devastating for patients' quality of life after surgery. It is therefore essential that adequate systems are developed within clinical care in order to better plan and mitigate these instances of severe perioperative complications.

Trying to approach that problem, some studies have been carried out to assess how EuroSCORE I and II predict combinations of postoperative complications [23, 28–31]. Looking at the AUC, our AB model outperforms both versions of EuroSCORE in all aforementioned studies. Our RF model has a similar performance to EuroSCORE and EuroSCORE II in one study (AUC = 0.72 for both) [31]. Even though our GBM model has the lowest performance out of these three in our study, it still outperforms the commonly used risk models in most mentioned studies, apart from EuroSCORE and EuroSCORE II in one study [31].

Performance Measures. Even though the AB model has the highest AUC, the performance of sensitivity and negative predictive value are the most important for the purpose of developing a decision support application for severe complications. A model with a high specificity can be used to rule out patients who do not need specific treatment [32]. However, our aim is to develop a model which can identify which patients are more likely to develop severe postoperative complications, in order to improve care planning, management and monitoring. Having a higher negative predictive value, meaning the patient probably does not have the disease when the test is negative, reassures the provider of the treatment to do no harm.

Some of the previously mentioned papers evaluating the commonly used preoperative risk tools predicting complications have similar results based on AUC as our models. However, these studies have not reported other performance measures such as sensitivity, specificity, PPV and NPV.

Current Challenges in Predicting Postoperative Complications. At present a major obstacle in predicting postoperative complications is that there is currently no single nomenclature of surgical complications; unlike for clinical diagnosis (i.e. the International Statistical Classification of Diseases, ICD-10[4]). Due to that, when comparing our results with the literature, all of the aforementioned studies have a different definition for "morbidity", which includes a different set of combined complications. The reporting of different complication outcomes in the scientific literature therefore prevents the objective comparison of the performance of these predictive risk models.

It is also worth mentioning that prediction models appear to have very good performance when applied at the population level [9, 10], i.e. their prediction accuracy generally performs well when applied to broad group or categories of patients. However, the prediction performance of these models at the 'individual' level is in fact far less satisfactory [33].

Conclusion and Future Work. In this study, we have highlighted how the use of machine learning techniques could be applied to the problem of predicting postoperative complications and compared the performance of several approaches.

Through our analysis we found two machine learning models suitable for predicting severe postoperative complications: random forest and gradient boosting model based on sensitivity (0.852 and 0.875, respectively) and negative predictive value (0.923 and 0.920, respectively). Either of these models could help a clinician to identify patients who are at risk of having severe postoperative complications in order to allocate resources or avoid high-risk treatments. In order to develop a usable clinical decision support system that relies on the models developed in this study, a further validation study needs to be undertaken.

References

1. Findlay, G.P., Goodwin, A.P.L., Protopapa, K., Smith, N.C.E., Mason, M.: Knowing the risk: a review of the peri-operative care of surgical patients, London (2011)
2. Bouamrane, M.-M., Mair, F.S.: Implementation of an integrated preoperative care pathway and regional electronic clinical portal for preoperative assessment. BMC Med. Inform. Decis. Mak. **14**, 93 (2014)
3. Bouamrane, M.-M., Mair, F.S.: A study of clinical and information management processes in the surgical pre-assessment clinic. BMC Med. Inform. Decis. Mak. **14**, 22 (2014)
4. Moonesinghe, S.R., Mythen, M.G., Das, P., Rowan, K.M., Grocott, M.P.: Risk stratification tools for predicting morbidity and mortality in adult patients undergoing major surgery: qualitative systematic review. Anesthesiology **119**, 959–981 (2013)
5. SCTS: Blue Book Online. http://bluebook.scts.org/
6. Pinto, A., Faiz, O., Davis, R., Almoudaris, A., Vincent, C.: Surgical complications and their impact on patients' psychosocial well-being: a systematic review and meta-analysis. BMJ Open **6**, e007224 (2016)
7. Knapik, P., Ciesla, D., Borowik, D., Czempik, P., Knapik, T.: Prolonged ventilation post cardiac surgery - tips and pitfalls of the prediction game. J. Cardiothorac. Surg. **6**, 158 (2011)

[4] http://apps.who.int/classifications/icd10/browse/2010/en.

8. Eappen, S., et al.: Relationship between occurrence of surgical complications and hospital finances. JAMA **309**, 1599–1606 (2013)
9. Roques, F., Michel, P., Goldstone, A.R., Nashef, S.A.M.: The logistic EuroSCORE. Eur. Heart J. **24**, 1–2 (2003)
10. Nashef, S.A., et al.: EuroSCORE II. Eur. J. Cardio-Thorac. Surg. **41**, 734–744 (2012)
11. Barnett, S., Moonesinghe, S.R.: Clinical risk scores to guide perioperative management. Postgrad. Med. J. **87**, 535–541 (2011)
12. Chawla, N.V., Bowyer, K.W., Hall, L.O., Kegelmeyer, W.P.: SMOTE: synthetic minority oversampling technique. J. Artif. Intell. Res. **16**, 321–357 (2002)
13. Galar, M., Fernandez, A., Barrenechea, E., Bustince, H., Herrera, F.: Review on Ensembles for the Class Imbalance Problem: Bagging-, Boosting-, and Hybrid-Based Approaches. IEEE Trans. Syst. Man Cybern. Part C Appl. Rev. **42**, 463–484 (2012)
14. Kuncheva, L.I.: Combining Patterns Classifiers: Methods and Algorithms. Wiley, New York (2004)
15. Bischl, B., Mersmann, O., Trautmann, H., Weihs, C.: Resampling methods for meta-model validation with recommendations for evolutionary computation. Evol. Comput. **20**, 249–275 (2012)
16. Freund, Y., Schapire, R.E.: Experiments with a new boosting algorithm. In: The Thirteenth International Conference on Machine Learning, pp. 148–156 (1996)
17. Friedman, J.H.: Greedy function approximation: a gradient boosting machine. Ann. Stat. **29**, 1189–1232 (2001)
18. Walker, S.H., Duncan, D.B.: Estimation of the probability of an event as a function of several independent variables. Biometrika **54**, 167–179 (1967)
19. Ho, T.K.: The random subspace method for constructing decision forests. IEEE Trans. Pattern Anal. Mach. Intell. **20**, 832–844 (1998)
20. Zhang, H.: The optimality of naive Bayes. In: FLAIRS2004 Conference (2004)
21. Breiman, L.: Bagging predictors. Mach. Learn. **24**, 123–140 (1996)
22. Valverde-Albacete, F.J., Pelaez-Moreno, C.: 100% classification accuracy considered harmful: the normalized information transfer factor explains the accuracy paradox. PLoS ONE **9**, e84217 (2014)
23. Wang, T.K., Li, A.Y., Ramanathan, T., Stewart, R.A., Gamble, G., White, H.D.: Comparison of four risk scores for contemporary isolated coronary artery bypass grafting. Hear. Lung Circ. **23**, 469–474 (2014)
24. Reis, C., Barbiero, S.M., Ribas, L.: The effect of the body mass index on postoperative complications of coronary artery bypass grafting in elderly. Braz. J. Cardiovasc. Surg. **23**, 524–529 (2008)
25. Kaul, P., Naylor, C.D., Armstrong, P.W., Mark, D.B., Theroux, P., Dagenais, G.R.: Assessment of activity status and survival according to the Canadian Cardiovascular Society angina classification. Can. J. Cardiol. **25**, e225–e231 (2009)
26. Ji, Q., Zhao, H., Mei, Y., Shi, Y., Ma, R., Ding, W.: Impact of smoking on early clinical outcomes in patients undergoing coronary artery bypass grafting surgery. J. Cardiothorac. Surg. **10**, 16 (2015)
27. Sharif-Kashani, B., et al.: Smoking and wound complications after coronary artery bypass grafting. J. Surg. Res. **200**, 743–748 (2016)
28. Geissler, H.J., et al.: Risk stratification in heart surgery: comparison of six score systems. Eur. J. Cardio-Thorac. Surg. **17**, 400–406 (2000)
29. Hirose, H., et al.: EuroSCORE predicts postoperative mortality, certain morbidities, and recovery time. Interact. Cardiovasc. Thorac. Surg. **9**, 613–617 (2009)

30. Pitkänen, O., Niskanen, M., Rehnberg, S., Hippelainen, M., Hynynen, M.: Intra-institutional prediction of outcome after cardiac surgery: comparison between a locally derived model and the EuroSCORE. Eur. J. Cardiothorac. Surg. **18**, 703–710 (2000)
31. Wang, T.K.M., Harmos, S., Gamble, G.D., Ramanathan, T., Ruygrok, P.N.: Performance of contemporary surgical risk scores for mitral valve surgery. J. Card. Surg. **32**, 172–176 (2017)
32. Lutkenhoner, B., Basel, T.: Predictive modeling for diagnostic tests with high specificity, but low sensitivity: a study of the glycerol test in patients with suspected meniere's disease. PLoS ONE **8**, e79315 (2013)
33. McEvoy, J.W., et al.: Risk and the physics of clinical prediction. Am. J. Cardiol. **113**, 1429–1435 (2014)

Mining Compact Predictive Pattern Sets Using Classification Model

Matteo Mantovani[1](✉), Carlo Combi[1], and Milos Hauskrecht[2]

[1] Department of Computer Science, University of Verona, Verona, Italy
{matteo.mantovani,carlo.combi}@univr.it

[2] Department of Computer Science, University of Pittsburgh, Pittsburgh, USA
milos@pitt.edu

Abstract. In this paper, we develop a new framework for mining predictive patterns that aims to describe compactly the condition (or class) of interest. Our framework relies on a classification model that considers and combines various predictive pattern candidates and selects only those that are important for improving the overall class prediction performance. We test our approach on data derived from MIMIC-III EHR database, focusing on patterns predictive of sepsis. We show that using our classification approach we can achieve a significant reduction in the number of extracted patterns compared to the state-of-the-art methods based on minimum predictive pattern mining approach, while preserving the overall classification accuracy of the model.

1 Introduction

Past decade has witnessed an explosion in the number of medical and healthcare datasets available to researchers and healthcare professionals. However, the analyses and utilization of these datasets still lack the data collection efforts. This prompts the development of appropriate data mining techniques and tools that can automatically extract relevant information from data and consequently provide insight into various clinical behaviors or processes captured by the data. Since these tools should interact with medical experts, it is important that all the extracted information is represented in a human-friendly way, that is, in a concise and easy-to-understand form.

One way to present knowledge to humans is to use if-then rules, that relate a condition defining a subpopulation of instances (or patients) with observed outcomes. The strength of this relation can be expressed using various statistics, such as precision and support. This human-friendly form facilitates the exploration, discovery and possible utilization of these patterns in healthcare. For example, consider a rule mining algorithm that identifies a subpopulation of patients that respond better to a certain treatment than the rest of the patients. If the rule clearly and concisely defines this subpopulation, it can be validated and potentially utilized to improve patient management and outcomes.

Many strategies to mine 'if-then' rules from the data exist. One is association rule mining [1,2]. It gained a lot of popularity in data mining research

D. Riaño et al. (Eds.): AIME 2019, LNAI 11526, pp. 386–396, 2019.
https://doi.org/10.1007/978-3-030-21642-9_49

[14], including medical data mining [8,18]. The key strength of association rule mining is that it searches the space of rules completely by examining all patterns that occur frequently in the data. Its disadvantage is that the number of association rules it finds and outputs is often very large. This may hinder the discovery process and the interpretability of the results. Hence, it is desirable to reduce the mined rule set as much as possible while preserving the most important relations (rules, patterns) found in the data. Various rule interestingness statistics and constraints based on such statistics have been proposed to address this problem [13].

The objective of this work is to study new ways of improving association rule mining that can lead to a smaller set of rules that are sufficient to capture the essential underlying patterns in the data. This requires analyzing relations among the mined rules and defining criteria for assessing the importance of individual rules w.r.t. other rules. The key principle studied and applied in this work for filtering the rules is rule redundancy. Our approach builds upon the minimum predictive pattern mining idea proposed by Batal and Hauskrecht [6] to eliminate spurious and highly redundant rules, and attempts to improve it by reducing the set of mined minimum predictive rules using an auxiliary classification model that combines the rules into one model. Since in general the search for the optimal set of rules is equivalent to the optimal subset selection problem [17], we propose and experiment with a more efficient greedy rule selection algorithm that avoids the need to explore and evaluate all possible rules subsets.

We have tested our method on data from MIMIC-III [15] EHR database. More specifically, our goal is to discover patterns that are associated with sepsis and its treatments. We compare our method to the original one [6] and show that the number of rules found by our method is significantly smaller than the original set. Moreover we show that the performance of the classification model that is based upon our rule set is close or better than classification models built by Batal's rule sets.

2 Related Work

Association rule mining [1,2] is a method for identifying strong relations in a dataset based on some measure of interestingness (e.g., confidence/precision, support or lift [13]). Typically, such relations are expressed in terms of if-then rules consisting of different rule antecedents (conditions) and consequents (targets). The majority of association rule mining algorithms rely on Apriori algorithm [2]. The algorithm searches the pattern space defining the condition of the rule by starting with more general patterns with the highest support before inspecting more specific patterns with a lower support. The process is bottomed-out by the minimum support parameter.

When the rule mining process is focused on a specific target class, we refer to it as to predictive pattern (rule) mining [16]. The task of identifying all important predictive patterns from a large pool of frequent patterns is similarly to association rule mining time-consuming, and may lead to a huge number predictive rules. One important contribution in limiting the size of the rule set is the minimal predictive rule mining approach proposed in [6] to eliminate spurious predictive patterns. Briefly, a pattern is called spurious when it is predictive when evaluated alone, but is redundant given one of its subpatterns. Spurious patterns may be formed by adding irrelevant items to other simpler predictive patterns. Approach in [6] eliminates spurious patterns using statistical test based on binomial distribution. Later the same authors proposed a more robust Bayesian criterion to perform the spurious pattern elimination [3]. The minimum predictive rule mining approach has been successfully adapted and applied to mine temporal clinical data [4,5,7].

Predictive pattern mining process can be used for knowledge discovery when the goal is to extract a set of rules describing patterns that are important for a specific target class. Alternatively, it can be used to define a classifier [6]. In such a case, predictive patterns can be viewed as nonlinear features helping to improve overall performance of a classification algorithm. This complementary use of predictive patterns raises an interesting question. Is it possible to reduce the set of extracted predictive rules with the help of a classification model? That is, are there any rule redundancies that can be eliminated when we combine the rules into a classification model? Research in this work is centered around this interesting question. More specifically, we use mined set of minimum predictive rules to define features of the linear classification model based on Support Vector Machines (SVM). Then, feature selection methods are applied to further reduce the rule set, aiming to extract the set that optimizes the classification performance of the classification model.

3 Method

3.1 Definitions

Assume a dataset with only categorical features (attributes): all numeric features should be first discretized. Each (feature, value) pair is mapped to a distinct item in $\Sigma = \{I_1, ..., I_l\}$. A pattern is a conjunction of items: $P = I_{q_1} \wedge ... \wedge I_{q_k}$ where $I_{q_j} \in \Sigma$. If a pattern contains k items, we call it a k-pattern (an item is a 1-pattern). Assume an item $I = (fea, val)$, where fea is a feature and val is a value. Given a data instance $\boldsymbol{x}$, we say that $I \in \boldsymbol{x}$ if $fea(\boldsymbol{x}) = val$ and that $P \in \boldsymbol{x}$ if $\forall I_j \in P : I_j \in \boldsymbol{x}$.

Given a dataset $D = \{\boldsymbol{x}_i\}_{i=1}^n$, the instances that contain pattern P define a group $D_P = \{\boldsymbol{x}_j | P \in \boldsymbol{x}_j\}$. If P' is a subpattern of P ($P' \subset P$), then $D_{P'}$ is a supergroup of D_P ($D_{P'} \supseteq D_P$). Note that the empty pattern Φ defines the entire population. The support of P is defined as: $sup(P) = |D_P|/|D|$.

In this paper we are interested in mining patterns that are predictive of class c. So for pattern P, we can define a predictive pattern (or a rule) R: $P \Rightarrow c$ with respect to class label c. The confidence of R is the precision (or posterior probability of c in group D_P). Note that confidence of $\Phi \Rightarrow c$ is the prior probability of c. We say that rule R': $P' \Rightarrow c'$ is a subrule of rule R: $P \Rightarrow c$ if $c' = c$ and $P' \subset P$.

Let $\Omega = \{P_1, ..., P_m\}$ be a set of patterns predictive of c. Given a dataset $D = \{\boldsymbol{x}_i, y_i\}_{i=1}^n$ defined in d-dimensional feature space and a set of patterns Ω the instances in D can be mapped into a new m-dimensional binary array D_Ω as follows:

$\boldsymbol{x}_i \rightarrow \{b_{i,1}, ..., b_{i,m}\}$ where $b_{i,j} = 1$ if $P_j \in \boldsymbol{x}_i$ and $b_{i,j} = 0$ if $P_j \notin \boldsymbol{x}_i$.

We refer to new $D_\Omega = \{\boldsymbol{x}'_i, y_i\}_{i=1}^n$ as to the pattern induced projection of the dataset D based on patterns in Ω. The pattern induced dataset D_Ω and its instances can be used to define and also learn a binary classification model $f : \boldsymbol{x}'_i \rightarrow y_i = c$ that distinguishes instances with the target class c from other classes. Effectively, this classification model combines a set of patterns predictive of c into a unified model for predicting the same class.

3.2 Problem

Our objective is to identify a small set of predictive patterns (rules) for the target class c from the data. To achieve this we propose a new two-step pattern mining process.

First, the number of predictive rules one can define by considering just the rule support and its precision can be enormous and may include a large number of spurious patterns. Hence we restrict our attention only to non-spurious rules. We mine these rules using Apriori algorithm proposed by [6] that includes binomial test when selecting more specific rules.

Second, to further limit the number of predictive rules we combine the minimal predictive patterns into a unified classification model to search for the optimal minimal pattern set Ω^* predictive of the target class c. We define the optimal pattern set to be the minimal pattern set that leads to the best combined generalization performance discriminating class c from the rest of the classes.

In the following we first describe the idea behind the minimum predictive patterns, and the unified classification models. After that we propose a greedy search algorithm that combines the two ideas into one search mechanism for identifying small sets of predictive patterns.

3.3 Minimum Predictive Patterns

Our solution builds upon the concept of minimum predictive patterns (MPRs) proposed by Batal and Hauskrecht [6].

Definition: A predictive pattern $R : A \rightarrow c$ is a called minimal, if and only if, R predicts class c significantly better than all its subpatterns.

The gist of this definition is that every item in the condition of the predictive pattern R is an important contributor to its prediction, that is, removal of any of the items in the condition would cause a significant drop in its predictive performance. The significance of the pattern R is determined using a statistical test derived from the binomial distribution. Let us assume we are interested in testing the significance of rule $R : A \rightarrow c$. Assume that pattern

A consists of N instances, out of which N_c instances belong to class c. Let P_c represents the highest probability achieved by any subpattern of R, that is, $P_c = \max(A' \subset A)Pr(c|A')$. To test, if the pattern R is significantly different, we hypothesize (null hypothesis) that N_c is generated from N according to the binomial distribution with probability P_c. If we cannot reject the hypothesis at some significance level, then, R is not significantly different from the subpattern with P_c. However, we say that pattern R is significantly different when we can reject the above hypothesis and show that the probability that generated N_c class x instances out N is significantly higher than P_c. We can perform this test using a one sided significance test and calculate its p-value. If this p-value is significant (smaller than a significance level α), we conclude that R significantly improves the predictability of c over all its simplifications, and hence R is a MPR. The mining algorithm to mine minimal predictive patterns relies on the Apriori algorithm that uses a minimum support parameter. The algorithm generates all patterns starting from more general patterns to more specific that satisfy the minimum support, but only the patterns that satisfy the binomial test (the minimality condition) are retained. As shown by studies in [6] such an algorithm retains significantly smaller subset of predictive patterns.

3.4 Combining Predictive Patterns via Classification Model

Our second solution attempts to reduce the number of minimum patters mined by considering their combinations. Briefly, we are interested in retaining only a subset of minimum predictive patterns that are critical for predictive performance of the classification model defined on the pattern induced dataset.

There are many classification models one can define on the binary dataset induced by the predictive patterns. In this work, instead of considering all possible classification models, we restrict our attention to linear support vector machines (SVM) models with shared discriminant functions (discriminating class c from the rest of the classes) that are defined by a linear combination of predictive patterns. To judge and compare the quality of such models across many features we use the area under the ROC curve (AUROC) statistic.

In general the problem of finding the optimal subset of minimum predictive patterns that leads to the best performing classification model is intractable. In order to make the search more efficient we resort to greedy pattern search approach. To make the choices of patterns we rely on the wrapper approach that tests, and selects patterns by considering the internal validation approach. That is, in order to compare two distinct sets of patterns Ω and Ω', we use the internal train and test splits of the data to evaluate the AUROC performance of the two sets in combination with the SVM model. The model and its patterns set with better AUROC performance is preferred. In the following we describe the specific algorithm we use to search a subset of minimum predictive patters to identify the best set.

3.5 Greedy Pattern Subset Selection Algorithm

Our approach starts by splitting dataset D into the training and test sets. All pattern selection and learning is always done on the training set. We use the test set only for the final evaluation.

Since our algorithm searches and compares many different subsets of predictive patterns, we use internal validation process to measure their quality and choose better subsets. Briefly, in order to evaluate and compare the goodness of a specific set of patterns Ω to other candidate sets, we use a classification model based on the linear SVM that is run on the data induced by Ω. We use multiple internal validation splits of the training data to make the comparison. The training dataset is divided as follows: first we randomly pick 30% of the data rows and use them as the test set, the remaining rows are reshuffled 10 times and for every reshuffle 80% of the data are used as the internal training set and the remaining 20% as the internal validation set. The goodness of Ω is then estimates by averaging the AUROC score for all internal splits obtained through reshuffling.

While our ultimate goal would be to find a set of predictive patterns that are optimal in terms of the quality of the predictive performance of a classifier that combines them, the full search is infeasible. To avoid the full pattern subset search, we adapt a greedy approach that generates, examines and selects the patterns level-wise, where a level k covers all k-patterns. More specifically, our method uses a two-stage procedure. First, using an Apriori algorithm with the minimum support threshold and the binomial test proposed by Batal et al., we generate a set of minimum predictive patterns for each level k. Second, we use these minimum level-wise patterns to construct greedily the final set of patterns. We implemented two procedures to conduct the greedy search. One that searches and constructs the subset of patterns starting from the most general (level 1) patterns and gradually adds new more specific (higher level) patterns. We refer to this procedure as the top-bottom greedy procedure. The other procedure starts from the most specific patterns (the highest level minimum predictive patterns) and greedily adds to the set more general patterns of lower complexity. We refer to this method as to the bottom-up greedy procedure.

Let us assume that Ω' is our current set of patterns (selected in the previous steps). Our greedy search algorithm on level k works by first trying each minimum pattern on level k in combination with Ω'. Each of these combinations are ranked in terms of the AUROC score based on the internal validation. This order defines a greedy order in which all k-level minimum patterns are sequentially tried and if successful (in terms of AUROC improvement) they are added (one-by-one) to the resulting set of patterns. The same procedure for greedily adding the patterns on level k is applied whether we build the patterns in the top-down fashion (from level 1 patterns) or from the bottom-up (from highest level patterns). The reason for using the bottom-up greedy search process is that it tends to retain a greater number of the more specific patterns.

4 Experiments

4.1 Data

To test and validate our method, we analyze clinical data derived from MIMIC III dataset [15] with the goal of identifying patterns predictive of sepsis diagnosis. Briefly, MIMIC is a publicly available database that contains EHR data for patients treated in intensive care units between 2001 and 2012. The data are de-identified and associated with 46000+ patients and ~60000 admissions. The data consist of multiple clinical data sources: measurements of hourly vital signs (heart rate, blood pressures, oxygen saturation, and so on), administered drugs, labs and diagnosis for every patient. However, before analysis, it is necessary to transform the MIMIC-III raw data in a form that we could mine. This was accomplished through an E.T.L. (Extract, Transform, Load) process. One source of our data was CHARTEVENTS, that is the vital signs table. We used it to extract specific measurements of hearth rate, diastolic and systolic blood pressure, white blood cells, and body temperature across the admission. For each of these variables we created two attributes, one containing its maximum value during the hospitalization of a patient, and the other one containing its minimum value. Instead of numerical values, all these measurements were discretized to low, normal and medium ranges, using the thresholds shown in Table 1. Other information we selected from the records came from "PROCEDURES_ICD" table which we used to determine whether a patient had a procedure or not (true\false attribute is created) during the hospitalization. We applied the same transformation to table "DIAGNOSES_ICD" to identify all the patients diagnosed with sepsis. "INPUTEVENTS_MV" table consists of medication administration records. We used it to extract some medications administered to the patient, such as vancomycin, piperacillin/tazobactam, ciprofloxacin, epinephrine, norepinephrine, vasopressin, dopamine, metoprolol, potassium chloride, phenylephrine, omeprazole (prilosec), and pantoprazole (protonix). Let us note that while some of these medications are commonly used for treating patients with sepsis, whereas other medications such as metoprolol, potassium chloride, phenylephrine, omeprazole (prilosec), and pantoprazole (protonix) are more general. These were included to test the effectiveness of our method when mining patterns related to sepsis. At the end of the E.T.L. process we obtain data for 21880 patients, 2806 of them with sepsis.

Table 1. Thresholds used to discretize the considered vital signs in low, medium, high.

	Heart rate	Diastolic BP	Systolic BP	White blood cells	Body temperature
Low	<60	<60	<90	<4.0	<36.0
High	>90	>90	>140	>12.0	>38.0

4.2 Results

Table 2 shows the results we obtained on MIMIC-III data for the minimum predictive rule mining approach by Batal and Hauskrecht [6], and two versions of our greedy classification model driven subset selection approach. The main statistics we use to evaluate the quality of the predictive rule set is the area under the ROC (Receiver Operating Characteristics) curve (AUROC) [12]. All AUROC statistics listed in the table are obtained on the test data. In addition to AUROC performances, we list the number of patterns found by the different methods. For example, the minimum predictive pattern (MPR) baseline used 85 patterns and reached AUROC performance of 0.8580. As we can see, both greedy methods outperformed (in terms of the AUROC classification performance) the baseline. Moreover this improvement is accompanied by a significant reduction in the total number of patterns used in the set compared to the baseline. We note that while there is nearly no difference in the AUROC performance among the two versions of our greedy method, the number of patterns found and used by the two is significantly different. In particular, we observe that the majority of the patterns in the bottom-up approach are more complex patterns while the majority of patterns in the top-down approach are 1-patterns. This shows that bottom-up approach tends to keep more detailed patterns compared to the top-down approach.

Table 2. Comparison between the results for our method and Batal et al.'s predictive pattern mining method.

Method	AUROC	Number of patterns
MPR (Batal et al.)	0.8580	85
Our method (bottom-up)	0.8643	33
Our methods (top-down)	0.8635	19

5 Discussion

Sepsis is the systemic response to infection, and there are many conditions that would indicate its occurrence during the admission or hospital stay, such as: temperature >38 °C or <36 °C; heart rate >90 beats per minute; systolic blood pressure <90 mm Hg, and white blood cell count >12,000/cu mm or <4,000/cu mm [9]. Moreover, patients with sepsis are usually treated with antibiotics such as vancomycin, piperacillin/tazobactam, ciprofloxacin, and drugs treating episodes of hypotension such as epinephrine, norepinephrine, phenylephrine, vasopressin, and dopamine [11].

Table 3 lists all minimal predictive patterns that we mined using the bottom-up greedy procedure. The table entries include the absolute weight the rule was assigned by the final classification model, the rule support and the rule precision. By analyzing the results with respect to sepsis symptoms and treatments we

Table 3. The mined set of minimal predictive patterns with their absolute weight, support and precision

Pattern	Rule weight	Support	Precision
Norepinephrine = true	0.4667	0.1453	0.5571
Norepinephrine = true & Vancomycin = true	0.2628	0.1114	0.4865
Piperacillin/Tazobactam = true	0.2476	0.143	0.4449
MaxSystolicBloodPressure = low	0.1981	0.4311	0.5267
Ciprofloxacin = true	0.1750	0.1157	0.2671
Vancomycin = true	0.1619	0.3890	0.7897
Pantoprazole (Protonix) = true & MaxSystolicBloodPressure = low	0.1418	0.1437	0.3128
Norepinephrine = true & Piperacillin/Tazobactam = true	0.1159	0.0566	0.2854
MaxWhiteBloodCells = high	0.1017	0.5669	0.7486
MinWhiteBloodCells = low & MaxHeartRate = high	0.0870	0.0653	0.1412
Vancomycin = true & MinWhiteBloodCells = high	0.0856	0.0940	0.2026
PotassiumChloride = true & MaxWhiteBloodCells = high	0.0738	0.3275	0.4566
Vancomycin = true & MaxHeartRate = high	0.0628	0.3015	0.6729
MinWhiteBloodCells = low	0.0601	0.0913	0.1640
MinDiastolicBloodPressure = low & MaxWhiteBloodCells = high	0.0533	0.3102	0.4454
Vancomycin = true & MaxWhiteBloodCells = high	0.0527	0.2821	0.6135
Pantoprazole (Protonix) = true & Piperacillin/Tazobactam = true	0.0512	0.0662	0.2285
MaxWhiteBloodCells = high & MaxSystolicBloodPressure = low	0.0500	0.3255	0.4535
Piperacillin/Tazobactam = true & MaxWhiteBloodCells = high	0.0471	0.1106	0.3646
Pantoprazole (Protonix) = true & Ciprofloxacin = true	0.380	0.0574	0.1533
MinTemp = low	0.0369	0.0618	0.0813
Vancomycin = true & *MaxDiastolicBloodPressure = high*	0.0324	0.1089	0.2559
Ciprofloxacin = true & MaxHeartRate = high	0.0241	0.0930	0.2351
Ciprofloxacin = true & MaxWhiteBloodCells = high	0.0117	0.0873	0.2214
Pantoprazole (Protonix) = true & Norepinephrine = true	0.0094	0.0679	0.2899
MaxWhiteBloodCells = high & MaxHeartRate = high	0.0075	0.4129	0.6338
Pantoprazole (Protonix) = true & Vancomycin = true	0.0058	0.1433	0.3814
Pantoprazole (Protonix) = true & *PotassiumChloride = true*	0.0053	0.1616	0.3068
Pantoprazole (Protonix) = true & MinDiastolicBloodPressure = low	0.0047	0.1369	0.3067
PotassiumChloride = true & MaxDiastolicBloodPressure = high	0.0047	0.1320	0.2240
MaxDiastolicBloodPressure = high & MaxHeartRate = high	0.0001	0.1451	0.2570
MaxSystolicBloodPressure = low & MaxHeartRate = high	0.0015	0.3109	0.4692
Pantoprazole (Protonix) = true & MaxWhiteBloodCells = high	0.0001	0.1851	0.3646

see 21 patterns (out of 33) that match exactly sepsis related symptoms and/or treatments, and 9 more with the sepsis related patterns but in conjunction with Pantoprazole (Protonix). Pantoprazole is a proton pump inhibitor (PPI) and, even though it is not used to treat sepsis, PPIs are used for stress-related mucosal damage (SRMD). SRMD is an erosive gastritis of unclear pathophysiology, which can occur rapidly after a severe insult such as trauma, surgery, *sepsis* or burns [10]. In other words, it is still reasonable to mine patterns with Pantoprazole, because it is weakly related to sepsis. Finally, we have only 3 patterns, indicated in Table 3 in italic, that include items we would consider to be weakly related to sepsis: 2 patterns have *MaxDiastolicBloodPressure* = *high* and one that includes *PotassiumChloride* = *true*. This demonstrates our algorithm is able to select a much smaller subset of patterns compared to MPR method and that the majority of the patterns predictive of sepsis are reasonable.

6 Conclusion

In this work we have developed and tested a new framework for mining predictive patterns that compactly describe a class of interest. It uses a greedy algorithm to mine the most predictive patterns level-wise and including only those that improve the overall class prediction performance. We tested our approach on intensive care data from MIMIC-III EHR database, focusing on patterns predictive of sepsis. The results preserve the overall classification quality of state-of-the-art methods based on minimum predictive pattern mining approach, but with a significant reduction in the number of extracted patterns.

Acknowledgement. This work was supported by NIH grant R01-GM088224. The content of this paper is solely the responsibility of the authors and does not necessarily represent the official views of the NIH.

References

1. Agrawal, R., Imielinski, T., Swami, A.: Mining association rules between sets of items in large databases. In: Proceedings of SIGMOD (1993)
2. Agrawal, R., Srikant, R.: Fast algorithms for mining association rules in large databases. In: Proceedings of VLDB (1994)
3. Batal, I., Cooper, G., Hauskrecht, M.: A Bayesian scoring technique for mining predictive and non-spurious rules. In: Flach, P.A., De Bie, T., Cristianini, N. (eds.) ECML PKDD 2012. LNCS (LNAI), vol. 7524, pp. 260–276. Springer, Heidelberg (2012). https://doi.org/10.1007/978-3-642-33486-3_17
4. Batal, I., Cooper, G.F., Fradkin, D., Harrison, J., Moerchen, F., Hauskrecht, M.: An efficient pattern mining approach for event detection in multivariate temporal data. Knowl. Inf. Syst. **46**(1), 115–150 (2016)
5. Batal, I., Fradkin, D., Harrison, J., Moerchen, F., Hauskrecht, M.: Mining recent temporal patterns for event detection in multivariate time series data. In: Proceedings of the International Conference on Knowledge Discovery and Data Mining (SIGKDD) (2012)

6. Batal, I., Hauskrecht, M.: Constructing classification features using minimal predictive patterns. In: Proceedings of the 19th ACM International Conference on Information and Knowledge Management, pp. 869–878. ACM (2010)
7. Batal, I., Valizadegan, H., Cooper, G.F., Hauskrecht, M.: A temporal pattern mining approach for classifying electronic health record data. ACM Trans. Intell. Syst. Technol. (ACM TIST) **4**(4), 63:1–63:22 (2012). Spec. Issue Health Inform
8. Bellazzi, R., Zupan, B.: Predictive data mining in clinical medicine: current issues and guidelines. Int. J. Med. Inform. **77**(2), 81–97 (2008)
9. Bone, R.C., et al.: Definitions for sepsis and organ failure and guidelines for the use of innovative therapies in sepsis. Chest **101**(6), 1644–1655 (1992)
10. Brett, S.: Science review: the use of proton pump inhibitors for gastric acid suppression in critical illness. Crit. Care **9**(1), 45 (2004)
11. Dellinger, R.P., et al.: Surviving sepsis campaign: international guidelines for management of severe sepsis and septic shock, 2012. Intensive Care Med. **39**(2), 165–228 (2013)
12. Fawcett, T.: An introduction to ROC analysis. Pattern Recogn. Lett. **27**(8), 861–874 (2006)
13. Geng, L., Hamilton, H.J.: Interestingness measures for data mining: a survey. ACM Comput. Surv. (CSUR) **38**(3), 9 (2006)
14. Han, J., Pei, J., Kamber, M.: Data Mining: Concepts and Techniques. Elsevier, New York (2011)
15. Johnson, A.E., et al.: MIMIC-III, a freely accessible critical care database. Sci. Data **3** (2016)
16. Jovanoski, V., Lavrač, N.: Classification rule learning with APRIORI-C. In: Brazdil, P., Jorge, A. (eds.) EPIA 2001. LNCS (LNAI), vol. 2258, pp. 44–51. Springer, Heidelberg (2001). https://doi.org/10.1007/3-540-45329-6_8
17. Koller, D., Sahami, M.: Toward optimal feature selection. Technical report, Stanford InfoLab (1996)
18. Lavrač, N.: Selected techniques for data mining in medicine. Artif. Intell. Med. **16**(1), 3–23 (1999)

Unsupervised Learning

Inferring Temporal Phenotypes with Topological Data Analysis and Pseudo Time-Series

Arianna Dagliati[1,2(✉)], Nophar Geifman[1], Niels Peek[2,3], John H. Holmes[4], Lucia Sacchi[5], Seyed Erfan Sajjadi[6], and Allan Tucker[6]

[1] Centre for Health Informatics, University of Manchester, Manchester, UK
arianna.dagliati@manchester.ac.uk
[2] Manchester Molecular Pathology Innovation Centre, University of Manchester, Manchester, UK
[3] NIHR Manchester Biomedical Research Centre, cUniversity of Manchester, Manchester, UK
[4] Department of Biostatistics, Epidemiology, and Informatics, Penn Institute for Biomedical Informatics, University of Pennsylvania Perelman School of Medicine, Philadelphia, USA
[5] Department of Electrical, Computer and Biomedical Engineering, University of Pavia, Pavia, Italy
[6] Department of Computer Science, Brunel University London, London, UK

Abstract. Temporal phenotyping enables clinicians to better under-stand observable characteristics of a disease as it progresses. Modelling disease progression that captures interactions between phenotypes is inherently challenging. Temporal models that capture change in disease over time can identify the key features that characterize disease subtypes that underpin these trajectories. These models will enable clinicians to identify early warning signs of progression in specific sub-types and therefore to make informed decisions tailored to individual patients. In this paper, we explore two approaches to building temporal phenotypes based on the topology of data: topological data analysis and pseudo time-series. Using type 2 diabetes data, we show that the topological data analysis approach is able to identify trajectories representing different temporal phenotypes and that pseudo time-series can infer a state space model characterized by transitions between hidden states that represent distinct temporal phenotypes. Both approaches highlight lipid profiles as key factors in distinguishing the phenotypes.

Keywords: Type 2 diabetes · Unsupervised machine learning · Longitudinal studies · Electronic phenotyping

D. Riaño et al. (Eds.): AIME 2019, LNAI 11526, pp. 399–409, 2019.
https://doi.org/10.1007/978-3-030-21642-9_50

1 Introduction

Electronic temporal phenotyping is the identification of clinically meaningful event sequences from patient data that have been collected over time. The identification of temporal phenotypes that are specific to subgroups of patients can assist researchers in identifying useful cohorts and could also be used to generate hypotheses for precision medicine research. What is more, they help experts to better understand the disease in question and how it progresses over time, while ensuring that existing guidelines and care plans are appropriate. Existing methods for temporal phenotyping include the use of temporal graphs extracted from electronic health records [1, 2]. Unlike most previous research that is based on extracting phenotypes from longitudinal electronic health records, we are interested in the construction of temporal phenotypes based on the overall structure of data (that is not necessarily longitudinal) and the identification of realistic trajectories through this structure in time.

Topological Data Analysis (TDA), enables structural phenotype discovery from large, complex data by creating networks of individuals and linking those who display demographic, clinical, and biomarker similarities. TDA provides an analytic method for complex clinical and -omics data to identify shape characteristics that are robust to changes by rescaling distances resulting in a qualitative description of the data. Leveraging methods adapted from topological mathematics, which studies the characteristics of shapes that are not rigid, TDA approaches consider fundamental properties: like coordinate invariance, deformation invariance and compression [3, 4, 7]. TDA works by clustering related data points, representing these as a non-dimensional network graph. This allows for visualization of a "disease space", the underlying shape of the data, and the identification of relevant groupings. Relevant features of TDA include the possibility of studying patients' conditions as a continuum, where subjects can fluctuate over the disease space, moving through the nodes of the network graph. Furthermore, TDA provides intuitive representations of results, which are drawn from simple linear algebra steps and geometric parameters. Its simplicity and ease of interpretation responds to a current compelling challenge of artificial intelligence: to translate research results into transparent and accessible tools based on data visualization and interactive data exploration [5]. Algorithms underpinning TDA are well defined [3–5].

A pseudo time-series (PTS) [9, 10] exploits the characteristics of disease progression so that realistic trajectories can be constructed from cross-sectional data. It uses known labels that determine the beginning and endpoints of a trajectory so that a time-series can be created to better understand the metabolism or cell cycles in genomic data [11, 12], or the different variations of progression in diseases such as glaucoma or cancer [13]. PTS has also been used to integrate longitudinal studies with cross-sectional data [14].

In this paper we explore both TDA and PTS for building different trajectories from health record data in order to better understand the temporal phenotypes that can identify different subtypes of type 2 diabetes mellitus (T2DM).

2 Methods

In the following we describe our approach to discover T2D temporal phenotypes. First, we use TDA to identify subgroups of disease characteristics from cross-sectional record-level data, not ordered in time; we consider these as "sub-phenotypes". Then, TDA is used to identify an overall, complex structure with multiple trajectories by applying a minimum-spanning-tree filter, which identifies a number of feasible trajectories representing different temporal phenotypes. Second, we explore pseudo-time approaches which involve using a combination of distance metrics and graph theory to reconstruct transitions among the phenotypes and infer realistic trajectories through the data space from early disease stages through to advanced ones (Fig. 1).

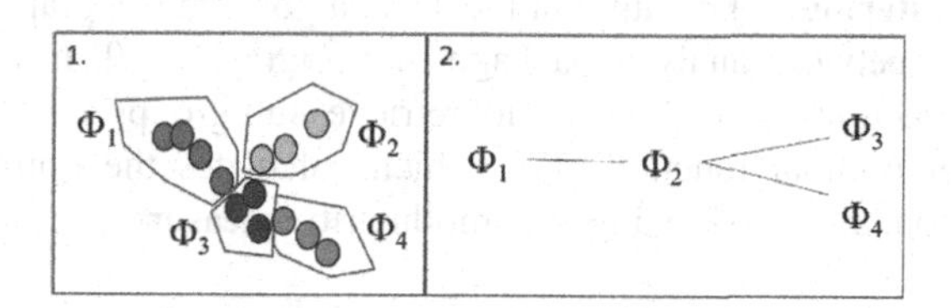

Fig. 1. Methodological steps: 1. TDA finds sub-phenotypes, 2. Pseudo-time reconstruct transitions and trajectories to derive temporal phenotypes.

2.1 Topological Data Analysis

Topological Data Analysis (TDA) allows modelling complex data with an organization principle focused on capturing data shapes. While topology is a mathematical formalism for measuring and representing shapes, TDA uses topology in order to visualize and explore high dimensional and complex real-world data sets and represent them as a graph. The mathematical tools to identify shape characteristics of data sets with topology are called topological mappers [15, 16] and they work by identifying the shape of a data set along specific filter functions, as follows: (1) The points in the dataset are represented with a similarity metric that measures the distance between points in the space; (2) The filter functions (lenses) project the points into a coordinate space and describe the distribution of data in that space; (3) The projections are partitioned into overlapping bins. The bins are defined by resolution, which sets the number of bins that are created within the projections' range of selected lens values, and by gain, which defines the amount of overlap between bins; (4) a clustering step is carried out within each of these bins. This step defines the geometric scale of the shape and is defined by the number of clusters in each bin; (5) finally, the graph is generated by plotting clusters as the graph nodes where shared samples (between bins) are connected by an edge. Once the graph is generated it is possible to colour nodes and edges with the average value of filter functions or to generate a specific function that represents variables of interest (e.g. number of observations in the bins, average age of the subject in bins etc.). We used the Topological Data Mapper implementation described in [4] to perform our analysis, and perform the analysis using the function mapper2D from Topological Data Analysis using Mapper R package [https://github.com/paultpearson/TDAmapper].

Parameterization of TDA. We used cosine distance in conjunction with singular-value decomposition (SVD) and L1-infinity centrality (which assigns to each point the distance to the point most distant from it) as filter functions to build the topology. This is based on the same pipeline adopted in [6] and has been found to provide a more detailed and succinct description of the data than typical scatterplots. We explored the effect of varying resolution parameters (i.e. number of bins and their overlapping) and the geometric scale (i.e. the number of clusters within bins) and using a grid search. It's important to tune parameters and scale in order to insure a shape granularity fine enough to detect temporal behaviors (i.e. repeated observations in time of individual patients aren't restrained within the same node). Too-coarse granularity would result in state changes within nodes, which might impede trajectory discovery.

Topology and Clustering. The output of the TDA algorithm is a graph object that can be analysed with a network analysis package [http://igraph.org/]. In order to identify distinct topology sections that allow us to retrieve sub-groups of observations, we applied the cluster optimal function [17], which calculates the optimal community structure of a graph, by maximizing the modularity measure over all the possible partitions.

Minimum Spanning Tree on Topology. In order to identify specific trajectories from the overall topology, we applied a minimum spanning tree filter to detect the shortest paths within the topology. The weights were based on the average time of the observations represented in the topology's edges. While temporal features were not used to retrieve the original topology, the minimum spanning tree was guided by time to illustrate disease temporal pathways.

2.2 Pseudo-Time Trajectories

Pseudo-time methods can be used to infer state-space models that are characterized by transitions between explicit hidden states, representing distinct temporal phenotypes. Specifically, the outcome we considered is the development of complications during disease progression. This is one of the main indicators of the progression of the disease for T2DM patients [18]. For the PTS analysis, a distance matrix was constructed using cosine distance as was used for the topological analysis. A number of data points were sampled by bootstrapping the data and the weighted graph was constructed based on the cosine distance. This was implemented in MATLAB using the Bayes Net Toolbox [19]. The EM algorithm was used to infer parameters and the junction tree algorithm to perform inference.

2.3 MOSAIC Data

Data for this study was previously collected for clinical and management purposes during the MOSAIC project funded by the European Commission under the 7th Framework Program, (Theme Virtual Physiological Human, 2013–2016) [17–19]. Health records were accumulated from 356 pre-diagnosed T2DM patients, which resulted in 3959 instances in our data set. Risk factors found to influence T2DM [20] include: body mass index (BMI), systolic blood pressure (SBP), diastolic blood pressure (DBP), high-density lipoprotein (HDL), triglycerides, glycated haemoglobin (HbA1c), total cholesterol and smoking habits. Accordingly with previous studies on the MOSAIC project [20], the experimental results were mined for microvascular comorbidities (diabetic nephropathy, neuropathy, and retinopathy). The following variables were used to build the topology and pseudo time-series: age, smoking habit, HbA1c, BMI, SBP, total cholesterol, and triglycerides. Continuous variables were standardized on a –1 to +1 scale. While we did not exploit the temporal nature of this data for phenotype identification, we used the fact that many of these patients had varying follow-up measurements to evaluate our trajectories. In particular, we used time-since-first-visit to assess whether the trajectories correctly model patient progression.

3 Results

3.1 Topological Data Analysis

The graphs shown in Figs. 2 and 3 illustrate the result produced by the TDA algorithm. Each node represents a cluster of data points as observations in time (i.e., an encounter in the MOSAIC data set). The nodes are coloured with the time (days) from the first visit of each encounter. Figure 4a reports the distribution of the value on a continuous scale from blue (time = 0 days from the first visit) to red (time = 4000 days from the first visit).

First, we explored the effect of varying the number of clusters within each bin, which defines the geometric scale of the topology (Fig. 2a). In general, a lower value results in very small clusters (sometimes individual data points), and for higher values the network starts to become extremely sparse or loosely connected. In both cases, edges, which are based on shared samples, are impossible to extract and resulting shapes don't show any relevant topological features. Figure 2a demonstrates a relatively stable topology for between 8 and 12 clusters per bin. For the remainder of the analysis we chose a value of 10. Secondly, we explored the resolution scale while also varying the degree of cluster overlap (gain) when determining the topology (Fig. 2b). In general, higher gain results in more edges. Increasing the resolution of a graph increases the number of bins. In Fig. 2b the horizontal axis represents the number of overlapping intervals and the vertical axis represents the percentage overlap. Note that while the percentage doesn't affect the shape considerably, the interval sizes between 6 and 14 enable a stable shape. For higher values, the network becomes too unstable and it is more difficult to recognize any characteristic shape or trajectories within the network.

Figure 3 illustrates a stable topology generated with seven bins, 60% overlap, and a geometric scale of 8; this is the one used in the following analysis steps. Figure 3a reports the topology enriched by time from the first visit, whose distribution is given in the bottom panel. It is possible to identify a clear temporal direction from the blue bottom node towards the red nodes. Figure 3b reports the topology enriched by the five clusters that we obtained applying the optimal community structure cluster.

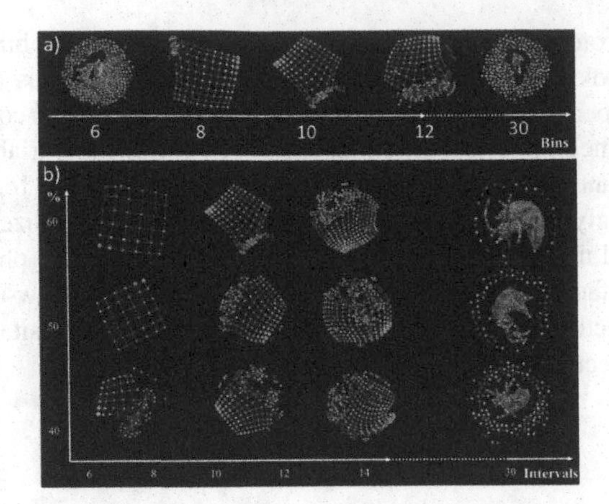

Fig. 2. Topologies (a) varying Geometric Scale and (b) varying Resolution Scale and Percentage Cluster Overlap (gain).

The minimum spanning tree identified seven distinct trajectories (Fig. 3c); all of which start from the central blue cluster which accounts for the first observations in time. We grouped the trajectories as follows: (A) the two trajectories that lead to the red clusters, (B) the two trajectories that lead to the orange clusters and (C) the three trajectories that lead to the yellow clusters past the green clusters. These three groups represent disease progression phenotypes, which we refer to as temporal phenotypes.

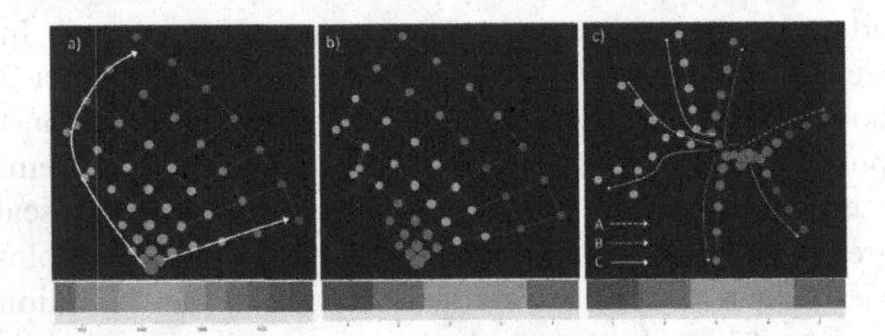

Fig. 3. The network retrieved via TDA and displayed with igraph. In (a) nodes are coloured by time from the first visit, in (b) with the cluster membership. In (c) The Minimum Spanning Tree identifies trajectories of patients. The node colouring is based upon the clustering membership.

3.2 Pseudo-Time Trajectories

Figure 5a reports the weighted graph constructed on the basis of a cosine distance. This graph was used to construct the minimum spanning tree (Fig. 4b). Randomly predefined points were identified as the start-point and endpoint of a trajectory. Here we accepted any sampled patient that has no microvascular complications as a potential start and any patient with microvascular complications as a potential endpoint. The shortest path was identified between the start-point and the endpoint within the minimum spanning tree, resulting in a single pseudo time-series (Fig. 4c). The entire resampling procedure was repeated 1000 times to generate multiple pseudo time-series. Figure 5 illustrates the cosine distance plot enriched with the information about having developed or not a micro vascular complication during the observation period. The pseudo time series (10 samples and all of them) have been plotted upon the graph showing the correlation between trajectories of disease and complications. Having constructed 1000 pseudo-time series, we used an Autoregressive Hidden Markov Model (ARHMM) with five discrete hidden states to build a model to capture the dynamics of the different trajectories through the data.

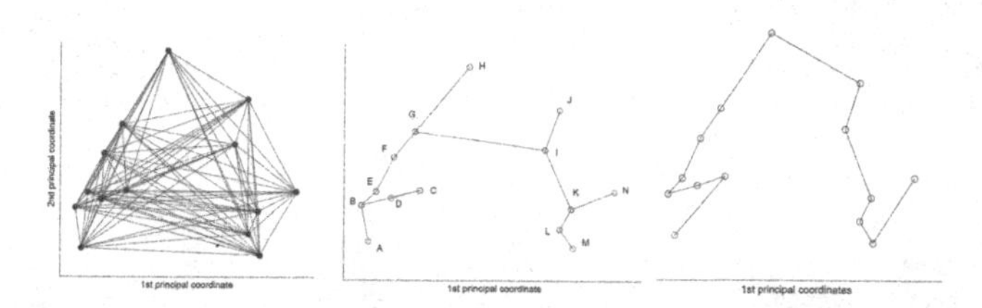

Fig. 4. Generation of pseudo-time series from left to right: (a) the weighted graph of a sample of data (b) the minimum spanning tree of the weighted graph and (c) the Pseudo Time-Series (Color figure online)

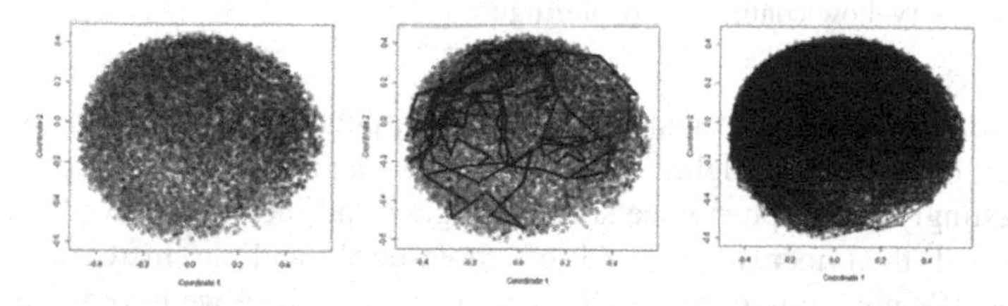

Fig. 5. Left: Multidimensional Scale plot of Cosine Distance where red represents patients with at least one microvascular complication, and black represents none. Middle: Cosine Plot with 10 sample Pseudo-time Series trajectories plotted, right: Full 1000 Pseudo-time Series Generated

3.3 Clinical Assessment

Using data for T2DM patients, we created a topological data network, selecting that with the most stable topology, and enriched the topology with time-from-the-first-visit information. This process revealed potential trajectories for disease progression (Fig. 3a) and sub-groups of observations from the topology clustering (Fig. 3b). Having identified the most suitable topology, the graph was used to build a minimum spanning tree in order to identify pseudo-time-based trajectories (Fig. 3c). Using this approach, seven potential trajectories were identified. These trajectories have been grouped in three temporal phenotypes: A, B and C (Fig. 3c), which show the progression of each trajectory (each one representing a T2DM temporal phenotype) towards the disease's deterioration and distinct outcomes. When the temporal phenotypes are compared in terms of microvascular complications, we found that patients in the C phenotype develop higher prevalence of complications (61%, num. patients = 159), compared to B

(43%, num. patients = 574) and A phenotypes (23%, num. patients = 191). Therefore, minimum spanning tree paths can identify groups of patients less (A phenotype) or more (B and C phenotype) exposed to the development of T2DM-related complications over time. We characterize these phenotypes using relevant clinical features values as they develop in time (Fig. 6).

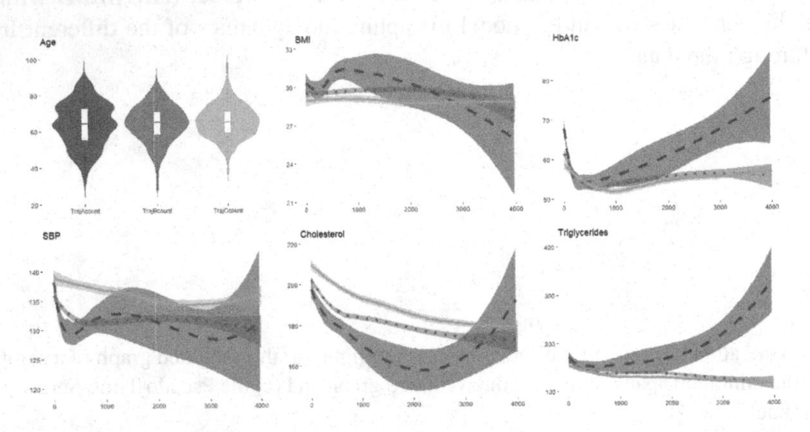

Fig. 6. Clinical characteristics over time of subjects in the A (red-dashed), B (orange-dotted) and C trajectories (yellow-continuous) (Color figure online)

In addition to the high prevalence of micro-complications, patients belonging to the C phenotype demonstrated higher cholesterol levels and systolic blood pressure over time. Interestingly, the A phenotype shows a higher and increasing level of HbA1c, a decreasing and then increasing trend of cholesterol, and an increasing trend of triglycerides. We now turn to the pseudo-time approach where we have inferred a five-state Auto Regressive Hidden Markov Model from the 1000 pseudo time-series generated from the original data. Table 1 illustrates the expected values for the key features of the data for each of the five hidden states.

Table 1. Expected values for the 5 hidden states, where t2d represents time-since-first-visit, TotChol represents total cholesterol and Trigl represents triglycerides.

State	1	2	3	4	5
% Female	0	0	0	50	59
% Male	100	100	100	50	41
Age	59.16	**69.41**	63.7	**67.78**	56
t2d	3.77	9.76	**13.4**	**11.86**	5.42
HbA1c	47.66	50.3	**62.6**	53.54	60.7
BMI	28.1	27.58	30.07	30.31	**31.02**
SBP	129.59	129.5	**136.08**	134.8	132.73
TotChol	187.51	167.28	183.7	188.86	**207.62**
Trigl	126.98	108.13	136.71	124.38	**232.46**

Looking at the expected statistics in Table 1 (highest values in bold), State 1 represents younger patients who have the shortest period of time since their first visit, State 2 represents the oldest patients, State 3 represents people with the highest Hba1c and SBP values and are the patients who have been visiting for the longest of time since their first visit, State 4 represents older patients who have been visiting for a relatively long period, while State 5 represents the youngest patients with the highest BMI.

Table 2. State transition matrix

	State 1	State 2	State 3	State 4	State 5
State 1	0.733	**0.145**	0.023	**0.084**	0.015
State 2	0.036	0.866	0.049	**0.049**	0
State 3	0.055	**0.127**	0.678	**0.132**	0.007
State 4	**0.159**	0.113	**0.157**	0.542	0.029
State 5	**0.134**	0	0.107	**0.140**	0.618

Table 2 illustrates the transition probabilities between these states. The transition probabilities in Table 2 demonstrate that most states are relatively stable with higher probabilities of remaining the same than of changing. Highlighted in bold are the two highest *transition* probabilities from one state to another. This is presented as a diagram in Fig. 7a, which captures a natural flow from State 5 to two potential end-States 3 and 4. This flow is supported by a general increase in the expected time-since-first-visit (t2d) shown in the diagram, as well as increasing age. End State 3 represents patients with very high hb1ac and relatively lower cholesterol whereas end State 4 captures older patients with relatively higher cholesterol but lower hb1ac and very low triglyceride levels. Figure 7b shows two potential trajectories in the form of state transitions based on the HMM model: State transitions 5-1-4 and 5-1-2-3, for patients' triglycerides (left) and cholesterol (right). It is interesting that the lipid profiles were discovered as a defining characteristic of the two trajectories, similar to the TDA results in Fig. 6.

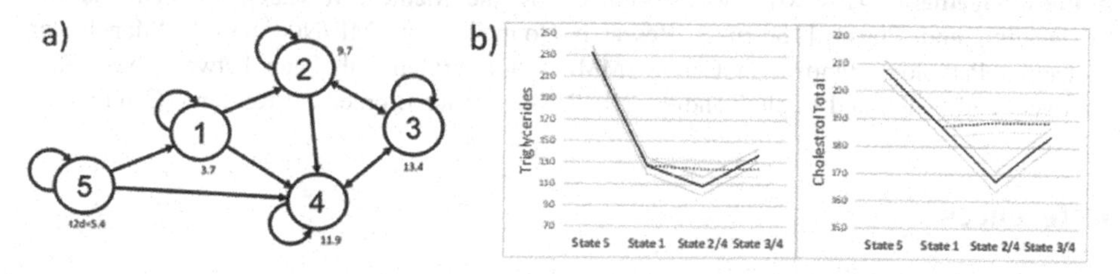

Fig. 7. (a) Transition Diagram with expected time since first visit. (b) Mean statistics for two trajectories 5-1-4 (dashed) and 5-1-2-3 (solid) for Triglycerides (left) and Cholesterol (right)

4 Discussion

In this paper we present a comparison of two approaches to automatically building temporal phenotypes from health records. TDA has been used to capture the overall shape of the data and a minimum spanning tree filter was applied to identify different trajectories. This approach highlighted subcategories of T2DM including one sub-cohort that displays different levels of cholesterol and initial Hba1c from the rest of the population. We also explored the use of PTS methods where different trajectories have been bootstrapped from the data and a state-space model was learned with five hidden states. This approach has identified only two trajectories; however, these are clinically relevant and support the findings made using TDA. Neither TDA nor PTS relied on temporal features of the data in the health records to build these models. As a result, both approaches could be used to construct temporal phenotypes from cross-sectional data if appropriate disease staging information is included. Here we used micro-vascular comorbidity data, but any data that helps to stage a disease could be used. Another important step of the analysis of temporal trajectories regards their comparison among individual patients and their use for predicting disease deviations or adverse outcomes [22–25]. Future steps will employ similarity measures to compare individual trajectories to modelled ones, as in [21], and to predict deviation in the disease space drawn from probabilities of moving forward into nodes with higher density of complications.

Our approach is somewhat limited by the degree of validation. The limitation of PTS is that we only investigated only a 5-hidden-state ARHMM, as suggested by the topology clustering. It is likely that for large datasets, the number of hidden states could be much higher and as a result more complex trajectories can be discovered. Further validation of the joint utility of the two methods could also be carried out. This could be achieved using data simulated via pseudo-time approaches to derive topologies in order to provide a more robust evidence for approaches that combine the two methods. Furthermore, we will compare the presented approach to baseline methods, including simplified version of the analysis pipeline, standard clustering methods and time series methods based on dynamic time warping.

Acknowledgement. This work was co-funded by the Medical Research Council and the Engineering and Physical Sciences Research Council grant MR/N00583X/1 "Manchester Molecular Pathology Innovation Centre (MMPathIC): bridging the gap between biomarker discovery and health and wealth" and the NIHR Manchester Biomedical Research Centre.

References

1. Dagliati, A.: Temporal electronic phenotyping by mining careflows of breast cancer patients. J. Biomed. Inf. **66**, 136–147 (2017)
2. Hripcsak, G., Albers, D.J.: Next-generation phenotyping of electronic health records. J. Am. Med. Inform. Assoc. **20**(1), 117–121 (2012)
3. Offroy, M., Duponchel, L.: Topological data analysis: a promising big data exploration tool in biology, analytical chemistry and physical chemistry. Anal. Chim. Acta **910**, 1–11 (2016)

4. Carlsson, G.: Topology and data. Bull. Am. Math. Soc. **46**(2), 255–308 (2009)
5. Shortliffe, E.H., Sepúlveda, M.J.: Clinical decision support in the era of artificial intelligence. JAMA – J. Am. Med. Assoc. **320**(21), 2199–2200 (2018)
6. Li, L.L., et al.: Identification of type 2 diabetes subgroups through topological analysis of patient similarity. Sci. Transl. Med. **7**(311), 311ra174–311ra174 (2015)
7. Nielson, J.L., et al.: Topological data analysis for discovery in preclinical spinal cord injury and traumatic brain injury. Nat. Commun. **6**, 8581 (2015)
8. Torres, B.Y., Oliveira, J.H.M., Thomas Tate, A., Rath, P., Cumnock, K., Schneider, D.S.: Tracking resilience to infections by mapping disease space. PLoS Biol. **14**(4), e1002436 (2016)
9. Tucker, A., Garway-Heath, D.: The pseudotemporal bootstrap for predicting glaucoma from cross-sectional visual field data. IEEE Trans. Inf. Technol. Biomed. **14**(1), 79–85 (2010)
10. Magwene, P.M., Lizardi, P., Kim, J.: Reconstructing the temporal ordering of biological samples using microarray data. Bioinformatics **19**(7), 842–850 (2003)
11. Campbell, K.R., Yau, C.: Uncovering pseudotemporal trajectories with covariates from single cell and bulk expression data. Nat. Commun. **9**(1), 2442 (2018)
12. Gupta, A., Bar-Joseph, Z.: Extracting dynamics from static cancer expression data. IEEE/ACM Trans. Comput. Biol. Bioinform. **5**(2), 172–182 (2008)
13. Li, Y., Swift, S., Tucker, A.: Modelling and analysing the dynamics of disease progression from cross-sectional studies. J. Biomed. Inform. **46**(2), 266–274 (2013)
14. Tucker, A., Li, Y., Garway-Heath, D.: Updating Markov models to integrate cross-sectional and longitudinal studies. Artif. Intell. Med. **77**, 23–30 (2017)
15. Nicolau, M., Levine, A.J., Carlsson, G.: Topology based data analysis identifies a subgroup of breast cancers with a unique mutational profile and excellent survival. Proc. Natl. Acad. Sci. **108**(17), 7265–7270 (2011)
16. Lum, P.Y., et al.: Extracting insights from the shape of complex data using topology. Sci. Rep. **3**, 1236 (2013)
17. Brandes, U., et al.: On modularity clustering. IEEE Trans. Knowl. Data Eng. **20**(2), 172–188 (2008)
18. Teliti, M., et al.: Risk factors for the development of micro-vascular complications of type 2 diabetes in a single-centre cohort of patients. Diabetes Vasc. Dis. Res. **15**(5), 424–432 (2018). p. 1479164118780808
19. Dagliati, A., et al.: A dashboard-based system for supporting diabetes care. J. Am. Med. Inform. Assoc. **25**(5), 538–547 (2018)
20. Dagliati, A., et al.: Machine learning methods to predict diabetes complications. J. Diabetes Sci. Technol. **12**(2), 295–302 (2017)
21. Dagliati, A., Tibollo, V., Cogni, G., Chiovato, L., Bellazzi, R., Sacchi, L.: Careflow mining techniques to explore type 2 diabetes evolution. J. Diabetes Sci. Technol. **12**(2), 251–259 (2018)
22. Batal, I., Fradkin, D., Harrison, J., Moerchen, F., Hauskrecht, M.: Mining recent temporal patterns for event detection in multivariate time series data (2012)
23. Batal, I., Valizadegan, H., Cooper, G.F., Hauskrecht, M.: A temporal pattern mining approach for classifying electronic health record data. ACM Trans. Intell. Syst. Technol. **4** (4), 63 (2013)
24. Moskovitch, R., Shahar, Y.: Fast time intervals mining using the transitivity of temporal relations. Knowl. Inf. Syst. **42**(1), 21–48 (2015)
25. Moskovitch, R., Shahar, Y.: Classification of multivariate time series via temporal abstraction and time intervals mining. Knowl. Inf. Syst. **45**(1), 35–74 (2015)

Unsupervised Mitral Valve Segmentation in Echocardiography with Neural Network Matrix Factorization

Luca Corinzia[1(✉)], Jesse Provost[1], Alessandro Candreva[2], Maurizio Tamarasso[2], Francesco Maisano[2], and Joachim M. Buhmann[1]

[1] Institute for Machine Learning, ETH Zurich, Zurich, Switzerland
{luca.corinzia,jbuhmann}@inf.ethz.ch, jprovost@student.ethz.ch
[2] Department of Cardiology, University Hospital Zurich, Zurich, Switzerland
{alessandro.candreva,maurizio.taramasso,francesco.maisano}@usz.ch

Abstract. Mitral valve segmentation specifies a crucial first step to establish a machine learning pipeline that can support practitioners into performing diagnosis of mitral valve diseases, surgical planning, and intraoperative procedures. To this end, we propose a totally automated and unsupervised mitral valve segmentation algorithm, based on a low-dimensional neural network matrix factorization of echocardiography videos. The method is evaluated in a collection of echocardiography videos of patients with a variety of mitral valve diseases and exceeds the state-of-the-art method in all the metrics considered.

Keywords: Mitral valve segmentation · Echocardiography · Neural network matrix factorization

1 Introduction

The mitral valve (MV) of the heart regulates the blood flow between two heart chambers, namely the left atrium and the left ventricle. It is formed by two leaflets, the anterior and the posterior leaflet, that are attached to a fibrous ring known as the mitral annulus. In healthy patients, the left atrium contracts during diastole and the blood flows through the open MV into the left ventricle that is dilating. During systole the left ventricle contracts and pushes the blood into the aorta through the aortic valve, and the MV closes so that the blood does not flow back into the atrium. Various diseases concern the MV causing an alteration of healthy blood flow between left atrium and left ventricle. Briefly, two possible scenarios are possible: (i) *mitral stenosis* that is characterized by a narrowing of the mitral annulus and hence a decline of the blood flow to the left ventricle and (ii) *mitral regurgitation* that causes blood to flow back to the left atrium during systole. This last condition defines the second most common cardiac valvular defect amenable of surgical intervention [7].

L. Corinzia and J. Provost contributed equally to the work.

D. Riaño et al. (Eds.): AIME 2019, LNAI 11526, pp. 410–419, 2019.
https://doi.org/10.1007/978-3-030-21642-9_51

Echocardiography (echo) is a medical imaging technique that produces 2D and 3D pictures and videos using ultrasound waves generated by a transducer, scattered/reflected by biological tissue and read by a detector. Echo is the standard imaging tool in the clinical routine to perform the diagnosis of most of heart diseases and dysfunctions, including MV diseases [1,7,15] since it is inexpensive, non-invasive and it enables both qualitative and quantitative assessment of the myocardium and of the MV functions. The current clinical protocol requires practitioners to manually measure a plethora of diagnostic parameters of the cardiac valves as well as of the cardiac chambers. In this paper we propose NN-MitralSeg, an unsupervised MV segmentation algorithm that supports a systematic and fast evaluation of MV health status for the medical practitioners. Our method improves on the Robust Non Negative Matrix Factorization method (R-NNMF) proposed in [4] and it outperforms R-NNMF on a dataset of 38 patients affected with MV dysfunction and mitral regurgitation.

2 Related Work

MV segmentation in 2D and 3D echo enables automatic diagnosis and personalized prognosis and, therefore, it has received a lot of attention recently. Many early methods are based on active contour algorithms or on other methods that depend heavily on human-in-the-loop contributions. Active contour algorithms [2,9] require practitioners to initialize the segmentation algorithm, placing manually a contour close to the desired position in a given frame or on multiple frames [10,13]. Then the MV is segmented on the given frames optimizing a fixed energy function, and the mask is propagated over time with the support of the optical flow [11] and/or of a dynamical model of the MV [12]. In [3] the authors proposed a method that leverages both an active contour algorithm that segments the myocardial walls and a thin tissue detector that finds the valve leaflets. Also in [14] medical practitioners initialize the segmentation denoting multiple points that are then connected using J-splines.

The first attempts to design a fully automated MV segmentation algorithm are proposed in [4,16]. The 2D echo video is factorized using (robust) non-negative 2-rank matrix factorization. Every frame of the video is decomposed as a non-negative linear mixture of two frames and a sparse signal, where the frames, the mixture coefficients and the sparse signal are obtained optimizing an l_2 loss. The 2-rank factorization captures most of the myocardium wall motion, while the high dimensional sparse signal represents the echo noise and the MV movement. Then the MV is segmented using simple diffusion and thresholding of the sparse signal. Despite producing satisfactory results on high quality echos, these methods performs poorly on noisy low quality videos, due mostly to the misplacement of the region of interest (ROI).

We propose Neural Network Mitral Segmentation (NN-MitralSeg)[1], a method that improves on [4] with a two-fold contribution: (1) we use a neural network

[1] Code, hyperparameters and network specifications are available at https://github.com/jprovost14/NN-MitralSeg.

matrix factorization [5] (also know as neural collaborative filtering in [8]) to account for both linear and non-linear contributions of the myocardial wall motion, in combination with a parametrized threshold operator to learn the high dimensional sparse signal that captures the MV, and (ii) we leverage the information of both the sparse signal and of the dense optical flow to delineate the ROI.

3 Method

3.1 Model

Each echo is initially represented as a tensor $\mathbf{T} \in \mathbb{R}_+^{h \times w \times T}$, where h and w are respectively the height and the width of a single frame and T is the number of frames in the video. We reshape each frame of the echo into a column vector and then concatenate all the columns to get a matrix $\mathbf{X} \in \mathbb{R}_+^{N \times T}$ where $N = h \cdot w$. The matrix $\mathbf{X}$ is then embedded in a low dimensional space as follows. For each row (pixel) $n \in N$ and each column (frame) $t \in T$, we associate the latent feature matrices with non-negative entries $\mathbf{U}_n$, $\mathbf{V}_t, \in \mathbb{R}_+^{D \times K}$, where D and K are two hyperparameters of the model. Let $f_{\theta_{LD}}$ denote the *low dimensional* network with weights θ_{LD} and f_{θ_T} denote the *threshold* network with weights θ_T. The low dimensional network reconstructs the inputs as

$$\hat{X}_{n,t} = f_{\theta_{LD}}(\mathbf{u}_{n,1} \cdot \mathbf{v}_{t,1}, \ldots, \mathbf{u}_{n,D} \cdot \mathbf{v}_{t,D}).$$

where $\mathbf{u}_{n,j} \cdot \mathbf{v}_{t,j}$ is the inner product between the j-th row vectors of $\mathbf{U}_n$ and $\mathbf{V}_t$. It is easy to see that the input of the $f_{\theta_{LD}}$ is equivalent to the diagonal of the product matrix $\mathbf{U}_n\mathbf{V}_t^T$, hence it is a D-dimensional latent feature vector. Notice that K-rank non-negative matrix factorization is obtained enforcing $f_{\theta_{LD}} = \mathbf{1}$ and $D = 1$, where $\mathbf{1}$ is the identity function (see [8]), hence K can be interpreted as the generalized rank of the factorization. The non-negativity of the latent features is imposed using non-negative activation functions. Given the reconstruction $\hat{X}_{n,t}$, the difference between $X_{n,t}$ and $\hat{X}_{n,t}$ serves as the scalar input to the threshold network and is transformed to get the scalar output:

$$\hat{S}_{n,t} = f_{\theta_T}(X_{n,t} - \hat{X}_{n,t}).$$

The threshold network is composed by just one node and a ReLu activation function and acts as a parametrized threshold operator. A diagram for the general architecture is given in Fig. 1.

3.2 Learning

The training of this model occurs in three stages. In the first stage the *low-dimensional* network is trained to provide an accurate approximation $\hat{\mathbf{X}}$. Subsequently, both the *low-dimensional* network and the *threshold* network are trained iteratively such that the threshold network fully reconstructs the error $\mathbf{X} - \hat{\mathbf{X}}$. The final stage consists of imposing the sparseness on $\hat{\mathbf{S}}$ using a ℓ_1 regularizer.

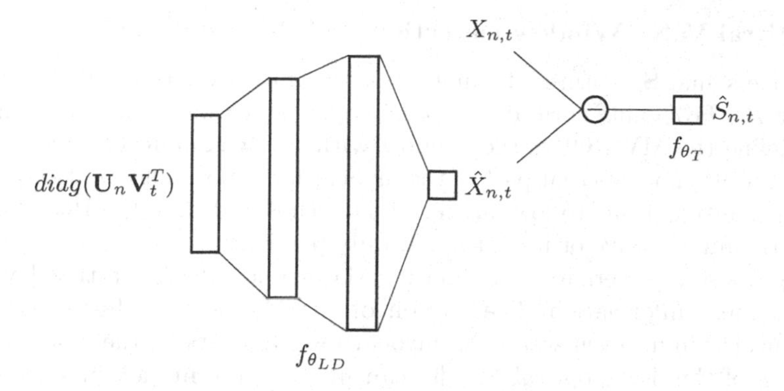

Fig. 1. Diagram of the model used in NN-MitralSeg. The network $f_{\theta_{LD}}$ maps the pixel and frame latent features $\mathbf{U}_n$, $\mathbf{V}_t$ to the reconstruction $\hat{X}_{n,t}$. The input of the network is the diagonal of $\mathbf{U}_n\mathbf{V}_t^T$ denoted as $diag(\mathbf{U}_n\mathbf{V}_t^T)$. The threshold operator f_{θ_T} is then applied to $\hat{X}_{n,t} - X_{n,t}$ to give the sparse signal $\hat{S}_{n,t}$.

Pre-training the Low-Dimensional and Threshold Networks. Pre-training the parameters θ_{LD} and $\{(\mathbf{U}_n, \mathbf{V}_t)\}_{n,t}$ of the low-dimensional network ensures that the network can produce an accurate approximation of $\hat{\mathbf{X}}$, which is used as input into the threshold operator. The pre-training of the low-dimensional network is performed as in [5]; freezing the latent (pixel and frame) features $\{(\mathbf{U}_n, \mathbf{V}_t)\}_{n,t}$ while updating θ_{LD}, and then freezing the low dimensional network θ_{LD} while updating $\{(\mathbf{U}_n, \mathbf{V}_t)\}_{n,t}$. The objective that is optimized during this stage is given by:

$$\|\mathbf{X} - \hat{\mathbf{X}}\|_F^2 + \beta \left[\sum_n \|\mathbf{U}_n\|_F^2 + \sum_t \|\mathbf{V}_t\|_F^2 \right],$$

where β is a regularization parameter and $\|\cdot\|_F$ is the Frobenius norm. In the second stage also the threshold network is trained in an iterative fashion: updating θ_T while freezing θ_{LD} and $\{(\mathbf{U}_n, \mathbf{V}_t)\}_{n,t}$; then updating θ_{LD} and $\{(\mathbf{U}_n, \mathbf{V}_t)\}_{n,t}$ as described above while freezing θ_T according to the loss function given by:

$$\|\mathbf{X} - \hat{\mathbf{X}} - \hat{\mathbf{S}}\|_F^2 + \beta \left[\sum_n \|\mathbf{U}_n\|_F^2 + \sum_t \|\mathbf{V}_t\|_F^2 \right].$$

Training on the Full Objective. The goal of pre-training is to obtain two networks that can fully reconstruct the echo. The low-dimensional network captures the myocardium movement and the threshold operator captures the echo noise and the mitral valve movement. Sparsity is enforced by regularizing the loss function with the ℓ_1-norm:

$$\|\mathbf{X} - \hat{\mathbf{X}} - \hat{\mathbf{S}}\|_F^2 + \beta \left[\sum_n \|\mathbf{U}_n\|_F^2 + \sum_t \|\mathbf{V}_t\|_F^2 \right] + \lambda \|\hat{\mathbf{S}}\|_1,$$

where λ is the sparsity coefficient and $\|\cdot\|_1$ denotes the ℓ_1-norm.

3.3 Mitral Valve Window Detection and Segmentation

The sparse signal $\hat{\mathbf{S}}$ captures the motion of the mitral valve. In [4] the authors compute the Frobenius norm on all possible 3D windows of the sparse matrix $\hat{\mathbf{S}}$ and define the MV ROI as the window with the maximum Frobenius norm. However, it often occurs that part of the myocardium movement is also captured in the sparse signal due to low quality of the echos and then the ROI does not contain the mitral valve or it captures it only partially.

We propose an alternative method for MV window detection that leverages also movement information. The motion of the MV is much faster compared to the myocardium, even when the myocardium appears in the sparse signal. The norm of the dense optical flow [6] can measure the motion in a video and a large norm is indicative of fast motion. First the sparse signal $\hat{\mathbf{S}} \in \mathbb{R}_+^{n \times \times T}$ is reshaped into a 3D array of the same shape of the original video $\mathbb{R}_+^{h \times w \times T}$ and then thresholding is applied in order to retain only the p percent high intensity pixels. The dense optical flow is then computed for every frame of $\hat{\mathbf{S}}$ and is denoted as $optical\_flow(\hat{\mathbf{S}})_t$. Similar to the window detection method in [4,16], the ROI of the MV is then identified as the window with largest sum among the frames of the optical flow norms. The selection is made between windows spanning the whole 2D frame, with a fixed stride. Denoting by $\mathbf{W}_l \in \{0,1\}^{w \times h}$ the windows as binary masks, the ROI selection can be summarized as

$$l^* = \arg\max_l \sum_{t=1}^{T} \|optical\_flow(\hat{\mathbf{S}})_t \cdot \mathbf{W}_l\|_2^2$$

The segmentation is consequently performed on the sparse signal enclosed in the ROI similarly to [4] using simple isotropic 2D diffusion on each frame.

4 Experiments and Results

4.1 Dataset Description

A total of 38 transthoracic echos were obtained from the MitraSwiss Registry, a Swiss-wide prospective registry which includes patients undergoing percutaneous mitral valve repair using the MitraClip system. All patients had moderate-to-severe (3+) or severe (4+) mitral regurgitation of functional or degenerative origin as graded according to current recommendations of the American Society of Echocardiography [18]. Imaging data were processed in an anonymized way and all patients provided written informed consent to be entered into the database. Only 4-chamber echo views are used, and for every echo, a rectangular window around the MV and three selected frames were annotated by an expert medical doctor.

4.2 Window Detection

A comparison of the sparse signal according to R-NNMF [4] and our method NN-MitralSeg is showed in Fig. 2 for a R-NNMF failure case. As it can be seen

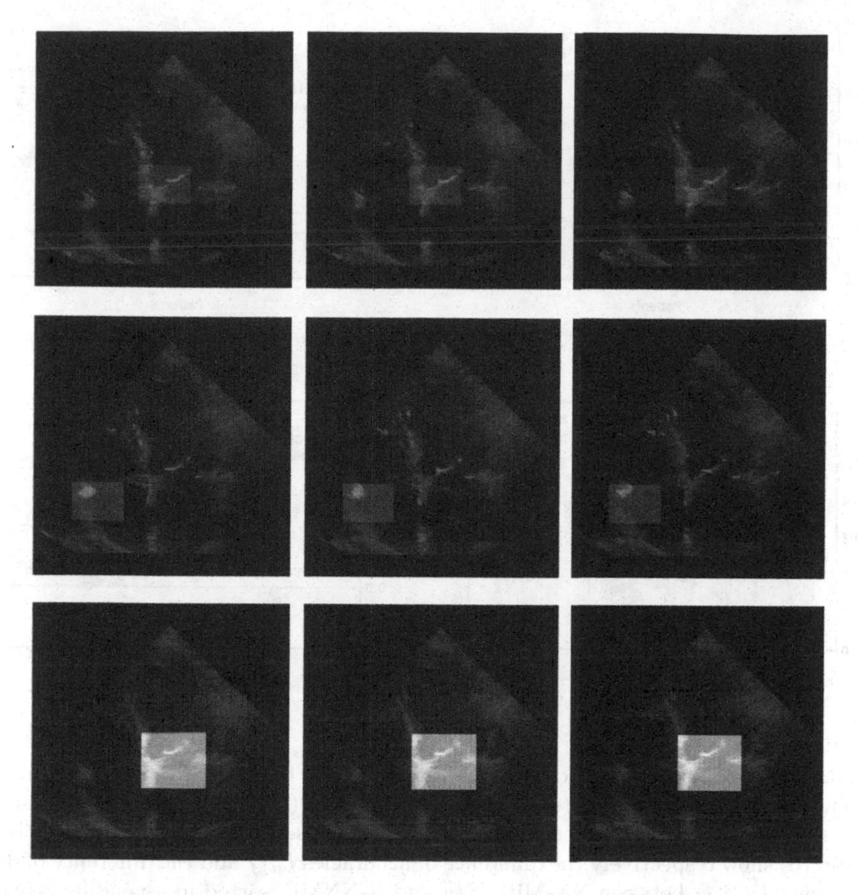

Fig. 2. A failure case for the window detection method of R-NNMF [4]. The sparse signal (in blue) is given for both our method (NN-MitralSeg, top row) and R-NNMF (middle row) with reference to the original frames (bottom row) for three consecutive frames. The mitral valve region is always highlighted as the shaded area. The region is misplaced by R-NNMF due to a strong myocardium movement contribution in the sparse signal. (Color figure online)

the failures of the R-NNMF window detection method are due to a strong presence of the myocardium movement in the sparse signal, as a consequence of the low expressiveness of the linear model used in R-NNMF. We compare the performance of the mitral valve window detection according to accuracy (I), defined as the percentage of pixels in the computed ROI that intersect the gold standard window. Note that in this specific task, the window sizes are fixed and not inferred by the model, hence the accuracy is a reliable measure of performance. I is sorted in descending order according to our method in Fig. 3a. In Fig. 3b the difference between the accuracies of our method and of R-NNMF I_{diff} is sorted in ascending order, alongside the average I_{diff} over all echos μ and the p-value of the one-sided t-test. In Table 1 we also report the number of success

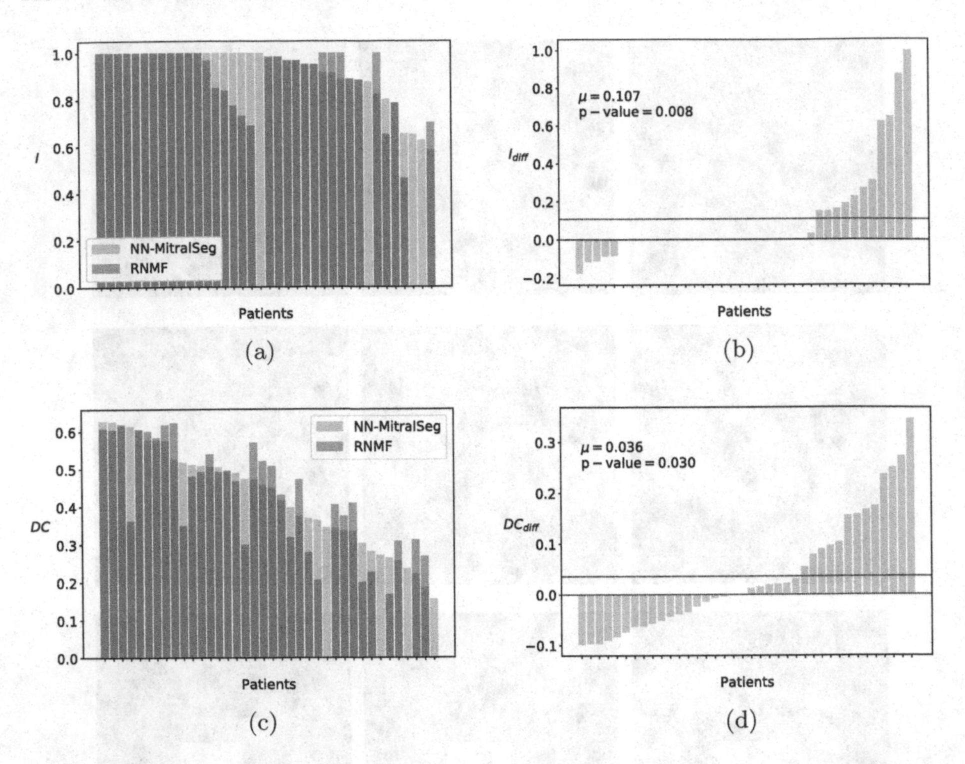

Fig. 3. A comparison of our method NN-MitralSeg and the R-NNMF [4] according to window detection accuracies I (a-b) and Dice coefficient (c-d). (a) and (c) show respectively the accuracy and the Dice coefficient sorted in descending order by our method. (b) and (d) show respectively the difference of accuracies I_{diff} and the difference of dice coefficients DC_{diff} between NN-MitralSeg and R-NNMF, sorted in ascending order.

cases where the accuracy I is higher then a given threshold, and the average Intersection over Union score (IoU).

4.3 Mitral Valve Segmentation

The output of the segmentation algorithms are compared with the ground truth in Figs. 3a and d according to the Dice coefficient (DC). The DC is reported for every echo and it is sorted in descending order according to the score of our method. The DC difference DC_{diff} between the two methods is also reported in Fig. 3d sorted in ascending order. We observe that NN-MitralSeg outperforms the state-of-the-art in both window detection and in the dense MV annotation by a statistically significant margin. A detailed comparison of the MV segmentations produced by the two algorithms is documented in Figs. 4a and b where we show the masks and the ground truth respectively for the highest and lowest five scoring echos (according to our method).

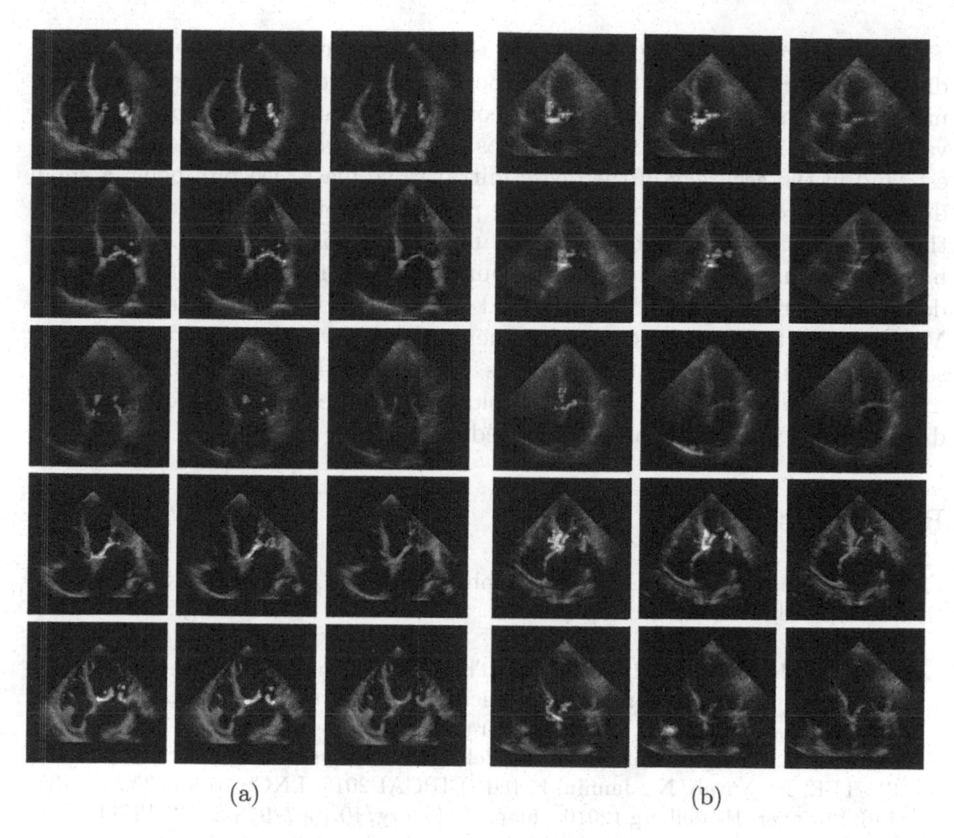

Fig. 4. The mitral valve segmentation masks for the echos with the (a) five highest and (b) lowest Dice coefficients according to NN-MitralSeg. From left to right: NN-MitralSeg (yellow), R-NNMF (green) and ground truth (red). (Color figure online)

Table 1. Number of success cases and average Intersection over Union score for the window detection algorithm. The total number of echo is 38.

	$I > 0.65$	$I > 0.85$	IoU
NN-MitralSeg (Ours)	**35**	**31**	**0.35132**
R-NNMF [4]	32	25	0.30883

5 Conclusion and Future Work

We proposed NN-MitralSeg, a fully automated and unsupervised mitral valve segmentation algorithm based on non-linear matrix factorization. An echocardiography video is decomposed into a low dimensional signal that captures the linear and non-linear myocardial wall motion, and a high dimensional sparse signal that accounts for the echocardiography noise and mitral valve movement.

The mitral valve is then segmented from the sparse signal using thresholding and diffusion algorithms. This method outperforms the state-of-the-art fully automated algorithm in a data-set of 38 videos with patients suffering various mitral valve dysfunctions, in both the task of positioning the rectangular region of interest, and in the accuracy of the dense mitral valve mask. Despite being a small dataset, due to the health status of the patients it contains a larger variability then an healthy control dataset of the same size. A possible future development includes the use of both the sparse ground truth segmentation masks and the dense (inaccurate) annotation generated by unsupervised algorithms (like NN-MitralSeg) to train segmentation deep networks in a weakly-supervised-learning scenario [17] in an online fashion. This would provide practitioners with segmentation algorithms that could be deployed in the real-time echocardiography during mitral valve intraoperative procedures.

References

1. Baumgartner, H., et al.: Echocardiographic assessment of valve stenosis: EAE/ASE recommendations for clinical practice. J. Am. Soc. Echocardiogr. **22**(1), 1–23 (2009)
2. Blake, A., Isard, M.: Active Contours: The Application of Techniques from Graphics, Vision, Control Theory and Statistics to Visual Tracking of Shapes in Motion. Springer, London (2012). https://doi.org/10.1007/978-1-4471-1555-7
3. Burlina, P., et al.: Patient-specific modeling and analysis of the mitral valve using 3D-TEE. In: Navab, N., Jannin, P. (eds.) IPCAI 2010. LNCS, vol. 6135, pp. 135–146. Springer, Heidelberg (2010). https://doi.org/10.1007/978-3-642-13711-2_13
4. Dukler, Y., et al.: Automatic valve segmentation in cardiac ultrasound time series data. In: Medical Imaging 2018: Image Processing, vol. 10574, p. 105741Y. International Society for Optics and Photonics (2018)
5. Dziugaite, G.K., Roy, D.M.: Neural network matrix factorization. arXiv preprint arXiv:1511.06443 (2015)
6. Farnebäck, G.: Two-frame motion estimation based on polynomial expansion. In: Bigun, J., Gustavsson, T. (eds.) SCIA 2003. LNCS, vol. 2749, pp. 363–370. Springer, Heidelberg (2003). https://doi.org/10.1007/3-540-45103-X_50. http://dl.acm.org/citation.cfm?id=1763974.1764031
7. Hayek, E., Gring, C.N., Griffin, B.P.: Mitral valve prolapse. Lancet **365**(9458), 507–518 (2005)
8. He, X., Liao, L., Zhang, H., Nie, L., Hu, X., Chua, T.S.: Neural collaborative filtering. In: Proceedings of the 26th International Conference on World Wide Web, pp. 173–182. International World Wide Web Conferences Steering Committee (2017)
9. Isard, M., Blake, A.: Contour tracking by stochastic propagation of conditional density. In: Buxton, B., Cipolla, R. (eds.) ECCV 1996. LNCS, vol. 1064, pp. 343–356. Springer, Heidelberg (1996). https://doi.org/10.1007/BFb0015549
10. Mikic, I., Krucinski, S., Thomas, J.D.: Segmentation and tracking of mitral valve leaflets in echocardiographic sequences: active contours guided by optical flow estimates. In: Medical Imaging 1996: Image Processing, vol. 2710, pp. 311–321. International Society for Optics and Photonics (1996)

11. Mikic, I., Krucinski, S., Thomas, J.D.: Segmentation and tracking in echocardiographic sequences: active contours guided by optical flow estimates. IEEE Trans. Med. Imaging **17**(2), 274–284 (1998)
12. Schneider, R.J., Tenenholtz, N.A., Perrin, D.P., Marx, G.R., del Nido, P.J., Howe, R.D.: Patient-specific mitral leaflet segmentation from 4D ultrasound. In: Fichtinger, G., Martel, A., Peters, T. (eds.) MICCAI 2011. LNCS, vol. 6893, pp. 520–527. Springer, Heidelberg (2011). https://doi.org/10.1007/978-3-642-23626-6_64
13. Shang, Y., Yang, X., Zhu, L., Deklerck, R., Nyssen, E.: Region competition based active contour for medical object extraction. Comput. Med. Imaging Graph. **32**(2), 109–117 (2008)
14. Siefert, A.W., et al.: Accuracy of a mitral valve segmentation method using J-splines for real-time 3D echocardiography data. Ann. Biomed. Eng. **41**(6), 1258–1268 (2013)
15. Zamorano, J., et al.: Real-time three-dimensional echocardiography for rheumatic mitral valve stenosis evaluation: an accurate and novel approach. J. Am. Coll. Cardiol. **43**(11), 2091–2096 (2004)
16. Zhou, X., Yang, C., Yu, W.: Automatic mitral leaflet tracking in echocardiography by outlier detection in the low-rank representation. In: 2012 IEEE Conference on Computer Vision and Pattern Recognition (CVPR), pp. 972–979. IEEE (2012)
17. Zhou, Z.H.: A brief introduction to weakly supervised learning. Natl. Sci. Rev. **5**(1), 44–53 (2017)
18. Zoghbi, W.A., et al.: Recommendations for evaluation of the severity of native valvular regurgitation with two-dimensional and doppler echocardiography. J. Am. Soc. Echocardiogr. **16**(7), 777–802 (2003)

Unsupervised Learning from Motion Sensor Data to Assess the Condition of Patients with Parkinson's Disease

Teodora Matić[1], Somayeh Aghanavesi[2], Mevludin Memedi[3], Dag Nyholm[4], Filip Bergquist[5], Vida Groznik[1,6], Jure Žabkar[1], and Aleksander Sadikov[1](✉)

[1] Faculty of Computer and Information Science, University of Ljubljana, Ljubljana, Slovenia
aleksander.sadikov@fri.uni-lj.si
[2] Computer Engineering, School of Technology and Business Studies, Dalarna University, Dalarna, Sweden
[3] Informatics, Business School, Örebro University, Örebro, Sweden
[4] Department of Neuroscience, Neurology, Uppsala University, Uppsala, Sweden
[5] Department of Pharmacology, University of Gothenburg, Gothenburg, Sweden
[6] Faculty of Mathematics, Natural Sciences and Information Technologies, University of Primorska, Koper, Slovenia

Abstract. Parkinson's disease (PD) is a chronic neurodegenerative disorder that predominantly affects the patient's motor system, resulting in muscle rigidity, bradykinesia, tremor, and postural instability. As the disease slowly progresses, the symptoms worsen, and regular monitoring is required to adjust the treatment accordingly. The objective evaluation of the patient's condition is sometimes rather difficult and automated systems based on various sensors could be helpful to the physicians. The data in this paper come from a clinical study of 19 advanced PD patients with motor fluctuations. The measurements used come from the motion sensors the patients wore during the study. The paper presents an unsupervised learning approach applied on this data with the aim of checking whether sensor data alone can indicate the patient's motor state. The rationale for the unsupervised approach is that there was significant inter-physician disagreement on the patient's condition (target value for supervised machine learning). The input to clustering came from sensor data alone. The resulting clusters were matched against the physicians' estimates showing relatively good agreement.

Keywords: Unsupervised learning · Motion sensor · Parkinson's disease · Objective evaluation · Patient monitoring · Bradykinesia · Dyskinesia

1 Background and Experimental Setup

A clinical trial conducted at the Uppsala University Hospital, Sweden [5], recruited 14 males and 5 females with advanced Parkinson's disease (PD) and

D. Riaño et al. (Eds.): AIME 2019, LNAI 11526, pp. 420–424, 2019.
https://doi.org/10.1007/978-3-030-21642-9_52

experiencing motor fluctuations. Their mean age was 71.4 and mean years with levodopa treatment was 10. Each patient received a 150% of their individual levodopa-carbidopa equivalent morning dose and was followed up until the medication wore off. Patients were observed once within 50 min before the morning dose was administered, once at the time the dose was administered, and approximately every 20–30 min afterwards. During each observation the patients performed a set of standardised motor tasks as specified in the UPDRS-III (Unified Parkinson's Disease Rating Scale; motor section) including hand rotation tests, leg agility, and walking. Three movement disorder specialists provided ratings for finger taps, rapid alternating of hands, leg agility, arising from chair, gait, and body bradykinesia/hypokinesia, each on a scale from 0 (normal) to 4 (extremely severe). They also rated the severity of dyskinesia on a scale from 0 (no dyskinesia) to 4 (severe dyskinesia) and the overall mobility of the patients on a Treatment Response Scale (TRS) on a scale from −3 (very bradykinetic) to 0 (normal) to 3 (very dyskinetic) [4].

In addition to clinical-based measures, 3D accelerometer and gyroscope data was available for each observation as the patients wore motion sensors on each ankle and wrist during all observations. For leg agility tests, the patients were asked to sit on straight-back chair and stomping each foot on the floor 10 times as fast as possible. During rapid alternating movements of hands, subjects were seated on a chair and performed hand rotation tests for 20 s, starting with the right hand and then with the left hand [6]. During walking tests, the patients walked for about 4 m at a self-selected pace. [1]

This paper presents an unsupervised learning approach applied on the data from the clinical trial with the goal to check how well motion sensor data alone can indicate the patient's motor state. The decision for the unsupervised approach was due to significant inter-physician disagreement on the patient's condition (target value for supervised machine learning).

2 Unsupervised Learning Results

Altogether there were 178 features from motion sensors, including 24 from leg agility tests, 88 from hand rotation, 37 from gait, and 29 from arm swing. The principal component analysis (PCA) was applied on them in order to reduce their number before the unsupervised learning. It was performed in R, using the *princomp* function with the correlation matrix as the input. It yielded 33 components with eigenvalues over the Keiser criterion of 1 and in combination explained 87% of variance. These 33 components were used as the *only* input for cluster analysis. The algorithm was blind to the patients' condition, time of dose and their other medical data (e.g. levodopa concentration in blood, medical history, etc.) as the goal was to check whether and/or how well motion sensor data alone can indicate the patient's motor state.

The cluster analysis was run using the trimmed k-means algorithm, using R's package *tclust* [3]. Trimmed approach was chosen in order to remove the influence of the most outlying cases that can skew the results of the analysis.

Ten per cent, or 21 cases, were trimmed, leaving 183 out of 204 individual patient measurements to be clustered.

Figure 1 shows the resulting two clusters. The clusters are roughly similar in size, one containing 104 (56.8%) and the other 79 (43.2%) measurements. The size of the individual recordings (dots) represents TRS ratings (discussed below) and the numeric labels are patient IDs. The visualisation shows that the large majority of the discrepancy between TRS ratings and the obtained clusters are due to just a few patients: these are patients with IDs #14 and #8 in the red cluster and patient #41 in the blue cluster. Some recordings of patient #33 are puzzling (red ones on the bottom).

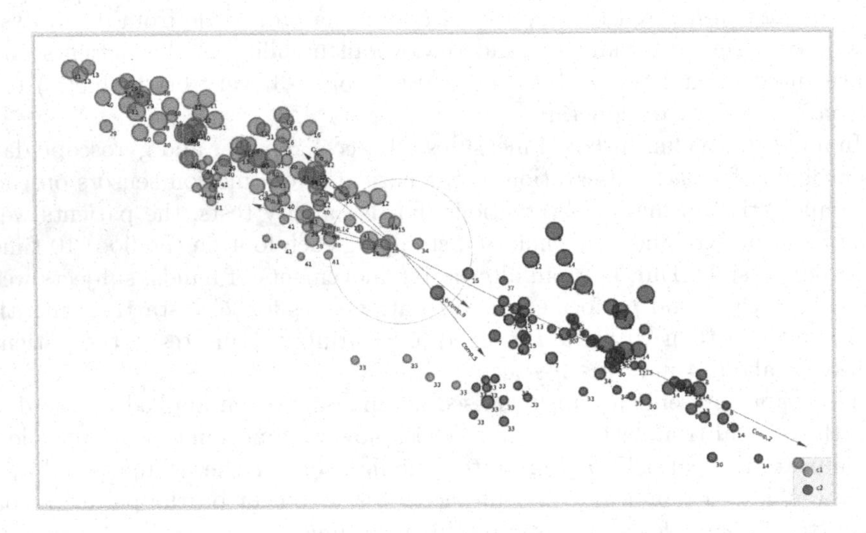

Fig. 1. Visualisation of the two clusters. Dot size represents the TRS score and numeric labels are patient IDs. (Color figure online)

The clustering results were visualised using FreeViz [2] with the aim to separate the instances of different classes; in our case clusters 1 and 2. FreeViz projection in Fig. 1 shows that clusters 1 and 2 are well separated. In contrast, Fig. 2 shows a different projection, where the clusters overlap, however the intent there is to show the relation of clusters to TRS values.

The two clusters were compared with the TRS ratings of the individual recordings (Fig. 2). These ratings would represent the class/target value in a supervised setting. However, as stated earlier, we used the unsupervised setting since there is relative disagreement (first order intra-class correlation coefficient of 0.79) between individual physicians when rating the patients' recordings thus making the class value potentially unreliable. The y-axis represents TRS ratings and the x-axis represents discrete timestamp of the recording with the first recording being the baseline one before the patients received the medication. We see that there is a relatively good correlation between TRS ratings and the two

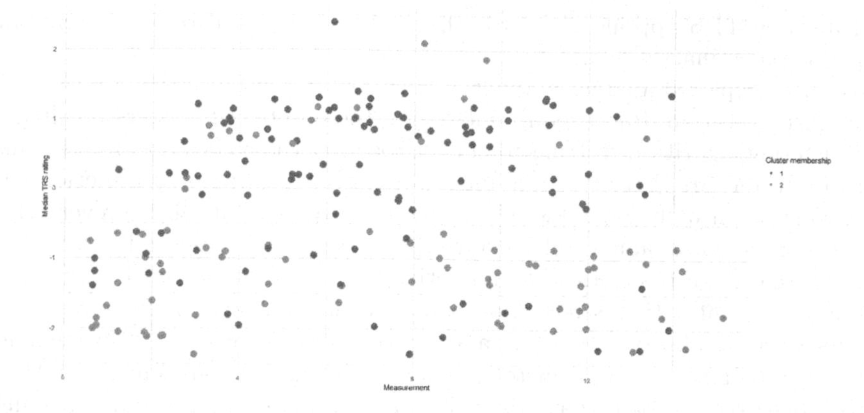

Fig. 2. Comparison between TRS ratings and the two clusters. The x-axis represents recording time (the first measurement is before medication), and y-axis represents TRS.

clusters, one representing high TRS values and the other representing low TRS values. The TRS is usually interpreted in three levels with an additional layer around zero. However, three cluster analysis did not correspond to that – this is further debated in the discussion.

We treated the TRS ratings as a continuous variable (which it is) for the purpose of this analysis even though in clinical practice it is usually discretised. Just to give a general feel as to the correspondence between learned clusters and the average TRS scores in a discretised setting, we put a threshold at TRS rating of zero, thus dividing it into positive and negative bands. The clusters (if interpreted as low and high) correctly classify 132 out of 183 cases; that is 72.1% accuracy. We excluded the 21 outliers, even though all but one is in the lower band and classified correctly as we wanted to report the worst-case scenario. The majority class is the lower band containing 51.9% of the recordings.

3 Discussion

This paper presents the initial results of an unsupervised analysis of the motion sensor recordings of PD patients during standardised tests to quantify their motor states. It compares the obtained clusters with the TRS ratings representing the clinicians' assessment of the patients.

We visualised the clusters using FreeViz best linear projection to show the quality of separation. Furthermore, we treated the TRS rating as a continuous variable and presented the agreement between TRS ratings and the two clusters graphically as this best shows the recordings in discrete time after the baseline (pre-medication). To give a numeric feel for this agreement we divided the TRS ratings to those above and below zero – the worst-case agreement is 72.1% compared to 51.9% majority class. It has to be noted, however, that since there is noteworthy inter-rater disagreement on TRS ratings, the "misclassified" recordings might actually be correct, and these could facilitate the search for better guidelines on TRS appraisal. This is the major goal of future work stemming from this preliminary analysis.

We have explored a larger number of clusters than just two that we reported in the results section. Purposefully increasing the number of clusters resulted in the clustering algorithm starting to cluster together the recordings of individual patients. It appears that each individual patient has his or her own motion patterns which eventually get detected by the clustering algorithm. However, when clustered in a small number of clusters these seem to represent the pathology (patient's condition) as testified by a relatively good correlation with TRS scores and by individual patient's recording being spread over several clusters.

Moreover, we observed that just a few patients cause most of the discrepancy between TRS and the two clusters. Since these patients are clustered "as a whole", one explanation is that it is their unique motion patterns and not change due to the medication that the algorithm is primarily detecting. Further work should focus to re-observe these patients' video recordings to carefully observe for specifics in their movement to either confirm or reject this hypothesis.

Acknowledgements. The research is supported by the Slovenian Research Agency (ARRS) under the Artificial Intelligence and Intelligent Systems Programme (ARRS No. P2-0209), Swedish Knowledge Foundation, and Swedish Agency for Innovation.

References

1. Aghanavesi, S., Filip, B., Nyholm, D., Senek, M., Memedi, M.: Feasibility of a multi-sensor data fusion method for assessment of Parkinson's disease motor symptoms. In: International Congress of Parkinson's Disease and Movement Disorders (MDS), Hong Kong, 5–9 October 2018 (2018)
2. Demšar, J., Leban, G., Zupan, B.: Freeviz–an intelligent multivariate visualization approach to explorative analysis of biomedical data. J. Biomed. Inform. **40**(6), 661–671 (2007)
3. Fritz, H., Garcıa-Escudero, L.A., Mayo-Iscar, A.: tclust: an R package for a trimming approach to cluster analysis. J. Stat. Softw. **47**(12), 1–26 (2012)
4. Nyholm, D., et al.: Duodenal levodopa infusion monotherapy vs oral polypharmacy in advanced parkinson disease. Neurology **64**(2), 216–223 (2005). https://doi.org/10.1212/01.WNL.0000149637.70961.4C
5. Senek, M., et al.: Levodopa/carbidopa microtablets in Parkinson's disease: a study of pharmacokinetics and blinded motor assessment. Eur. J. Clin. Pharmacol. **73**(5), 563–571 (2017). https://doi.org/10.1007/s00228-017-2196-4
6. Thomas, I., et al.: A treatment-response index from wearable sensors for quantifying Parkinson's disease motor states. IEEE J. Biomed. Health Inform. **22**(5), 1341–1349 (2018). https://doi.org/10.1109/JBHI.2017.2777926

Correction to: The Minimum Sampling Rate and Sampling Duration When Applying Geolocation Data Technology to Human Activity Monitoring

Yan Zeng, Paolo Fraccaro, and Niels Peek

Correction to: Chapter "The Minimum Sampling Rate and Sampling Duration When Applying Geolocation Data Technology to Human Activity Monitoring" in: D. Riaño et al. (Eds.): *Artificial Intelligence in Medicine*, LNAI 11526, https://doi.org/10.1007/978-3-030-21642-9_29

Unfortunately the first author's name was spelled incorrectly. In the contribution it read "Yan Zheng" but it should have read "Yan Zeng".

The updated version of this chapter can be found at
https://doi.org/10.1007/978-3-030-21642-9_29

D. Riaño et al. (Eds.): AIME 2019, LNAI 11526, p. C1, 2020.
https://doi.org/10.1007/978-3-030-21642-9_53

Single-Cell Data Analytics in ScOrange (General Machine Learning)

Martin Stražar[1(✉)], Lan Žagar[1], Jaka Kokošar[1], Vesna Tanko[1], Pavlin Poličar[1], Aleš Erjavec[1], Ajda Pretnar[1], Anže Starič[1], Vilas Menon[2], Rui Chen[3], Gad Shaulsky[3], Andrew Lemire[2], Anup Parikh[4], and Blaž Zupan[1,3]

[1] University of Ljubljana, Ljubljana, Slovenia
martin.strazar@fri.uni-lj.si
[2] Howard Hughes Medical Institute, Ashburn, USA
[3] Baylor College of Medicine, Houston, USA
[4] Naringi, San Francisco, USA

Single-cell RNA sequencing (scRNA-seq) is an emergent technology that enables the discovery and tracking of cell types, tumor progression, differentiation factors, and pathogen identification. Single-cell Orange is an interactive visual programming tool that uses visual programming to assemble data analysis workflows (Fig. 1). Built on top of the general purpose data analysis platform Orange [2], scOrange implements data preprocessing, filtering, and batch effect removal techniques designed for scRNA-seq data. It includes a rich library of data visualization, modeling, and bioinformatics methods that support combinations of data management, visualization, and machine learning.

ScOrange aims for balance between easiness of use and the capacity to answer complex questions. End-to-end scRNA-seq analyses often consist of working with multiple data sets driven by biological and technical variability (Fig. 2). The principal advantages of interactive workflows are control over data flow and methods, effects of tuning modeling parameters, and visualization-based inspection at any step. Statistical analysis enables characterization of cell types, revealing *de novo* markers, or use a provided marker library and access to resources such as Gene Ontology or KEGG pathways, combined on production quality plots. Comparing to providing code or method descriptions in publications, workflows are a complementary medium that guarantees results sharing and reproducibility.

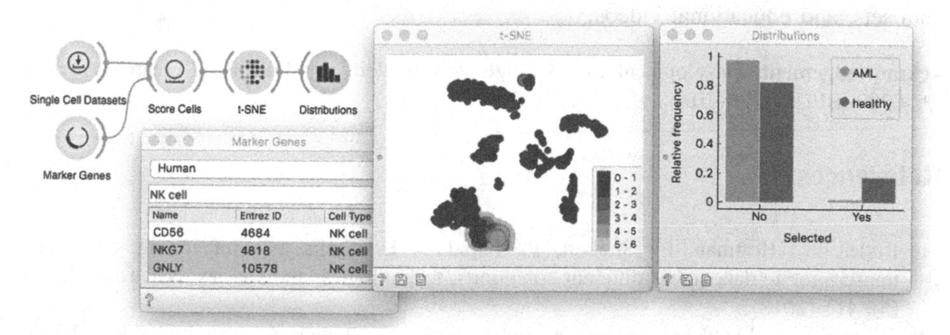

Fig. 1. An example workflow. Cells are scored with *Marker Genes* to identify natural killer (NK) cells in the *t-SNE* plot. Interactively selecting the found population reveals the abundance of NK cells in healthy and acute myeloid leukemia (AML) patients.

D. Riaño et al. (Eds.): AIME 2019, LNAI 11526, pp. 425–426, 2019.
https://doi.org/10.1007/978-3-030-21642-9

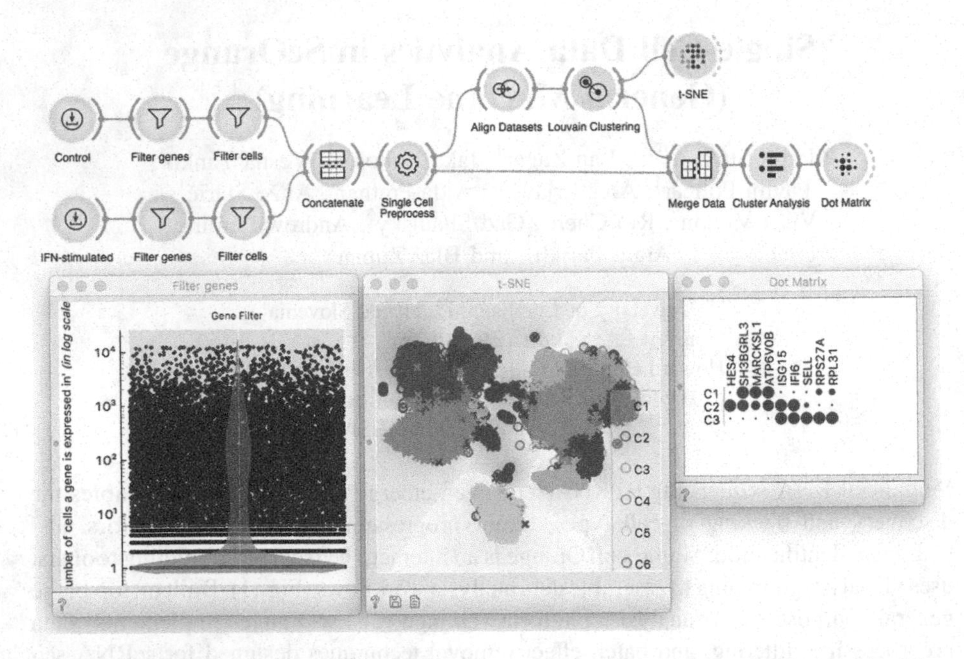

Fig. 2. End-to-end analysis of two peripheral blood mononuclear cell data sets. The combined data undergoes filtering, normalization, and batch effect removal via Canonical correlation analysis [1]. The resulting clustering is unbiased with regard to IFN-β treatment (crosses) and control cells (circles). Cluster analysis reveals cell types and gene markers, as well as cell type-specific response to IFN-β stimulation.

Computational techniques for single cell analysis and their access through R or Python are widespread [3], but add to the steep learning curve for the users and creates a barrier between biologists and their data. We aimed to design a tool that simplifies access to the data and allows construction of analysis pipelines within minutes after minimal training. The tool enables life scientists to focus on content, rather than on issues of programming. Finally, interactive visualizations support exploratory data analysis and focus on the most interesting data and patterns. scOrange is released in open source (http://singlecell.biolab.si), including documentation, workflow examples, data sets, and educational videos.

Acknowledgement. Development of scOrange was funded by Slovenian Research Agency (P2-0209, BI/US-17-18-014).

References

1. Butler, A., Hoffman, P., Smibert, P., Papalexi, E., Satija, R.: Integrating single-cell transcriptomic data across different conditions, technologies, and species. Nat. Biotech. **36** (5), 411 (2018)
2. Demšar, J., et al.: Others: orange: data mining toolbox in Python. J. Mach. Learn. Res. **14**(1), 2349–2353 (2013)
3. Wolf, F.A., Angerer, P., Theis, F.J.: SCANPY: large-scale single-cell gene expression data analysis. Genome Biol. **19**(1), 15 (2018)

Author Index

Printed in the United States
By Bookmasters